DUMBARTON OAKS STUDIES

XLIV

A CRITICAL COMMENTARY ON THE TAKTIKA OF LEO VI

A CRITICAL COMMENTARY ON THE TAKTIKA OF LEO VI

JOHN HALDON

DUMBARTON OAKS RESEARCH LIBRARY AND COLLECTION

WASHINGTON D.C.

LIBRARY OF CONGRESS CATALOGING-IN-PUBLICATION DATA

Haldon, John F.
A critical commentary on the Taktika of Leo VI / John Haldon.—First paper.
pages cm.—(Dumbarton Oaks Studies XLIV)
Includes index.
ISBN 978-0-88402-391-3 (first paper : alk. paper)
1. Leo VI, Emperor of the East, 866–912. Tactica. 2. Military art and science—Early works to 1800.
3. Tactics—Early works to 1800. 4. Military art and science—Byzantine Empire.
5. Byzantine Empire—History, Military—527–1081. I. Title.
U101.L43H35 2014
355.4'2—dc23
2013026005
www.doaks.org/publications

Design and composition: Melissa Tandysh

Cover design: Kathleen Sparkes

CONTENTS

PREFACE

I am grateful to a number of friends and colleagues for advice and assistance in completing this volume. I owe a debt of gratitude to Mark Bartusis, who read through and commented on an early draft of the manuscript. Since Leo's *Taktika* is based heavily on the late sixth-century *Strategikon*, which in turn derives a great deal of its material from earlier Greek and Latin sources, the reader is referred for detailed discussion of the latter to the commentary and translation of Philip Rance, *The Strategikon of Maurice: Translation and Commentary* (Aldershot, 2014). I am especially grateful to him for allowing me access to the manuscript of this important work prior to publication, and in particular for his generosity in sharing his expertise on the manuscript tradition of the ancient military treatises, his deep knowledge of both Maurice's *Strategikon* and Leo's *Taktika*, and his valuable critical remarks on an earlier draft of this volume. I am equally indebted to Everett Wheeler, who took great pains in reading the whole manuscript through, whose expertise in the field of the ancient tactical handbooks saved me from a number of errors and omissions, and who offered both advice and bibliographical guidance, together with numerous suggestions for improving aspects of the commentary. To Dr. Jonathan Shepard and Professor Mary McRobert I am grateful for invaluable advice on issues of Old Slavonic, to Professor Emmanuel Bourbouhakis for advice and help in respect of the corpus of Greek proverbs, to Dr. Éva Bakay for assistance with the Hungarian archaeological literature, to Dr. Adam Bollók for his insights on Magyar archaeology, and to Professor John Pryor for his helpful remarks on the commentary to Constitution 19 on naval warfare. I am similarly indebted to the anonymous peer-reviewers for their helpful, detailed, and constructive criticism and suggestions. I am grateful to my friends and colleagues in the History Department at Princeton for their good-natured and collegial support throughout, to that same department and other sources at Princeton University for financial support while working on various stages of this undertaking, as well as to the Dutch Institute of Archaeology and to the Research Center for Anatolian Civilizations of Koç University, both in Istanbul, for hospitality during the initial stages of the project in 2009. I owe a special debt of thanks to Anna Patej, who labored as my research assistant during the summer and fall of 2011, while I was working on a very different sort of project in Turkey, and who carried out with patience and exactitude some of the more tedious but necessary cross-referencing. To Joel Kalvesmaki at Dumbarton Oaks my sincere thanks for painstaking and careful editing and guidance of the volume through the production process. Finally, of course, I want to express my debt, both for myself and on behalf of colleagues in the field, to the achievement of Father George Dennis, whose lifelong interest in Leo's *Taktika* and whose scholarly rigor and diligence in producing the new edition have made this commentary possible.

Note on References and Translations

References to the text of the *Taktika* are to the Dennis edition, and cross-references are given by Constitution number with section and line number (e.g., 11§38.213) in order to render the lemmata as precise as possible. Within the commentary the main entries or lemmata are given by line number or section number for that constitution only. For the older paragraphizations see the concordance (pp. 455–66).

Translations of Greek, Latin, and Arabic texts into English are always from the published versions employed and cited in this volume, where available (whether accompanying an edition of the text in question or published as separate works). All other translations, where no modern translation is cited in the text or bibliography, are the author's.

PART I

Introduction to the *Taktika*

INTRODUCTION

The *Taktika* attributed to the emperor Leo VI "the Wise" (886–912) is perhaps one of the best-known and least understood texts of its type. Composed early in the tenth century ostensibly to guide his provincial generals, Leo's treatise was intended to serve as both a corrective and a reminder of earlier "good practice," as well as a statement of imperial authority and power. It embodied Leo's divinely founded legitimacy as ruler over the Romans, the Chosen People in Byzantine eschatalogical thinking, and it reflected at the same time a particular moment in the evolution of medieval Christian views of Islam. That it was originally composed specifically with the threat from Islam in mind, a point of view that has recently prevailed, can be shown to be incorrect (see chapters 1 and 2 and the commentary at Const. 18), even if the anti-Islamic element was soon introduced by Leo himself in a revised and expanded version of Const. 18. Yet for the most part Leo's treatise has been seen as a somewhat antiquarian exercise, a Byzantine revival of an established genre of *strategika*, handbooks designed to advise military officers on aspects of tactics and military organization, a genre whose roots lie ultimately in the writing of authors such as Aeneas Tacticus (fl. 4th century BCE).[1] Because Leo's *Taktika* is based largely upon the text of the late sixth-century *Strategikon* ascribed to the emperor Maurice (582–602) (itself a departure in many important respects from the older tradition) and a handful of earlier Roman or Hellenistic writers, its relevance for its own times has been repeatedly called into question. As we will see in chapters 2 and 3 and in the course of the commentary to the twenty *diataxeis* or constitutions of the treatise, this view is both partial and misguided, although some have recognized the value and relevance of some of Leo's material for the period in which he was writing and for his reign and work more generally. Yet the *Taktika* was greatly valued by much later generations of

---

1 See Whitehead, *Aineias the Tactician.*

military thinkers in western Europe, and indeed part of the inspiration for the radical reforms in tactical organization and discipline introduced by leaders such as Maurice of Nassau in the late sixteenth and early seventeenth century was their reading of translations of ancient treatises such as Vegetius's *Epitoma rei militaris* and Aelian's *Tactica theoria*, together with Leo's *Taktika*. Vegetius had been known and valued since the twelfth century, of course, and became in part the basis for medieval theories of the state, as exemplified in John of Salisbury's *Policraticus* (1159) or Giles of Rome's *De regimine principum* (1275–77). Vegetius also lay behind some aspects of the reform of the Burgundian army in the 1470s, and was later employed by Machiavelli in his 1521 *Arte delle guerra*.[2] Although the *Epitoma rei militaris* was first translated into French in the late thirteenth century, it experienced real popularity only in the late fifteenth century, succeeded by a French translation of Aelian in 1536, the latter from an earlier Latin version of 1455–56.[3] The first edition of the Greek text of Aelian was published in 1552, and the first Greek editions of one or two other texts that had been known in Latin for a century or more, such as that of Onasander, soon followed.[4] Leo's *Taktika* was translated into Latin by John Cheke in 1544 and published in 1554, in a volume dedicated to the English king, Henry VIII;[5] Johan Meurs (Meursius) edited it on the basis of three late manuscripts and published it in Leiden in 1612, with the Latin translation from Cheke and with a dedication to the commanders of the armies of the United Provinces;[6] and it was certainly used by later sixteenth- and early

2 See Allmand, *Vegetius*; Richardot, *Végèce*; John of Salisbury, *Policraticus*; Nederman, *John of Salisbury*; Briggs, *Giles of Rome*.

3 See Allmand, *Vegetius*, 152–59; Hahlweg, *Heeresreform*, 154–64, 194–95, and esp. 304–5; for the texts: Löfstedt, *Vegesce Flave René*; older ed.: de Meun and de Pisan, *L'art de [la] chevalerie*; Volkyr de Sérouville, *De l'ordre*. See also Wijn, *Het krijgswezen*, from 480. Aelian and Onasander were among the earliest to be translated, in the fifteenth century, first into Latin and then into Italian; see Eramo, "'Certo tractatello'." For the manuscripts: Hoffmann, *Lexicon bibliographicum*, 1:15–17.

4 For further details, see the notes accompanying the entries for the ancient military writers in Dain and de Foucault, "Stratégistes byzantins," 319–32.

5 Cheke, *De bellico apparatu liber*.

6 Meursius, *Tactica*, probably composed at the behest of Willem Lodewijk of Nassau, who particularly valued Leo's treatise; see Parker, "Limits to Revolutions" 338 and n. 11. A year later the Meursius text was reissued, alongside Cheke's Latin translation, and bound with an edition and Latin translation (a revised version of that published in 1487) of Aelian by Sixtus Arcerius (which had been published separately, earlier in 1613): Meursius, *Claudii Aeliani*.

seventeenth-century writers on military matters, although it was not translated into French (and German) until the second half of the eighteenth century, when it was very influential in discussions of military theory.[7]

The translation or, at least, the use and interpretation of such texts was not just a military matter, however, since the late sixteenth century saw a widespread debate across western and central European societies about the moral and organizational superiority of classical and Roman institutions, of which the military was only one aspect.[8] But since the editions of the *Taktika* by Meursius and then Lami, reproduced by Migne in volume 107 of the Patrologia Graeca, Leo's treatise has received frequent attention as a source for Byzantine military affairs, always usefully but often somewhat uncritically, as in Oman's *Art of War*[9] or Delbrück's *Geschichte der Kriegskunst*.[10] Although there existed the Latin, French, and German translations,[11] there has been no English translation and none in any other language until the appearance of the new edition and translation by George Dennis.[12]

Byzantium was not a warlike society, at least in its general composition and in any traditional sense, yet the language and vocabulary of warfare permeated both secular and religious literature as well as oral culture in various ways. At the same time, warfare was seen as an evil, often even by the soldiers most actively involved. But Byzantium was also a culture that knew, in differing ways at different levels of social experience, what it was defending, and why. Herein is to be found the psychological aspect of its success—indeed, one may legitimately ask how the east

---

7 See Joly de Maïzeroy, *Institutions* and, based on the French translation, von Bourscheid, *Strategie und Taktik*.

8 See the article "Rome," in Grafton, Most, and Settis, *Classical Tradition*; Schulten, "Nouvelle approche"; Oestreich, *Neostoicism*, esp. 39–117; Rothenberg, "Maurice of Nassau"; and note also Grafton, "Availability," 767–91. See in general the papers in Colson and Coutau-Bégari, *Pensée stratégique*; and Ilari, *Imitatio*.

9 When it first appeared, of course, the book was groundbreaking in a number of ways.

10 Now accessible in English as Delbrück, *History of the Art of War*, 2:339–83 and 3:189–202 (for Byantium). Other less important commentators who employed the *Taktika* include Kalomenopoulos, Στρατιωτικὴ ὀργάνωση.

11 Joly de Maïzeroy, *Institutions*, was reprinted in Liskenne and Sauvan, *Bibliothèque historique*, 3:437–552.

12 Dennis, *Taktika*. A modern Greek translation based on the version in PG 107 appeared in 2001. It is a relatively free rendering, accompanied by no critical apparatus, notes or analysis: Potamianos, Τακτικά. For a Russian translation of Consts. 3–6: Kučma, "Taktika"; and of Const. 7: idem, "Metodika."

Roman state could have survived the territorial collapse and economic crisis of the second half of the seventh century at all, the more so in light of its precarious strategic situation. And while part of the answer lies in economic organization, logistical sophistication, diplomacy, and international politics, the strength of the imperial ideology, in the various forms through which it was effective in society as a whole, was undoubtedly crucial. The certainties which this system of beliefs and values presented to the literate cultural and political elite must be allowed a good portion of the credit for the survival of the east Roman empire. In particular the close relationship between the Church, as the formal representative of Orthodox Christianity—firmly rooted in the hearts and minds of the ordinary population— and the emperors, and the ideological motivation thus generated to maintain the existence of the state, played a key role.

Being more or less continually at war across several centuries, even if not on all frontiers or in all regions at once, and even if interspersed with periods of relative peace, had its effects on the mass of the ordinary people of the empire, in particular on the rural population of farmers and peasants who constituted the greater part of the population. It was they who bore the burden of maintaining the armies, whose lives were in part regulated by the timetable and often exceedingly oppressive, if not ruinous, demands of the state's fiscal apparatus, quite apart from the effects of warfare and raiding on the provincial economies of the empire at different times. Byzantine society was thus molded in its institutional forms and in its evolution and development by factors associated with warfare, and this gives the study of its military and corresponding fiscal organization particular importance. Leo's treatise was not especially concerned with this institutional context, of course—although, following tradition, he pays lip service to the need to protect the civil population of the state from the potentially damaging depredations of the empire's own soldiers. Yet the general situation, in conjunction with the self-conscious role of the emperors as protectors of all their subjects inscribed within the imperial ideology and the east Roman symbolic universe or thought-world, can account in part for the revival of the tradition of tactical manuals in late ninth- and tenth-century Byzantium.

The commentary is preceded by three chapters dealing with the context, motives, and date of the *Taktika*, together with its structure, sources, and language. For ease of reference I have repeated or expanded, and on occasion corrected, the few editorial comments or references to Leo's sources accompanying the edition of the text by Fr. Dennis, simply to ensure that all the relevant information is accessible in one volume. Since his edition includes a glossary of technical Byzantine terms, however, I have elected not to repeat that exercise in the present volume, also because much longer and more detailed discussions and definitions of technical language occur as part of the line commentary. Fr. Dennis established a paragraphization of the text based on M and his own reading of the way the text deals with its various subjects, and this arrangement differs frequently from those of Meursius and Vári. Since older discussions of the *Taktika* naturally employ these earlier paragraphizations, a concordance between the three editions seemed an essential aid to scholars who need to be able to locate with ease references to the older versions (see pp. 455–66).

Leo's treatise was for many years assumed to represent the first handbook of its kind since the time of the emperor Maurice. It has now become clear that Maurice's treatise was paraphrased at some point in the middle years of the seventh century (in a text known as the *De militari scientia*), and that a group of three texts (on strategy, on naval warfare, and on military harangues), originally thought to belong to the sixth century, should probably be dated to the ninth century, and to the pen of an otherwise unknown Syrianos *magistros* (although mentioned in the treatise on imperial expeditions compiled at the order of Constantine VII),[13] at least in the form in which they have come down to us. Although his conclusions can certainly be modified and corrected in several respects, Alphonse Dain reconstructed a range of possible corpora of military writings that may belong to this period also, from some of which Leo may well have drawn information or anecdotes for his own treatise. The so-called *Corpus perditum* and the *Apparatus bellicus*, two collections of extracts from ancient or later authors (including sections of the *De re strategica* of Syrianos), may likewise belong in the late ninth century.[14] Thus it is possible that Leo's treatise does not mark quite such a radical revival as was once thought. Be that as it may, it certainly deserves more detailed and careful analysis, both in terms of its structure and sources, as well as with respect to the factual, cultural, and historical information it contains. This commentary is intended to provide the basis for such an analysis.

13 See *CPTT*, 210–11 and C.196–98. No details of this Syrianos are known, although he was clearly seen as an authority in the tenth century—he is also cited in *scholia* to the *Taktika* as well as to the treatise of Nikephoros Ouranos.

14 Dain and de Foucault, "Stratégistes byzantins," 353, 360–61.

CONTEXT AND MOTIVES FOR COMPOSITION

Leo VI "the Wise":
Writings and Reputation

The emperor Leo VI was born in September 866, the second son of the emperor Basil I, founder of the so-called "Macedonian" dynasty.[1] By 869 he had been crowned coemperor, and after the death of his father, in 886, he became sole emperor, ruling until his death, in May 912. In the older literature he is often characterized as a representative of the metropolitan bureaucratic elite and even of the upper levels of craftsmen and artisans in Constantinople, an opponent of the senate who set in motion the governmental regulation of industry and commerce in the capital city, a view no longer fashionable.[2] His reign has been the subject of many articles and several books, and we do not need to pursue the complex political history of the period in detail here.[3] Suffice to say that he has been judged very differently by different scholars, partly because he was a "noncampaigning" emperor (thus contrasting with his father in particular, but with most emperors before him as well), partly because of his difficult relationship with his father,

---

1 At some point between the 1st and 19th of the month; see *PmbZ* 24311; *PBE* Leo 25; for Basil I: *PmbZ* 832 and 20837; Ostrogorsky, *History*, 233 and nn. 1 and 2; also Vogt, "Jeunesse" and Grumel, "Léon VI," 5–42. In fact, there are good grounds for believing that Leo was in reality the illegitimate son of Michael III and his mistress Eudokia Ingerina, although the issue remains unresolved; see *PmbZ* 21754; Mango, "Eudocia Ingerina"; Kislinger, "Eudokia Ingerina"; and against this position Toul, "Περὶ τῆς νοθογενείας."

2 Ostrogorsky, *History*, 245–46.

3 Detailed and extensive account with literature now in *PmbZ* 24311; also Tougher, *Leo VI*; the survey in Ostrogorsky, *History*, 242–55; and among Mango's many contributions his essential study, "Legend." The now outdated but pioneering work of Popov, *Imperator Lev*, retains some interest. The literature is extensive; specific contributions will be noted where relevant. There is a useful short survey in Antonopoulou, *Homilies*, 3–16.

who had him imprisoned at one point, partly because of his difficult relationship with the patriarch Photios, whose pupil he was,[4] and partly because of the range of his literary and legislative interests, as a result of which he was given his epithet "*sophos*," "the wise."[5] Much of Leo's legislative activity took place in the first half of his reign, including probably the collection of 113 novels, presented in the same format as the novels of the emperor Justinian I and explicitly intended to connect with the legislative activity of Basil I. Although previously thought to have been issued as a single collection, it now appears that there were in fact at least two redactions or partial collections issued during Leo's reign.[6] The collection represents a mixture of topical revisions of older laws, the ratification of customary practice, and rulings connected with ecclesiastical matters. Thus novels 2–17 and 75 are addressed to Leo's younger brother, Stephen, who occupied the patriarchal throne from 886 until 893; many of the others are addressed to Stylianos Zaoutzes, for whom Leo established the position of *basileopator* in 888–89, and who was himself the likely author of many of the novels.[7] The great work of codification known as the *Basilika*, an extensive 60-book collection based on the Justinianic corpus, was promulgated during Leo's first years, probably in 888, although commissioned and largely completed during the reign of his father, following the preparatory work undertaken at the instigation of Photios from the 870s on.[8]

---

4 For Basil I see the literature cited in the preceding notes; Blysidou, Ἐξωτερική πολιτική, esp. 164–89; and Tougher, *Leo VI*, 73–74. For Photios: *PmbZ* 6253 and 26667.

5 See Ostrogorsky, History, 241–42, and Tougher, "Thought-World." On his education, see Vogt, "Jeunesse," 403–5. For a useful survey of some of Leo's writings, see also Moravcsik, *Byzantinoturcica*, 1:400–409; and Antonopoulou, *Homilies*, 16–23 with extensive literature. For the internal politics of Basil's reign and the quarrel between him and Leo, see Blysidou, Ἐξωτερική πολιτική, esp. 171–208 and Tougher, *Leo VI*, 35, 57–60.

6 Troianos, Νεαρές, 19–25; for the older view: Noailles and Dain, *Novelles*, viii–xiii. For discussion of the date and inception of both Leo's and Basil's legislative and codificatory activity, see van Bochove, *To Date*, who takes issue with the dating proposed by Schminck, *Studien*; Schminck, "Probleme"; idem, "'Novellae extravagantes'"; and Troianos, Νεαρές, 29–36.

7 See Oikonomidès, *Listes*, 307 on Zaoutzes and the title of *basileopator*; with *PmbZ* 27406; and *PmbZ* 27208 (Stephen). On the legislation and novels, see in particular Troianos, Νεαρές, loc. cit.; idem, "Novellen Leons VI."; Fögen, "Legislation und Kodification"; and eadem, "Leon liest Theophilos." Note also Fledelius, "Woman's Position."

8 For discussion of Leo's legislative activity, with earlier literature, see Troianos, Νεαρές, esp. 17–35 and idem, "'Ἐκκλησιαστικές' Νεαρές." See further, Schminck, "'Frömmigkeit'" and Fögen, "Reanimation." See the extensive introduction in Scheltema and Van Der Wal, *Basilicorum*, with the discussions in Dagron, "Lawful Society" and Simon, "Legislation."

Apart from his legislation, of course, Leo's reputation was founded on his concern for matters of ecclesiastical administration (as evidenced in the *diatyposis*, which listed the autocephalous archbishoprics and metropolitan sees of the empire, and which was at a later date expanded into a full episcopal *notitia* for the empire),[9] his interest in military matters, and a wide range of other texts generally ascribed to him, whether or not he was actually directly responsible for them. In many cases the attribution can be neither proved nor disproved. In the secular sphere Leo is attributed with anacreontics addressed to his brother as well as to the general Andronikos Doukas, along with a series of epigrams on diverse subjects (such as the Roman names of the month); he is also attributed with a series of palindromes (Krumbacher gives as an example: ὦ γένος ἐμόν, ἐν ᾧ μέσον ἐγώ), known to the Byzantines as "crabs."[10] Among the better-known works ascribed popularly to Leo (who is also confused in the tradition with the ninth-century polymath Leo the mathematician) are a series of compositions relating to prediction: books interpreting thunder (*brontologia*) or the movements of sun and moon as well as one dealing with earthquakes and another on using biblical texts for prediction. It is not without interest that in his account of how to manage an imperial military expedition the emperor Constantine VII recommends that the imperial *vestiarion* should take along tactical and technical manuals relevant to warfare, as well as books on the interpretation of chance events, of the weather, of thunder, and of earthquakes—were these the texts commissioned by his father?[11] But perhaps best known, and certainly enormously influential, is the oracle book ascribed to Leo, representative of a popular genre and surviving in many versions, as it was repeatedly recopied and formed the base for similar works well into the later medieval period. Such books were widely used and seem to have received less censure from the church than, for

9 See Beck, *Kirche*, 151 and 547; Darrouzès, *Notitiae Episcopatuum*, Not. 7 (a. 901–7); Grumel, *Regestes*, 133–35 (no. 598).

10 See Krumbacher, *Geschichte*, 721 and Hunger, *Profane Literatur*, 2:105–6; and for a bibliography on Leo's literary oeuvre: *PmbZ* 24311 (34–43). For Andronikos Doukas: *PmbZ* 20405.

11 Beck, *Volksliteratur*, 204–5 and Hunger, *Profane Literatur*, 2:223–24, 237–39. On prediction and oracular literature within the context of apocalyptic and eschatological writing of the Byzantine period, see Magdalino, "History of the Future"; Brandes, "Kaiserprophetien," 183–84; Magdalino, "Year 1000"; the essays in Magdalino and Mavroudi, *Occult Sciences*; and also Dagron, "Saint." Using the Bible for prediction was frowned upon by the church, although it never seems to have issued a formal ruling. See Klingshirn, "Divination"; idem, "Christian Divination"; Michel, "Sort"; and von Dobschütz, "Sortes." For Constantine VII's recommendation: *CPTT* C.196–202.

example, astrology, although the two genres were in various ways closely associated.[12] In the religious sphere Leo authored homilies and theological works, many of which were brought together shortly after his reign in a collection, of which one version contains 34 and the other 37 separate compositions,[13] dealing with feasts of the liturgical calendar as well as special occasions, such as the consecration of churches or the appointment of his brother Stephen as patriarch. Among the better known of these compositions is the funeral oration for his father Basil I.[14] He also compiled, or has had attributed to him, a collection of 190 maxims, the so-called *Pattern of Guidance for Souls* (Οἰακιστικὴ ψυχῶν ὑποτύπωσις), now shown to have been influenced by the writings of Hippocrates.[15] More doubtful is Leo's authorship of a group of liturgical prayers for Easter, an apologetic text about accusations that the emperor was a secret apostate, some iambics about theological subjects, and a number of texts addressing issues of biblical interpretation or morality.[16] And he was also, quite incorrectly, along with the emperor Leo III, attributed with an interesting polemical letter addressed to the caliph ʿUmar II (717–20), which survives only in a Latin translation of a Greek original. While the authorship is problematic, the text refers to aspects of Islam which evolved only after the time of ʿUmar II and has been dated to the time of Leo VI. We will return to this issue in the commentary on Constitution 18 below.[17]

The fact that so much was attributed to Leo, even within a generation of his death, says a great deal for his reputation. In the *Vita* of Theophano, his first wife, he is referred to both as "the wise" as well as "all-wise" (*pansophos*); the text, composed probably during Leo's own lifetime, generally corroborates agreement that his reputation as wise was well established before his death.[18] His poor treatment of Theophano during their marriage seems to have been ignored in light of

---

12 See Berger, "Das apokalyptische Konstantinopel," esp. 151–53. For the text: Brokaar et al., *Oracles*; also Vereeken and Hadermann-Misguich, *Oracles*.

13 See Antonopoulou, *Homilies*; Beck, *Kirche*, 546. Edition: Antonopoulou, *Homiliae*.

14 Vogt and Hausherr, *Oraison funèbre*.

15 See Grosdidier de Matons, "Hippocrate et Léon VI" and Papadopulos-Kerameus, *Varia*, 213–53.

16 Beck, *Kirche*, 547.

17 Text: PG 107:315–24; see Beck, *Kirche*, 547; esp. idem, *Vorsehung und Vorherbestimmung*, 43–46; Gaudeul, "Correspondence"; and Jeffery, "Ghevond's Text."

18 See Kurtz, *Zwei griechische Texte*, 4 and Magdalino, "Bath of Leo." Theophano: *PmbZ* 28122.

his pious treatment of her memory after her death.[19] Whether this was the case or not, his reputation as a wise (or "most wise," as he is frequently described)[20] and responsible ruler, an emperor who fulfilled his duties as God's vicegerent on earth, came to be firmly embedded in the historiography of the tenth century and later. Partly, of course, this lay in the fact that the history-writing of the period was to a large degree commissioned or written during the reign of Leo's son Constantine VII. While Solomon is nowhere mentioned explicitly in the *Taktika* (although the Solomonic attributes do appear; see on Const. 20.7–8), Leo certainly understood the power of the images of David and Solomon, images skillfully deployed by his father in efforts to legitimate his rule and in his advice to his son.[21] Similar imagery is used by Leo in his self-presentation,[22] and by his son Constantine VII.[23] Indeed, it is not without significance that in both the opening paragraph of the prooemium as well as in the acrostic built into Constitution 20, Leo describes himself as *eirenikos*, "peaceful," which, as has been pointed out, is the literal meaning of the name Solomon. There are many other parallels.[24]

Traditional scholarship ascribed to Leo a series of reasons for his lack of an active military career, but it has recently been argued quite persuasively that none of these is really satisfactory. Indeed, it is quite clear that Leo was deeply interested in military matters, and that a much better explanation for his absence from the battlefield is quite simply that he was attempting to emulate the emperor Justinian, whose golden age learned Byzantines could look back on as a time of military glory and of peace and prosperity. Basil I had taken his eldest son Constantine, whom he assumed would succeed to his throne, on campaign with him. But he never took Leo, and this may also have been an important factor. Yet in his legislative activity,

19 See, for example, Karlin-Hayter, "Vita Euthymii," commentary, 167; eadem, "Mort de Théophano"; Tougher, *Leo VI*, 134–40; and Strano, "Teofano."

20 See Tougher, "Wisdom of Leo VI," 178.

21 Magdalino, "Basil I, Leo VI"; see also Dagron, *Emperor and Priest*, 192–200; Markopoulos, "Constantine the Great," 161; and idem, "Laudatory Poem"; and see also Brubaker, *Vision and Meaning*, 185–93 and 265–66 on visual imagery linking Basil and David, and Basil and Solomon.

22 See the evidence collected by Tougher, "Wisdom of Leo VI"; and Magdalino and Nelson, "Introduction," esp. 22–25; also Antonopoulou, "Homiletic Activity," 320. Constantine VII: *PmbZ* 23734.

23 See Mango, "Legend"; on the Solomonic imagery, see Tougher, "Wisdom of Leo VI," 172–78; the literature in n. 20 above; and on Const. 20.7–9 below.

24 Kartsonis, *Anastasis*, 192 and in general, Tougher, "Wisdom of Leo VI."

in his active interest in the war against Islam, and in his paternal relationship with his generals, Leo's model was the great emperor of the sixth century, and in his style of rule as well as his presentation of self Leo emphasized his own power and authority while recalling the glories of the past.[25] Significantly, Leo was to establish a trend which lasted well into the second half of the tenth century—neither his brother nor his son and grandson campaigned, and indeed their military achievements were represented for them both symbolically and in practical terms through the activities of their leading generals, soldiers such as John Kourkouas or Nikephoros Phokas. This was a model which was not without precedent, although none of these tenth-century emperors would have been aware of it so explicitly. From the late fourth century right through into the early seventh century, hardly any western or eastern emperors were campaigning rulers, in stark contrast to those who went before them and those from Heraclius onward (although Maurice had made forays into Thrace, against the advice of his senior commanders).[26] In the fourth and fifth centuries this was in part because many of them succeeded to the throne as infants, and their rule was dominated for much of their lives by powerful generals such as Stilicho, Flavius Constantius, or Aetius in the west; but it also reflected important shifts in the ways the imperial office was understood and perceived.[27] Similar changes, still needing further investigation, were happening to the imperial office in Leo's time, and such changes contributed to this noncampaigning aspect of his and his successors' rule.

Whether the *Taktika* was ever really effective as a guide and handbook for field operations, it was certainly read by some officers, as is clear from the reference to it in the treatise on skirmishing or guerrilla strategy ascribed to Nikephoros Phokas, and as is implied in several other texts of the period.[28] In this respect, it

---

25 See *PmbZ* 23742; and especially the discussion of Tougher, "Thought-World," 56–60; see also Karlin-Hayter, "Military Affairs" and Antonopoulou, "Manuels militaires."

26 John Kourkouas: *PmbZ* 22917; Nikephoros Phokas: *PmbZ* 25545. See Whitby, *Emperor Maurice*, 156–57. The few exceptions were the short-lived western rulers Avitus (455–56), Majorian (457–61) and the son of the western emperor Anthemius (467–72). The eastern emperor Marcian (450–57) may have led a campaign after his accession, but this is uncertain.

27 See McEvoy, *Child-Emperor Rule*. We should differentiate additionally between those emperors who had been effective military commanders before their elevation to the throne (such as, for example, Constantius III, i.e., Flavius Constantius), and those who had never campaigned at all.

28 *Skirmishing*, 20.1.9–10; see Darkó, "Glaubwürdigkeit." Nikephoros II Phokas: *PmbZ* 25535.

should be recalled that the *Book of the Eparch*, also compiled during the reign of Leo and with an apparently official status, seems never to have been formally promulgated, survived in a single manuscript only, and became in effect a work of literature rather than a practical piece of active legislation.[29] Leo's *Taktika* clearly had a greater circulation. As such, and as a very explicit statement of the emperor's own central position in the Byzantine state and society, it may well have had a greater impact than some of Leo's legislation.[30]

The Taktika: Context

The *Taktika* is one of the best-known, although certainly not the best-understood, middle Byzantine texts, particularly in the genre of military treatises. It was not Leo's first attempt at compiling a book of guidance on military matters. The so-called *Problemata*, ascribed to Leo in the single manuscript in which it survives (Laurentianus 55-4; see chapter 2 below on the manuscript tradition), was probably also compiled by him, and some time before the *Taktika*. Much has been written on this type of literature, and this is not the place to repeat this material.[31] Leo's *Taktika* is important for several reasons. It represents the first of a series of military treatises that distinguish the tenth century—later texts such as the so-called *Sylloge tacticorum* (which may or may not have been produced during Leo's reign, see below) seem to have taken their inspiration at least in part from the *Taktika*. And while it is true that the treatise ascribed to a certain Syrianos *magistros* titled *On Strategy*—Περὶ στρατηγίας—may well be a mid-ninth-century

---

29 See Schminck, "'Novellae extravagantes'," 201–9.

30 We should also be sensitive to the ways in which the *Taktika* evolved as a work both of literature and of reference to military matters, both during and after the Byzantine period as, for example, in the later Arabic translations; see, e.g., Christides, "Naval Warfare." For some comments on the approach to the evolving history of a medieval text, see Baun, *Tales*, esp. 34–75; and on the compilatory literature of the tenth century in particular, see Trombley, "*Taktika*" and Holmes, "Political Culture."

31 The two standard works of reference are Dain and de Foucault, "Stratégistes byzantins"; Hunger, *Profane Literatur*, 323–40: "Kriegswissenschaft"; see also McGeer, "Military Texts"; Sullivan, "Military Mauals." For additional comment on the genre from the Roman period onward and in particular on its continuity after the sixth century, see Kučma, "Militärische Traktate"; Cosentino, "Writing." For the older tradition, see Burliga, "Aeneas Tacticus"; Wheeler, "Military Treatises"; idem, "Polyaenus," with detailed discussion and older literature; also Tejeda, "Warfare, History and Literature" and Whitehead, "Fact and Fantasy." See also below, p. 41 n. 8.

work,[32] Leo's treatise is far more detailed in its descriptions of military organization, armament, and campaigning, and it directly takes up the tradition inaugurated in the *Strategikon* of Maurice, whereas *On Strategy* remains somewhat closer to Hellenistic and Roman antecedents.

This does not necessarily mean that no interest in such works continued through the late seventh and eighth centuries, although the evidence suggests that if it did, it was limited to minor revisions and summaries of the more practical work of Maurice—as exemplified, for example, in the so-called *De militari scientia*, a revision and summary of the *Strategikon*, with some additional material from another source or sources, produced at some point during the seventh century, most likely during the first phase of the wars with the Arabs, ca. 633–40. The text omits the information on the Avars and Turks (although making frequent reference to "Scythians"), but does mention the Saracens, for example.[33] It is notable that whereas the classical treatises by certain authors, such as Aeneas the Tactician or Polyaenus and several others, are known only from an extremely limited manuscript tradition, Maurice's *Strategikon* was transmitted through a more extensive range of manuscripts, illustrative of its persistent popularity.[34] And if there were other military compilations made during the period, they appear largely, if not entirely, to have been based on Roman and Hellenistic texts, such as the lost *Apparatus bellicus*.[35] Syrianos's tripartite compilation certainly broke new ground in some ways, in particular by its incorporation of a guide to naval warfare, based in part on classical antecedents but including some new elements and material.

---

32 Although this is still debated; see below and notes 45–46. For the structure and content of *Περὶ στρατηγίας*, see Eramo, "Composition" and idem, "Compendio."

33 Edited by Müller, "Griechisches Fragment"; see Dain and de Foucault, "Stratégistes byzantins," 346. A translation of the text, accompanied by a copy of Müller's edition, has been prepared by Bertazzoli, *Il De militari scientia*. I am indebted to Prof. S. Cosentino for obtaining a copy of this work for me. See the detailed discussion and commentary in Rance, *Strategikon*, app. 1a.

34 For Leo's sources, see chapter 2, below; and for the manuscripts of the *Strategikon*, see the detailed analysis in Rance, *Strategikon*, chapter 1.

35 While there are no doubt substantial emendations to be made to the original analytical survey by Dain, completed by de Foucault, in 1967, as a result of more recent work, the broad outlines of the tradition as it appears in the manuscripts of the tenth century and later remains substantially valid. For the *Apparatus bellicus*, see Dain and de Foucault, "Stratégistes byzantins," 359–61 (although placed by Dain and Vieillefond in the tenth century it may very probably be somewhat earlier) and Zuckerman, "Chapitres peu connus."

His detailed presentation of the form, content, and function of military harangues and rhetoric was likewise a novelty for military treatises. The section on strategy, based in many respects on Roman and Hellenistic antecedents, includes material on frontier fortresses and on defending towns, which reflects the realities of warfare in the eastern border regions in the ninth century, even if there are obvious debts to Hellenistic antecedents such as Philo of Byzantium, and thus ultimately Aeneas Tacticus.[36] Since it seems Leo had access to some, perhaps all, of this substantial text, it would be resonable to ascribe some of the inspiration for Leo's own compilation to this work.

The model that was followed by the tenth-century military writers was that established, or reestablished, by Leo. Although the first fifty-five chapters of the *Taktika* of Nikephoros Ouranos repeat or paraphrase substantial parts of Leo's treatise, other texts, such as the *Sylloge tacticorum*, the *De velitatione bellica* (*Skirmishing*), the *Praecepta militaria Nicephori imperatoris*, and an anonymous treatise on encampments (*Campaign Organization*), while certainly less dependent upon the *Taktika* of Leo, were nevertheless influenced in their composition and their rationale by its existence and format.[37] This includes also the basic philosophy of such treatises, that the good general will avoid battle where possible and employ stratagems and guile to defeat an enemy, thus avoiding unnecessary slaughter and bloodshed. Frequently picked out by modern commentators as a peculiarly Byzantine and Christian approach,[38] such views can be traced back to writers such as Polybius, Aeneas Tacticus, and Asklepiodotos; and many of their prescriptions, filtered through Polyaenus, for example, can be found in Constitution 20 and the Epilogue.

The tenth century saw an efflorescence of interest in all aspects of warfare and war making, such as siege warfare and artillery. Thus while Leo's *Taktika* has a section on sieges, several texts are devoted specifically to poliorcetics.[39] Others

---

36 See Dain and de Foucault, "Stratégistes byzantins," 323–24; below and nn. 44–45.

37 See the bibliography of sources under the titles *Sylloge tacticorum*; *Skirmishing*; *Campaign Organization* (this anonymous treatise dates probably from the reign of John Tzimiskes or Basil II); Nikephoros, *Praecepta*; and Nikephoros Ouranos, *Taktika*. See also Dain, *"Tactique" de Nicéphore Ouranos*, 19–20 (and 123–25); *PmbZ* 25617.

38 As in, for example, Oman, *Art of War*, 43 and more recently Luttwak, *Grand Strategy*, for example (see 80, 257, 281–303).

39 Sullivan, *Siegecraft*, 26–113 (text and trans.) and 153–248 (comm.); van den Berg, *Anonymous de obsidione toleranda*, with the English translation by Sullivan, "Instructional Manual."

concerned warfare at sea, to which Leo also dedicated a chapter (for which he claimed he found no ancient sources) in his *Taktika*, a version of which also circulated independently.[40] During Leo's reign too, possibly associated with the compilation of the material which later became the *Taktika*, and certainly at Leo's behest,[41] another treatise was compiled by the *magistros* Leo Katakylas. This served as the basis for a reworked and expanded account of imperial military expeditions assembled during the reign of Leo's son Constantine VII, very likely with Constantine's own input.[42] Somewhat later, during the reign of Constantine VII, interest in military records and the organization of campaigns became especially evident in the form of the accounts of a series of military undertakings dating from the reign of Leo himself, of Romanos I, and of Constantine VII, all preserved as miscellaneous items in the second part of the *Book of Ceremonies*, and probably associated with the politics of financing such (extremely costly) efforts.[43]

This literary and compilatory activity seems to reflect what we may reasonably call the spirit of the times: older treatises, such as the late sixth-century *Strategikon* ascribed to the emperor Maurice, had had no immediate successors, if one sets the shortened paraphrase *De militari scientia* to one side.[44] As noted already, the anonymous *On Strategy*, traditionally thought to be a sixth-century text, has now been shown convincingly to be but one part of a tripartite compilation, probably (although there is still disagreement on the date) of the ninth century, consisting of the section on strategy, a section on military rhetoric, and a treatise on naval warfare.[45] All

---

40 For naval warfare (including the section on naval warfare in the *Taktika* of Leo, the treatise dedicated to Basil the *parakoimomenos*, and the naval sections of the *Taktika* of Nikephoros Ouranos), see the edited collection in Dain, *Naumachica* and esp. Pryor and Jeffreys, *Age of the Δρόμων*. Syrianos magistros, *Naumachiai*, in: Dain, *Naumachica*, 43–55 and Pryor and Jeffreys, *Dromon*, 455–81; Leo's *Naumachika*: Dain, *Naumachica*, 19–33 and Pryor and Jeffreys, *Dromon*, 483–519; Basil, *Naumachika*, in: Dain, *Naumachica*, 61–68 and Pryor and Jeffreys, *Dromon*, 521–45. These texts are also edited, with critical notes and translation into modern Greek, in Dimitroukas, *Ναυμαχικά*.

41 See *CPTT* C.26–29.

42 *CPTT*; *PmbZ* 24329.

43 Holmes, "Political Culture," esp. 60–69; Haldon, "Chapters II, 44 and 45," 265–68; Trombley, "*Taktika*."

44 Dennis (ed.) and Gamillscheg (trans.), *Strategikon des Maurikios*; for the *De militari scientia*, see n. 33 above.

45 That the *Rhetorica militaris* was one element in a larger compendium was recognized already by Lucas Holsten in the seventeenth century, followed by Köchly and Rüstow,

three works were part of a single treatise compiled by a certain Syrianos *magistros*, although nothing is known of this author. If—as the evidence of the *Rhetorica militaris* might suggest[46]—the context was that of a new awareness of the nature of Islam as a direct challenge to the existence of Christianity, as antithetical to Christian values and mores, as hostile to the very existence of the Christian Roman empire, then this would place the date of composition of the treatise as a whole at some point in the last years of Michael III or during the reign of Basil I, where it also fits well with

---

Griechische Kriegsschriftsteller, 2.2:14–15; see Eramo, *Siriano*, 14; and taken a step further by Müller, *Seekrieg*, 46–47 and Lammert, "Älteste erhaltene Schrift," 279–81; followed by Dain and de Foucault, "Stratégistes byzantins," 344, and others such as Hunger, *Profane Literatur*, 2:328. The general consensus now is that this is certain, although there is continued disagreement over the date of composition. See Eramo, "Retorica militare"; eadem, *Siriano*, 14; eadem, "Ῥωμαῖοι e Ἄραβες"; eadem, "Composition" (neutral on date); Zuckerman, "Military Compendium" (a sixth-century date); and Cosentino, "The Syrianos's 'Strategikon'." The three sections are published separately, and are now available in modern editions. Syrianos, *Strategy*: ed. and trans. Dennis, in *Three Byzantine Military Treatises*, 1–136; Syrianos Magistros, *Naumachiai*, ed. and trans. Pryor and Jeffreys, *Age of the Δρόμων*, 455–81 (and see n. 39 above). The third section, the *Rhetorica militaris*, is edited by Eramo, *Siriano*; and see eadem, "Retorica militare." The older edition is Köchly, *Rhetorica militaris* (and the first three chapters of the *Rhetorica* are reproduced in Köchly and Rüstow, *Griechische Kriegsschriftsteller*, 2.2:15–20). This guide to military rhetoric may be usefully compared with two tenth-century harangues ascribed to Constantine VII, both of which clearly draw on the *Rhetorica* in terms of structure and content: see Markopoulos, "Ideology of War," 49–52 with further recent literature. The harangues were originally edited by Ahrweiler, "Discours inédit" and Vári, "Zum historischen Exzerptenwerke," both now translated with detailed commentary by McGeer, "Two Military Orations." On imperial propaganda in military contexts more generally, see Koutrakou, *Propagande*, esp. 354–79.

46 See, for example, *Rhetorica militaris*, 10.1: Οἱ πολεμοῦντες ἡμῖν βάρβαροι διὰ τὴν πίστιν ἡμῖν πολεμοῦσιν ("those barbarians who fight us do so on account of our faith") and 26.1: Θαυμάζω, εἴ τις ὁρῶν τὴν τῶν πολεμίων ἀσέβειαν οὐ πάσῃ δυνάμει κατ' αὐτῶν διανίσταται ("I wonder that if, when a person observes the impiety of the enemy, he does not stand against them with all his might"); 36.9: εἰ δὲ μὴ οὕτως δεῖ λαμβάνειν τὸν θεῖον νόμον, ἐπεὶ καὶ τὸν Πέτρον τὴν μάχαιραν σπασάμενον ὁ Χριστὸς διεκώλυσεν, ἀλλ' οὖν βιάσασθαι χρὴ διὰ τὸ πολιτικὸν συμφέρον καὶ τὸ κατεπεῖγον τοῦ πράγματος ("but if one ought not to understand God's law thus, since Christ prevented Peter from drawing the sword, yet one must appeal to force on account of the common interest and the pressure of circumstances") (author's trans.). The enemy of the faith is not named explicitly (which one would not expect in a rhetorical guide of this sort), and it could thus be argued that the text could refer to any enemy of Christianity. Yet the enemies of the empire are not seen in this light until the wars of Heraclius against the Persians. Although it is possible that the text was composed at this time—ca. 610–28—a mid-to-late ninth-century date seems more probable, given the writing of Niketas Byzantios at this time (see below with nn. 73–77).

the intellectual currents then prevailing at the imperial court.[47] Beginning during Basil's reign, interest in the late Roman (or even earlier) roots of some elements of medieval Byzantine culture underwent a strong revival; and with the empire's reassertion of its military strength and effectiveness from the last years of the ninth century onward (despite some substantial defeats at the hands of both Arabs and Bulgars), this interest grew ever stronger. It is possible that it was at this time that some of the fragmentary texts that include extracts or paraphrases of older Roman or Hellenistic works, but which cannot be dated with any degree of precision, were compiled. Interest in military rhetoric was also represented in the form of a couple of harangues to the troops that have survived, and again an association between these tenth-century texts and the *Rhetorica militaris* of Syrianos has been identified.[48] The dramatic series of victories and the transformation of the hitherto defensive east Roman empire into an aggressively expansionist force from the 960s added further stimulus to this tradition, but now with a renewed interest in and emphasis upon the present, upon contemporary ways of doing things rather than on updated recapitulations of earlier texts. It is difficult to know to what extent this interest, reflected, for example, in the encyclopedism of the emperor Constantine VII and his circle, reflects a broader cultural trend, of course, and we should beware of generalizing too readily from the work, however prolific, of a relatively small group of literati and learned men clustered around the society of the imperial palace. But it is clear that the revival of interest in both the history and the culture of the past that can be seen in this semiofficial and official literature has its roots in the late years of Basil I and in Leo's reign. By the same token it reflects both the cultural politics and literary tastes of a group of courtiers and senior churchmen who wielded influence at court—men such as Photios are among the most obvious representatives of

---

47 For a probable ninth-century date and context, see esp. Cosentino, "Syrianos's 'Strategikon'," 262–80 and Rance, "Date." Note also Baldwin, "On the Date" (after the sixth century); but for a sixth-century date: Zuckerman, "Military Compendium," 216. For analysis and discussion of the contents of all three sections, see the articles of Cosentino and Zuckerman; Eramo, *Siriano*, 14–23; and eadem, "Composition."

48 See Dagron, "Byzance et le modèle islamique," 227 and n. 34; Eramo, *Siriano*, 21–23; eadem, "Retorica militare"; eadem, "Ὦ ἄνδρες στρατιῶται." The texts: see n. 45 above. That these speeches were in fact compositions of Constantine VII has been shown to be unlikely. Along with several other works ascribed to him, they were probably commissioned by the emperor but composed by others; see Ševčenko, "Re-reading Constantine Porphyrogenitus," esp. 186–87 and n. 49, and Taragna, "Λόγος e Πόλεμος."

this tendency.[49] Leo's treatise marks in this respect the first step in a renaissance of military writings, which stimulated both the preservation and recopying of older classical, Hellenistic, and Roman texts and the writing of new texts that had a more immediate, practical, and highly contemporary purpose.

While the impact of warfare on literature was most obviously reflected in military writings, it also marked the chronicle and historiographical literature, partly because it was the stock-in-trade of the traditional historical models. Whether we are speaking of Thucydides or Procopius, Theophanes continuatus or Leo the Deacon, the deeds of military leaders and emperors in war, the bravery of the soldiers, the hard-fought campaigns, all played a central role in the construction of the narrative, and this remained true to the end of the Byzantine period. Since being at war was the usual situation of the Byzantine state for much of its history, the fact that wars were fought in order to achieve a state of peace meant that the wars of the Romans could usually be presented in a positive light. They served not just the ends of individual war-leaders and rulers, but of the Roman empire and its people as a whole, hence also of God in his struggle with the forces of evil on earth.[50] The vocabulary of warfare permeated theological and religious literature, too, so that monastic communities were described as regiments of spiritual fighters for the faith, and the struggle of the Church against evil was phrased in terms of a military campaign, in which the weapons of prayer, contemplation, and spiritual purification were part of the armory of the east Roman church. Such motifs can be traced to the very beginnings of Christianity, of course. Indeed, early Christians who argued against Christian participation in fighting had argued precisely that the Christian community fought a spiritual war for the benefit of the Roman state.[51] Equally, however, the benefits of peace were emphasized, both in letters to foreign rulers and in political discourse in general since the emphasis upon either war or peace, and the appropriate characterization of the enemy at any given time,

49 See Flusin, "*Excerpta* constantiniens"; Tougher, "Thought-World," 58–60; and Ševčenko, "Re-reading Constantine Porphyrogenitus." For the older literature and discussion on this theme, see Cohn, "Bemerkungen" and de Boor, "Suidas."

50 There existed a tension between Christian and Roman identities in Byzantine concepts of their empire, with first one, then the other dominating according to context and the ideological demands of the political or cultural moment. See in particular Stouraitis, "Byzantine War," 98–100.

51 See Haldon, *Warfare, State and Society*, 13–16 and the detailed discussion in Stouraitis, *Krieg und Frieden*, 190–208.

depended upon pragmatic political demands and priorities as much as it did upon abstract and theoretical arguments.[52] Peace was always the ultimate aim, even if warfare was necessary to secure it.[53]

Motives

The *Taktika* is interesting not simply because it represents one stage of a revival of military literature. It has also generally been taken to represent Leo's particular response to the military and political situation with respect to the Caliphate more broadly, since it has been argued that Leo's sections on fighting the Saracens and on their methods of recruitment and supporting warfare against the empire reflected a new awareness and consciousness in the Byzantine world, at least at court and among the educated, of the nature of Islamic beliefs and social organization, however distorted or inaccurate in parts such views may have been, and aimed to present a specifically Christian response to this challenge.[54] Not only this, but it has also been argued, and is now widely accepted, that the treatise as a whole was compiled specifically with the threat from Islam and the Saracens in mind.[55]

Now, while it is certainly the case that in the epilogue (§71) Leo states that this was his own reason for compiling the *Taktika* (referred to as "the present book"), it is important to note that he does not mention this motive at all in the prologue, which derives largely from the *Strategikon*, although the opportunity was certainly there, as evidenced by Leo's many other changes and additions to his exemplar.[56] By the same token, the very similarly worded statement in Constitution 18.692–93, usually taken to say that Leo has composed the *Taktika* as a whole

52 See in general the discussion in Kolia-Dermitzaki, "Byzantium at War."

53 Haldon, *Warfare, State and Society*, 17–33 for discussion and sources.

54 See esp. Dagron, "Byzance et le modèle islamique"; idem, "Ceux d'en face"; idem, "Byzance entre le djihad et la croisade"; and further literature in the commentary to Const. 18, below. This position is taken up and argued at length in Riedel, "Fighting," esp. 106–52 (I am grateful to the author for permitting me to read the dissertation shortly after its completion); see also the brief summary of these arguments in Riedel, "Christian Philosophy of Warfare?"; and the commentary to Const. 18, below. For the place of the *Taktika* in the compilatory literature of the period, and the range of motives that may have inspired such work, see in general Holmes, "Political Culture" and Magdalino, "Knowledge."

55 See the articles of Dagron in the preceding note and Kolias, "*Taktika*," esp. 130; see also Riedel, "Fighting."

56 See Epilog. 319–21 and *pace* Sullivan, "Military Manuals," 153, who suggests that Leo also mentions the Saracens in the prooemium. He does not.

because of the Saracens, actually refers just to that constitution—*diataxis*—alone (or only to that particular section within Constitution 18), and clearly not to the treatise as a whole. Indeed, it seems more likely, both in view of the wording of these two references to the Saracens, and in view of the way in which Constitution 18 is put together, that such comments in fact represent an afterthought, inserted into the body of the *Taktika* at appropriate points, and reflecting Leo's particular concerns at that time—probably soon after the first version of the *Taktika* had been more or less completed.[57] As will be suggested in the commentary to Constitution 18, it is probable that the *Taktika* was commissioned and completed in its original form without these comments, even if it did include much of the material on the Saracens—this may be reflected in the independent duplicate section on the Saracens extant in M (the earliest manuscript witness, cod. Mediceo-Laurentianus Plut. 55.4, dated ca. 950–55),[58] after the main text of the *Taktika*, equivalent to 18§§103–25 (see chapter 2, p. 57 below). This section would then have been filled out and completed (from §126 on), and the comments on the purpose of Constitution 18, and of the treatise as a whole, added at a secondary stage (as noted already, likely very soon after the first draft was completed). Such a view is strengthened by the fact that these two explicit statements on the subject are buried in the middle of the constitution (probably because the passage in question was originally a concluding paragraph; see the commentary to 18.686–95) and in the penultimate paragraph of the epilogue, respectively, and these are hardly points at which any author would be particularly well-served if the intention was to highlight motives for composing the whole work and to lend it legitimacy or indeed urgency. This relatively obscure placing of such a purportedly key motive is only underlined by the fact, already noted, that there is not a word on this concern with Islam in the prooemium.

On this argument, then, the *Taktika* was not—in its first version, at least—composed with a view to alerting the generals or the world of literate Byzantines to the Islamic threat at all, but was much more probably a general, straightforward attempt to legislate for warfare as an activity which fell under imperial supervision and authority, just as he had done in other spheres—although it is quite clear that the *Taktika* cannot be considered in any technical sense to be imperial legislation.[59]

---

57 See Const. 18.686–95, with commentary to this section.

58 For the manuscript tradition of the *Taktika*, see chapter 2 below.

59 See below, commentary to Const. 18.686–95.

Yet it is also quite clear that the nature of the Islamic enemy, and of Islamic concepts of religiously motivated warfare and divine reward for fighting for the faith, were a major concern for Leo. Indeed, in Constitutions 12, 13, 14, and 18 in particular, as well as in the Epilogue, Leo is very concerned to emphasize the essential differences between Muslim and Christian reasons for fighting, underlining above all the struggle to defend their faith that should encourage all Romans to strive for the common cause, fighting for God, emperor, and their Christian Roman brethren, not simply as a response to his own understanding of Islamic beliefs, but following the precepts set out in the *Rhetorica militaris*. Muslim fighting techniques, armament, and strategy would hardly be new to the majority of the military commanders and generals whom he is nominally addressing, no more than would the threat posed to the empire be a subject with which they were unacquainted.

But while there was undoubtedly a preoccupation in imperial and military circles during Leo's reign with the threat from its Islamic foes in the east and the duties of a Roman empire in the face of this threat, it is not always clear that there was a desire to do anything more than defend and maintain the status quo. On the one hand, in an oration delivered (probably) in 901, Arethas of Caesarea could describe Leo's strategy in the east as aimed at the recovery of a lost inheritance, while the former eastern provinces, which included substantial numbers of Christians as well as the holy places associated with the life of Christ, could still be regarded as imperial lands to which the empire could quite legitimately lay claim, even if the tensions between the conquering Byzantines and the indigenous Christian populations in northern Syria in the late tenth and early eleventh century belie the universal appeal of such views among Byzantines.[60] On the other hand, through the first half of the tenth century the Byzantine response to Islamic military activity was largely defensive, at least until the real threat from the Hamdanid emir Sayf ad-Daula had to be countered.[61] The urgency with which Leo addresses the issue of the fight for Christianity and Roman identity is apparent, and there is no reason to doubt that he wanted to bring out this issue in the body of the treatise.[62] But that the first version, which certainly must have included these elements, was specifically written with this in mind seems doubtful, even

---

60 See Arethas, 2.33.18–19, 2.34.1–3; for discussion: Talbot, "Byzantine Pilgrimages"; Pahlitzsch, "Zur ideologischen Bedeutung"; and Patlagean, "Byzantium's Dual Holy Land."

61 Shepard, "Equilibrium to Expansion," 498–99, 516–17; see *PmbZ* 26998.

62 See commentary to Const. 18.508–14 and 620–25.

if Leo quickly made this his motive in restructuring the text and incorporating the additional material on the Saracens which became the "complete" version as represented by the text in W (cod. Vindob. phil. graec. 275, ca. 960–90, the oldest witness to a version of the *Taktika* that incorporates all twenty constitutions).

The key point is that the treatise is throughout addressed to an imperial general—in other words, to all imperial generals, who were drawn almost exclusively by this time from a recognized social elite, an elite originating in a *noblesse de robe* that was rapidly becoming an aristocracy—and that in Const. 1§13 a clear emphasis is laid on loyalty to the imperial throne. As a whole, the *Taktika* can hardly be plausibly taken as an instructional manual for active field commanders, who certainly had much greater technical and tactical experience than the emperor himself. Rather, it should be seen as an attempt to establish or to define a clearly Christian moral framework for the conduct of war, regardless of the enemy in question, in which loyalty to God and the divinely chosen emperor was the key motif.

Leo's Authorial Contribution

Leo's treatise derives both its form and much of its content from Maurice's *Strategikon*, and represents a book of advice focusing on issues associated with the specific concerns of the time as perceived by the emperor. It was certainly an officially sanctioned document. We can reasonably assume that most of the work of copying out the relevant parts of Maurice's text, or the passages from Onasander, Aelian, Polyaenus, and other texts was not done by the emperor himself but that he probably selected the material, arranged for it to be marked up for copying, and decided on the order in which the material was to be presented, as well as the way in which material from different parts of the *Strategikon* and his other sources was to be reorganized and integrated. Leo's personal perspective is reflected in the frequent comments occurring throughout the text, as well as in the opening and concluding paragraphs of each constitution. He is most likely responsible for selecting much of the moralizing and gnomic material and for deciding on placement, so that in many respects, even if most of the *Taktika* derives from other sources, Leo can be regarded as the author, not just as the person who commissioned the work.[63] Of course, it is clear that we can never prove this. It is not impossible that he employed a literate and articulate redactor

---

63 For discussion of Leo's thought on law and society, see Troianos, "Λέων ϛ′ ο Σόφος."

for much of this work, including the task of selecting appropriate material for Constitution 20 and the Epilogue.

Whether or not the *Taktika* was intended to have the force of imperial law (even if unofficially it would have carried great weight and influence), it was meant to enshrine the God-granted authority of the Roman emperor in matters concerned with warfare and to codify both ancient tradition and contemporary practice, just as his legislative activity was grounded in and drew inspiration from his divinely inspired wisdom and used both ancient and more recent knowledge.[64] The emphasis on divine authority, divine sanction, the need for the commander and his soldiers to place their trust in God, the need to observe Christian morality and ethics in military affairs, and the God-given authority of the emperor which gives the *Taktika*, along with other official imperial legislation, its legitimate authority, is in itself not new—the *Strategikon* voices the same sentiments ("we urge upon the general that his most important concern be the love of God and justice; building on these, he should strive to win the favor of God, without which it is impossible to carry out any plan, however well devised it may seem, or to overcome any enemy, however weak he may be thought"), and similar ideas can be found in the prooemium to the *Ecloga* of Leo III and Constantine V, as well as in other texts.[65] And the *Strategikon* is certainly the first treatise in the tradition of military literature to emphasize divine favor and the intercession of the mother of God; indeed, unlike Leo's short and straightforward opening invocation, Maurice begins with a prayer, concluding with the "Amen."[66]

Yet in Leo the appeal to divine protection and support, the role of piety, and the emphasis on reliance upon God is given very particular expression throughout the rest of the treatise, and he quite deliberately moves the amen from the beginning of the prologue to the very end of the epilogue (l. 338). For the most part, while Leo is by no means anonymous as the author of the *Taktika*, he adopts the neutral role of adviser, merely laying out in an orderly manner the information

---

64 See Dagron, "Lawful Society," 39 and 46, and Magdalino, "Knowledge," 194–97.

65 For Maurice, see *Strategikon*, pr.36–40. See also *Ecloga* (ed. Burgmann), prologue and Eng. translation in Barker, *Social and Political Thought*, 84–85: "Since God has put in our hands the imperial authority . . . we believe that there is nothing higher or greater that we can do than to govern in judgment and justice . . . and that thus we may be crowned by His almighty hand with victory over our enemies (which is a thing more precious and honourable than the diadem which we wear) and thus there may be peace. . . ." See also Eramo, *Siriano*, 175–76.

66 *Strategikon*, pr.1–9.

relevant to each theme or topic.[67] Yet in respect of the role of God in granting victory to the Romans, or punishing them with defeat, Leo the emperor becomes particularly apparent as an author with a specific, personal vested interest in the advice he is giving and its reception. He is the writer who wishes to persuade his readership both that the treatise offers the best and safest way of dealing with the matter in hand, and that he speaks with the approbation of God in his role as divinely sanctioned and blessed emperor. This powerful motif recurs in other contexts in Leo's reign—Leo's wisdom was presented as an attribute fostered by divine support—and is reinforced both in texts and in material visual culture and symbolism. Indeed Leo presents himself elsewhere as a mediator between God and the Roman people as well as the shepherd guiding the flock or, as in the prooemium to the *Taktika*, as the helmsman guiding the ship of state.[68] The message is repeated at frequent intervals through the various parts of the treatise, with the implication that the whole of the *Taktika,* not simply the opening sentences, is an invocation of divine support. The distribution of explicit and implicit references or allusions to the divine—to divine support, to the relationship between piety and military success, to that between God and humankind more generally, as well as to the fate of those who fail to recognize the importance of divine support—is indicative of this emphasis. The structure of the prooemium combines all these elements, the formal legislative act as well as the attention to divine support and the fundamental need to place one's trust in God. In the constitutions that follow, where he is dealing with general issues of command, of generalship, and of the place of the eastern Roman empire in the order of things, Leo makes the greatest number of references to the need for both piety and the recognition of God's support. Where he is dealing with very practical matters—weaponry, marches, sieges, and so forth—this motif appears less frequently, except in Constitution 14,

67 This is not the place to enter the complex and ramified debate about the nature and impact on the writing of medieval notions of the self, of the author, and of the relationship between authors and their readers in various social and cultural contexts. For some orientation see Pohl, "Introduction: Ego Trouble?" and Lyubarskii, "Writers' Intrusion"; more generally, see the essays in Mondrain, *Lire et écrire*; Cavallo, *Lire à Byzance*; and Pizzone, "Rhythm."

68 See, e.g., the description of Leo as an emperor who is "neither ignorant nor lacking in wisdom but, on the contrary, who enjoys great wisdom and is favored by the grace of heaven in both his thoughts and his deeds" (Philotheos, *Kletorologion*, 85.6–8); and in his own *Homilies* (ed. Antonopoulou), 73, 78–79. For the visual material, see esp. Corrigan, "Ivory Scepter." The motif of the good captain guiding the ship of state was, of course, already ancient, and appears in Onasander and earlier authors (see the commentary to the prologue below).

on battle itself. While many of these references occur already in his main source, the *Strategikon*, the concentration on this motif in parts of Constitution 18 and throughout Constitution 20 and the epilogue is striking, as the following selection of references makes clear:

CONSTITUTION 18

§19 "... all ... who are engaged in the struggle for Christ our God. ..."

§40 "Justice pursued them for breaking their oath to Christ our God, the emperor of all. ..."

§42 "We now refer any military action against them to God ..."

§105 "They [the Saracens] cannot bear to call Christ God, <although he is indeed> true God and savior of the world." "Fighting, therefore, against such impiety by means of our own piety and orthodox faith and observing divine and civil laws as all the more inviolate, we wage war against them."

§123 "... against those people who blaspheme the emperor of all, Christ our God, and they must strengthen those waging war on his behalf against the nations by every means, by arms, gifts, and processional prayers. ..."

§124 "... it will easily, with God's help, be crowned with victory over the barbarian Saracens."

§125 "... if we have the divinity as our ally in everything, we will easily achieve victory over those peoples."

§127 "... with God fighting along beside us ... we place our hopes in God."

§133 "... as our most blessed father and autokrator of the Romans once did in his days by his sacred command."

§150 "... with God as your support, because of your faith in him and your love and good disposition toward Our Majesty."

CONSTITUTION 20

§3 "... he will reverence God and obey orders and he will carry them out as though he had received them from God."

§14 "They will dispel cowardly thoughts and will stir up manly ones, taking the very word, victory, as a good omen, and expecting to receive the reality from God."

§28 "When, with God's favor, an enemy city is taken by you. ..."

§39 "Inasmuch as God has been invoked, it is essential that what has been agreed on should remain firm ..."

§40 "But, after God, place your confidence in your weapons."

§42 "When God has granted you victory in open battle. . . ."

§47 "It is necessary to worship the Divinity at all times. . . ."

§58 "The person who first begins injustice has his victory taken away by divine justice itself."

§70 ". . . it is good to honor and revere the holy temples of God. . . ."

§72 "The bodies of the soldiers who have been killed in battle are sacred, especially those who have been most valiant in the fight on behalf of Christians."

§77 "In time of war it is necessary to offer prayers to God and to invoke him as an ally."

§79 "With God's help, you must hope that the rumor of victory will become a reality."

§82 "When God grants you the favor of routing the enemy in battle. . . ."

§97 "If the enemy decides to break the agreement, they will incur, along with disfavor and threats from God, <the reputation> of faithlessness. You, however, will remain safe and, with the help of God, will enjoy <the reputation> of being true to your word."

§112 "When the enemy, after God has granted you victory. . . ."

§140 "You must know, O general, that it is a religious deed and an extremely useful one. . . ."

§149 "You will make your soldiers stronger on the day of battle if you get up early and spread it about that you have had a dream ostensibly sent by God. . . ."

§156 "He made a counterargument by saying 'The thunder did not come because of us, but because of the enemy; God has sent the thunder on them. . . ."

§169 "For then God will become benevolent and will fight along with our armies."

§170 "Then entrust the entire undertaking to God."

§179 "They will think that the dream related by you is something positive, coming from God. . . ."

§191 "When, with the help of God, you have been favored with victory in battle. . . ."

§199 ". . . they will once more, as they attack, have the divine assistance."

§202 "When God has granted you victory, do not act in a precarious manner. . . ."

§213 "It is particularly incumbent upon you, truly reverent as you are toward the Divinity. . . ."

§218 "You will do this when, with God's help, you see that, at the first charge, you
gain an advantage over the enemy."

§221 "Above these, cherish and without hesitation obey the divine and wholly
true laws by which reverence is fortified. It is from these that the best
beginning proceeds, and the general will be most pleasing in the sight of
the God and emperor of all and to our Lord Jesus Christ and to us who rule
through him."

EPILOGUE

§2 "First, everything you intend to say or do should take its beginning from
God."

§3 "Thus, I judge it to be necessary for all things to take their beginning from
God." "Wherefore, we ought to do nothing apart from his will."

§4 "We should obey him to the extent that a private soldier obeys his commanding officer, as slaves a good master, and as officials the emperor."
"While the entire irrational flock is pastured and led by us, we are pastured
by God, the good shepherd, who for our sake out of love for mankind put
on our nature."

§5 "Nobody might ever deny the existence of God, except perhaps one who
has <already> destroyed his own soul. For everything is full of God and
he brings forth all things from nonbeing and completes everything and
finishes everything and provides for and manages all."

§6 "Proof of this truth is what he has made, heaven and all that is in it, the earth
and what is in it, indeed, the sea and everything in it. All that is helpful in
them bear witness to God's solicitude for us and manifest his providence."

§7 ". . . a person must not be invested with command before prayer and con-
verse with him, as though he were to consecrate himself to God. And he
should entrust the ordering of his own life to his providence."

§8 "Because of this, it is necessary for you, O general, to handle and observe all
things, above everything else, for the service of God. In particular, honor
and reverence his priests and bishops, and keep his holy temples as places
of asylum . . ."

§9 "To put it simply, preserve everything consecrated to God unmolested,
honored, and respected, for they are all holy and dedicated to God, who
is holy."

§10 ". . . but all, as belonging to God, must be free from all harm and abuse."

§11 "... who are intermediaries for us with God. ..."

§12 "What is done to them is referred to God, from whom they have received the high-priestly office. ..."

§13 (and preceding sections) "In this way, then, be reverent, orthodox, and well disposed toward God and those who serve him."

§14 "... you will conduct yourself reverently, in a manner pleasing to God. ..."

§15 "Your pious life having encompassed these things, be well assured that, along with righteousness, you will have God himself campaigning along with you."

§16 "The belief that one is not acting unjustly but is being treated unjustly will bring <you> the Divinity as your general and leader. ..."

§17 "Just as it is impossible for the unjust person not to suffer at some time the penalty for his injustice from God the judge, so it is impossible for one who has warded off and fought against injustice not to obtain victory from God."

§22 "You must especially <attend to> your own responsibilities by way of skillful commands, anticipation of what is to be done, concern for the Divinity. ..."

§41 "Be generous, respectful, and honorable regarding those who have died in battle, especially the men <cited for> bravery, assuring them of glory by blessing and an honorable burial."

§68 "Having been sanctified themselves, they are to bring to perfection and offer up the consummation of their own activity to God."

§73 "And so it is always necessary for you, O general, in a fitting, dutiful way, to devote yourself to prayer to God and to observe his commandments. ..."

It is not surprising that the greatest density of such material occurs in these three constitutions in particular: in Constitution 18, where Leo is dealing with the empire's enemies, and therefore stressing the orthodoxy and piety of the Romans, whose empire is protected by God; in Constitution 20, which sums up the advice and information offered in the preceding constitutions; and in the Epilogue, where a series of general exhortations rubs shoulders with specific pieces of advice, and where the crucial importance of divine support is underlined. In this, Leo is not setting a new agenda in respect of the relationship he makes explicit between the emperor, his commanders, military success, and enemy impiety, although he certainly expands it in relation to the Saracens. At the same time, he is building quite logically and consistently on the advice he had himself received from his father Basil I (although largely the work of the patriarch Photios). The young Leo was advised to place his faith in God in all things, and in doing so to emulate and honor his own father: Αὕτη ἡ πίστις πάντων τῶν ἀγαθῶν τὸ κεφάλαιον. αὕτη τῶν ἀρετῶν ἁπασῶν ἡ τελείωσις. Ταύτην οὖν ἔχε παρακαταθήκην ἄσυλον, ἥ σοι γέγονε τροφὸς ἀπὸ σπαργάνων αὐτῶν. Ἐγώ σου ταύτης ὑπῆρξα διδάσκαλος, μή με καταισχύνῃς πατέρα φιλόστοργον. ὀφείλουσι γὰρ οἱ ζωγράφοι τοὺς χαρακτῆρας ἀναστηλοῦν, οἱ δὲ τῶν βασιλέων παῖδες ἔμπνοοι εἰκόνες τῆς τῶν πατέρων ἀναδείκνυσθαι ἀρετῆς ("Faith itself is the culmination of all that is good, it is the summit of all virtues. It should, then, be your imperishable treasure. You have been brought up with it from the cradle itself, and it was I who taught it to you, so do not embarrass me, the loving father. For as it is the task of painters to capture and represent the distinctive features of things, so is it that the offspring of emperors must rise to become living portraits of their fathers' virtue").[69]

In the early tenth-century context in which he was writing, it seems that Leo was explicitly underlining the vast differences between the law-respecting and pious Romans on the one hand and the empire's barbarian enemies, and in particular the impious Saracens, on the other. Roman reliance on faith and divine support is a constant refrain throughout the *Taktika*: the emperor's military commanders are enjoined implicitly to emulate Leo in his piety and thus to act as though they too had a paternal responsibility for their soldiers, modeling the relationship with their troops on that which pertained between themselves and their emperor. The motif is not new—it appears in the ninth-century *Rhetorica militaris*, a text which, as we will see, Leo certainly must have known, and which no doubt reinforced his approach.[70]

Leo was also very probably drawing upon the ideals of filial devotion he had been urged to show toward his own father, and the precepts laid down for him in his youth, encapsulated in the book of advice referred to already. A particularly significant passage notes: Ἐπὰν δὲ τέλειος γένῃ τὴν φρόνησιν, κἀμὲ τὸν βασιλέα καὶ πατέρα καὶ παιδευτὴ εὐφρανεῖς, τότε καὶ αὐτὸς ἑτέρους πάλιν νουθετεῖν

---

69 Basil I, Κεφάλαια παραινετικά (ed. Emminger), 50 (ll. 17–22) (author's trans.). On the *paraineseis*: Markopoulos, "*Chapitres parénétiques*"; see also Tougher, *Leo VI*, 54–55.

70 Ed. Eramo, *Siriano*, 36.2–11; and cf. 45.1–2; see discussion in the commentary to Const. 2, below.

ἀξιωθήσῃ ("And when you reach the pinnacle of wisdom, hereby giving great pleasure to me, emperor, father, teacher, you yourself will be worthy of offering advice to others").[71] Indeed, considered in the light of such texts, the whole of the *Taktika* constitutes the emperor's advice to his commanders, which they themselves, when they follow it and have correctly understood it—and in emulation of their emperor—can apply in their own behavior, both as commanders of soldiers and in their private lives. The *Taktika* can be seen in many respects as a pamphlet on the Christian moral qualities of the general, and the moral conduct of warfare, as well as the proper management of an army. It mirrors the ideology of Christian imperial rulership, so that good rulership should be reflected in good generalship.[72]

Leo, Faith, and Warfare

While Leo's exemplar, the *Strategikon*, likewise emphasizes the role of God and the importance of piety and faith in the military commander, Leo's repeated insistence on this throughout the *Taktika* is striking, and in the context of the times, as has been pointed out, would suggest a greater awareness of the acute differences between Christianity and Islam. Leo's notions of the basics of Islam contained in Constitution 18.508ff. may well, as Dagron pointed out, derive their inspiration from the polemical writings of Niketas Byzantios, writing in the reign of Michael III, but the general sentiments about enemies of the faith might just as easily have been inspired also by the *Rhetorica militaris*, as I have suggested.[73] This and the passages that follow are especially important, since they represent the first explicit imperial recognition that the warfare along the eastern frontier is a war characterized, at least in terms of how contemporaries understood its origins, by religious belief and motivations (see the commentary to 18.620–25 below). The polemics of Theodore Abu Qurra or Niketas Byzantios, as well as other texts (martyrologies in particular) recognized the fundamental antipathy of Islam and Christianity to one another, while many other texts (including chronicles) included passages or comments denouncing Islam as heretical,

71 Ed. Emminger, 73 (ll. 24–26) (author's trans.).

72 On the ideal qualities of a general, see Antonopoulou, "Manuels militaires."

73 Dagron, "Byzance et le modèle islamique," 223–24 n. 21; Cosentino, "Syrianos's 'Strategikon,'" 265–72, arguments expanded and pursued in greater depth by Riedel, "Fighting." On the notions of just war in the *Taktika*, see Stouraitis, "Jihād and Crusade" and further literature and discussion in the commentary to Const. 18.508–16, 592–98, 620–25, below.

misguided, and false. The mid-ninth-century *Rhetorica militaris* is really the first sustained attempt to present warfare with a particular (but unnamed) enemy as religiously motivated and, followed by Leo in the *Taktika*, this is the first time in east Roman history that religion (both that of the enemy and that of the Romans) is understood as the particular motive for the enemy's attitude toward the Romans and Christians.[74]

Christianity (and more exactly, orthodox belief) had, of course, always been linked with the fate of the empire, and this was not the first time that an enemy was quite specifically identified through their religion.[75] But it is not easy to find explicitly religious motivations in the wars of the later Roman period, wars that involved primarily political and strategic concerns. Roman writers regularly referred to their own side as "the Christians" as well as "the Romans," and this awareness of difference is especially pronounced when dealing with enemies of a distinctly different faith. Indeed, awareness of difference in religion as one element among many in the accounts of war between the Christian Roman state and its enemies is, in itself, hardly surprising—Byzantine theologians as well as writers of miracle collections and saints' lives raise the issue of Jewish or heretical hostility to Orthodoxy throughout the seventh century, for example; religious debate and theological argument became the language through which politics and theories of power and authority were expressed. This can be seen increasingly from the late sixth century, but was stimulated after the defeats suffered by the Romans at the hands of Islam and the Arabs in the 630s and 640s.[76] And even if the wars which were fought against the Persians by the emperor Heraclius, culminating in the complete defeat of the Sassanid forces in 626–27, had an ideological quality that, as has often been pointed out, differentiates them from earlier conflicts, these were nevertheless understood to have been

---

74 In contrast, Polyaenus has been shown to be the first military author to present "barbarians" as a particular category of enemy (see *Strategika*, 7, pr.), whereas earlier treatises saw a generalized and unspecified enemy; see Wheeler, "Polyaenus," 48–50.

75 See Haldon, *Byzantium in the Seventh Century*, 355–75 for background and further discussion; and in general Kolia-Dermitzaki, Βυζαντινός "ἱερός πόλεμος" and Holum, "Pulcheria's Crusade."

76 See in particular Haldon, *Byzantium in the Seventh Century*, 37–40, 281–375; and for a broader and detailed discussion of Byzantine attitudes to warfare, peace, and the justification of fighting, see Stouraitis, *Krieg und Frieden*, 260–327; Theis, "Ist Frieden darstellbar?" with further literature; and the survey in Kolia-Dermitzaki, "Holy War."

righteous and just wars in defense of Christianity, rather than "holy wars" in any crude sense.[77] Correctly understood in the larger context of Byzantine history, Heraclius's war against the Persians was one of many justifiable wars waged to defend Christianity and the Roman empire, God's empire on earth. Certainly, as far as the sources represent the situation, the religious element was unusually strongly emphasized, but it seems that this was a somewhat isolated moment in the broader picture.

So, for the first time since the early seventh century—or so it would appear—and at an official level, Leo's treatise, perhaps inspired by the example of the *Rhetorica militaris* now attributed to Syrianos *magistros*, makes religion the key identifying sign par excellence of both sides. But whereas in the *Rhetorica militaris* religion is one of several key motives, in Leo's *Taktika* religious belief appears to be the main motive (along with greed) for the enemy's hatred and violence against Christians.

The issue of what exactly educated Byzantines knew and understood of Islam as a system of belief, both in its theology and in its practical day-to-day observance, has been the subject of considerable discussion. On the whole, the evidence suggests that in the latter respect some knowledge was reasonably widespread, if frequently inaccurate, but that in terms of the former Islam was understood as an aberrant form of a Christian—miaphysite—heresy until, during the course of the late eighth and ninth century, a somewhat more sophisticated, but still deeply hostile, understanding evolved, the result of an increasing willingness of theologians on both sides to engage in discourse, itself reflecting a combination of both political and philosophical-religious interests.[78] One—possibly but not certainly the earliest—redaction of the supposed correspondence between the emperor Leo III and the Caliph ʿUmar II was completed around the turn of the eighth and ninth centuries, for example;[79] the (Nestorian) patriarch Timothy reportedly debated on theological issues with the caliph al-Mahdi in the 780s.[80] But the view of Islam

77 See Stoyanov, *Defenders and Enemies*; McCormick, *Eternal Victory*, 69–71, 193–95; Speck, *Das geteilte Dossier*, 328–41; and note also in this context the issue of emperors who actively campaigned in the field in contrast with those who did not: see Nichanian, "De la guerre 'antique'." For Heraclius, see Kaegi, "Heraclians."

78 See the survey in Magdalino, "Road to Baghdad."

79 Meyendorff, "Byzantine Views of Islam," 125–27; Hoyland, *Seeing Islam*, 490–501; Gaudeul, "Correspondence"; Jeffery, "Ghevond's Text."

80 See Putnam, *L'église et l'Islam*.

as essentially a heresy retained its potency until well into the eighth century and in some circles beyond. John of Damascus viewed it in this light, for example, whereas his successor, Theodore Abu Qurra, writing in the early ninth century saw it rather for what it was, or had become—a coherent, powerful, and established religious system which posed a fundamental challenge to Christianity ideologically, culturally, and socially.[81]

During the middle and later years of the ninth century the evidence suggests that the volume of both diplomatic and cultural activity between the imperial court and the caliphal court at Baghdad (and maybe the courts of the local *amirs* who dominated the frontiers of Syria and Mesopotamia) was on the increase; and while this certainly involved debates over "secular" knowledge, contacts between the two worlds seem likewise to have fostered in the imperial court a far greater interest in Islam itself.[82] By the 860s the court of Michael III had encouraged the philosopher theologian Niketas Byzantios to compile a series of texts directed against key aspects of Islamic belief, including a letter on the morality of holy war,[83] writings which, as Dagron has reasonably surmised, seem to lie behind some of Leo's ideas in the *Taktika*, although it is unlikely that the writings ever reached a Muslim readership.[84] As well as the more open atmosphere of cultural contact, an added stimulus for this writing may well have been the perceptible shift in the fortunes of war from the early 860s onward, when in spite of regional setbacks, some of them serious, Byzantine forces began to achieve greater success on campaign and established a greater degree of equilibrium, which lasted—again with some substantial setbacks—beyond Leo's reign. Leo's father, Basil I, was particularly successful in this respect, an aspect of the reign which Leo could hardly fail to recognize and which must surely have influenced his thinking. Hagiography

81 On John's writings in general see Beck, *Kirche*, 476–86 and for Theodore: ibid., 488–89. See Meyendorff, "Byzantine Views of Islam," 116–18; in general Sahas, *John of Damascus*; Hoyland, *Seeing Islam as Others Saw It*, 480–89, with broader discussion 454–518; and on the background see Khoury, *Polémique byzantine*; Griffith, "Constantinople," esp. 186–94. See also, with further literature on Byzantine views and misunderstanding of Islam, Simelides, "Byzantine Understanding."

82 See Magdalino, "Road to Baghdad."

83 In PG 105:821–42; see Beck, *Kirche*, 530–31; Krausmüller, "Killing at God's Command"; Meyendorff, "Byzantine Views of Islam," 121–22; and see Khoury, *Théologiens byzantins*; Argyriou, "Perception de l'Islam."

84 See Dagron, "Byzance et le modèle islamique," esp. 221 n. 14, 223–24, n. 21, and esp. Stouraitis, *Krieg und Frieden*, 333–35.

may have played a role in generating a greater awareness of the differentness of the Muslim "other"—the acts of the forty-two martyrs of Amorium, for example, where aspects of Islamic belief and practice, including the issue of predestination, become themes within the narrative, themes which had already become topoi in the polemical literature.[85] Whether or not Leo was responding to this general situation or to a specific event or events, there can be little doubt that something stimulated this very obvious focus on Islam in Constitution 18. And just as significantly, there seems little doubt that this marks the beginning of a far more explicit religious message embodied in later texts in the tenth century, the most obvious examples of which are represented by the military harangues usually ascribed to Constantine VII, texts which connect with both the *Rhetorica militaris* and Constitution 18 of the *Taktika*.[86]

Leo's *Taktika* reflects another trend at this period, quite apart from its relevance for attitudes to technical matters, legal and administrative as well as military. For it also embodies a late ninth-century interest in rediscovering the roots of the east Roman state, and in reasserting the identity of the eastern empire as the "real" Roman empire, in contrast to the western empire established by the crowning of Charles the Great in St. Peter's on Christmas Day in the year 800. The *Taktika* is an important piece of evidence for this movement—even if it was confined largely to the court and to the elite at Constantinople—fitting in as it does with the situation following the bad-tempered exchange of letters between Pope Nicholas I and the emperor Michael III, or between John VIII (or Anastasius Bibliothecarius) and Basil I and his patriarch Photios in the 860s and 870s.[87] It is not a coincidence that imperially sponsored codifications of law—the *Eisagoge/Epanagoge*, the *Procheiros Nomos*, the *Basilika*, and Leo's own compilation of *Novellae*—were inaugurated at this time.[88] Leo had no compunction about

---

85 Meyendorff, "Byzantine Views of Islam," 129–30; and compare the comments of Niketas Byzantios on Islamic fatalism (in Khoury, *La polémique byzantine*, 243–45), with *De XLII martyribus Amoriensibus*, 73–74, for example.

86 See literature and references cited in notes 38 and 45 above, the texts and discussion in McGeer, "Two Military Orations"; and in general on the attribution of a range of literary works to Constantine VII, Ševčenko, "Re-reading Constantine Porphyrogenitus"; see also the useful discussion of Taragna, "Λόγος ε Πόλεμος" and comments of Eramo, "Retorica militare."

87 See in particular Wickham, "Ninth-Century Byzantium," 253–54 and esp. Fögen, "Reanimation of Roman Law," 17–22.

88 On which see Schminck, *Studien*, 1–15; see also *PmbZ* 24311 (at 39).

changing older legal prescriptions in his novels, for example, the better to represent his own views of his role as divinely appointed and divinely guided ruler.[89]
The *Taktika*, while certainly occupying a somewhat different place in the range of
imperially sponsored literary activities, nevertheless deserves to be understood
from this point of view.[90]

89 Fögen, "Leon liest Theophilos," see 96–97.

90 The revival or renaissance of Byzantine secular literature and learning which begins
in the late eighth century but has become closely identified with the "Macedonian" dynasty
founded by the emperor Basil I, and within the context of which these developments can be
understood, has received a great deal of scholarly attention. For a range of views and approaches
to its origins and course, see: Speck, "Weitere Überlegungen"; Fögen, "Reanimation of Roman
Law"; Speck, "Byzantium: Cultural Suicide?"; and Auzépy, "Manifestations." See the insightful
discussion in Magdalino, "Non-juridical Legislation."

2

SOURCES, DATE, LANGUAGE, AND STRUCTURE

Sources

The *Taktika* is based heavily on the earlier *Strategikon* of Maurice, with substantial extracts or paraphrases of material from the mid-first-century CE *Strategikos* of Onasander, from the early second-century CE *Taktike theoria* of Aelian, from the compilation of historical examples of stratagems and tactics on land and at sea by Polyaenus, written in the second half of the second century CE,[1] and from the ninth-century treatises on strategy, military rhetoric, and naval warfare of Syrianos *magistros*. Leo appears also to have drawn inspiration for some of his gnomic advice from the *paraineseis* or "books of advice" composed for him in the name of his father Basil I, probably by the patriarch Photios. Leo himself is credited with the composition of a collection of 190 maxims, a "pattern of guidance for the soul," addressed to an unnamed senior cleric and composed after his accession to the imperial throne; he may also have drawn on a collection of strategic *gnomika* independent of the material in Polyaenus, especially for the epilogue and parts of Constitution 20,[2] as well as on the Old and

---

1 On all of whom see Dain and de Foucault, "Stratégistes byzantins," 327–30, 333–35; Ambaglio, "Onasandro"; with discussion by Dain, *Manuscrits d'Onésandros*, 151–54, and with Kučma, "'Strategikos' Onasandra"; and the useful analytical commentary and discussion in Petrocelli, *Onasandro*; Dain, *L'histoire*, 77–151; and for Polyaenus the literature cited in n. 16 below. See also Loreto, "Generale e la biblioteca," but with the critical remarks of Wheeler, "Hugo Grotius," 145 and n. 19. While Leo probably did not know Aeneas Tacticus (mid-fourth century BCE), the latter's text on defending fortified places sets the tone for much of the strategic literature that followed; see Whitehead, *Aineias the Tactician*, who summarizes the older literature and notes that already by the late Roman period much of Aeneas's original composition had been lost, fragmented, or excerpted; Rance, "Aineias' *Poliorketika*"; the introduction in Bettali, *Enea tattico*; and Dain and de Foucault, "Stratégistes byzantins," 319–21.

2 Such collections were popular in middle Byzantine times, and it would be reasonable to suppose that Leo knew of one or several. Kekaumenos advises his son to read them, for example:

New Testaments, which—in the forms available and accessible to them—had a particular resonance for Byzantines, along with the (imperial) imagery of David and Solomon.[3] He was undoubtedly familiar with some of the considerable corpus of gnomic literature available to educated Byzantines, including the collections commented upon by his mentor Photios. He seems to have had access to some of the military texts (such as Syrianos *magistros*) directly, not only via the *Strategikon*, and he used these in assembling his material, as we will see.[4] This is clear with Onasander, for example, as well as with Aelian and Polyaenus, and he refers on a number of occasions to the "ancient authors" as a group whom he has consulted (but whose detailed accounts he does not always understand; see, e.g., 7§67.471–73).[5] It is true that substantial passages are taken verbatim from the *Strategikon* in particular, while very many more passages from the same source are only lightly reworked or partially paraphrased, and this has led many commentators to assume that, since both the historical situation in which the empire

---

Kekaumenos, *Strategikon* (Spadaro), 2:54 (88.13–31) (Litavrin, 154.23–156.8). For Leo's maxims, in part at least influenced by Hippocrates, see chapter 1, n. 15 above; and for their addressee and date of composition (both unspecified), see Leo VI, *Guidance* (ed. Papadopoulos-Kerameus), 213.5–10, 214.15–22. See also Odorico, *Prato e l'ape* and Searby, *Corpus Parisinum*, esp. 1:1–8 and 22–23. For general comments on Leo's ancient (military) sources only: Kourakes, *Taktika*, 27–37.

3 For Byzantine approaches to the Old and New Testaments, how they were understood, and how they were employed or excerpted, see Miller, "Prophetologion"; and see Magdalino and Nelson, "Introduction," 11 and Markopoulos, "Constantine the Great," 161.

4 See *ODB*, art. "Gnome"; Odorico, "Gnomologium Byzantinum"; Hose, "Das Gnomologion des Stobaios." For Syrianos, see chap. 1 above, pp. 18–19 n. 45. The *Strategikon* draws on a much wider range of sources, both directly in Greek and indirectly, through Greek translations or paraphrases of earlier Latin sources, such as Vegetius, *Epitoma rei militaris*, on which see Charles, *Vegetius in Context* and the critical review by Wheeler in *Bryn Mawr Classical Review* 2008.06.42. For detailed analysis of Maurice's sources, see Rance, *Strategikon*, chapter 5 ("Sources"). Until recently the standard view was that Leo was merely an antiquarian copier of older material with only limited contemporary relevance, except in the field of literary classicism. There were exceptions in particular issues, as with Leo's view of lineage and noble or aristocratic family, for example; see the comments of Ostrogorsky, "Observations," 4–5. For the best recent contextualization of the *Taktika*, see Magdalino, "Non-juridical Legislation"; and for the broader view, see Riedel, "Fighting."

5 Although he refers to Onasander by name only twice, at pr.57 and 14.677, he borrowed from him on many occasions, and not just via the *Strategikon*, as the commentary below makes clear. See Dennis, *Taktika*, index fontium, 671. For discussion, see Vári, *Bölcs Leó*, 56–59 and Dain, *Manuscrits d'Onésandros*, 151–54. For Leo's debt to Aelian, see Dain, *L'histoire*, 134–47; for Polyaenus, see below with nn. 16–18.

found itself had changed so dramatically between the late sixth and the late ninth century, and since the military organization and armament of the army had also changed, any details that Leo includes could in fact tell us very little about the real situation of his own times. As will be clear from the commentary, this is certainly not the case.

Leo thus exploited a variety of written sources and clearly drew on oral reports, in some cases possibly from eyewitnesses. The parainetic literature mentioned already was nearest to Leo in time (although it remains unclear as to what form it had taken by the time Leo was using it), yet it is apparent that this work did not influence substantially the structure of the *Taktika*, which was very clearly based on the *Strategikon*, albeit reorganized at various points to suit Leo's aims. The *Taktika* was the end result of a longer development of Leo's thoughts on the matter of compiling a guidebook. The so-called *Problemata* (ascribed in its single manuscript to Leo), if it is indeed by that emperor, or at least commissioned by him, seems to represent an earlier stage of his ideas about such a project, and is presented as a series of questions and answers extracted from Maurice's *Strategikon*.[6] That Leo progressed from this rather simple model to the more sophisticated *Taktika* is evident in the fact that on occasion he retains elements of the question-and-answer format of his original summary, as in Constitution 12 (see commentary to 12.637–90), even though this section of the *Strategikon* (7.B) is not included in the *Problemata*.[7]

Leo appears to have been less burdened than Syrianos *magistros* or the later author of the *Sylloge tacticorum* by the weight of the earlier tradition, in particular by the weight of earlier Hellenistic and Roman treatises, preferring instead to take his inspiration for a practical manual from the *Strategikon*, although he employed a relatively narrow range of sources in compiling the *Taktika*.[8] He copied wholesale

6 Leo excerpted the *Strategikon* verbatim in order to construct this text, in frequent contrast to the *Taktika*, where the original is regularly altered or paraphrased.

7 Dain, *Problemata*. The text survives only in M (the oldest witness, ca. 950–55), which suggests, as we will see, that it probably did not circulate outside the imperial palace.

8 The weight of this tradition was substantial: for the Byzantine aspect, see Hunger, *Profane Literatur*, 2:323–40; Dagron and Mihăescu, *Traité sur la guérilla*, 139–60; McGeer, "Military Texts"; Sullivan, "Byzantine Military Manuals"; and esp. Loreto, "Generale e la biblioteca." The tradition preceding Maurice in both Greek and Latin was yet more burdensome; see in particular Meißner, *Die technologische Fachliteratur*; Campbell, *Greek and Roman Military Writers*; and the literature cited above, chap. 1, p. 15, n. 31. For the sources exploited by, or having influence upon, the author of the *Strategikon*, see Rance, *Strategikon*, chapter 5.

and often verbatim large sections of Maurice's text, including, for example, the regulations on military discipline, which makes it inherently unlikely that this "military code" or set of "military laws" was available to him in an independent form (although the so-called *Nomos stratiotikos* appear to have been in existence and associated with the manuscript tradition of the *Ecloga* of Leo III and Constantin V by ca. 886 and thus, in their turn, cannot be said to derive from the *Taktika*).[9] Whether or not Leo was aware of this collection, there is no evidence that he used it in his section in the *Taktika* on discipline and punishments.

But while using Maurice as a model, Leo was in effect doing something relatively new, or at the least revitalizing a genre which had been moribund for almost three centuries, and this constituted in itself a novelty—even if we accept that one or two compilers, including Syrianos *magistros*, had produced treatises on the classical model before Leo. Syrianos *magistros*, the assumed author of the treatise on strategy, produced a far more generic work of far less detailed technical instructional value than Leo, who opted to follow the highly practical model offered by the *Strategikon*, and the last time such a detailed practical handbook had been prepared (whether we take this to have been the *Strategikon* or its slightly later summary version, the *De militari scientia*) had been before the great transformation of the late seventh and eighth centuries. Leo had taken on a substantial task, therefore, in both revising and updating older material, some details of which were clearly lost on him (for example, technical references to the various corps of *boukellarioi, phoideratoi [foederati], optimatoi [optimates]*, and the *Illyrikianoi* in *Strategikon*, 2.6.20–35; see on Constitution 12§§30.218–34.260), as well as in incorporating the structural information on provincial military organization or on thematic military capabilities that would make his treatise useful in the way he sets out in his prologue.

If the generally accepted reasons for much of his literary and codification activities are valid, it is not hard to see why the *Strategikon* was chosen as the inspiration for the *Taktika*, nor why the latter was given the form it was, since Maurice offered both an explanation (in the prologue) and a method, in the form

---

9 See introductory commentary to Const. 8 below. Vári, "Zur Überlieferung," 83–87. The "military code" is associated with the *Appendix Eclogae*, which was compiled at some point before 886: Burgmann and Troianos, "Appendix Eclogae," esp. from 89. For arguments that the military code derives from Leo's writing, see Schminck, "Probleme," esp. 176–78. See Burgmann, "Nomoi," who argues instead (60), and more plausibly, for a probably seventh-century date.

of the treatise itself, that admirably suited Leo's purpose as an emperor who was both asserting his own absolute authority in yet another matter directly relevant to effective rule and governance, and who wished seriously to address day-to-day issues affecting his subjects and the health of the state over which God had set him and from whom his divinely inspired wisdom was drawn. Leo had a particular interest in legislating for the future of his empire, both in terms of personal perspective, in respect of asserting and underlining his position and authority as emperor and autocrat, as well as with regard to his appreciation of the empire's political situation.[10]

That Leo was aware of comparable military writings from the Islamic tradition is most unlikely, although should not be entirely discounted. Such texts did exist and in the context of the mid-ninth-century cultural and diplomatic traffic between the courts of Constantinople and Baghdad it is not impossible that this was the sort of text which might have attracted the attention of a learned Byzantine visitor to the latter. A military treatise known as the *Stratagems* (al-Hiyāl) ascribed to a certain al-Sha'rānī al-Harthamī, and surviving in a much later abridged form under the title *The Organization of Warfare* (Siyāsat al-Hurūb), was written for the caliph al-Ma'mūn, for example.[11] In its structure it follows none of the individual Hellenistic or Roman treatises, but, like Leo's *Taktika*, seems to be an edited compilation from several of them. Beginning with a section (pp. 15–22) on the qualities and necessary virtues of a commander, the importance of taking advice, of sensible delegation of authority, of humility, and of devotion to and reliance upon God (cf., e.g., Onasander), it goes on to discuss in order the value and management of spies (23–25), camp security (26), tactical organization and function of different types of unit (26–27), the structure of different types of unit (28–29), marches and the order of march (29–30), camps and camp sites (31–32), scouting and marching in enemy territory (32), battlefield tactics (33–55), fortifications and sieges (56–64), and general advice on training, discipline, sedition in the ranks, and tactical organization, concluding with a reminder that all victory comes from God (71). The themes chosen

10 See the discussion in chapter 1 above, and Grosdidier de Matons, "*Constitutions tactiques*"; Vernadsky, "'Tactics' of Leo the Wise," 333–35; and, most recently, emphasizing the legislative aspects of Leo's writing, Magdalino, "Non-juridical Legislation," 174–78.

11 *Mukhtasar siyāsat al-Hurūb*; see also the discussion in Kennedy, *Armies of the Caliphs*, 111–14.

for discussion and the language bear some similarities to the material found in Onasander, Aelian, and Arrian, but whether the original was compiled from or took inspiration from these authors, either from translations or directly, must remain unknown. It is unfortunate that the abridged version omits all reference to its sources, but a Hellenistic text or texts, or even a later Sassanian source, is entirely possible.[12] As we have said, there is no evidence that Leo was aware of such writings, but it is worth noting that, paradoxically, the tradition upon which he drew for some of his work was a heritage shared with the archenemy described in Constitution 18, against the threat from whom Leo himself claims, in a revised version of his treatise, the *Taktika* was particularly intended to alert his generals and soldiers.

As we have seen already, one of the stimuli for his treatise seems to have been the text, now generally believed to have been compiled by Syrianos *magistros*, known under the title *Rhetorica militaris* which, while concentrating largely on the strategies to be deployed by a commanding officer to encourage his soldiers, includes also the advice that practical measures are an absolutely essential part of military preparedness, and that practical guidance and experience for the soldiers is the foundation of military success (*Rhetorica militaris*, 15.1, 15.40, 15.41). This is a text that was certainly exploited also by Constantine VII.[13] Leo's debt here (as well as to the other parts of the original collection, *On Strategy* and the *Naumachiai*, as the line commentary will show) seems clear, especially as regards the justification for warfare and the motives for Christians to defend their faith against Islam, a motif that recurs here and there in the treatise and especially in parts of Constitution 18. It is likely also that Leo's Constitution 19, on naval warfare, owes

12 Further discussion on the textual background, with detailed account of the literature and relevant textual material, in Rance, *Strategikon*, chap. 5 ("Sources"), showing the likelihood of a Sasanian source or sources; and Hamblin, "Sassanian Military Science." See also Shatzmiller, "Crusades and Islamic Warfare," esp. 253–57. General, if somewhat outdated, comments: Zaki, "Military Literature."

13 Cf. *Rhetorica militaris*, 15.1: Κατασκευάζεται δὲ ἀπὸ τῶν εἰρημένων κεφαλαίων οὐ μόνον ὁ πόλεμος, ἀλλὰ καὶ τἄλλα, δι᾽ ὧν πόλεμος κατορθοῦται. πέντε δέ ἐστι ταῦτα τὸν ἀριθμόν· ὅπλων παρασκευή, γυμνασία τῶν τακτικῶν, ἀνδρεία, καρτερία πόνων, καὶ ἡ πρὸς τοὺς ἡγεμόνας ὑπακοή· The text then goes on in caps. 40–43 to specify in detail what each of these entails. It is well-established that the two military harangues of Constantine VII likewise derived a great deal from, and were to some extent following the advice in, the *Rhetorica*. See the careful analysis in Markopoulos, "Ideology of War," esp. 49–53; McGeer, "Two Military Orations," 114–15, 122–24; and Karapli, *Κατευόδωσις στρατού*, 207–9 and 235–37.

its origins, at least as a topic worthy of inclusion, to Syrianos, for in the absence of any specific independent treatise on the subject, much of Constitution 19 seems to be based on advice or instructions taken from Syrianos or on anecdotes from Polyaenus, as the commentary below demonstrates.

As far as Leo's access to older authorities is concerned, his "ancient" material is in many instances simply drawn from Maurice, whom he frequently then supplements or expands by drawing on the few ancient texts he chose to exploit or which were available to him. A good example of the way he integrated material from several sources occurs in Constitution 20, the first section of which is derived largely from book 8 of Maurice, the later parts of which were drawn almost entirely from Polyaenus in particular, with a number of paragraphs also drawing on Onasander.[14] Where Leo clearly has the ancient author in front of him and can include material which Maurice simply omitted or paraphrased, this is largely from a manuscript tradition that appears to have grouped Maurice together with a number of ancient treatises, including Onasander and Aelian, a grouping which Dain suggested took place at some indeterminate point after the early seventh century. Leo drew directly upon this version of Maurice for the *Problemata* and (although often rephrasing or paraphrasing the original) for the *Taktika*, and this manuscript tradition can be differentiated from others by the variant readings contained in the works of Leo himself, although no witnesses are extant. Thus it is probable that the older manuscripts in which the tactical treatises are found (see chapter 1 above, and discussion of the manuscripts and dating of Leo's *Taktika*, below) reflect, with some exceptions (and of course the later compositions themselves), a good deal of the material that was available to Leo when he began planning his treatise, a corpus identified by Dain that comprised Maurice, Aelian, and Onasander. It is probably not simply coincidence that the ancient military treatises that survive with the most extensive and varied majuscule traditions also became the three main sources of Leo's *Taktika*.[15]

14 For the tradition from which such collections of sayings and gnomic advice were drawn, see Wheeler, *Stratagem*; also Krentz and Wheeler, *Polyaenus*, 1:, vi–ix; Meißner, *Fachliteratur der Antike*, 79–82 and 185–88; and Kučma, "'Strategikos' Onasandra."

15 Dain, *L'histoire*, esp. 122–27, 134–47 and 204–7 (and see also idem, *Manuscrits d'Onésandros*, 15–18 and 167–71); see Dain and de Foucault, "Stratégistes byzantins," 380–85 for the contents of the earliest manuscripts of the tenth century. For further comment on the version of Maurice employed by Leo, see Rance, *Strategikon*, chap. 1, ("The Manuscript Tradition").

As noted above, Leo's sources seem to have been limited, and he seems to have made little attempt to draw on a wider range of materials, although a number of the anecdotes Leo reproduces from Polyaenus's *Strategika* are found neither in Maurice nor in the *Hypotheseis* (or *Excerpta Polyaeni*), a collection of fewer than half of the original stratagems that could have been compiled at any point between the third and ninth centuries; nor in the derivative and abridged paraphrase of the Ambrosian tradition of Polyaenus. Leo seems thus to have had a fuller version of Polyaenus than that transmitted in the *Hypotheseis* (the earliest extant manuscript containing the original collection being cod. Laurentianus 56-1, of the thirteenth century).[16] Altogether Leo incorporates, paraphrases, or takes ideas from at least forty anecdotes from Polyaenus, excluding those on which he may have drawn but for which other possible sources, such as Syrianos, also exist. Of these, at least ten do not appear in the *Hypotheseis*, and Leo certainly used four and maybe all ten.[17] Interestingly, he tends to anonymize the tales he uses, removing

16 Polyaenus wrote around the year 162 CE and dedicated his *Strategika* to coemperors Marcus Aurelius and Lucius Verus. The Byzantine manuscript tradition is slender compared with that of the other treatises, and is largely disconnected from them; see Schindler, *Überlieferung*; Schettino, *Polieno*; and Hunger, *Profane Literatur*, 2:325–26. The original collection, culled from a range of earlier writers, is arranged in eight books, and included some 900 stratagems, of which only 830 survive, as a result of the loss of folios in the MSS; see Melber, *Polyaeni Strategematon*, xiii–xx; and Krentz and Wheeler, *Polyaenus*, 1:vi–xx; Dain and de Foucault, "Stratégistes byzantins," 333–35. For the *Hypotheseis*, see Krentz and Wheeler, *Polyaenus*, 1:xx–xxi and Dain and de Foucault, "Stratégistes byzantins," 337. Most scholars who have discussed the issue have argued that Leo had access only to an abridged version of Polyaenus: for example Schettino, *Polieno*, 21 and n. 2, and Trombley, "*Taktika*," 267, 272. The *Hypotheseis* formed the basis for several further sets of excerpts, the two earliest being the so-called *Strategemata Ambrosiana*, compiled at some point after Leo's *Taktika*, probably around the middle of the tenth century, and consisting of some 238 excerpts; and the so-called *Leonis imperatoris strategemata*, of some 27 sections and including 116 anecdotes, included in the (probably mid-tenth-century) *Sylloge tacticorum* (chaps. 76–102). See also Polyaenus, *Strategika*, xxi–xxii. On these and later versions of the *Strategemata Ambrosiana*, see Dain and de Foucault, "Stratégistes byzantins," 364–65 and Krentz and Wheeler, *Polyaenus*, 1:xx–xxiii; also see Wheeler, "Polyaenus"; Schindler, *Überlieferung*; and Dain, "Cinq adaptations."

17 See Polyaenus, *Strategika*, 8.16.1–2, for example, and compare with Const. 20§80, where no such detail is found in the *Hypotheseis*; or Polyaenus, *Strategika*, 3.9.38 and Const. 20§196, a tale not in the *Hypotheseis*. Additional material drawn from Polyaenus which does not appear in the *Hypotheseis* includes 2.3.11 (at Const. 20§158) and 3.9.47 (at Const. 20§197). Material not in the *Hypotheseis*, but on which Leo may also have drawn: 1.32.2 (Const. 20§156), 3.9.10 (Const. 20§76), 3.9.50 (Const. 11§21; 20§21, but see Maurice, *Strat.*, viii, 1.27), 4.18.2 (Const. 11§21), 6.12 (Const. 20§144, possibly), and 7.11.4 (Const. 11§21), although his debt to these is less obvious.

the names of the classical protagonists of the original text and retelling the story in a neutral form, a procedure found, for example, in late antique Christian literature that makes use of pre-Christian material as well as in Byzantine adaptations and reuse of classical texts.[18]

Whether Leo in fact used Arrian remains unclear, since on all occasions except one when this author might have been a source, the passages in question can also be found in the *Tactica theoria* of Aelian (in turn taken in several cases from Maurice). Arrian is barely mentioned by Byzantine writers of this period, although in the manuscript tradition his writing was certainly available. Aelian, however, was widely read. Anna Komnene notes in passing (*Alexiad*, 15.3.6) that her father Alexios had read Aelian, for example. Dain showed that the recension of Aelian that Leo used was different from the two recensions that can be reconstructed in the MS tradition (including that in M). On the basis of Leo's use of Aelian he further argued that Leo had a version of Arrian's *Techne taktike*, which he used to supplement certain mutilated or incomplete passages in his text of Aelian, notably at Constitution 6§§25–32 and 7§69.[19] A comparison of the relevant paragraphs is inconclusive, since it will be seen that while Leo's text is indeed identical to that of Arrian after the first lines of the passage, the latter reflect more closely the Aelianic text, which is clearly corrupt, as Dain saw.

Melber (*Polyaeni Strategematon*, xxi–xxii) believed that Leo also used the *Hypotheseis*, but this is uncertain—the textual variations in Leo may just as likely derive from an intermediate version of Polyaenus's complete work. For detailed demonstration that Leo used Polyaenus directly and not through the *Hypotheseis/Excerpta Polyaeni*, see Wheeler, "Notes," esp. the table at 162–63.

18 Compare, for example, Const. 14§97 and Polyaenus, *Strategika*, 2.1.3; Const. 17§19.117–19 and Polyaenus, *Strategika*, 2.1.4; Const. 17§89.517–33 and Polyaenus, *Strategika*, 3.13.1; Const. 20§144 and Polyaenus, *Strategika*, 5.16.3; 6.12; Const. 20§146 and Polyaenus, *Strategika*, 3.9.2; 1.40.7; Const. 20§154 and Polyaenus, *Strategika*, 2.1.17; 4.4.3. In one or two cases Leo retains the original name, as in the example of Scipio in Const. 20§80 and Polyaenus, *Strategika*, 8.16.1–2. See Hunger, *Profane Literatur*, 2:323–24; a similar redactional approach is found in late antique Syriac texts; see, e.g., Brock, "Syriac Attitudes" and Rigolio, "Sacrifice," to whom I am indebted for valuable discussion.

19 Dain, *L'histoire*, 134–47. Arrian did not copy Aelian, rather both authors drew on a common source, dubbed by Dain the "*Techne* perdue," which both follow, section by section, in almost identical language. For the relationship between Aelian, Arrian, and the "*Techne* perdue": Dain, *L'histoire*, 26–40; also Dain and de Foucault, "Stratégistes byzantins," 329, 331–32, 383–84; for Arrian in the Byzantine tradition more generally: Hunger, *Profane Literatur*, 2:325.

ARRIAN 32.1

ἄγε εἰς τὰ ὅπλα. <ὁ> ὁπλοφόρος <μὴ> ἀπίτω ἀπὸ τῆς φάλαγγος. σίγα
καὶ πρόσεχε τῷ παραγγελλομένῳ. ἄνω τὰ δόρατα. κάθες τὰ δόρατα. ὁ
οὐραγὸς τὸν λόχον ἀπευθυνέτω. τήρει τὰ διαστήματα. ἐπὶ δόρυ κλῖνον.
ἐπ' ἀσπίδα κλῖνον. πρόαγε. ἐχέτω οὕτως. ἐς ὀρθὸν ἀπόδος. τὸ βάθος
διπλασίαζε, ἀποκατάστησον. τὸν Λάκωνα ἐξέλιττε, ἀποκατάστησον. ἐπὶ
δόρυ ἐκπερίσπα, ἀποκατάστησον.

AELIAN 42.1

ἄγε εἰς τὰ ὅπλα. παράστητε παρὰ τὰ ὅπλα. ὁ σκευοφόρος ἀποχωρείτω
τῆς φάλαγγος. σίγα καὶ πρόσεχε τῷ παραγγελλομένῳ. ὑπόλαβε, ἀνάλαβε.
διάστηθι. ἄνω τὰ δόρατα. στοίχει, ζύγει, παρόρα ἐπὶ τὸν ἡγούμενον. τὸν
ἴδιον λόχον ὁ οὐραγὸς ἀπευθυνέτω. συντήρει τὰ ἐξ ἀρχῆς διαστήματα. ἐπὶ
δόρυ κλῖνον, πρόαγε, ἔχου οὕτως, εἰς ὀρθὸν ἀπόδος. ἐπ' ἀσπίδα κλῖνον,
πρόαγε, ἔχου οὕτως. ἐπὶ δόρυ μεταβαλοῦ, πρόαγε, ἔχου οὕτως. ἐπ' ἀσπίδα
μεταβαλοῦ, πρόαγε, ἔχου οὕτως.

LEO, *TAKTIKA*, 7§69.487–92

ἄγε εἰς τὰ ὅπλα. παράστητε παρὰ τὰ ὅπλα. <ὁ> ὁπλοφόρος μὴ ἀπίτω τῆς
φάλαγγος. ὁ σκευοφόρος ἀποχωρείτω τῆς φάλαγγος. σίγα καὶ πρόσεχε
τῷ παραγγελλομένῳ. ἄνω τὰ δόρατα. κάθες τὰ δόρατα. ὁ οὐραγὸς τὸν
λόχον ἀπευθυνέτω. τήρει τὰ διαστήματα. ἐπὶ δόρυ κλῖνον. ἐπὶ ἀσπίδα
κλῖνον. πρόαγε. ἔχ' οὕτως. εἰς ὀρθὸν ἀπόδος. τὸ βάθος διπλασίαζε,
ἀποκατάστησον. τὸν Λάκωνα ἐξέλισσε, <ἀποκατάστησον>. ἐπὶ τὸ δόρυ
ἐκπερίσπα, ἀποκατάστησον.

While it cannot be proven beyond doubt, this does not seem to me to be sufficient evidence for Leo's using Arrian here instead of Aelian, since we do not know
what form the nonextant recension of Aelian took at this point. Leo does note
that Aelian and Arrian "speak as though from one mouth" (7.469–70), which
suggests that he knew this much about Arrian's treatise (although Arrian did not
copy and extend Aelian; rather both writers were drawing on the same exemplar,
which Arrian extended to incorporate more explicitly Roman material). But it
seems more likely that the version of Aelian he had before him already included
these missing elements or alternative readings, since otherwise one might have
expected a more obvious and liberal use of Arrian's treatise, which—along

with Aelian and Onasander, as well as Leo's *Taktika*—was later included in M, the Laurentianus collection of military treatises compiled in the middle of the tenth century.[20]

In support of this position we should note that while Aelian deals only with ancient Greek and Hellenistic military organization, Arrian supplements their common source with frequent reference to the Roman practice of his own day. At no point does Leo draw on Arrian for this information, which seems odd given his particular interest in the Roman past in the relevant constitutions (esp. Const. 4§§58–62, but based in the first instance on Maurice, *Strategikon*, 12.B.8.3–36, who follows Aelian; Const. 6§§25–34, following Aelian, *Tactica theoria*, 2.11–13; and Const. 7§§67–69, selectively following Aelian, *Tactica theoria*, 24–31). In the last case Leo might have added comments on later Roman practice after Arrian, *Techne taktike*, 3, 4, 9, 10, and especially 35 onward, which deals specifically with Roman cavalry tactics and formations. The absence of an Arrianic element is especially obvious at Constitution 7§§54–55, where Leo discusses the defensive "Roman"—i.e., Byzantine—heavy infantry formation described as a *phoulkon* (Lat. *fulcum)*, yet makes absolutely no reference to a passage in which Arrian describes the comparable Roman *testudo* formation (*Techne taktike*, 11.4–5, see ll. 15–16: καὶ ἀπὸ τοῦδε τοῦ συνασπισμοῦ τὴν χελώνην Ῥωμαῖοι ποιοῦνται. . . .). For while it seems likely that Leo in fact partly misunderstood this passage in Maurice, and that the contemporary term for such a formation—as suggested by a passage in the slightly later *Sylloge tacticorum*—was *syskouton*, this presented a perfect opportunity, one might have thought, for Leo to contrast ancient and modern, as he claims to do.[21] And by the same token, at Constitution 5§2.13–15 (where Leo describes briefly the Macedonian pike, 16 cubits in length) the source must have been Aelian 14.2, since Arrian 12.7 writes 16 feet, in error (and cf. Polyaenus, *Strategika* 2.29.2: likewise 16 cubits). Leo incorrectly claims that the (ancient) Romans used this weapon—had he read Arrian he would have known that this was not the case! Nor is there any evidence in Leo's text at all that he was familiar with (or at least, that he used) Arrian's *Ektaxis kata Alanon*, relevant, for example, to Leo's Constitutions 9 and 10, on marches and the baggage train. This text, too, was copied, in an often garbled form, into M (ff. 196–97), the

20 Dain and de Foucault, "Stratégistes byzantins," 331–32, 383–84.

21 For discussion of this tactic and Leo's misconstrual, see commentary to Const. 7§54.370, below.

only witness, and might thus have been available in the palace before that manuscript collection was compiled in the middle of the tenth century (although its
presence in that collection was more likely a product of the researches commissioned during the reign of Constantine VII). There is thus no persuasive trace of
any independent use of Arrian in the *Taktika*, even if the *Ektaxis* formed the basis
for Maurice, *Strategikon*, 12.A.7 (a section of the *Strategikon* which is in fact not
exploited by Leo in the *Taktika*).[22] We may conclude that, in view of the fact that
Leo does not show any other dependence on Arrian, in spite of his awareness of
the similarities between the two texts, the question of whether he actually used
Arrian at all must remain open.

Dennis identifies four occasions where Leo supposedly quotes or paraphrases Aristotle. Three of these (Consts. 14§22.157–58, repeated at 14§64.469–70;
and 20§90) appear in Maurice's *Strategikon* (7.B.12 and 8.2.60), and the fourth
(Const. 2§15.96–102) is found in a series of passages borrowed from Onasander
(*Strategikos*, 2.131–34 [1§21]). None in fact derive from Aristotle. Indeed, the context and language are quite different and the connection seems to reflect simply the
common use of a single word. As for Leo's other ancient sources, the lines purportedly from Demosthenes at Constitution 19§37.215–16 appear in fact to be derived
from Syrianos, *Naumachiai*, 9.10–12, while those from Isocrates (Const. 3§8.41),
Menander (Const. 14§21.148–49), and Polybius (Const. 20§27) are found already in
Maurice (cf. respectively Maurice, *Strategikon*, 8.2.31, and cf. 8.1.5; 7.B.12.4; 8.1.26),
or in Basil I's Κεφάλαια παραινετικά addressed to Leo (see Const. 20§2.25 [τὸ γὰρ
μέλλον ἀόρατον] and cf. Κεφάλαια παραινετικά, 63, §38.24–25).[23] Again, none of

22 See Rance, *Strategikon*, chapter 5 ("Sources"); also idem, "*Fulcum*," 295–99; and
Nefedkin, "Mavrikii i Arrian" (I am grateful to Philip Rance for this reference); but with a dif-
ferent view of the origins of the *fulcum*: Wheeler, "Legion as Phalanx," part 1, esp. 357–58. For the
Byzantine tradition: Dain, "Manuscrits"; on the original text: Stadter, "*Ars tactica* of Arrian";
Wheeler, "Arrian's *Ars Tactica*"; idem, "Legion as Phalanx," part 1, 309–13; and part 2, 152–59;
more generally: Bosworth, "Arrian and the Alani," 217–55.

23 The maxim is at Isocrates, *Ad Demonicum*, 29.3 and is not found in the *Strategikon*.
Whether the *paraineseis* were composed by Photios remains unknown, although likely; like
the author of the *paraineseis* he was certainly very familiar indeed with the work of Isocrates;
see Wilson, *Photius, Bibliotheca*, 12, 16; and see cod. 159 (101b–102b) and cod. 260 (486b–487b),
and for a detailed list of sources used by the author of the *paraineseis* (including, extensively,
from Agapetos and Isocrates, as well as from biblical texts), see Emminger, *Studien*, 3:42–49; for
the manuscript tradition, see ibid., 23–42. See also Hunger, *Profane Literatur*, 1:160–61 for the
tradition of *Fürstenspiegel*, in which the *paraineseis* are to be understood. That Leo was himself
directly familiar with Isocrates is a moot point, but he certainly knew the *paraineseis*.

these has any demonstrable direct connection with the named ancient authors. As noted already, Basil's *paraineseis* clearly inspired Leo for some of his maxims (cf., e.g., Const. 20§70 to Basil I, Κεφάλαια παραινετικά, 51, §3; Const. 20§213 to Κεφάλαια παραινετικά, 60, §30). A supposed association made by Dennis between Euripides and Constitution 14§9.48–50 (Maurice, *Strategikon*, 7.B.7), is spurious (again an association based merely on the presence of some common vocabulary), as is that between Constitution 14§15.101 (Maurice, *Strategikon*, 7.B.11.1–9) and Thucydides 7.61. Of the three tales or maxims taken from Plutarch, one comes through Polyaenus (Const. 20§162 = Polyaenus 2.10.5) and one (Const. 2§32.220–21, repeated at 20§128.633–34) is found in both Maurice, *Strategikon*, 8.2.79 and Basil I, Κεφάλαια παραινετικά, 67, §46. The other (Const. 20§40.216–18, see comm.) seems to be taken from Plutarch's *Moralia* 3 (*Apophthegmata Lakonika*) 210E, although it is just as likely that Leo took it from another later source as yet unidentified, maybe from one of the other texts which could have been available at the time, as (probably) reflected in the contents of the earliest manuscripts to include the various military and poliorcetic literature. Alternatively, it may have been a text that he recalled from his earlier reading and education.[24] We should bear in mind that Leo was educated as a young man by Photios, and that there is good evidence for the availability of a wide range of classical texts to which Leo probably had access at this time and from which, either directly or through memory, he was able to quote. It is possible that Isocrates numbered among these, for example, although—as we have seen—the examples of Isocratic material in the *Taktika* can all be found in other sources that Leo certainly used.[25] Nevertheless, there is really no reliable indication that Leo drew directly on the nonmilitary classical authors named above.

It seems fairly clear, therefore, that in respect of access to ancient texts Leo worked from a relatively limited manuscript base, although we cannot know whether or not he simply ignored or was unaware of some texts that he would have found in the palace library, or other libraries, had he looked. More likely, he did not feel he needed to consult any other texts apart from these, since it appears that the palatine library was reasonably extensive, at least according to an oblique reference

24 See Dain and de Foucault, "Stratégistes byzantins," 380–85. For aspects of the Byzantine reception of Plutarch, see the essays in Garzya et al., *Moralia di Plutarco*; Ziegler, *Überlieferungsgeschichte*; and Tartaglia, "Saggio."

25 See Mango, "Availability of Books"; Wilson, *Photios, Bibliotheca*, 6–17; and esp. Lemerle, *Byzantine Humanism*; note also Dain, "Transmission."

to the availability of literature at the emperor's disposal in an oration of Arethas of Caesarea.[26] As noted above it is certain that he had copies of Onasander, Aelian, and Maurice, probably from the same manuscript corpus; as well as Polyaenus (or a derivative compilation); and very probably the three texts of Syrianos *magistros*.[27] Exactly how extensive a resource the palatine archive, or any other library or collection to which Leo might have had access, actually was must remain unknown. Apart from the patriarchal library, collections of literary and theological works were, as far as the evidence can inform us, mainly to be found largely in monastic contexts until at least the second half of the ninth century, for example, and it was probably only under Leo VI, and certainly during the reign of his son Constantine VII, that the palatine resources were expanded. The range of material was apparently wide, but access seems to have been relatively limited.[28] Presumably the palace archives included legal texts and imperial legislative material, as evidenced from the codifying activities of both Basil I and Leo VI and their advisors (although since the patriarch Photios was also instrumental in these efforts such material may well have been found in the patriarchal library also). Certainly in respect of material to do with the military, or with the imperial aspect of the military, it does seem that such resources were quite limited: it is worth recalling that when Leo VI commissioned the *magistros* Leo Katakylas to search out material on imperial expeditions, the latter found nothing in Constantinople (assuming that he had access to palatine archives), but had recourse instead to a monastic library.[29]

To argue that Leo was merely a compiler and imitator both misrepresents and misunderstands Leo's text and to some extent his intentions. In the first place, he was necessarily bound—by literary convention, precedent, and his own cultural values—to use as much as he could of the earlier treatises, precisely on the grounds

---

26 Commented on by Tougher, *Leo VI*, 115. See Westerink, *Scripta minora*, 2:46.

27 See esp. Mazzucchi, "Basilio parakimomenos." For a detailed reassessment of the manuscript tradition of Maurice's *Strategikon*, which has implications also for understanding the sources of Leo's *Taktika*, see Rance, *Strategikon*, chap. 1, "The Manuscript Tradition."

28 On the availability of manuscripts in imperial and patriarchal circles at this time, see Mango, "Availability of Books" and Wilson, "Libraries"; see also Mango, "Discontinuity"; and for the military treatises in particular, Loreto, "Generale e la biblioteca," is useful, although should be used with care due to excessive hypothesizing about ancient authors. For Leo's education, see Vogt, "Jeunesse," 403–4 and chapter 1 above. For range of material and access, see, e.g., Canfora, "'Cercle des lecteurs'" and Irigoin "Survie et renouveau."

29 *CPTT* C.24–30.

that innovation for its own sake was regarded not simply as unnecessary but positively a bad thing. But Leo also believed firmly that the words of the ancients had a contemporary relevance if they were appropriately framed, as is clear from his legislation.[30] Second, as Dain pointed out long ago,[31] the *Taktika* was intended as a summary of the older "tactical" tradition, with Leo's own additional remarks and advice, and set in the "modern" context of Leo's own time and its concerns. Third, the notion that things had changed sufficiently to render the *Taktika* merely an exercise in antiquarianism can be challenged. Many technical terms had changed their meaning, of course, but Leo is aware of this and says so, and where an older word might no longer be understood he offers his own "modern" gloss or equivalent. Further, the fundamental assumptions of a disciplined, nonmercenary "national" army, if we may use the term, which the middle Byzantine social and political elite regarded as part of its Roman inheritance, still applied, so that much of what Leo takes from Maurice continued in his eyes and in the eyes of his contemporaries to have relevance.

Indeed, where Maurice notes in his prooemium that he will use contemporary Latin words and phrases, especially for the commands to be given to the soldiers in order to carry out maneuvers on the drill ground and in actual combat, Leo explicitly reverses the sentiment, noting that he has replaced older, unintelligible Latin terms with contemporary Greek words or phrases, just as he had done in the prooemium to the *Procheiros Nomos*.[32] While this is also a topos and follows the model he found in Maurice's prologue,[33] there is enough corroborative material to show that a considerable part of Leo's writing did indeed reflect, or was made to reflect, contemporary realities as well as ideas, and this makes the *Taktika* an important source for a range of topics related to the empire's military, political, and social history. His tactical arrangements for a small provincial force of four thousand soldiers outlined in Constitution 18§§136–49 are in accord with his earlier account in Constitution 12, derived from Maurice, of the standard Roman battle formation, yet the detail is quite original to Leo and carefully tailored to the sort of small-scale

---

30 The last point is emphasized by Dagron, "Lawful Society," 39, 46; for Byzantine contempt for "innovation" (νεοτερισμός), see esp. Hunger, "On the Imitation," with idem, "Reconstruction," 510.

31 Dain and de Foucault "Stratégistes byzantins," 355.

32 Pr.72–73; cf. *Procheiros Nomos*, pr.52–53.

33 See below, p. 126.

fighting (along the Byzantine-Arab frontier in particular) with which this section is concerned and which can be corroborated from narrative sources. This is especially the case, as has been argued by Dagron, for example, with respect to Leo's sections in Constitution 18 on the Arabs of Tarsos, and in other constitutions where he compares Islamic approaches to fighting and the reward granted to soldiers for their sacrifices in battle.[34] As we will also see below, it is likewise the case in the prooemium, where Leo draws not just on earlier tactical writings but on contemporary imperial legislation as well. By the same token Leo draws on information collected by eyewitnesses who fought for his father Basil I, possibly in some cases from Basil himself (see on Consts. 9§14, 11§§21–22, 17§65, 18§40, as well as the information in 18§§136–39, already noted).

The sources or inspiration for most sections of the *Taktika* can be traced, therefore, but there remain some important gaps. The origins of the list of weaponry and equipment in Constitution 5, as well as of certain passages in Constitutions 3 and 4, remain uncertain. Dain suggested a lost work from which Leo may have drawn;[35] but it seems in fact far more likely that Leo simply used material he had readily available: the list of armaments in Constitution 5 seems in fact to be Leo's (or his redactor's) own composition, with the information simply compiled from Constitution 6, which Leo had in turn based on Maurice, *Strategikon*, 1 and 12, but updated and emended to conform with what he understood as contemporary practice (although it nevertheless includes a number of anachronisms, such as the σωληνάρια or arrow guides listed at Const. 5§3.27). The description of the various grades and ranks in a provincial army in Constitution 4 is again based on Maurice, but with a substantial number of emendations, editions, glosses, or omissions to make it relevant for a contemporary reader. It is possible that parts of Constitutions 4 and 5 together were originally part of a separate document, although the fact that 4§§6–33 was taken to form §§1–26 of a separate text excerpted from the *Taktika* in the late tenth century does not necessarily support this contention, and given Leo's editorial interventions in both the structure and the language of the *Taktika* there seems little reason for doubting that he might have been responsible for these sections himself.[36]

34 See Dagron, "Byzance et le modèle islamique."

35 Dain and De Foucault, "Stratégistes byzantins," 356.

36 This text is known under the title Ἐκ τῶν τακτικῶν τοῦ βασιλέως Λέοντος τοῦ σόφου, see Dain, *L'"extrait tactique,"* 9–12, and text at 84–88.

Finally, we may consider the sources of Leo's information from eyewitnesses or contemporaries. He himself as much as states that one of these was Nikephoros Phokas, or one of his close associates, since this senior officer is referred to on several occasions in this light (Consts. 11§121, 15§32.202, 17§65.374, and commentary, below). Likewise, his father, Basil I, is referred to on two occasions, although we cannot know whether Leo heard these stories from him directly (9§14.60, 18§95.454). But from a different and slightly later but quite reliable text we know that Leo certainly consulted with and probably derived information from another senior officer, Leo Katakylas (or Katakalon), an experienced general whom, as already noted, Leo commissioned, according to Constantine VII, to research a treatise on imperial military expeditions, the basic text for which survives in a short text found in M and in the Leipzig manuscript of the *De cerimoniis*.[37] This is all we can securely say about the sources of Leo's information, but it is illustrative of the sort of people with whom he discussed his interest in military affairs, as well as the range of his interests (see below, on "Structure").

Manuscripts and Date

The earliest versions of the *Taktika* are found in three groups of manuscripts, all of the period ca. 950–ca. 1050.[38] The first group includes the earliest surviving manuscript, cod. Mediceo-Laurentianus Plut. 55.4 (= M), dated ca. 950–55. This is a collection of sixteen Hellenistic, later Roman, and Byzantine tactical handbooks, including the major ancient authorities, along with works of Leo VI and Constantine VII. The manuscript is missing a number of folios,[39] and must be supplemented, therefore, from the second manuscript in this group: a Vienna codex

---

37 For detailes, see *CPTT* 180–81.

38 For surveys and discussion, see Vári, *Leonis imperatoris tactica*, 1:xi–xxxv; Dain, "Inventaire raisonné," with corrections and emendations in Andrés, "Nota"; Dain and de Foucault, "Stratégistes byzantins," 354–57, with the stemma textuum at 376.

39 Detailed description: Dain, *L'histoire*, 183–202, 375–77; for the missing folios: Vári, *Leonis imperatoris tactica*, 1:xi–xii; and for the related, later, manuscripts: Dain, "Inventaire raisonné," 34. M is perhaps the best known of the manuscripts containing classical, Hellenistic and Byzantine military writers; it was certainly the product of an imperial scriptorium and has been shown to have close affinities with a small group of manuscripts all dated to the 950s and representative of the encyclopedism of the court of Constantine VII: Irigoin, "Centres de copie," 178–81; Dain and de Foucault, "Stratégistes byzantins," 382–85; Haldon, ed., *CPTT* 38–39 with literature. This date has been challenged by Schindler, *Überlieferung*, 216–17, but the proposed alternative date (in the 980s) has not met with general acceptance. For the context and

(W = cod. Vindob. phil. graec. 275), of a slightly later date, otherwise a near exact equivalent to the text in M, although not necessarily copied from it, and with a somewhat different disposition of some elements of the text of the *Taktika* (see below). Again, this manuscript lacks a number of folios, including those with the opening sections of the prooemium, and several from the final part of the treatise, including most of Constitution 20 and all of the epilogue (from 20§45.242 to the end of the treatise); it contains only works of Leo VI.[40] The place of production of these two manuscripts is unknown, but an imperial scriptorium has been hazarded for M; W still lacks a detailed codicological study and analysis, although a palatine location is not unlikely, given its contents.

A second manuscript tradition is represented by cod. Ambrosianus B 119 sup. (= A, dated 959–63, commissioned by Basil the *parakoimomenos*), and includes a slightly reworded version of the *Taktika* (clarifications, technical terms updated, in some respects a more demotic vocabulary) as well as a separate version of Constitution 19 on naval warfare, which is placed after the main body of the *Taktika* and at the head of a small corpus of naval texts (see introductory notes to the commentary to Const. 19, below), as a result of which A's Constitution 20 is numbered 19. As has been pointed out, this version occasionally includes readings which are more accurate than those of M or W, but is on the whole of only limited help in reconstituting the probable original full version of the treatise.[41] These three early manuscripts represent some of the stages through which the treatise went as it was composed, and reflect several revisions, the earliest of which at least, as I will suggest, were carried out by Leo himself, or on his instructions.[42]

The prooemium, which is common to all three versions, lists the full twenty constitutions. But the redaction which survives in M has what later became

---

tenth-century encyclopedism, see Dain, *L'histoire*, 183–85; idem, "Encyclopédisme"; idem, "La transmission"; and Wilson, *Scholars*, 142–45.

40 See Vári, *Leonis imperatoris tactica*, 1:xii; Dennis, *Taktika*, ix–xii; Irigoin, "Centres de copie," 178–79; Dain and de Foucault, "Stratégistes byzantins," 355; Dain, "Inventaire raisonné," 35, and 35–40 for the dependent later MSS.

41 Vári, *Leonis imperatoris tactica*, 1:xiii (with incorrect date attribution). For discussion of the MS tradition, see Leoni, *La parafrasi Ambrosiana*, 18–20; Mazzucchi, "Basilio parakimomenos" 267–84, 292–316; Dennis, *Taktika*, xi, xiii; with Dain, "Inventaire raisonné," 40 and Dain and de Foucault, "Stratégistes byzantins," 385.

42 Vári, *Leonis imperatoris tactica*, 1:xi–xxxv; idem, "Überlieferung," 47–87, see 47; summary in Dain and de Foucault, "Stratégistes byzantins," 354–57 and 376; Andrés, "Nota," 261–63; Dennis, *Taktika*, ix–xiii; and the commentary.

Constitutions 15, 17, and 19 merely appended to the main text of the *Taktika*; whereas these Constitutions were retained in the body of the treatise copied into the Vienna codex (W) at a slightly later date. Thus in M they had an independent existence, and were each preceded by the heading Λέοντος ἐν Χριστῷ βασιλεῖ αἰωνίῳ βασιλέως Ῥωμαίων, suggestive of this autonomous status (see also below).[43] Unlike Constitutions 1–18, the constitution on naval warfare also has independently numbered sections in all three early manuscripts. But since Const. 20 and the epilogue likewise have numbered sections, it is not clear whether this is further evidence that it was at some point handled somewhat differently from the other constitutions, or not. By the same token (and as noted in chapter 1 above), a substantial section (§§103–25 [=PG §§109–31]) of the original Constitution 16 (in M; in W and later versions Const. 18) on how to deal with different nations, was duplicated, after the three additional constitutions appended to the *Taktika* in M, as a separate text (fols. 401–2, 404), with the title: Λέοντος ἐν Χριστῷ βασιλεῖ αἰωνίῳ βασιλέως Ῥωμαίων πῶς δεῖ σαρακηνοῖς μάχεσθαι. There seems also to have been an intermediate recension of this particular section, of which no independent version survives, but which can be detected in the structure of this part of Constitution 18 as a whole.[44] As we have noted, all three early versions of the *Taktika* share the same prooemium, but also have what should be the third and fourth constitutions (according to Leo's original list at the end of the prooemium) in reverse order. Vári follows Meursius, who first edited a version of the text in 1612, in restoring this assumed original order, and Dennis maintains this.[45]

A third group of manuscripts is represented by three codices, all dated ca. 980–1050. The oldest, probably written ca. 1020, is cod. Vat. graec. 1164 (V), although substantial sections of the *Taktika* are missing. The remaining two codices are represented by four manuscripts which were divided at a date some time

---

43 Of these separate chapters, the naval warfare treatise was edited and published by Dain (see n. 4 above) and has now been reedited with an English translation in Pryor and Jeffreys; the section on the Saracens by J. Lami in 1745: *Johannis Meursii opera omnia*, 6:4–6, from p. 1414. See further on the excerpts from the text Hunger, *Profane Literatur*, 2:333–34.

44 Detailed discussion below, introductory comments to Const. 18 and commentary to 18.686–95.

45 Dain and de Foucault, "Stratégistes byzantins," 362; Dain, *L'"extrait tactique,"* 9, and see in detail the discussion in the commentary to Const. 18.686–95, below. On the manuscript placement of Consts. 3 and 4 see Vári, *Leonis imperatoris tactica*, 1:42, note; 48, note; Dennis, *Taktika*, 38, note.

after their production (P = cod. Paris. Graec. 2442; B = cod. Barberinianus graec. 276 [II 97]; N = cod. Neapolit. Graecus 284; E = cod. Scorialensis graecus Y-III-11). These originally formed just two codices (P with B, N with E), incomplete versions of the *Taktika* being found only in B and E.[46] All three codices were produced in Constantinople, in the scriptorium of Ephraim, with B (+ P) probably the closest to V, produced at about the same time, and E (+ N) somewhat later, perhaps ca. 1050. They have been shown to derive from a single common exemplar (dubbed by Dain the "Mazoneus"), produced ca. 980–1000. The versions in V, B, and E have in common the fact that the last sections of Constitution 18 (from §120, middle of l. 584 [= PG §126]), all of Constitution 19, and Constitution 20 up to and including §180 (= PG §180), are missing. B and E also share the fact that most of the remainder of Constitution 20 (from §187 on) and the Epilogue have been separated from the main text of the treatise by a misbinding in their exemplar, and are found inserted between chapters 1 and 2 of the treatise entitled *Parekbolai*, which precedes Leo in both manuscripts (the same was true of V, but this MS has lost the folios containing the *Parekbolai* as well as Consts. 1–5 of the *Taktika*).[47] Cod. A has been shown to represent a parallel tradition to that of the hypothesised Mazoneus, and thus to the recension represented in VBE, both traditions deriving from a common (the "Ambrosian") ancestor, having no affinity with the branch represented by M. But A shares some features with W, occasionally retaining what is probably a more accurate reading. This may be because the paraphrast has merely tacitly corrected an error he found in his exemplar.[48] Dennis's edition is founded chiefly on M and W, supplemented by A and by VBE, although a small group of four other later

---

46 See Vári, *Leonis imperatoris tactica*, 1:xiii–xiv; Dain and de Foucault, "Stratégistes byzantins," 385–87 with literature; Dain, "Inventaire raisonné," 35–40 for these MSS and their later dependents.

47 For details of the folios concerned, see Dain, "Inventaire raisonné," 41–42; Irigoin, "Centres de copie"; and Gamillscheg and Harlfinger, *Repertorium*, 82–83. For their common source, the Mazoneus, so-called by Dain after a colleague, see Dain, *L'histoire*, 203–40 and Dain and de Foucault, "Les stratégistes byantins," 387. For the *Parekbolai*, a tenth-century compilation, see Dain and de Foucault, "Stratégistes byzantins," 368. Dennis did not use BE for the text from 20§187 to the end of the Epilogue, but since the reading in A is normally the same as in VBE, this hardly affects his edition.

48 For the complications of the relationship between A and the other early manuscripts, see Vári, *Leonis imperatoris tactica*, 1:xxx–xxxiii; Dain, *Manuscrits d'Onésandros*, 36–42, 44–47; idem, "Inventaire raisonné," 40–42; esp. Mazzucchi, "Basilio parakimomenos," 285–90; and Leoni, *Parafrasi Ambrosiana*, xvi–xvii.

manuscripts deriving directly from W should also have been adduced in respect of some of W's readings (in particular the earliest of these, Paris. Gr. 1385, 13th c.).[49]

A number of different dates have been proposed for the compilation of the *Taktika*. Thus Kulakovskij argued for an early date of 890–91; most recently Dagron suggested ca. 895 for an original version, with some later additions.[50] Other dates suggested have been ca. 900,[51] between 904 and 908,[52] some time after 904,[53] and after 906–7, on the grounds that a reference to an attack on Cyprus referred to in Constitution 20§212 took place in 906.[54] The three firm chronological references in the text itself, assuming they are not later interpolations (for which there are no real grounds in this instance), are to an expedition, probably in 895/96,[55] led by Nikephoros Phokas (see 11§21 and commentary to 11§21.121–27), and another campaign, probably that of the year 902 to recapture Theodosioupolis (see Const. 18§134, and commentary to 18§134.675–78). In addition, Leo also refers in 18§40.212–21 to the conflict with the Bulgarians in 894–96.

Three sets of evidence provide the framework for a chronology. First, the composition of the *Taktika* has been associated with the date at which the *Procheiros Nomos* (see below) was promulgated and, in particular, attention has been drawn to the titulature used by the author in the various versions.[56] Now it is worth noting that the title to the whole work as it is preserved in M refers to the emperor not as *basileus*, the usual title for an emperor by this time, but as *autokrator*. The preference for using the title *autokrator* has been noted by several scholars in connection with Leo's reign and that of his brother and successor Alexander.[57]

---

49 See Dain, "Inventaire raisonné," 35.

50 Kulakovskij, "Lev Mudryj," 400–401; Dagron, "Byzance et le modèle islamique," 219 n. 1.

51 Kučma, "'Taktika L'va'," 77. See also Ostrogorsky, "L'expédition du prince Oleg," 51.

52 Vári, "Desiderata," 226.

53 Moravcsik, *Byzantinoturcica*, 1:404.

54 Dain and de Foucault, "Stratégistes byzantins," 355, based ultimately on "a forgotten Byzantine conquest." In fact, this does not seem to have entailed a conquest of Cyprus at all, but is rather a reference to Cyprus as a convenient location off which a combined fleet might assemble in order to preempt an attack on Byzantine territory. See commentary on 20§212.1103 below.

55 And not ca. 900 as the editor (Dennis, *Taktika*, 203 n. 11) suggests; see Cheynet, "Phocas," 293–95.

56 Schminck, *Studien*, 92–96.

57 For example, Ostrogorsky, "Mitkaisertum," 166–78; Rösch, Ὄνομα βασιλείας, 36; and Schminck, *Studien*, 92 and n. 232, 93–94 with sources and literature.

·

The reasons for this preference appear to be associated with Leo's wish to distance his own title from that of a junior emperor. It is well established that the title of *basileus*, introduced by Heraclius, became the standard form to designate the emperor thereafter.[58] In 899, in the title of the *Kletorologion* of Philotheos, Leo is as usual referred to as *basileus*;[59] yet in an inscription dated to 904 he is referred to not as *basileus* but as *autokrator*,[60] a title he clearly preferred and used thereafter, as did his brother (although in the acrostic in Const. 20, from which Alexander's name was later expunged, the latter's original title is given as *basileus*).[61] Use of the title *autokrator* in the *Taktika* might in consequence suggest a date of composition after this had become the official preference, thus from at least the year 904 (and the appearance of the title *basileus* thus indicating, perhaps, an original date of composition before this point).

Second, and as noted already, reference is made at Constitution 18.210ff. to the Bulgars' breaking of the peace with the empire in 894, the subsequent transporting by Byzantine vessels of the Magyars across the Danube, the initial imperial victory, and the peace that was later concluded, following the Byzantine defeat at Bulgarophygon, in 896 (Const. 18§42.227ff.)—a peace that was, according to the text, still in force.[62] This places the compilation of the original version between ca. 898 and the end of Leo's reign. The absence of references to historical events after this date means nothing, of course—Leo's vague comments on the current threat from Islamic enemies may date to any period in the latter part of his reign.

Third, it is notable that at two points in his prooemium Leo refers to the *Taktika* with the words *procheiros nomos*, a point which, it has been argued in light of the uniqueness of this term, must surely refer back to the existence of the legal codification of the same name. But scholars disagree fundamentally about the date of issue of this code, ranging between a traditional date between 870 and 879, on the one hand, to a more radical redating to the year 907, so that this is of

---

58 See Chrysos, "Title *Basileus*."

59 Philotheos, *Kletorologion*, 81.4; Schminck, *Studien*, 92–94.

60 Spieser, "Inventaires," 162, no. 12.

61 See introductory note to Const. 20 below. Alexander used the title also on his coinage; see Grierson, *Catalogue*, 1:179; 2:523, 525. Leo and Alexander were acclaimed *autokratores* in 886; see Christophilopoulou, Ἐκλογή, 95–96, but the title seems not to have been regularly employed in documents until ca. 903–4.

62 See Tougher, *Leo VI*, 175–80.

little help in determining when the *Taktika* was initially compiled.[63] And finally, it is interesting that in the final paragraph of Constitution 1, which introduces the theme of the virtues and attributes of a *strategos*, Leo is at pains to stress that the ultimate aim of the *strategos* who is worthy of his title and position is to cultivate good relations with the emperor, rather than, "by paying little account to suitable and fitting matters, to arrive at the opposite" (1§13.44–45). This is an odd thing to write, and may reflect the fact that between 905 and 908 Leo was having to deal with the desertion to the Arabs of the general Andronikos Doukas and his son, but that in the final year Constantine Doukas (Andronikos had died in the interim) returned to an enthusiastic welcome by the emperor.[64] The comment may suggest both Leo's own disappointment at the events surrounding these developments as well as his worries about future such provincial elite opposition. If this suggestion is correct, it might point to a date of composition of this paragraph in the period ca. 908 or shortly thereafter.

We may therefore draw the following, tentative, conclusions. First of all, it seems fairly certain that the extant version of the *Taktika* in M must represent a version which was produced between 904 and Leo's death in 912, because it employs the title *autokrator* (and possibly after the issue of the *Procheiros Nomos* in 907, if we accept Schminck's still problematic redating). But in view of the use of the title *basileus* in the three separate sections which were extracted from the original *Taktika* as transmitted in M, as well as its appearance in the title of the separate section on how to fight the Saracens, we can suggest either that these may have

---

63 For 907: Schminck, *Studien*, 98–101, based upon the dating of two sermons of Arethas of Caesarea and upon a reference to Leo's legislation on fourth marriages, and the relationship of this evidence to the contents of the prooemium to the *Procheiros Nomos*. See Oikonomidès, "Leo VI's Legislation"; but note also Goria, review of Schminck, *Studien*, which draws attention to unresolved issues in this respect. The traditional dating of the *Procheiros* to the period 870–79 has recently been defended by van Bochove, "Some Byzantine Law Books" and esp. idem, *To Date*, 29–56. See also Signes Codoñer and Santos, *Introducción al derecho*, 147–60 for a similar position to that of van Bochove. On the whole, Schminck's argument for the date of 907 has found little support. Magdalino, "Non-juridical Legislation," 177 notes additionally that Leo also refers to his treatise as an "Introduction" (*eisagoge*: pr.62), recalling another of the legal manuals associated with his name and that of his father Basil I, although this is taken from Maurice, *Strategikon*, pr.23–24, so has little value in this context (even if Leo himself may have drawn the comparison).

64 See Tougher, *Leo VI*, 208–10.

been written before ca. 904, thus in their original independent form were probably composed by Leo before his preference for the title *autokrator* became standard; or were excerpted from the original *Taktika* and retitled by a later scribe or redactor using the standard imperial title of *basileus*. It has been argued that the hostility to Islam and the Saracens in Constitution 18 reflects the changed political relations between the Caliphate, in particular the emirates of northern Syria and Cilicia, and the empire after the year 900, when, following a period of relative peace, war broke out again and when the empire suffered some significant reverses.[65] But in fact, since hostilities had been worsening since at least 897, and were already bad before then, this argument has little purchase for the date of composition. The date of composition of the additional Saracen material in the *Taktika* cannot be fixed with any exactitude, and the general context alone might quite readily render the production of separately composed independent chapters on naval warfare, surprise attacks, and sieges in the same period plausible. This partly reiterative process may well reflect the working methods of the scribes whom the emperor employed to collate and recopy the material.

In consequence, it must remain uncertain whether or not the shorter version of the *Taktika* (in M) should be understood as a reflection of an earlier version, incompletely or inadequately updated after 904, and intermediate between an original draft of the *Taktika* and the version extant in W. It includes the reference in Constitution 16 (in M; = Const. 18 in W) to the Saracen danger as the motive for Leo's composing (part of) Constitution 18, and that in the epilogue to this motive for composing the *Taktika*; but as noted, it does not incorporate into the main treatise the three independent sections on sieges (Const. 15), surprise attacks (Const. 17), and naval warfare (Const. 19), which employ the imperial title *basileus*. What is now Constitution 17 must, despite its existence as an addendum to the main text of the *Taktika* in M, nevertheless have been conceived, or revised at some stage, by Leo as part of a single work, since the text refers explicitly on two occasions within this constitution to the earlier ones on marches and on camps (see below on Const. 17§50.283–84 and §55.314), and 18§102.489 refers back to Constitution 17; similar considerations apply to Constitution 19.

Given these cross-references to other parts of the treatise in the separate sections, we can suppose that the final version of Leo's original treatise, that which became the basis for the tradition upon which M, W, and A were based, included

---

65 Vasiliev, *Byzance et les Arabes*, 2.1:137.

all nineteen constitutions, regardless of the original order of Consts. 3 and 4. Possibly under the direction of Constantine VII the redactor of M, or of the exemplar on which M was based, was asked to extract the chapters on surprise attacks, siege warfare and naval warfare, as well as the section on how to fight the Saracens (which, however, may have existed independently already), perhaps preparatory to turning them into independent treatises. In the process, the constitutions after 14 were renumbered accordingly (or even given an enumeration for the first time), although the prooemium was not altered to reflect these changes.[66]

The constitution on naval matters may already have existed at an earlier stage as an independent text. A, as we have seen, represents a parallel tradition to that of the hypothesized Mazoneus (from which VBE are derived), and both of these traditions derive from the common ("Ambrosian") ancestor. They have no direct affinity with the branch represented by M, although A shares some features with W. Yet all three manuscripts, MWA, have an internally numbered version of the naval constitution as well as of Constitution 20 and the epilogue. On the assumption that such a numbering in only three constitutions is unlikely to have been introduced independently by two different redactors, this can only mean that the section numbers must have been present before the creation of the exemplars from which both the tradition behind M as well as that behind the "Ambrosian" family were derived. The most likely explanation is that the last three sections—19, 20, and the epilogue—were prepared by a different copyist, who introduced the numbering for clarification. It must also mean that the constitution on naval matters was fully integrated into the treatise at this point, the numbering being retained when the text was appended to the extant redacted versions of the *Taktika* (first into M, and then, independently, A).

In contrast to the process through which the version of the *Taktika* in M came into being, the copyist of W was almost certainly working from a somewhat later redaction, one on which M was based but that was created after the split between the "Ambrosian" tradition and that represented by M and W as indicated by the closeness of the latter to one another. It is impossible to determine whether W was based directly on the same manuscript as M or on another exact copy no longer

---

66 That the *Taktika* in its extant versions was certainly the product of several redactional phases is widely agreed, although the stages themselves remain obscure. See Vári, "Überlieferung," 47; Hunger, *Profane Literatur*, 2:331 and n. 34; Dain and De Foucault, "Stratégistes byzantins," 355.

extant, which still included these three constitutions as well as the full version of the Saracen section in Constitution 18, but retained the inversion of Consts. 3 and 4.[67] On this argument, M represents a redacted version of an earlier manuscript of the *Taktika*, deliberately extracting the four above-named sections or constitutions, whereas W represents a copy from the same earlier manuscript, but retaining the original structure, Leo's final plan for the treatise, and including in the body of the work what, in M, became the "additional" excerpted constitutions or sections. Thus the version in M may not necessarily represent a "pre-edition,"[68] but rather an excerpted version of the original, which is probably more accurately represented by W.

It seems likely that we are dealing with at least two missing stages in the process by which W was eventually produced. This may be indicated also by the fact that the section on the Saracens in Constitution 18 (16 in M) appears to show at least three separate stages or elements in its composition, and may thus reflect Leo's initial draft, or an early version of a separate text. In addition, we know that the acrostic composed from the first letters of each paragraph in Constitution 20 was altered to erase the name of the emperor Alexander, Leo's brother, possibly— but not necessarily—at the order of Constantine VII.[69]

To summarize: M probably reflects the work of a scribe or redactor working from a version which reflected Leo's original plan (and which already included the additional Saracen material), but who took the constitutions on sieges, ambushes, and naval warfare out of his exemplar and gave them their own titles and separate identity. The same redactor probably also excerpted the material on how to fight the Saracens, although the possibility that behind this lay an original text, perhaps a product of Leo's own hand, should not be excluded. W probably comes from a copyist who, following the plan set out in Leo's prooemium, was reproducing what he found in the common ancestor of both W and M (including the inversion of Consts. 3 and 4). The version in A, a codex slightly later than M (its relationship to W, however, remains to be determined), which can be dated to 959–63,

---

67 This scheme agrees with the *stemma codicum* proposed by Vári, *Leonis imperatoris tactica*, 1:xxxii (and cf. discussion at 1:xxx–xxxv), and diverges from that proposed by Dennis, *Taktika*, x–xi, who suggests that W may have been copied directly from M (although allowing the possibility of a copy from a manuscript closely linked to it). See also Dain, *L'histoire*, 34–35 and Dain, *Problemata*, 7–9.

68 See previous note and Dain and de Foucault, "Stratégistes byzantins," 356.

69 See below on Const. 20 for discussion and literature.

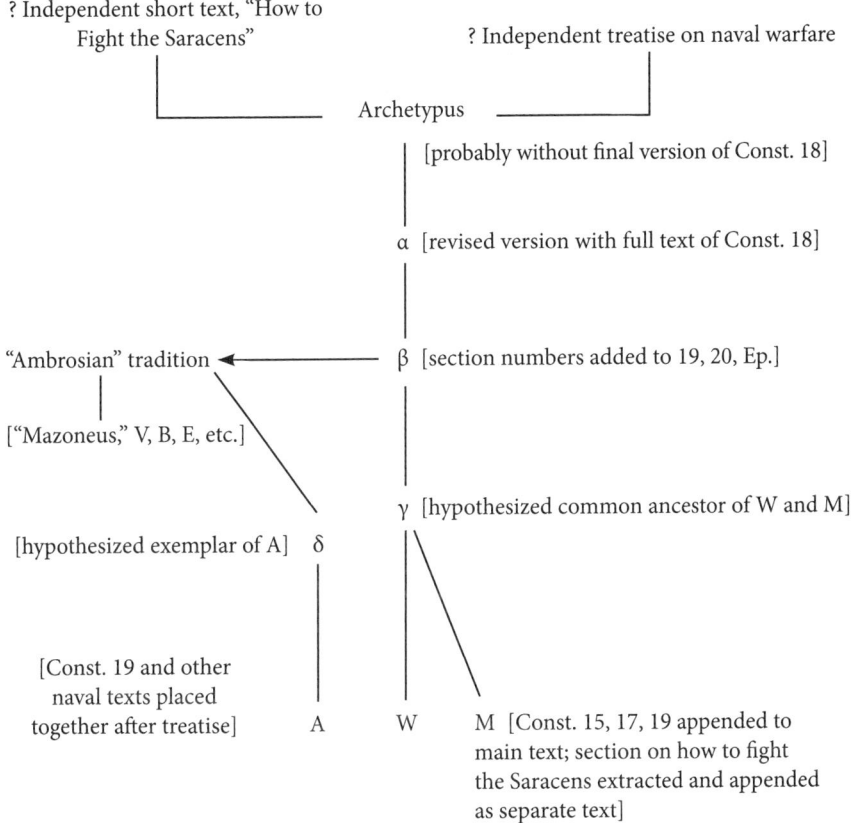

? Independent short text, "How to Fight the Saracens"

? Independent treatise on naval warfare

Archetypus

[probably without final version of Const. 18]

α [revised version with full text of Const. 18]

"Ambrosian" tradition ← β [section numbers added to 19, 20, Ep.]

["Mazoneus," V, B, E, etc.]

[hypothesized exemplar of A]   δ

γ [hypothesized common ancestor of W and M]

[Const. 19 and other naval texts placed together after treatise]   A   W   M [Const. 15, 17, 19 appended to main text; section on how to fight the Saracens extracted and appended as separate text]

Hypothetical *stemma* for versions of the *Taktika*

has likewise inverted Consts. 3 and 4, as in M and W, but has far fewer similarities with either in respect of errors and omissions, although in some cases it more accurately reflects what was probably the original reading of the archetype, or a copyist who knew the material better. For example, at Constitution 14.476, both M and W have the reading ὀφθαλμός whereas in Maurice, *Strategikon* 12.B.17.22, which Leo follows, and in the paraphrase of the *Taktika* in A, the term is in fact ὄμφαλος (a technical term taken directly from Hellenistic and Roman texts; see Constitution 14§86.591 and commentary), as noted by the editor in the critical apparatus. As we have remarked, the redactor of A took Constitution 19 on naval warfare out of the main body of the *Taktika* and appended it as a separate text together with other texts on naval warfare. Since A is, however, to some extent a paraphrase, and as we have seen, based on a different tradition to that of WM (that is to say, that represented by VBE), it is difficult to establish its relationship with the putative exemplar(s) of M or W.[70] Whether the earlier version of the section in Constitution 18 on the Saracens, as well as the separate versions of the constitutions on siege warfare, ambushes, and naval warfare, were also originally composed by Leo before the rest of the treatise, and before he incorporated them into a larger work; or whether they were simply taken out of the main text and retitled by a later redactor, probably under Constantine VII (this seems more likely), remains uncertain. The evolution of the text may be illustrated graphically in the accompanying figure.

What emerges from this discussion is that Leo VI was intensely interested in the codification of military matters from at least ca. 900 onward. Already at a somewhat earlier point he had compiled, or ordered the compilation of, a series of questions and answers, *erotapokriseis,* the answers being drawn verbatim from the *Strategikon* of Maurice.[71] But Leo was not satisfied with this, it seems. Influenced by the style, structure, and purpose of Maurice's *Strategikon,* which is obviously the model for his own longer treatise, and the various factors already noted above, he decided eventually to produce something more useful, something organized by specific topics and that would be helpful in the practical conduct of warfare. An initial version was then finished, perhaps by the year 904 or shortly thereafter. Somewhat earlier he might also have prepared sections on siege warfare, ambushes, naval warfare, and maybe something on dealing with the Saracens. Additional material was produced, perhaps very shortly afterward in view of its content and the context described above, but remained independent of the main treatise.

One other treatise remains particularly problematic for this period, namely the so-called *Sylloge tacticorum.* This text is not found in the majority tradition in any of the tenth-century manuscripts which survive, although it is ascribed by its fourteenth-century copyist to the year AM 6412 (= CE 903–4), and therefore to

70 See above; Leoni, *La parafrasi Ambrosiana,* xviii–xxviii; and Mazzucchi, "Basilio parakimomenos," 310–16.

71 Dain and de Foucault, "Stratégistes byzantins," 354. The *Problemata* is divided into twelve chapters, which reflect in order the twelve books of the *Strategikon.*

the reign, if not the pen, of Leo VI. The imperial title *autokrator* appears in two places in the heading, which may in fact indicate a post-904 date (at least for this particular ascription), as we have seen.[72] It draws on the *Taktika* of Leo in several places, along with the work of Onasander, Aelian, Polyaenus, Julius Africanus, the *Strategikon* of Maurice, and Syrianos, which were all mined for appropriate material. As Dain has shown, much was taken from collections which are no longer extant but which can be more or less completely reconstituted from later manuscripts.[73] The *Sylloge* thus draws on a broader range of sources than does Leo, and this, quite apart from some of the contents and its very different style, indicates that Leo was not the author.[74] Sections 1–75 present ancient, Roman, and Byzantine methods of tactical organization and armament, while sections 76–102 constitute a body of parainetic literature drawn from the *Excerpta Polyaeni*, misleadingly referred to in the literature as the "Stratagems of the Emperor Leo."[75] The treatise contains virtually no material on the Saracens (although it does mention "Arabs" in chap. 24, on ambushes) or on naval matters, nor does it have more than three short paragraphs on sieges, at a time when all these topics were still highly relevant.[76] Some of the contents would appear to date from a slightly later period—reference to the *menaulion* as a heavy infantry pike or spear (as opposed to a javelin or throwing spear, as in the *Taktika*), or to trapezoidal or triangular shields, for example, as well as to particular cavalry formations such as the *allagion*, which do not appear in the *Taktika* and which appear to be a development of the years after Leo's reign, would suggest this, although it may be that an original version again dates to Leo's own time and was later brought up-to-date, and that only this revised version survived. It was itself the source of some material for the

72 See *Sylloge tacticorum*, pr.13 and 19, with Dain and de Foucault, "Stratégistes byzantins," 357–58; discussion with review of literature in Mecella, "Kestoi."

73 Material from the *Taktika* appears in *Sylloge tacticorum*, 35, 36, 38, 39, and 44, for example (see critical apparatus in the edition of Dain); and material from Onasander represents a different MS tradition of this text from that excerpted by Leo; see Korzenszky and Vári, *Sylloge*, xvii, all of which makes it highly unlikely that the compiler of the *Sylloge* was associated with Leo and those working under his direction.

74 Dain and de Foucault, "Stratégistes," 350–53, 357, 388 (on the only manuscript in which the *Sylloge tacticorum* was transmitted, Laurentianus 75-6, 14th c.).

75 Edited with Engl. trans. in Krentz and Wheeler, *Polyaenus*, 2:850–1075. The inaccurate name of the treatise reflects earlier assumptions that the author of the *Sylloge* was, in fact, Leo.

76 *Sylloge tacticorum*, 11, 53, 54.

Praecepta of Nikephoros, so certainly predates that treatise.[77] Like Leo's *Taktika*, the *Sylloge tacticorum* presents both ancient as well as contemporary material, often explicitly comparing the two and clarifying the differences. In general, it would seem far more likely (even if the initial version was begun in Leo's reign or under his auspices) that those parts of the treatise which display a clear mid-tenth-century character (such as the descriptions of the "contemporary" Roman *bandon* or *allagion*, for example) were part of a now-lost source, and were assembled, together with extracts from a range of ancient and Byzantine authors, by a later redactor.[78]

Structure, Style, and Language

The *Taktika* announces in the prooemium a clear plan, yet this plan is not followed beyond Constitution 14 in the earliest manuscript (M, dated ca. 950–55): Constitutions 3 and 4 of the *Taktika* are reversed, so that this version of the treatise includes seventeen constitutions only, plus the epilogue; while the chapters on siege warfare, surprise attacks, and naval warfare appended in M to the main body of the work at fols. 379, 387, and 394 appear as chapters 15, 17, and 19 respectively only in the second manuscript, the slightly later Vienna codex (W), otherwise a near exact equivalent to the text in M, although not necessarily copied from it, as we have seen.[79] In the third version of the *Taktika* (in Cod. Ambrosianus B 119 sup. = A, ca. 959–63), commissioned by Basil the *parakoimomenos*, Consts. 3 and 4 are again inverted, as in M and W, and while both incomplete and in many respects a slightly condensed version of the original, its existence suggests the contemporary importance of the text to those interested in these treatises, whatever their purpose in making the copies.[80] There seems thus to have been relatively soon after

77 See *Sylloge tacticorum*, 35.4 and 5; 38.1 and 3, for example. Further discussion in McGeer, "*Syntaxis armatorum quadrata*"; and for the *Praecepta*, McGeer, *Sowing the Dragon's Teeth*, 184–88. The *Sylloge* is a fascinating text and merits a great deal more scholarly scrutiny.

78 Krentz and Wheeler consider the text to have been "the work of a forger," meaning presumably that the compiler exploited the name of Leo VI in assembling his text: Krentz and Wheeler, *Polyaenus*, 1:xxi–xxii.

79 For the particulars of the tradition behind these three constitutions, see the relevant introductory remarks in the commentary below, as well as Dennis, *Taktika*, ix–xii; Irigoin, "Centres de copie," 178–79; and Dain and de Foucault, "Stratégistes byzantins," 355.

80 The Ambrosian manuscript includes paraphrases of Maurice, *Strategikon*, as well as that of Onasander, both of which include a substantial number of contemporary terms which

its composition some considerable interest in Leo's treatise. That interest extended into the early eleventh century. As we have seen, three manuscripts produced in Constantinople ca. 1020–50 (VBE) contain the *Taktika*, as well as the anonymous treatise *Campaign Organization* and that on *Skirmishing*. Interestingly, in these three MSS the treatise *Campaign Organization* is copied directly after Leo's *Taktika* (and without attribution, so that it appears as part of Leo's work), but the latter ends midway through Constitution 18 (from §120.584, which breaks off after the opening phrase, and is completed with the text of Const. 20§186), omitting sections which Dagron has plausibly argued were by then no longer seen as relevant or appropriate, because the "Saracens" were no longer perceived as a problem in the same way as had been the case a century earlier.[81] Such a clear attempt to render the *Taktika* relevant and part of a broader corpus of tactical writings is significant. Of course, its later widespread copying in the early modern period tells us nothing about its popularity or circulation in Byzantine times (Vári counted some 88 manuscripts and Dain found some 100 manuscripts containing some or all of the *Taktika*).[82] But the manuscript tradition of the *Taktika*, given the hypothesized number of prototypes behind MWA and VBE, does suggest that it was reasonably well known and reproduced in a palatine or "high command" context at least; indeed, in the glosses and emendations introduced into the original version of the *Taktika*, the redactor of the tradition preserved in VBE demonstrates a high level of technical understanding and familiarity with the Roman and Hellenistic tradition.[83]

Leo's aim in writing what would in effect be a book of moral guidance in the form of a handbook for his senior officers raises the question of the function of such technical and practical treatises, an issue which remains the focus of debate

---

aid in explaining technical language and point up the connections between the original form of both Maurice's or Leo's texts and other military treatises to which Leo had access. See the original discussion in Vári, "Überlieferung," 47–68, as well as Dain and de Foucault, "Stratégistes byzantins," 385. More recent discussion, with detailed analysis and literature: Leoni, *La parafrasi Ambrosiana* and esp. Mazzucchi, "Basilio parakimomenos." On Basil see Brokaar, "Basil Lecapenus" and Featherstone, "Further Remarks," esp. 117–21, with earlier literature. See additionally Korzenszky and Vári, xvi–xviii, and the parallel text of the paraphrase accompanying their edition of Onasander.

81 See Dagron and Mihăescu, *Traité sur la guérilla*, 157–60.

82 See Andrés, "Nota"; Vári, *Leonis imperatoris tactica*, 1:xi–xxix; and Dain, "Inventaire raisonné."

83 For discussion, see Rance, "*Fulcum*," 318–20.

and disagreement.[84] It is in fact very hard to know whether the *Taktika* was turned to practical effect outside a small circle of senior generals, at least insofar as it was consulted and used, even if it clearly had no discernible impact on the arms, equipment, and administrative structure of the provincial armies. Between the 950s, when the first three surviving manuscripts were copied (although the date of W remains to be confirmed by detailed analysis), and the 1020s, at least two more copies were made, and a third ca. 1050, all in the same Constantinopolitan scriptorium, that of Ephraim, thus all within a few years of one another—a response to popular demand among senior officers, given the victories of the emperor Basil II and his generals (in particular Nikephoros Ouranos). Reference to the older as well as more recent tactical treatises occur in several texts.[85] We have already seen that a later treatise compiled at the orders of Constantine VII refers to the fact that when an emperor goes on a military expedition he should have with him a small library of tactical and strategical manuals, as well as other books which might be useful, and refers to both ancient and more recent compilations—while he names Polyaenus and Syrianos, it is possible that Leo's treatise was also what he had in mind, although it is not mentioned explicitly. Assuming the treatises ascribed to Syrianos *magistros* are indeed to be dated to the middle or late ninth century, he surely counts as one of the "recent" writers—unless, of course, his origins and identity were as mysterious to writers in the tenth century as they are to those of the twentieth and twenty-first![86] The author of the treatise *Skirmishing* (20.11–12), in contrast, does make specific reference to Leo's treatise.[87] Yet in view of the fact that at least six copies or versions of the *Taktika* were available by the 1020s and in Constantinople,[88] it seems not unreasonable to suppose that the manuscripts in which the *Taktika* is included were just the sort of collections that a senior officer, or an emperor, might have taken along with him on campaign. They were indeed highly regarded: apart from the reference to such authors in the treatise on imperial military expeditions already noted, there are several other indications that this was the case. Writing in the middle of the eleventh century, for example,

---

84 See Cosentino, "Writing," 95–99 and McGeer, "Military Texts," 907–14.

85 See, for example, *Sylloge tacticorum*, §45.17, 48.1, 53.8.

86 *CPTT* 210–11 and C.196–98.

87 See chapter 1 and n. 28 above. For Constantine's use of the *Taktika*, see Markopoulos, "Ideology of War," 51.

88 Dennis, *Taktika*, xi–xii; and see above, p. 69.

Michael Psellos remarks that the *caesar* John Doukas learned tactics and strategy from such collections, naming specifically Aelian and Apollodorus, and Alexios I was reportedly familiar with Aelian's work. But again, no mention is made of Leo's treatise,[89] even though a series of additional small texts, deriving from the tradition upon which A was based, were excerpted at some point in the second half of the tenth century (six paragraphs from the *Taktika* on naval warfare, titled Ἐκ τοῦ κύρου Λέοντος τοῦ βασιλέως—see commentary to Const. 19, below—and a selection of paragraphs from Const. 4 with additional material, referred to as Ἐκ τῶν τακτικῶν τοῦ βασιλέως Λέοντος τοῦ σόφου).[90]

The extent, therefore, to which Leo's *Taktika* may have achieved a degree of popularity in the course of the tenth century must remain uncertain. The fact that multiple copies were made suggests that it had a greater currency than many other compilations of the same period that were more limited in scope and purpose. The three treatises on imperial military expeditions, for example, survive complete in only two manuscripts, the Lipsiensis (L) and the Laurentian (our cod. M).[91] The former includes the *Book of Ceremonies*, was commissioned probably by the *parakoimomenos* Basil in the 960s, on the basis of a collection or file of other documents compiled during the reign of Constantine VII, and was probably never read outside the circle of senior court officers, if it was read at all. The Laurentian manuscript, M, contains most of the second of these treatises (B) only (an original work by Leo Katakylas), so that the Lipsiensis is the unique witness to the first and third of the texts, A (the so-called list of *aplekta*) and the longer text C, as well as a slightly different recension of B. Text B thus had an independent existence in the slightly earlier manuscript M, in which the *Taktika* of Leo is also found, although all three texts appear to have been unknown outside this palatine milieu.[92]

---

89 Psellos, *Chronographia* 7§16 (2:181.12–14 Renauld), who had himself composed a summary of Aelian's *Tactica theoria* (in Köchly and Rüstow, *Kriegsschriftsteller*, 2.2:232–38. For Alexios I: Anna Komnene, *Alexiad*, 15.3.6. For a general comment see McGeer, *Sowing the Dragon's Teeth*, 191–94, and Kučma, "Militärische Traktate," esp. 329. The classicizing elitism of an author such as Psellos would militate against his mentioning a more recent compilation or product such as the *Taktika*, in favor of a classical text or two.

90 Dain, *L' "extrait tactique,"* 9–12 (text and translation: 84–100).

91 On which see *CPTT* 36–39.

92 *CPTT* 45–65. See Featherstone et al., "Studien," 423–30 and Mazzucchi, "Basilio parakimomenos," from 279.

In this connection it is worth repeating the point made above that M—which includes the shorter version of the *Taktika*—appears to have no direct descendants, but is associated closely with a number of other MSS produced within and for a restricted palatine context, and was thus unlikely to have been accessible to those outside this closed context, even if associated with the palace.[93] The *Book of Ceremonies*, the *De thematibus*, and the *De administrando imperio* were likewise intended for a palatine readership only, and their manuscript tradition reflects this.[94] To what extent Leo's treatise can fairly be said to contrast with this pattern is difficult to judge, since the earliest manuscript tradition (MWA and VBE) was—so far as can be said in the current state of our knowledge—associated nearly exclusively with a palatine context. But—as noted already—this, and the fact that it was an imperial compilation, may suggest that very senior officers at least did consult this text, and maybe took a copy with them on campaign. Whether it really aided them in their campaigns and in battlefield situations is, of course, quite a different matter, although, as just noted, *Skirmishing* advises commanders to refer to the *Taktika* for advice on how to counter hostile attacks into Roman territory (see Consts. 11§21.116–27; 17§59ff., esp. §65.373–76). But in the end, perhaps, searching for its "practical" application misses the point: for as we have already suggested, the *Taktika* had as much to do with the moral universe inhabited by Byzantine generals as it had with tactics and strategy.

Not a great deal has been written about the structure or language of the treatise, primarily because of assumptions about its entirely derivative nature. As can be seen in particular from the prooemium, Leo understood his work as an official document which, while not having the force of imperial law, was nevertheless to be taken as having the backing of the emperor and to represent official "policy," insofar as an emperor could "legislate" on matters which clearly required a great deal of initiative and independent judgment from those who were using the treatise. It has already been remarked that the prooemium emphasizes the legal character of the *Taktika* by comparing it to the *Procheiros Nomos*, a point reinforced in Constitution 2, where Leo refers to it as an imperial constitution summarizing key aspects of a

93 See detailed discussion in Irigoin, "Centres de copie," 177–81 and literature cited in the preceding note.

94 See, e.g., *DAI* 1:32; Moravcsik, *Byzantinoturcica*, 1:367, 382 (on *De thematibus* and *De cerimoniis*).

general's conduct and practice, aspects which will be elaborated at greater length in the appropriate (later) constitutions.[95] The prooemium is constructed very much like an imperial *novella constitutio* (see p. 122 below), and it is hardly an accident of vocabulary that the "chapters" of the treatise are referred to as *diataxeis* or constitutions. Comparison of the prooemium with elements both of imperial novels as well as with the *proemia* to legal codifications, in particular the *Procheiros Nomos*, whether it was issued in 907 or earlier, makes this clear enough. The sentiments expressed in the prooemium to his collection of novels, in particular regarding the lack of understanding or ignorance of the older texts, on the one hand, and the resultant confusion and ignorance which characterized the current situation, on the other, while certainly reflecting standard topoi in such contexts, are closely paralleled in the prooemium to the *Taktika*.[96] As was long ago pointed out, a similar close parallel exists between the key terms employed by Leo to organize his material about tactics, strategy, and generals in Constitution 1, and the first three chapters of the *Eisagoge* (formerly referred to as the *Epanagoge*). In both, the terms employed to describe the aim (*skopos*), purpose (*telos*), and characteristic(s) (*idion/idia*) of emperor and patriarch, in the legal handbook, and of a general, in the *Taktika*, are the same.[97]

Unlike the *Strategikon*, which appears to have been aimed at a specific set of issues associated with cavalry warfare in the late sixth century, the *Taktika* does have pretensions to being a more comprehensive manual for generals, in its coverage dealing with all the elements of military organization, armament, tactical administration, leadership, and the ethics of warfare in a Christian state. Whatever the elements of archaism and antiquarianism in the *Taktika*, it was intended to be used in practical contexts (a point elucidated in the details of the commentary below) even if we cannot know whether this was ever the case in reality. For it must be admitted that the manuscript tradition in the late tenth and early eleventh century, as we have seen, barely supports the supposition of its use outside immediate imperial circles: indeed, Maurice and Leo are generally found in the same codices together (except for W), and are drawn on in equal measure by later compilers.[98]

95 See *Taktika*, Const. 2§33.222–24. Both here and at an earlier point in the constitution (§20.139–40) Leo refers to the following or corresponding (παράλληλα) books of the treatise, but this is not a reference to another treatise, *pace* Magdalino, "Non-juridical Legislation," 177.

96 Leo, *Novellae*, Pr. p. 40.14–25 (Troianos); p. 5.14–17.16 (Noailles et Dain).

97 Vernadsky, "Tactics of Leo the Wise," 333–35.

98 See Rance, *Strategikon*, chap. 1, "The Manuscript Tradition." For general considerations on the political context, see esp. Karlin-Hayter, "Military Affairs" and Tougher, *Leo VI*, 167–72.

The standard formulae, to the effect that the author has eschewed a flowery and/or rhetorically and sophisticated literary style, together with the usual protestations of humility, appear where we would expect them, in the prooemium, and as well-known topoi of ancient and Byzantine literature hardly need comment, appearing also in the prologues to works by Constantine VII, for example, as well as other middle Byzantine writers.[99] The title (if it is in fact part of the original work) refers to the whole compilation as a summary, *synopsis*, of instructions, again emphasizing the humility of the author and the limited nature of his claims—a wise comment in view of the remark of the anonymous author of the late tenth-century treatise *Campaign Organization* that "a long treatise usually brings forth unpleasant reactions..."![100]

Similar remarks occur also in the prologue to the *Strategikon*, although Leo alters the wording somewhat; the same considerations apply to his comments regarding vocabulary and the translation of ancient Latin terms into a contemporary Greek, more readily accessible to the reader of the time.[101] Leo intersperses some of his chapters with specific examples drawn from recent experience, so that we may reasonably accept that, whatever the limitations of his treatise, he was clearly concerned with the current situation of his empire militarily, both in terms of training and organization of the empire's armies as well as with issues of foreign policy and dealing with various threats.

The language of the *Taktika* is essentially that of the *Strategikon* and Leo's other sources, although where the older Roman and Hellenistic writers are concerned he generally paraphrases, or modifies technical phrases or terms to render them more contemporary. While he updates or glosses many words or expressions and employs others that had come into Greek since the seventh century, such as *stabarosai*, Slavic in origin (see on Const. 11§8.41), on the whole he does not employ a particularly demotic language, in great contrast to texts such as the *Praecepta militaria* of Nikephoros Phokas or the paraphrased versions of much of the *Taktika* as redacted by Nikephoros Ouranos. Indeed, comparison

---

99 See the brief summary on "humility" in Byzantine literary composition in Kazhdan and Ševčenko, *ODB* 3:1387; Ševčenko, "Levels of Style," 295–96; and Browning, "Language," esp. 103–4, 110–12, 118. For similar topoi in, for example, the works of Constantine VII, see Moravcsik, "Συγγράμματα."

100 *Campaign Organization*, chap. 32.18–22 (Dennis, 326). Note that Leo himself is keen "not to stretch out the discourse to too great an extent"; see Const. 7§67.470–71, for example.

101 See pr.66–76, commentary, with further sources and literature.

with the latter text, conveniently edited and published by Vári below his edition of the *Taktika*, demonstrates the difference very clearly, both in technical language and grammar. The process of bringing Leo's vocabulary up to date can be seen in a comparison of the older manuscript witnesses (MW) with the later tradition (AVBE), where the former retain Maurice's σκουλκάτωρες, or σκούλκα, whereas the later recensions replace this with the contemporary βιγλάτωρες and βίγλα (cf. Const. 17§72.430, §78.464, §80.471, §85.496), reflecting an ongoing editorial process. A similar process can be observed in the Ambrosian paraphrase of the *Strategikon* in A (ca. 959).[102] To what extent the military vocabulary Leo employs would be understood by a reader not associated with the army, and to what extent Leo was himself familiar with all the developments of his own age is a moot point, in light of his own lack of military and campaigning experience. The tactical administrative structures he describes—*tourma, droungos, bandon*—are merely the contemporary equivalents of the divisions described in the *Strategikon*, with the difference that Leo goes into greater detail about their make-up and interrelationship. The fact that his account is intended both to pay respect to the older traditions and the tactical manuals upon which he drew as well as to exemplify the "average" military arrangements of a provincial (thematic) army might also explain why he omits all mention of many developments which we know, from other sources, were already under way at the time of writing: the *droungarokometes*, for example. These were officers who commanded *banda* rather than anything larger, and whose appearance in the sources from the time of Basil I and after shows that the clear distinction between the *droungos* and its constituent subelement the *bandon* was beginning to disappear, with the result that the two ranks of *droungarios* and *komes* were initially associated together and eventually merged.[103] Yet he does describe very clearly the administrative structure of the *themata* under the overall supervision of the commander, the *strategos*, emphasizing the fact that he indeed intended his treatise to reflect contemporary realities as far as possible.

---

102 Vári, *Leonis imperatoris tactica*, 1:66–84; 89–143; 199–317; 2:1–160. I thank Philip Rance for pointing out, further, that the differences are even greater: Vári edited the text of Ouranos in its later, so-called Munich recension (Monacensis 452), a version characterized by a selective paraphrase edited to correct perceived vulgarisms; whereas an earlier version (the so-called Oxford recension, which remains unpublished) demonstrates more significant lexical and stylistic divergences.

103 For further discussion, see chapter 3 below and on Const. 4§12.76.

The *Taktika* appears to represent the fulfillment of a long-nurtured desire to address issues of military effectiveness, and while this interest may well have been stimulated by his early reading of Maurice and by his concerns for the empire's military effectiveness, it led to a series of works which progressively established the groundwork for the *Taktika*, beginning with the *Problemata*, followed by the first version of the *Taktika*, the plans for a treatise or chapters on imperial expeditions, researched by Leo Katakylas but never completed, it would seem—and this might explain the absence of any reference to imperial field armies in the *Taktika* (see below). Where the original version of the *Sylloge tacticorum* fits into this picture remains unclear.

The process of compiling a treatise such as the *Taktika* cannot have been simple. In the first place Leo and his redactors had to order their material from the major sources at their disposal, by subject and theme, and as far as possible avoid duplications. In the second place, they had to decide to what extent they should change any technical vocabulary to render it more intelligible to an early tenth-century reader, and whether or not they should add glosses to assist in this. In the third place, the material from different sources had somehow to be integrated logically so that each section as a whole made sense as it was read.

The agenda for this thematic approach is set out at the end of the prologue, where he lists the topics that were to be presented in each constitution. Whether this represents the original plan for the *Taktika* must remain unknown: it is interesting to note, for example, that what is currently Constitution 16 (W) and was Constitution 15 in M (coming before Const. 18 = 16 in M; Const. 20 = 17 in M, and the Epilogue), has the appearance in some respects of a concluding section, dealing briefly with matters to which the general should pay attention after the war was over. Given the fact that Constitution 18, on foreign enemies, itself seems to have had at least two stages of writing (see p. 57 above), it may be that an ur-version of the *Taktika* was a much simpler text, intended to consist of just the first fourteen constitutions, with what is now Constitution 16 as the final one, before Leo decided to write the current prooemium and add Constitution 20 and the Epilogue. But this remains hypothetical.

The order in which Leo presents his subjects is based closely on the *Strategikon*, and it was long ago suggested that this reflected the "military year," that is to say, the procedures, beginning with recruitment, training and equipping the troops, and following through to campaigning and facing an enemy force in open battle, which a late Roman army underwent in the course of a single year.[104] But while the *Strategikon* forms the basis for his treatise, Leo rearranges the material into a larger number of separate sections, or constitutions, expanding the material from the *Strategikon* with additional information from the older treatises and occasionally from more recent writings or from contemporary witnesses. It is true, of course, that Leo's text is full of repetitions, a reflection no doubt of his method, as Dain pointed out long ago.[105] This is, perhaps inevitably, especially the case with Constitution 20 and the Epilogue, as reference to the commentary below will quickly demonstrate. Leo's working method is not difficult to see if we summarize briefly the way each of his twenty constitutions is put together.

The introductory sections provide the tone and moral framework for the whole treatise. But while the prooemium is based on that of the *Strategikon* (together with parts of the prologue to *Strategikon* 7.A), Constitutions 1 and 2 are based on material about the capacities and requirements of a good general, and are taken from Aelian and Onasander. The following constitutions then largely follow Maurice, often verbatim but with occasional explanatory interventions from Leo, and with the emperor's own introductory and linking passages. Constitution 5 has no obvious source but seems rather to be a summary derived from Constitution 6 and compiled by Leo or the redactor(s). Constitution 6 is in turn taken from *Strategikon*, 1 and 12 (thus bringing material on cavalry and infantry together), as is Constitution 7, although with supplementary material, largely from Aelian. Constitution 8 reproduces Maurice almost exactly. Constitutions 9–11 are based on Maurice with additional material from Onasander, whereas 12 and 13 are largely from Maurice. Constitution 14 again supplements Maurice with Aelian and Onasander; 15 is drawn from Maurice, Onasander, and an unidentified text dealing with siege warfare; 16 is based almost entirely on Onasander, with some Maurician elements. Constitution 17, on "surprise attacks" is taken from the *Strategikon*, but shows a much greater degree of restructuring than usual, with passages from *Strategikon* 9–11 as well as from Aelian and Onasander. Constitution 18, based on *Strategikon* 11, with some material from *Strategikon* 6, but with the material rearranged, and with a substantial new section on the Saracens, is widely recognized to be Leo's most original contribution. Here he uses material from Maurice imaginatively, turning Maurice's Persians into Romans, for example, in

104 See Aussaresses, *L'Armée byzantine*, 5.
105 Dain and de Foucault, "Stratégistes byzantins," 355.

respect of planning and generalship, and updating the text to make it relevant to his contemporaries, mostly by changing names or omitting details from his exemplar. Constitution 19 is drawn from a range of sources, including Syrianos *magistros* as well as the relevant naval anecdotes in Polyaenus, and must have entailed more research than the other constitutions in view of the absence of a single treatise upon which Leo could draw—the very diverse nature of the material in this constitution would appear to support this. Finally, Constitution 20 is a collection of advice drawn in the first section from Maurice and in the second from Polyaenus and Onasander. Since much of the material from the last two authorities was already employed by Maurice, there is a great deal of repetition. The Epilogue represent Leo's more personal input, with a series of some seventy-three paragraphs referring back to the earlier parts of the *Taktika*, to biblical citations, or to the advice of the ancient tactical writers. Together, Constitution 20 and the Epilogue serve to close the treatise and balance the opening sections in the prooemium and Constitutions 1 and 2. Throughout the various constitutions, additionally, Leo peppers his text with comments drawn from a range of other sources, such as the parainetic literature already referred to.

In order to restructure the material he had from Maurice, and incorporate the additional information from the range of other sources upon which he drew, it is clear that the *Strategikon* offered the most helpful framework within which Leo or his redactors could work, adding and complementing Maurice's text at various points from the ancient authors at his or their disposal, and placing the material from the *Strategikon* within a moral framework taken largely from Aelian, Onasander, Polyaenus, and the Bible. It seems unlikely that the work proceeded by assembling a series of dossiers, each devoted to a specific topic, in which excerpts and extracts from his sources were collected and then collated. The way previous authors are incorporated into Leo's basic text appears to be the result of a process whereby once the relevant section(s) of Maurice had been copied, the redactor was instructed to go through the relevant older writers, according to the topic at hand and what was available to them, and incorporate key sections, presumably already marked out by Leo, with an appropriate explanatory linking comment (as, for example, at Consts. 4§57, 6§25, 7§67, 14§85), or gloss, as at 14§65.475–77.

Leo's thematic logic is fairly apparent, starting with what we might term the organizational and administrative aspects in the first eleven constitutions—the role and character of the commander, the strategic and tactical structure of the armies, and their armament, equipment, training, marches, baggage, and encampments. The next five constitutions (12–16) are all devoted to battle and field tactics in real warfare situations, with Constitution 15, on sieges, situated somewhat awkwardly between 14, "About the day of battle" and 16, "About matters after the war." The following constitutions, 17, on surprise attacks, 18, on the practices of various peoples, and 19, on naval warfare, all represent subjects which could not be easily accommodated within either of the two major groups. The placement of 15, 17, and 19 in W may explain why they were taken out of the original treatise by the redactor of M and tacked onto the end of the *Taktika*; it might equally reflect the fact that they already existed independently, although it remains impossible to say whether this also suggests that they were not integral to the original form of the *Taktika*.

A notable omission, it would appear, are the various detailed tactical diagrams or illustrations included in the original *Strategikon*, and it is difficult to know why these were not included in view of their value for explicating some of the more complex battlefield formations which Leo repeats from Maurice. Indeed, at Constitution 18§67.326–27 Leo emends the text of the *Strategikon* (11.2.87–88) specifically to delete the reference to such a diagram. There is no evidence that these were ever incorporated in or attached to the *Taktika*, and no version of the treatise includes any reference to them, whereas, in contrast, other treatises of the tenth century included a considerable number.[106]

Throughout the work, Leo (or the redactor[s] working under his direction) edited the material so as to differentiate it from the *Strategikon*. Whereas the *Strategikon* focuses heavily and quite explicitly on cavalry, although infantry are taken into account in book 12, Leo establishes from the start an arrangement of material that, we may assume, more accurately reflected either contemporary practice as Leo understood it (or thought he understood it), or the desirable state of affairs which Leo aimed to achieve. With this in mind, therefore, he moves much of the material from book 12 of the *Strategikon*, devoted to infantry, into several of his earlier constitutions, thus creating a more balanced and generally applicable account. This follows the precedent set by the author of the anonymous seventh-century *De militari scientia*, chaps. 2–4, a short paraphrase of the *Strategikon* found only in M (fols. 68r–76r), where material from *Strategikon*,

---

106 See Sullivan, *Siegecraft*, 153–54 and notes on ll. 3 and 5; on tactical diagrams in the military treatises, see Eramo, "Disegni di guerra"; Mazzucchi, "Καταγραφαὶ dello *Strategicon*."

12.B.16–17 is placed alongside material from *Strategikon*, 3.5,[107] although there is no evidence that Leo had access to a version of this text. He does the same with his material on several other occasions (in the constitution on ambushes, for example). By framing the whole within the introductory sections on generalship and command structures, on the one hand, and the gnomic advice in Constitution 20 and in the Epilogue, on the other, therefore, Leo in fact achieves a remarkably well-balanced and practical account—in spite of the flaws already noticed—of how to administer, train, equip, and go into battle with a tenth-century east Roman provincial army.

He thus makes a real break with Maurice's work. The *Strategikon* was designed to confront the particular interests of, and problems associated with, cavalry warfare and joint cavalry-infantry operations in the late sixth century, rather than to set out a generally applicable blueprint for the armies of the empire as a whole or to compose a treatise on the art of warfare as such. Leo's *Taktika*, on the contrary, whatever its admittedly derivative and imitative character, achieves something quite different and, in much greater detail than the older Roman treatises, embodies a complete handbook for generalship, tactics, and strategy, with all the necessary logistical and material infrastructural information necessary to realize this. His achievement in compiling and writing the *Taktika* is, contrary to the general view, entirely original.[108] And as we have seen, the tenth-century military handbooks take their inspiration from this model, rather than that of the ancients or from Maurice, however extensively they also borrowed from these sources too.

Application

It nevertheless remains the case that there is no obvious indication that Leo's prescriptions had any practical impact at all on Byzantine military organization, strategy, or tactics—and the effect of the *Taktika* on his senior generals cannot, unfortunately, be measured. We have already discussed the question of readership and have seen that there is little more to inform us about this than the fact that military commanders were assumed to take such manuals on campaign with

---

107 This is an important point, first noted by Rance, *Strategikon*, chapter 1, n. 80. See *De militari scientia*, chap. 2–4, at pp. 115–16; new (unpubl.) ed. by Bertazzoli, *Il De militari scientia*.

108 Most scholars are lukewarm regarding the question of Leo's originality. For a more positive assessment, see Cosentino, "Writing," 88 (but note also Dain and de Foucault, "Stratégistes byzantins," 356) and Tougher, *Leo VI*, 168–72.

them, or at least be familiar with them, that there were at least six manuscript versions of the text produced in palatine circles in the late tenth and early eleventh centuries, that the emperor Constantine VII was familiar with the text, and that the general Nikephoros Ouranos paraphrased it in his own treatise—very little evidence, therefore, to give us any idea of how practically efficacious the *Taktika* might have been.[109] An examination of the record of the Byzantine armies in the late ninth century, compared with their performance and achievements after the composition of the *Taktika*, or at least after Leo's involvement with the military and his expressed wish to improve matters, shows in fact very little change. The army was no less successful after 911–12 or thereabouts than it had been before; neither was it particularly unsuccessful from the 880s onward. And the remarkable successes of the second half of the tenth century are more readily explained by sociopolitical and tactical organizational developments that took place after Leo's time. The question arises, therefore, as to what Leo's real concerns actually were. It has been cogently argued that the *Taktika* reflects one element of a grand scheme through which the emperor would establish the administration of the empire—military, judicial, and fiscal—on a sound footing.[110] This scheme was embodied in texts such as the *Kletorologion* of Philotheos, on the one hand, or the codifications of the law, on the other, and his collection of *novellae*, which attempts a radical pruning of the legal inheritance, can readily be understood in the same way.[111]

To what extent can the *Taktika* be understood in this light? In considering the contents and shape of the treatise, and leaving to one side Leo's rhetorical flourishes regarding the decline in Roman military competence, three salient features stand out. The first is the emphasis Leo places on generalship, on the qualities of the general as a person, as a Christian, and as a leader who can win the trust and faith of his troops—this is obvious enough in Constitutions 2 and 3, but it recurs throughout the other constitutions and then once again is brought to the forefront in both Constitution 20 and the epilogue. The second feature is the fact that the *Taktika* is concerned only with provincial military matters: only a single (but dubious)

---

109 See Loreto, "Generale e la biblioteca." The issue is, of course, intimately associated with the question of the extent and degree of literacy in the Byzantine world at this time; see the essays in Holmes and Waring, *Literacy*.

110 Magdalino, "Non-juridical Legislation."

111 See Noailles et Dain, *Novelles*, xiii–xvii and Troianos, "Novellen Leons VI."

mention of the imperial *tagmata* is to be found;[112] no reference to the important role of the *domestikos* of the *Scholai*, by Leo's time the officer regularly given command of imperial campaign forces, occurs;[113] no suggestion that the imperial elite units might be found fighting alongside the provincial forces is found. The third feature is the fact that Leo's description of the tactical military organization of the provincial armies completely ignores developments about which we are tolerably well informed from other contemporary or near-contemporary sources, and appears to have had no visible impact on actual military organization or tactical practice, issues we will take up in chapter 3, below. And there is not a single reference to the fact that some of the troops fighting in Byzantine armies might not be Byzantines but foreign mercenary or allied soldiers. The *Taktika* thus appears to represent a partial and resolutely provincial context, concerned only with orthodox Christian subjects of the emperor, and with absolutely nothing to say of imperial forces or "modern" developments.

Yet this impression of Leo's military interests is, on further inquiry, only partially accurate. Leo was by no means concerned only with provincial military matters. It is a reasonable assumption that the researches of the *magistros* Leo Katakylas regarding imperial military expeditions, commissioned by Leo himself according to his son Constantine VII, were intended to fill this absence and, potentially, to provide an accompaniment to the *Taktika*. As we have seen, Katakylas was very probably one of the senior commanders whom Leo consulted on military affairs. Indeed, they must have discussed such issues when Leo gave Katakylas his commission to carry out this research assignment. The extant version of Katakylas's original text survives in M, and was also copied into the Leipzig manuscript of the *De cerimoniis*, commissioned—as was A—by the *parakoimomenos* Basil.[114] It survives in no other manuscript. It gives details precisely of those aspects of east Roman campaigning that the *Taktika* omits: how the imperial baggage train was to be assembled, what supplies were to be made available to the imperial column and how, the order of march of the imperial *tagmata*, the imperial encampment, and so forth. The work does not appear to have been completed in time, or maybe was felt to be unsatisfactory or inadequate by Leo, and it was

---

112 See commentary to Const. 6.69, below.

113 The brief allusions to the campaigns of Nikephoros Phokas the elder (11.121, 15.202, 17.374), who held this position, hardly count.

114 See *CPTT* 37, 66–68 with further literature. On Katakylas see *PmbZ* 24329.

left to Constantine to dig out Leo Katakylas's files so that he or his redactors could work them up into a more-or-less coherent text.[115] Part of the material for this treatise may be represented by another fragmentary text dealing with the ways in which expeditions involving several *themata* and the *tagmata* were to be conducted, and which used specific historical campaigns to exemplify strategy and tactics. It survives only in the text A of the so-called *Treatises on Imperial Military Expeditions*, in very fragmentary form, as notes intended to be expanded into a more coherent text.[116] Although the date of the original dossier remains uncertain (there is evidence that under Constantine VII some more up-to-date information was added: text A refers to the *thema* of Seleukeia, for example, established ca. 927–34), it is very likely that it was in fact originally part of the material collected by Leo Katakylas, or another person working under Leo VI's direction: it is in many respects more typical of Leo's military writings than those of his son Constantine VII. For as has also been shown, whereas Leo's work is often marked by an element of novelty and independent judgment, that of Constantine VII is much more concerned with the continuation and bringing to completion of the enterprise initiated by his father.[117]

This text has been dubbed a "list of *aplekta*," but it was not, in fact, a list of imperial assembly camps or bases, but rather the skeleton outline of a chapter or document about campaigning routes and bases in Asia Minor. The appearance of *ote* ("when") instead of [*isteon*] *oti* ("[know] that," used to introduce a prescriptive detail or instruction in texts such as the *Book of Ceremonies* and similar compilations) in the text, the repetition of the Anatolikon army in the list, the omission of several thematic armies, along with other anomalies, all suggest that in fact this list was made up originally of extracts taken from various narrative accounts of imperial campaigns, stuck together as examples and never properly edited to make sense. It is further evidence that Leo had originally intended to produce a second albeit shorter treatise to deal with those aspects of Byzantine military affairs not covered by the *Taktika*. We need, therefore, to reconsider how

---

115 This is text B in *CPTT* Text C, commissioned by Constantine VII, represents a greatly expanded and much more detailed account drawing on imperial archives and other sources.

116 Text A is transmitted in L only, and is generally assumed to have been about the major base camps in the provinces, the *aplekta*. In fact it was probably part of a somewhat different text. See Haldon, in *CPTT* 62–65.

117 See *CPTT* 62–65; and comment of Magdalino, "Non-juridical Legislation," 177–79.

we contextualize the *Taktika* and bear in mind that it was not by any means an isolated product of Leo's interests in military affairs.

In many respects, then, it is true that Leo was not very innovative in purely military terms and that he updated his material or added contemporary information only where he had a particular, personal interest in so doing—on the issue of the Byzantine response to Islam along the eastern frontier and as an existential foe of Christianity, for example. The picture painted by the *Taktika* of the provincial armies ca. 900 might almost be called "old-fashioned," and certainly neither takes account of recent administrative and structural developments, nor considers the possibility that Constantinople might be involved in thematic military affairs. Leo is, however, quite explicit that he wants to set out a template, an exemplar, to be followed by his generals, of organization, practice, and equipment. In other words, he wants his *Taktika* to be prescriptive as far as possible, rather than descriptive. He is not, with the one or two exceptions chosen in order to illustrate his points, describing things as they are, for if he were, as we will see in chapter 3, he would have had to introduce a much more complex and differentiated account of provincial tactical structures. On the contrary, using the model he found in the *Strategikon*, a template which certainly fitted the notional structure of the *themata* in a previous age, it seems that he wanted to reestablish the regular, symmetrical, and orderly arrangements which, he believed, had once prevailed and which had lent Roman arms their strength. From this perspective, the details of contemporary tactical arrangements were both unnecessary and indeed irrelevant. The fact that Leo's enterprise was probably doomed, and reflected an imperial wish far more than a practical blueprint, does not alter this.

Real issues are addressed, nonetheless. That the state of archery in the Byzantine army was a real concern is apparent from the fact that this is a topic he introduces himself, since it does not appear in Maurice in this strongly nuanced form, and it accurately reflects the devastating effectiveness of the Turkish horse archers who served in the Muslim armies.[118] The equipment with which the provincial troops were supplied or had to supply is clearly also an issue, because again Leo's concerns are his own in many cases and not a reflection of what the *Strategikon* has to say. Contemporary methods of supplying and maintaining the soldiers likewise constitute an issue, and here Leo addresses the thematic

118 See on Const. 6.30–33; and see 11.235; 20§81.

commander, the *strategos*, directly. The problems associated with supporting, maintaining, and recruiting soldiers is one which became increasingly acute during the first half of the tenth century, generating the legislation of emperors such as Romanos I and Constantine VII on military lands, on the "powerful," and on state control over its fiscal resources; but many of these issues are prefigured in some sections of the *Taktika*, as we shall also see, which gives the treatise a wider socio-historical relevance than has often been assumed. He exploited the *Strategikon* as a model through which to approach the nature of the empire's enemies, and although he simply reproduced much of what Maurice wrote, he made an effort to render his information relevant and set it in a contemporary context, even if, as a careful scrutiny of the text reveals, he was often less well-informed about technical issues and specifics, remaining at a level of generalization which frequently failed to move beyond caricature (see, for example, on Constitution 18§§42ff.; §§76–92 on Turks, Bulgarians, Franks, and Lombards).

And finally, as we have already remarked, Leo was genuinely concerned with the threat from the east—Constitution 18, while dealing with all of the empire's warlike or potentially dangerous neighbors, builds up to the section on the Arabs very deliberately, and it is clear from Leo's text that the danger from this quarter, both militarily as well as ideologically, appeared in his eyes as especially acute. While—as argued above—this may not have been the original motive behind the compilation of the *Taktika*, it soon became so, and it represents the most original part of the whole treatise, relying upon eyewitness accounts from senior military commanders, probably upon written reports also (as well as on tales from his father Basil I and his generals), on a careful expropriation of texts such as the *Rhetorica militaris*, and perhaps also upon the writings of Niketas Byzantios, which dealt with aspects of Islamic beliefs about warfare and fighting and which enabled Leo to demonstrate the inherent superiority of Christian beliefs and mores as both more civilized and ultimately bound to prevail.[119]

Given these considerations, it seems entirely reasonable to suggest that, from the point of view of Leo and his probable readership, the *Taktika* did possess a direct relevance to the situation of the early tenth century. Many of the conditions and

---

119 For good discussion see Kolias, "*Taktika*," as well as Dagron, "Byzance et le modèle islamique." For attitudes to the empire's enemies more generally in the military treatises, see Wiita, *Ethnika* and the discussion of Dagron, "Ceux d'en face."

much of the technological context of war-making, the funding and support of military activity, and the sociopolitical context had changed little in their fundamentals from the sixth century. The basic tactics followed by infantry or cavalry in a range of different battlefield situations were hardly different between the late sixth and the early tenth centuries—the enemies were different, styles of armament had changed, aspects of recruitment and training were different, but in most respects the descriptions of various tactical maneuvers in the *Strategikon*, the issues of tactical cohesion and orderly movement, of military discipline, and of the sanctions to be imposed for a range of military offences, remained just as relevant in Leo's day as they had been in the sixth century. Indeed in this respect the older treatises were also just as valuable as guides to managing and training infantry or cavalry. Where the strategic situation had changed, and where it suited his argument, Leo was quite able to provide up-to-date information to assist the general, most notably in Constitutions 17 and 18, where his account of the rapid to-and-fro of frontier warfare and raiding prefigured the much more detailed account of the treatise on skirmishing warfare written in the 950s or 960s. None of this means that Leo was not also consciously working within the literary tradition, that he did not incorporate bodies of material that we might consider, in some technical or utilitarian respects, to have been quite out-of-date, that he himself did not note on occasion that the information was included for the sake of completeness and possible value to the reader, rather than because it had a direct application, nor that his own entirely independent contributions represented only a fraction of the total text. Nor does it mean that the weaponry, tactics, and cultural values of the empire's enemies had not changed considerably in all sorts of ways which Leo did not explicitly take into account. Yet still, the fundamental technical, organizational, and strategic context meant that much of the older, especially the late Roman, material incorporated into the *Taktika* continued to have a purchase on the reality of the times, the more so in light of Leo's efforts to remind his readers of the contemporary situation.

A final point worth underlining is the fact that continuity of key cultural assumptions, in the context of a Christian moral universe, lent the much older *Strategikon* a contemporary value in the late ninth and tenth century which earlier, non-Christian treatises lacked, although it was not difficult to read into some of the moral judgments of a text such as Onasander's *Strategikos* a more "up-to-date" Christian ethos. The result was that an author such as Leo had little difficulty in making good use of even a Hellenistic treatise's key precepts governing campaigning and strategic planning, the behavior appropriate for a general or his soldiers upon taking a city, or the best ways to respond to overtures for peace from the enemy. Given that the aim was primarily the practical application of precepts intended both to achieve victory or, at worst, to save an army from destruction, the theory was less important than the practice, for the most part.[120]

But this was where Leo's *Taktika* achieved a unique status, precisely because it offers not only an empirical guide to making war. Far more explicitly than Maurice, Leo frames this within a clearly defined and self-consciously Christian symbolic universe, one which he, probably along with many others, perceived as directly and explicitly challenged by a rival and alternative belief system, a system which for Leo represented in every sense the antithesis of his own. His *Taktika* thus offered both a method and a theory of warfare as a just defense of the one true faith, and could be understood as a necessary corollary of the need to defend God's empire on earth.[121]

The *Taktika* raises issues of social, administrative, fiscal, and military history, along with many other topics, both directly and indirectly. Questions of popular belief and superstition, of literary allusion, classical knowledge, geography, and a whole range of topics, which are likewise embedded in Leo's text, will be dealt with at the relevant place in the line commentary. But we may conclude by emphasizing again that the *Taktika*, so often dismissed by scholarship as antiquarian rather than contemporary, is a great deal more than a mere copying out of an older work. It possesses an immediacy and relevance to its own time that is more often than not overlooked or denied, and it offers an invaluable insight into early tenth-century Byzantine culture in more ways than one.

120 See the useful comments of Kučma, "Militärische Traktate," 329.
121 For a detailed discussion, see Riedel, "Fighting."

3

ARMY AND SOCIETY
AT THE TIME OF LEO VI

To read Leo's introductory comments to the *Taktika*, one might think that Byzantine arms were in a truly parlous state at the time of composition. Of course, it is recognized that the rather grim picture he paints is partly rhetoric and partly based on Maurice, who made exactly similar claims, claims that form one of a series of topoi associated with the Graeco-Roman tradition of technical treatises.[1] Not only that, but as we have seen in chapter 2, Leo ignores the *tagmata* or imperial field units based in or near Constantinople, and makes no explicit mention of the great variety of types of unit in the army of his time, of the presence of mercenaries and foreign soldiers from outside the empire, such as the Khazars in the *Hetaireia*, for example, or (except indirectly and toward the end of Const. 18) of the differences in size and strength of the different provincial forces which his generals commanded.[2]

On the other hand, there were issues of administration and discipline which would have been familiar to Leo—or anyone who was familiar with the *Strategikon*—in terms of drawing comparisons with what was thought to have been the "old" Roman way of fighting and of tactical and strategic organization. By Leo's time the strategic organization of the armies was entirely different from that which Maurice had known, the most significant difference being the evolution of a highly provincialized "thematic" structure and the division between the more professional *tagmata* based in and around Constantinople and the provincial or thematic forces. The origins and the course of the evolution of these structures is now, with a few minor differences or emphases, more or less generally agreed, even if many points of detail continue to stimulate debate and disagreement. The

---

1 For example, Vegetius made similar complaints: *Epitome*, 1.28, 1.20.1–2, 1.21, 2.3.1, as did, nearer in time to Leo, Syrianos *magistros* (*Strategy*, 15.6–7).

2 See Const. 18§§146–149.

defeats and ensuing crisis of the 630s and 640s forced changes across the state administration. To the problems presented by the needs of defense as well those of maintaining its armies, the imperial government had responded by withdrawing the various field armies of the *magistri militum* back into the core territories between 637 and 640, mostly in central and western Anatolia. There, each army was billeted across groups of provinces, their number and thus total area being at least in part determined by the potential for the provinces in question being able to support them logistically, and a process was begun by which the groups of provinces occupied by each field army came collectively to be known by the name of that army. The field armies themselves were gradually transformed into provincialized militia-like forces, each with a central core of professional soldiers supported by both central and provincial revenues.

The groups of provinces evolved in turn into military regions or *themata*, a term which appears only in the early ninth century and probably reflects a series of fiscal administrative changes introduced by the emperor Nikephoros I, designed to alleviate pressure on the fisc as well as to ensure a regular source of recruits from the provinces.[3] Each army was commanded by a general, or *strategos*, who also had eventually—by the late ninth century—a supervisory authority over the civil and fiscal officials in his district, although there is little evidence to support the notion that this had been the case before the time of Nikephoros I. The earlier *themata* were thus named for the military divisions from which they evolved, whereas later *themata* received purely geographical names. The civil administration subsisted in an increasingly altered form until a series of measures to recognize the sort of changes which had occurred was undertaken, and the military and civil/fiscal arrangements were harmonized, a process which seems to have been completed by the time of the emperor Theophilos (829–42). The late Roman difference between mobile field units and stationary frontier forces seems to have vanished by the late seventh century (although the evidence for any sort of tactical structures at this time is negligible), until in the middle of the eighth century the emperor Constantine V reformed older elite units such as the *scholai* and *excubitores*, which by then were merely parade-ground

---

3 For the seventh-century origins of the middle Byzantine provincial administration, both fiscal and military, see Haldon, *Byzantium in the Seventh Century*, 173–253. The argument for the reforms of Nikephoros I and the introduction of the *themata* as a military-fiscal measure is set out in greater detail in Brubaker and Haldon, *Byzantium in the Iconoclast Era*, 744–55.

regiments. These served thereafter both as elite imperial mounted units and as a core of thoroughly professional soldiers bolstering the provincial armies when on campaign.[4]

By Leo's time there were over twenty *themata*, varying in size from the larger "original" districts of the Anatolikon or Armeniakon, to the much smaller and more recently established provinces of Sebasteia or Lykandos.[5] Along with the Arab geographers; the various lists of precedence; texts such as the later *De administrando imperio* or *De thematibus;* occasional snippets of information from narrative histories, hagiographies, and letters; and the evidence of lead seals in particular, the *Taktika* of Leo has been one of the main sources from which historians have attempted to reconstruct the military organization of the empire in the late ninth and tenth century. In terms of geography and extent, it is fair to say that we can reconstitute with a reasonable degree of accuracy the pattern of provinces across the empire at this time. We also have a good idea of how the fiscal administration of the empire operated, how armies were recruited and how they were paid.

Tactical Structures

It is difficult to say what sort of infrastructure existed to ensure the training of the soldiers in the provincial armies, or how they were equipped. Leo offers a great deal of detail on the latter, for example, yet because we still have virtually no archaeological evidence, it is very hard to establish whether or not his descriptions are merely idealized portraits and wishes and whether, and to what extent, they reflect any sort of reality, as will become apparent from the commentary on specific items.[6] At the organizational level, it is clear that each *thema* was by this time divided into a number of territorial *tourmai*, and that each such *tourma* was commanded—if not actually administered—by a *tourmarches*. An officer of similar rank, who may sometimes have been the senior *tourmarches*, probably based in

---

4 See Haldon, *Byzantine Praetorians* and Kühn, *Byzantinische Armee.*

5 See Oikonomidès, *Listes*, 346–54 and Haldon, *Warfare, State and Society*, 79–84, and table 3.1 with map 7. See also the surveys in Blysidou et al., *Μικρά Ασία των Θεμάτων* and Blysidou et al., *Βυζαντινά στρατεύματα στην Δύση.*

6 See the commentary to Consts. 5 and 6 below. In general on arms and armor, see Kolias, *Byzantinische Waffen*; Grotowski, *Arms and Armour of the Warrior Saints*; Dawson, "Kremasmata, Kabadion, Klibanion"; Dawson, "Suntagma Hoplôn"; Schreiner, "Zur Ausrüstung des Kriegers"; and Haldon, "Some Aspects of Technology" and idem, "Some Aspects of Arms and Armour" with various bibliographical resources in each.

the general's headquarters, was also known as the *merarches*, and may have provided the soldiers for the guards and retinue of the *strategos*.[7] Within each *tourma* were a number of smaller districts known by this time as *topoteresiai*, representing a fiscal administrative unit as well as the territory from which a particular unit within the *tourma*—a *bandon*—was drawn. There was not necessarily any neat equation between *bandon* and *tourma* as territorial and administrative districts, on the one hand, and the same terms applied to tactical bodies of soldiers, on the other. The commander of the army in the field could organize his troops into whatever divisions he felt appropriate, whether for the march, the camp, or the battle. Individual administrative *tourmai* could thus be brigaded together on campaign or in battle to make up a larger tactical *tourma* (and, conversely, large administrative *tourmai* might be broken up into smaller fighting units), so that tactical units or divisions may well not have overlapped exactly with the administrative districts from which the soldiers had actually been raised. This is why Leo states that the general should try to retain the cohesion of the units by keeping men from the same communities and districts together, retaining some local identities and solidarity.[8] It seems likely that the soldiers from each such *bandon* were generally billeted together, but that the units they formed might sometimes be grouped together or reorganized to produce units of different size or strength. This implies that soldiers who were placed in a particular territorial unit may generally have known one another, coming from the same or neighboring communities—according to the mid-eleventh-century *Life* of Lazaros of Galesion, the villagers recruited from the same community all went off to war together.[9] But because a *thema* might have three *tourmarchai* did not mean that they consisted of an equal number of smaller units, nor that the *tourmai* in the different *themata* were always of uniform size and structure. This is quite evident in the report of

7 In general on the Byzantine army from the tenth to late eleventh century, see Kühn, *Byzantinische Armee*. For the *merarches*, see on 3.8 and 4.69 below.

8 Const. 4§41; with §40 and §42. The *bandon* as a territorial subdivision of a provincial force probably originates in the seventh century, when the field armies of the various *magistri militum* were withdrawn into Asia Minor and distributed, according to a scheme carefully worked out in respect of the ability of the territory across which the units were based to support them. Although there is no specific evidence, the *bandon* had clearly been the basic administrative district within each *tourma* in the *themata* as can be seen in a number of territorial transfers from one *thema* to another in the time of Leo VI; see *DAI* 1: §50.83–132 and note 11 below.

9 For the importance of communal solidarities in battlefield contexts, see the commentary to 4§34.134–36, and §40.157–61, below; and see *Vita Lazari*, 529C (trans. Greenfield, 151).

the territorial transfers of *banda* included in chapter 50 of the *De administrando imperio*, where the newly established *tourma* of the *Kommata* in Kappadokia consisted of seven *banda*, whereas the new *tourma* of Saniana in the *thema* of Charsianon consisted of three *banda*.[10] The number of administrative *banda* in a *thema* may not necessarily have reflected a particular number of soldiers, or the strength of the thematic forces from that province at all. Indeed, as localization of recruitment, demographic changes, settlement patterns, and fluctuations in resource availability transformed the situation in each provincial field army from the second half of the seventh century onward, variation and difference will have increased, with the result that the number of infantry or cavalry soldiers needed for the tactical units—*banda*—could only rarely have matched the number of soldiers of the same type—infantry or cavalry—registered or actually available from each administrative *bandon* or *topoteresia*.

There existed also a purely tactical unit, known as a *droungos* and commanded by a *droungarios*, made up from a group of *banda*, but having no territorial identity, as far as the evidence permits us to say.[11] By Leo's time this notional unit seems still to have existed, if the contemporary information given by Leo in Constitution 18 (§§141 and 146) is reliable; but was already undergoing substantial change, one of the results of which was that the commanders of *banda* and *droungoi* were effectively assimilated to one another, commanding much smaller units within each *tourma*.[12]

Much of the military's regular structure described by Leo is corroborated for the middle or early part of the ninth century by the reports of Arab geographers. According to Ibn Khurradādhbīh, writing before 850,[13] each theme force

10 See *DAI* 1: chap. 50.92–105.

11 For the *topoteresia* and the *bandon*, see *CPTT* 257–58; also Haldon, "Military Service," n. 111 (45–46). That the *bandon* did indeed become the fundamental fiscal-territorial unit into which villages, as the basic fiscal unit of the empire, were grouped, is clear from much later evidence—in the empire of Trebizond, for example, where the *bandon* was the basic fiscal-administrative and military subdivision of the empire; see in general Bryer and Winfield, *Pontos*, vol. 1 and, more specifically, Bryer, "Rural Society in Matzouka."

12 See Haldon, "Chapters II, 44 and 45," 324–27.

13 His report was written in 846 and revised later, in 885: Miquel, *Géographie humaine*, 2:391–93. There are four such reports, all of which derive at least in part from information written down by a certain Muslim b. Abī Muslim al-Jarmī, a Muslim prisoner before 845, whose report was written in the period from the 830s to 846; see Brooks, "Arabic Lists" and

consisted of a *patrikios* who commanded 10,000 soldiers. Under him were two *tourmarchai* each with 5,000 men, and under each *tourmarches* were 5 *droungarioi* commanding 1,000 men each. The *droungarioi* were in turn in charge of 5 *kometes* with 200 men each, and each of these units of 200 were divided into 5 units of 40, each under a *kentarchos*, further subdivided into units of 10 under a *dekarchos*.[14] This broadly matches Leo's description in Constitution 4, and we may accept that the basic structure of thematic or provincial armies in the ninth century, derived certainly from that of the field armies of the *magistri militum* in the first half of the seventh century, was regular enough for Arab observers to derive this account from what they saw or heard. But even if there had been a relative degree of uniformity across the provincial field armies of the seventh century, and even if, as seems reasonably certain, this was reflected in the description preserved in Ibn Khurradādhbīh and in Leo's *Taktika* (which is why Leo felt he could lift Maurice's account with few changes other than to nomenclature), the situation had undoubtedly changed in several respects by the early tenth century. A difference between the older "Roman" *themata* and newer formations, especially the so-called "Armenian" *themata* along the eastern frontier, had certainly evolved, for example. This can best be exemplified by taking a concrete example or two, in order to illustrate both the basic structures common to a number of *themata*, as well as the range of variation that pertained.

In the documents which survived in the imperial archives concerning the expeditions to Syria and later to Crete, in the years 910–11 and 949, a number of scraps of information regarding the administrative structure of certain provincial armies are given. Some of these concern the *Thrakesion* army in particular. Originally the army of the *magister militum per Thracias*, this force had been transferred to Asia Minor by the 640s, where it was allocated as its resource-catchment area a number of the western provinces in that region.[15] Some of the divisions of which it was composed retained their late Roman names right up

---

Miquel, *Géographie humaine*, 1:xviii. For discussion of the relationship between the texts, see Winkelmann, "Probleme." The three later works are by: Ibn al-Fakīh, see E. W. Brooks, "Ibn al-Fakīh al-Hamadānī" (written in 903; see Miquel, *Géographie humaine*, 1:xxii); Kudāma ibn Ja'far (written in 928–32; see ibid., 1:xxviii); and al-Mas'ūdī, whose *Kitāb al-Tanbīh w'al-Ishraf* written in the middle of the tenth century, also used al-Jarmī, but with several other traditions represented (see Miquel, *Géographie humaine*, 1:xxix).

14 Ibn Khurradādhbīh, 84.

15 Lilie, "'Thrakien' und 'Thrakesion'," 22–23.

to the tenth century, and from the documents for 949 we learn that it included a *tourma* of the *Biktores*, a *tourma* of the *Theodosiakoi*,[16] a *tourma* of the *paralia* (that is to say, of the coast or shore), and a *tourma* of the *merarches*. For the 910–11 expedition, the Thrakesion army was supposed to provide some 3,000 cavalry soldiers (although in fact only 1,000 seem to have taken part); in 949 no specific total for the *thema* as a whole is given, but from the information that does appear we can estimate a figure per *tourma* of between 600 and 800,[17] so that the total number of cavalry soldiers from the theme might thus be in the order of 2,400–3,200, roughly consistent with the numbers from the 910–11 expedition.[18] The probable number of soldiers from the Thrakesion in 910–11 and 949 was thus approximately twice as great as that which can be derived for the (less wealthy) Peloponnese for the reign of Romanos I (between 1,500 and 2,000), based on figures given for the rate of commutation of their military service.[19]

The main point here is less that numbers and sizes of units could vary considerably than that the internal structure of the different *themata* could likewise vary. In 949 the *thema* of Thrakesion sent 64 *droungarokometes* on the expedition, and these probably represent all the permanent "officers" of the three corps present. If the total did indeed number some 2,400 or a little more, then each of these officers may have commanded a small unit of between 30 and 40 soldiers.[20] In the 910–11 expedition the troops from the *thema* of Sebasteia included 960 soldiers, 5 *tourmarchai*, 10 *droungarioi*, and 8 *kometes* in the 911 expedition. The proximity in the hierarchy of the *droungarioi* and the *komites* is evident, and the numbers suggest the relatively small units commanded by these officers—perhaps organized in

16 On the origins of these units, see Schmitt, "Untersuchungen," 211–16 and Rance, *Strategikon*, chapter 7 ("Cavalry Regiments and Corps in the *Strategicon*").

17 See *De cerimoniis*, 663, 666–67; Haldon, "Chapters II, 44 and 45," 217.200–217, 221.49–58.

18 On the *merarches/meriarches*: Haldon, "Chapters II, 44 and 45," 326 and n. 347 and *CPTT* 249–50 for further discussion. That a *tourma* originally numbered about 1,000 soldiers is perhaps suggested by two passages in the *Chronographia* of Theophanes: in his account of the events of the years 717–18, the future emperor Leo is made to inform the Arab commanders that 1,000 soldiers and their *tourmarches* had entered Amorion to defend it against attack; while he records that in 792/93, one thousand soldiers of the Armeniakon *thema* were punished for rebellion, along with their *tourmarches*, who was executed: *Theophanis Chronographia*, 389; 469.

19 See discussion in Oikonomidès, "Social Structure," 114–15. A different perspective: Treagold, "Standardized Numbers," 1–14.

20 Haldon, "Chapters II, 44 and 45," 325–28

banda of some 50 or so men, grouped into larger divisions of some 150–200 under the five *tourmarchai*.[21] Such operational unit sizes are corroborated by the *Sylloge tacticorum* (a text which, as we have seen, was compiled at some point between the end of Leo's reign and the middle of the century), which notes that *banda* could vary: the basic cavalry *bandon*, referred to also as an *allagion*,[22] numbered 50; but units of anything up to 400 were possible, especially for infantry. The text notes that thematic *banda* number anything from 50 to 150, whereas the "imperial" *allagia* can be as large as 320 or 350 or at most 400 in strength, "such as the present unit of the Thrakesion, numbering 320, or that of Charsianon, numbering from 300 to 400," while some of the western *tagmata* number 400 also. The "imperial" units are undoubtedly professional units retained on a full-time basis and based in the provinces, as opposed to the thematic soldiery proper, raised on the basis of their being registered for the *strateia*. In the case of the Thrakesion, for example, the documents for 949 tell us that 800 thematic soldiers commuted their service at a rate of 4 *nomismata* each, and the government then hired 600 Armenian mercenaries in their stead to guard the coast of the province.[23] The late ninth-century Arab geographer Yaqʿūbī notes that the total of cavalry in the empire amounted to 40,000, and that there was a great difference in numbers between smaller *themata* such as Charsianon (which provided 500 cavalry), on the one hand, and Thrace or Macedonia, on the other, which can provide 5,000 (although this very probably included the *tagmata* based there also).[24]

The fact that Leo passes over these anomalies and variations in silence should certainly give pause for thought in respect of the value of his information for thematic military structures. Even though it is likely that some of the changes reflected in the documents for the 949 expedition referred to had not yet taken place or were only evolving in Leo's reign, there is enough evidence to show that the neat and symmetrical system he describes was far from the reality of even the older, original thematic divisions such as the Thrakesion. In this respect, therefore,

---

21 *De cerimoniis*, 656.13–16.

22 Because it changed or rotated station or post with other similar units; see Haldon, *Byzantine Praetorians*, 275, 277.

23 *Sylloge tacticorum*, 35.4–5; Haldon, "Chapters II, 44 and 45," 305–6 and n. 265 on rates of commutation (varying 4 to 6 nomismata per annum, with a lower rate for the indigent registered persons).

24 Al-Yaqʿūbī, *Kitāb al-Buldān*, 168 (*BHG*, 7:232). The text was compiled ca. 889. For the *tagmata* of Thrace and Macedonia, see Haldon, "Chapters II, 44 and 45," 330–34.

we should accept Leo's testimony only where we have some corroboration from other sources. Yet he does bemoan the "prevailing lack of training . . . carelessness, and . . . the small number of soldiers these days" (Const. 18§149.842–43) so that it would be reasonable to suppose that at this point in the *Taktika*, as at earlier points in Constitution 18 (e.g., §146.798–802), his account reflects a desired regular arrangement, or the attempt to restore such arrangements, through the advice given to his thematic generals, rather than an accurate description of the current situation in all its variety and inconsistency. It is significant that the later practical treatises, in particular the *Praecepta militaria* ascribed to Nikephoros II, describe a somewhat different structure.[25]

Strategy

Byzantine tactics and strategy over the course of the late seventh and eighth centuries had evolved in the context of the warfare that dominated—raiding and hit-and-run attacks for the most part, with occasional large-scale campaigns against enemy strongholds or population centers.[26] The provincial field armies of the second half of the seventh and the eighth century were referred to retrospectively by Theophanes in the early ninth century as *kaballarika themata*—"cavalry armies"—illustrating the result. Infantry became a subordinate and second-rate arm, composed of the poorer and least well-equipped conscripts from the provincial registers. The few technical references to types of provincial soldiery dating to the period before the middle of the tenth century make virtually no mention of them. An account drawn from ninth-century information describing provincial field armies mustered to meet the emperor on their way to campaign in Syria assumes that the armies (or at least those elements thought worthy of mention) are composed of cavalry. Official and semiofficial regulations about the minimum property required for the maintenance of soldiers before the second half of the tenth century refer only to regular thematic cavalry or to sailors of the provincial fleets. Infantry barely figure in these texts,[27] indeed it seems unlikely, given what is known of the tactical use and social status of provincial infantry forces

25 McGeer, *Sowing the Dragon's Teeth*, 198–222; Kühn, *Byzantinische Armee*, 123–31, and 260–62, 278–80.

26 For an overview see Haldon, *Warfare, State and Society*, 34–106; also Luttwak, *Grand Strategy*, but with Kaldellis, "Review."

27 See *CPTT* C.443–50; *De cerimoniis*, 695.14–18; and Svoronos, *Novelles*, no. 5, 118 (A.1). For the longer-term history of the relationship between cavalry and infantry in Roman and

at this time, that Leo's prescriptions in their respect reflect the general situation. Some infantry forces may have been coherent and disciplined enough to carry out the tactical role he assigns or assumes for them, although in view of his explicit purpose, which is to legislate to reform and improve Byzantine arms and military science, he and his readers probably felt that his account represented an attainable goal. His father, the emperor Basil I, was attributed with reforming efforts in the army: improved discipline and more rigorous military training and exercises are mentioned,[28] although the source which reports this was commissioned by his grandson, the emperor Constantine VII, and aims explicitly to improve Basil's image in the eyes of later generations. The absence of specific references to infantry soldiers reflects several features, not just the evolution of Byzantine tactics up to the tenth century, but also the nature of Byzantine society by this time.

The provincial forces adhered to a pattern of largely seasonal campaigning, localized recruitment, and physical dispersal, none of which was conducive to the maintenance of regular infantry units and the tactical discipline and order required for battlefield maneuvers. Garrison duties and irregular warfare in broken country remained within the competence of poorly equipped foot soldiers, of course. We know nothing of the process of tactical degradation of the provincial field armies across the centuries from the 650s and 660s, so we have no idea of how large the more professional or more permanent elements were, at least until the *Taktika*, when the figure of 4,000 is mentioned, a figure borne out—very approximately—by a reference in Theophanes to groups of *epilektoi* from each provincial army, numbering 3,000, and from some references in Arabic historians of the period. The occasional references to such *epilektoi* make it clear that these are usually cavalry.[29] The evidence shows that even though the provincial cavalry were often also defective in respect of organizational discipline, the nature of warfare—raids, harassment of invaders, and so forth—inevitably gave them greater prominence. Some fifty or so years after Leo was writing, the author of *Skirmishing* notes on several occasions that the enemy cannot be defeated without an adequate force of Roman infantry to press home attacks on their encampments,

---

Byzantine armies, see Haldon, *Warfare, State and Society*, 193–97; Rance, "Combat"; and cf. *Menae patricii, de scientia politica dialogus*, bk. 4 (trans. Bell, 123–42).

28 *Vita Basilii imperatoris*, 36.1–13 (132–34).

29 Cf. Const. 18§§136–39, §§146–49; cf. *Theophanis Chronographia*, 452.10 (trans. Mango and Scott, 624), for 778/79.

occupy the defiles, ambush the withdrawing enemy columns, and so on.[30] This reflects the fact that the attacking forces were often composed of large numbers of infantry, which made their raids the more dangerous, since they were better able to pursue the rural population to their fastnesses, pillage and ravage their villages and homesteads, and resist Roman cavalry attacks on their encampments, a point certainly appreciated by Leo and explicitly noted in the *Taktika*.[31] Additionally, since the limited evidence suggests that in the course of the late seventh and certainly by the late eighth century many provincial soldiers came to be responsible for providing their own equipment, social differentiation must have been closely reflected in tactical differentiation: only those who could afford a horse (or two), invest resources in their equipment and weapons, as well as have their family or others work their land for them, could have served as cavalry.[32] The result of this transformation meant that cavalry tended to be light cavalry with fairly minimal offensive and defensive equipment and few specialist skills (such as mounted archery, which in the context of a largely agrarian society requires constant training and practice), and infantry must have been mustered from the simple peasantry, requiring even less equipment.

Finally, the nature of the warfare in the east, which entailed the avoidance of battle wherever possible, probably did not promote battlefield confidence, tactical cohesion, or discipline, especially among infantry units. Leo notes explicitly: Σφαλερὸν γάρ, ὡς πολλάκις ἡμῖν εἴρηται, τὸ πρὸς δημόσιον πόλεμον ἀποκινδυνεύειν τινάς, κἂν πάνυ δοκοῦσι τῶν ἐχθρῶν περιττεύειν τῷ πλήθει· τὸ γὰρ τῆς τύχης ἀόρατον.[33] This is an attitude not merely restricted to this period, of course—it was a standard and well-established motif in both Graeco-Roman as well as east Roman or Byzantine conceptions of strategy to avoid battle at all times where the outcome was uncertain: the *Praecepta* of Nikephoros enjoin the commander to "Avoid not only an enemy force of superior strength but also one of equal strength, until the might and power of God restore and fortify the oppressed

---

30 See *Skirmishing*, 3.2–4; 9.14; 10.19–20; 23.2–3; 25.1 and compare with *Taktika*, Consts. 18§128.626–31; 20§206; *Theophanis Chronographia*, 358–59 (trans. Mango and Scott, 498–99); discussion: Dagron and Mihăescu, *Traité sur la guérilla*, 190–93.

31 See *Taktika*, Const. 18§131; *Skirmishing*, 10.1; 16; 20.11; and note 7.2; 15.2 on the importance of finding out whether infantry were involved in the enemy raid.

32 Haldon, "Military Service," esp. 20–24.

33 Const. 18§121.

hearts and souls of our host. . . ."[34] Nevertheless, the strategy of avoidance which the hit-and-run raiding warfare of the eastern frontier imposed upon imperial armies must inevitably have contributed substantially to the downgrading of infantry as a battlefield arm, a strategy exemplified in the treatise on skirmishing warfare.

Although written in the second half of the tenth century, and therefore in a strategically and tactically somewhat different situation from that in which Leo's *Taktika* was compiled, the treatise *Skirmishing* reflects quite clearly the traditional form of warfare which had dominated the eastern frontier in the second half of the ninth century and up until shortly before the time of writing. From this text it is quite apparent that the regular provincial or thematic infantry were necessary to effective warfare, yet were regarded as potentially unreliable, undisciplined, and easily demoralized. This must certainly lie behind some of the more important defeats suffered by imperial forces across the course of the ninth and early tenth centuries. According to *Skirmishing*, infantry would attack an enemy camp when ordered to do so less because they were brave soldiers than because they were eager for booty;[35] they were slow moving and might hold up the commander's main operation;[36] in the line of battle, in an attack upon an enemy formation, for example, cavalry soldiers and officers were to be drawn up in their rear to ensure they pressed home the attack, maintained order, and did not try to flee. The commander had to be attentive to their morale, encouraging them before any combat with harangues, promises of rewards, and so forth, to keep them from melting away.[37] The mass of the infantry was slow and difficult to muster in time, and relatively poorly equipped.[38] The infantrymen seem to have had very low status compared with mounted troops, and the loss of his horse was a social as well as a disciplinary disaster for the cavalryman, especially if he then had to serve with the infantry.[39] But in spite of these characteristics, they did continue to play a

34 For the Graeco-Roman tradition inherited by late Roman and Byzantine military thinkers, see Krentz, "Deception" and Wheeler, *Stratagem*; for Nikephoros Phokas, see *Praecepta*, 4§19.197–202.

35 *Skirmishing*, 10.16–17, 24.6.

36 Ibid., 10.2, 5, 8; 14.7.

37 Ibid., 23.3–4, 24.4.

38 Ibid., 3.3, 10.16, 23.6, 24.5. Cf. *Taktika*, Const. 20§206.1060.

39 See Haldon, *Byzantine Praetorians*, 299 and n. 894; Dagron and Mihăescu, *Traité sur la guérilla*, 185 and n. 20; Kazhdan, "Hagiographical Notes," 544–45.

role, and the sources report Roman heavy infantry fighting effectively alongside the cavalry during a campaign in southern Italy in the 880s, for example;[40] while in the campaigns led by Basil I in the 870s against both Arabs and Paulicians in eastern Anatolia, the siege and capture of fortresses and other strongholds can hardly have been carried out without a substantial force of effective infantry.[41] And although they do not appear often in the passages which were by Leo himself, he does include an important role for infantry in holding passes and controlling key strategic locations, for example.[42]

What this implies for the provincial field armies, therefore, is that they were far closer in appearance, morale, and fighting techniques to the infantry and cavalry levies and retinues of their medieval western European contemporaries than has generally been admitted. Yet the structure of the east Roman armies clearly did retain a strongly "Roman" element, as noted above in respect of tactical organization. The majority of the soldiers listed in the thematic registers around the year 900 may well have been of little military value except as a peasant militia, but the effective element of each provincial army appears to have consisted of a core of several hundred full-time professional soldiers, around whom then the 3,000–4,000 or so *epilektoi*, the more carefully selected elite force of cavalry, where they were available, was organized. As far as tactics on the field of battle are concerned, Leo has little to say that is not already in Maurice's *Strategikon*. The establishment of the different units of the imperial *tagmata* as an elite mounted force from the time of Constantine V on had provided a key professional, well-trained, and tactically disciplined element which formed the backbone of many Byzantine armies. Together with the inherited Roman concepts of tactical order, discipline, and cohesion, and supported by a well-oiled logistical system, this gave Byzantine armies, when reasonably well-led, an important advantage over their foes. But in a few paragraphs toward the end of Constitution 18, Leo sets out what he knew or had been told about the typical line of battle in a provincial army of his own day. While in its fundamentals it is hardly different from that described in the *Strategikon*, and certainly influenced by what Leo had read and had written already in this respect, it offers some insight into the ways in which Byzantine

---

40 *Vita Basilii imperatoris*, 65–66 (228–30).

41 Ibid., 40 (142–46).

42 For example, *Taktika*, Const. 17§59; 18§§128 and 133; and 20§206.

commanders were able to manage their troops in actual conflict, and may also hint at some of the changes in cavalry tactics that were to become more apparent in treatises such as the *Sylloge tacticorum* and, later, in the *Praecepta* of the emperor Nikephoros II.[43]

Numbers

The size of Byzantine armies is very difficult to estimate with any degree of accuracy—many reports in chronicles and histories are either implausible or, if plausible, difficult to corroborate, while every modern historian has a different notion of what might be plausible or reasonable, so that many different suggestions have been made. As we have already seen in the discussion on tactical structures above, unit sizes often varied considerably, so that it is difficult to draw firm conclusions simply from the numbers of units named in a source. When many divisions could be brought together, however, it is reasonable to assume that large armies of up to 30,000 or so could be fielded for a short period. In 778 a combined force of 100,000 from the themes of Thrakesion, Boukellarion, Armeniakon, Anatolikon, and Opsikion is reported to have marched against Germanikeia.[44] The same chronicle records that shortly thereafter the emperor raised a force of 12,000 cavalry and a fleet for a combined offensive into Bulgar territory. The first figure is probably exaggerated; the accuracy of the second is more plausible but remains doubtful, indicative at least of the order of magnitude of regular campaign armies at the time.[45] Similarly inflated is the figure of 80,000 men supposedly in the army of Thomas the Slav in 821.[46] More reasonable is the figure of 20,000 men raised by Constantine VI to oppose the Arabs in 797 (although the fact that it seems "reasonable" is no guarantee that it is accurate, and reflects the subjective opinion of

---

43 See commentary to 18§§136.696–149.846.

44 *Theophanis Chronographia*, 451 (trans. Mango and Scott, 623).

45 Ibid., 447 (trans. Mango and Scott, 618). Such variations, of course, reflect the sources used by the original chronicler.

46 Theophanes continuatus, 55.22; Genesios, 2.5. Many descriptions of campaigns and battles involve the formulaic use of terms such as "x myriads" (i.e., multiples of 10,000) for both sides, as at Theophanes continuatus 177.18–22, where opposing forces of 3 and 4 myriads are mentioned; it is difficult to know whether these are purely formulaic, or reflect the actual numbers involved. For the whole problem of calculating numbers and the relationship they bear to contemporary logistical issues, see Haldon, "Why Model," with other essays in that volume; and idem, "Roads and Communications."

the historian).[47] For an expedition in 773, the chronicler Theophanes records that the emperor Constantine V marched with a force of 80,000 men, made up of all the *themata* and the *tagmata*. While this is an implausibly large figure for an expeditionary force, it has been reasonably argued that this figure actually reflected the nominal total of the provincial and Constantinopolitan units at the time, and there are quite good grounds for accepting this.[48]

The theoretical model of each *thema* which the *Taktika* presents is drawn directly from the *Strategikon*, and while it retains some value, neither the apparently regular structure of the earlier period nor the regular numbers of the earlier units (although Maurice too allows for very considerable variations in unit strengths) were retained in tenth-century practice. According to Leo's figures, for example, the larger themes could field as many as 10,000–15,000 soldiers, yet most of the evidence points to much smaller operational numbers. The solution, as already indicated above, may be to see the notional figures as reflecting the number of households entered into the military registers in each military district, from which soldiers could be drawn, but which rarely if ever reflected the number of actual serving soldiers. In Constitution 12.174–75 Leo describes a "standard" (*symmetron*) force of from 5,000–10,000 or 12,000 (although this is taken directly from the *Strategikon*, 2.4.82–83 and consequently may have little value for Leo's time), and much more to the point, in Constitution 18§§136–49, when discussing the frontier warfare in the east, Leo takes as his basic provincial army a cavalry force of no more than 4,000 selected men, separated out from the total available registered force in the *thema*, probably a much better reflection of the logistical and demographic realities of the period. He himself notes that the need to bring several thematic forces together in order to meet large enemy attacks has arisen as a result of "the prevailing lack of training, by carelessness, and by the small number of soldiers these days."[49] Of course large armies could be and regularly were brought into the field by combining the forces of several commands, as outlined at 18§149, but they seem to have been the exception rather than the rule—when coordinated operations were planned, as with the campaigns in the 860s and 870s against the Arabs and the Paulicians, for example. When the empire was able to muster all

47 *Theophanis Chronographia*, 471.21–27 (trans. Mango and Scott, 648).

48 Ibid., 447 (trans. Mango and Scott, 617). See Treadgold, *Byzantium and Its Army*, 64; and Haldon, *Warfare, State and Society*, 99–106 and 109–15.

49 Const. 18§149.841–43.

the thematic forces, then substantial forces could be assembled. The 40,000 troops ascribed to the army of Michael III in the 860s, drawn from many *themata*, is not unreasonable. At the battle of Poson in September 863 the general Petronas was able to bring three columns together to surround and defeat what the chroniclers clearly regard as a major Arab force, which is reported to have been some 40,000 in strength. This was a serious invasion on the Arabs' part, intended apparently to follow up the limited successes of previous years; and it was opposed by the combined forces of the themes of Armeniakon, Boukellarion, Koloneia, and Paphlagonia, which approached from the north; of Anatolikon, Opsikion, and Kappadokia, with the forces of the *kleisourarchiai* of Seleukeia and Charsianon, which approached from the south; and the commander-in-chief, with the *tagmata* and the armies of Thrace, Macedonia, and Thrakesion, from the west.[50] This was a very large undertaking, but it was planned and carried out within the few months between the invasion entering imperial territory (the Arab forces marched north through the Armeniakon region, sacking the town of Amisos on the Pontic coast) and September, illustrating the efficiency with which imperial forces could be moved when required. It was rarely possible to keep large numbers of men and animals together for long, because of supply issues, quite apart from issues of discipline, to both of which the *Taktika* (albeit following the *Strategikon*) makes reference.[51]

The figures offered for the Byzantine army during the first half of the ninth century by some Arab sources, notably the account of a certain al-Jarmī, whose report may include information from before the 840s and even earlier, may offer a broad guide to the total paper strength of the army.[52] Other Arab writers, mostly derivative of this first report, also provide figures, from which it has been calculated that the total number of soldiers theoretically available in the middle of the ninth century may have been 120,000. This depends on quite high round figures for some of the *themata*, on the assumption of a total of some 24,000 soldiers (of the *tagmata* and related units) based in or near Constantinople, and on the assumption of an entirely regular internal tactical organization with equivalent

---

50 For Michael III: Theophanes continuatus, 177.18–22; Genesios, 4.14 (although the account seems to be borrowed from an earlier description of the battle of Dazimon in 838; see Treadgold, "Chronological Accuracy," 180–82). For the battle of Poson: Theophanes continuatus, 179.13–183.13, see esp. 181.11–20.

51 Haldon, "Chapters II, 44 and 45," 305–29.

52 Treadgold, *Byzantium and Its Army*, 43–86; idem, "Remarks on Al-Jarmi"; Haldon, "Kudāma Ibn Dja'far"; idem, *Byzantine Praetorians*, 629–33; Winkelmann, "Probleme."

numbers in like units, which cannot be safely taken for granted. Nevertheless, if we take the figures given by these Arab commentators for individual *themata*, and aggregating those *themata* that were originally part of the same field command before the reforms and changes of the eighth or first half of the ninth century, we arrive at some plausible totals for the overall nominal strengths of the armies in question.[53] On this basis the figure of 120,000 would be a reasonable total for the nominal roll of the army in the military registers for all categories of soldier. The number of active troops of military value is a different matter. The figures from other sources—contemporary or near-contemporary accounts of battles and campaigns, some semiofficial and official documents, and the like—suggest very much smaller numbers for the *tagmata* (a total of some 4,000 for all four imperial *tagmata*), relatively small field armies, thus that the active soldiers on the military registers numbered considerably fewer than this.[54] As we will see, the registered soldiers of the provincial armies of the Peloponnese may have been as few as 2,000 in the early 920s.[55] One Arabic source gives a total of 40,000 for all the cavalry forces of the empire in the second half of the ninth century. As noted already, Leo's *Taktika*, as well as narrative accounts of the wars of the late tenth century, suggest that field armies from 3,000 to 4,000 were more usual than anything bigger. Such magnitudes may be corroborated by a ninth-century Arabic military treatise, *The Organization of Warfare*, probably in its original form composed during the reign of al-Ma'mūn by a certain al-Harthamī, and likely connected with the early ninth-century general Harthama b. A'yan. This notes that armies may consist of anything from 800 to 12,000 men. A force of up to 4,000 men counts as a good-sized force, while more than 12,000 is regarded as very large. Recommended numbers are 4,000 or, for major expeditions, 12,000.[56]

53 Thus the combined strengths given for the three ninth-century *themata* of Boukellarion, Opsikion, and Optimaton, all divisions of the original Opsikion army, produce a total of 18,000, which would be a force slightly smaller than the praesental field army based in the same region in the later sixth century. See Ibn al-Fakīh, 72–77 and Kudāma ibn Ja'far, 197–98.

54 For varying views, see Treadgold, *Byzantium and Its Army*, 66–77; Winkelmann, "Probleme"; and in detail, Haldon, "Chapters II, 44 and 45," 305–34; note also Whittow, *Making of Orthodox Byzantium*, 181–93.

55 See on Const. 20§205.1056–57 (and cf. on 4§2.15–16).

56 Although the value of the comparison may be vitiated by the fact that the same text quotes a *hadith* in which the optimal unit strength is set at 400 and that of the ideal army at 4,000. See al-Harthamī, 28 (and see also 26–27) and chapter 2 above, pp. 43–44.

But the active elements selected for battle could be much smaller—in the attack on the Paulician position at Bathys Ryax in 878, for example, the forces of the Armeniakon and Charsianon *themata* were divided into two, and a force of 600 selected from both armies, under the two *strategoi*, made the attack. The rest of the "numerous Roman force" took up position on the ridges around the enemy camp, and at a given signal made a great din to shock and terrify the enemy, while the 600 and the generals went in to the attack.[57] The totals for Byzantine provincial armies in the Arabic sources leave out of account the imperial fleet, of course (although they include the soldier-sailors of the so-called "maritime" *themata* in western Asia Minor), which may at times have been considerable. In the expedition to Syria and Crete in 910–11 the imperial fleet included some 13,500 sailors, and the thematic fleets some 17,540 soldiers and sailors altogether, so that in any consideration of overall military manpower the maintenance of several thousand more sailors/soldiers should be borne in mind.[58]

Strategic Organization and the Themata *in the Reign of Leo VI*

The Byzantine strategic response to the loss of territory and resources, as well as the appearance of new enemies and new tactics, seems to have been slow and piecemeal, and in order to provide a longer-term context for Leo's military writing, it will be worth sketching briefly some of the key aspects of this evolution. Strategy in the East, following the early disasters of the Arab conquests and raids into Asia Minor, came to rest upon three elements. First, and wherever possible, raiding forces were held and turned back at the various passes and entry points to imperial territory. Occasionally, when a unified command over the field armies was established, this policy worked quite well. At the end of the seventh century, for example, Herakleios, the brother of the emperor Tiberios Apsimar, was made *monostrategos* of the frontier cavalry forces, and in 697/98 was sent to Cappadocia, from which he was able to launch successful attacks into Syria in 699/700, and heavily defeat a series of major raids in 700/1, 702/3, and 703/4. This series of spectacular successes was brought to a premature end by the seizure of the throne by Justinian II, in 705.[59]

---

57 *Vita Basilii imperatoris*, 42 (150–54). The rest of the combined total of the two *themata* can hardly have been more than a few thousand.

58 For a detailed analysis of the figures for the army and fleets in both 910–11 and 949, see Haldon, "Chapters II, 44 and 45."

59 *Theophanis Chronographia*, 371–72 (trans. Mango and Scott, 517–20).

Second, where this policy did not work (generally the case), then local forces were supposed to harass and dog the invading forces, making sure to follow their every movement so that the location of each party or group was known. A key aspect of this strategy was the garrisoning of numerous small forts and fortresses along the major routes, on crossroads and locations where supplies might be stored, and above and behind the frontier passes through which enemy forces had to pass to gain access to the Byzantine hinterland. As long as these were held, they served to hinder any longer-term Arab presence on Byzantine soil, since they posed a constant threat to the invaders' communications, to the smaller raiding or foraging parties they might send out, and to their logistical arrangements in general. Small and feebly occupied though many of them were, they were usually sited on good sources of water, and as long as they had advance notice of the raiders, they could store supplies adequate for most eventualities. Such a strategy served to discourage smaller raids, because time in hostile territory was precious and the soldiers and warriors were interested in collecting as much booty as they could and getting out again. Although both small and large fortified places frequently changed hands, the Byzantines clearly understood the importance of maintaining their control as a means of preventing efforts at permanent settlement and of minimizing the extent and effect of the raids.[60]

Third, and from the late eighth and early ninth century, the armies in the eastern regions were complemented by a series of special frontier districts that constituted independent commands. These were known as kleisourarchies (*kleisourarchiai*), that is to say, commands controlling the passes (*kleisourai*) through the mountains, created from subdivisions of the military provinces from which they were detached. They represent the crystallization out of the previous strategy of a new policy: a locally focused defense, involving a "guerrilla" strategy of harassing, ambushing, and dogging invading raiders, designed to stymie all but the largest forces and to prevent both the pillaging of the countryside and the economic dislocation which followed, as well as to make raiding expeditions riskier and less certain, in terms of easy booty, than before. The change may be signaled by the report in the chronicle of Theophanes that, in the year 779, the emperor Leo IV ordered each *strategos* to select 3,000 elite troops to harass

---

60 For the significance of these strongholds, see in general the account in Lilie, *Byzantinische Reaktion*; and for a specific example from the Balkan context, see Stavridou-Zaphraka, "Vodena."

the Arab army which had reached Dorylaion, rather than attempt a direct confrontation. Theophanes states that this was specifically to prevent them sending out pillaging raids; and he also reports the emperor's order to destroy pasture and other supplies, with the result that the Arab force had to retire after fifteen days.[61] The creation of the *kleisourai* appears to mark the development of a third element in defensive strategy, specifically aimed at frontier provinces and the conditions there prevailing, and suggests an awareness in Constantinople of the need for greater autonomy at the local level. Some of these frontier regions, such as Kappadokia, Seleukeia, and Charsianon, had uncertain status before this time, and this may have reflected their exposed situation and the need to maximize their defensive possibilities. The first, called *mikre* or lesser Kappadokia, was originally a *kleisoura*, possibly formed from the areas later to become the *tourmai* of Koron and Nigde (between 806 and 813) in the Anatolikon *thema*, but by the 830s may have had a *strategos*, with the addition of other districts (and may have included Loulon in the Anatolikon and Kasse in the Armeniakon).[62] Seleukeia was a *kleisoura* from the late eighth century, but became a *thema* only during the reign of Romanos I.[63] Charsianon was in origin a *tourma* of the Armeniakon army, but may have become a *kleisoura* as early as 793/94. By 873 it was a *thema*.[64] Similarly the soldiers of the district of Chaldia, originally part of the Armeniakon, appear by the late eighth century to have attained a degree of independence: seals of *doukes* of the corps, as well as seals of officers normally associated with the military establishment (*tourmarches, komes tes kortes*), all from the late eighth and early ninth century, suggest an independent command. There are no references to officials who might yet be connected with a permanent civil administrative establishment. *Archontes* had also held (and may have continued to hold) office until the time of the compilation of the *Taktikon Uspenskij*

61 For the frontier at this time, and the strategies pursued by both sides, see Haldon and Kennedy, "Arab-Byzantine Frontier." For Leo IV's order: *Theophanis Chronographia*, 452.6–12 (trans. Mango and Scott, 624). See Lilie, *Byzantinische Reaktion*, 171–72.

62 See for what follows Brubaker and Haldon, *Byzantium in the Iconoclast Era*, 755–62.

63 Cf. Honigmann, *Ostgrenze*, 42–43; Oikonomidès, *Listes*, 350; and Ferluga, "Niže vojno-administrativne jedinice," 80. For further seals and literature, Winkelmann, *Rang- und Ämterstruktur*, 111.

64 Winkelmann, *Rang- und Ämterstruktur*, 114–15; Honigmann, *Ostgrenze*, 49–51; Ferluga, "Niže vojno-administrativne jedinice," 79–80; Blysidou et al., Μικρά Ασία των Θεμάτων, 299–305.

(thus either ca. 809–28 or in the early 840s),[65] when a *strategos* is first attested. He reappears in the year 867, and it is generally assumed that the regular thematic arrangements were then in place.[66] The units associated with the region Koloneia, also part of the Armeniakon division, similarly had a *doux* and an *archon* during the ninth century, before appearing as a *thema* in the 860s.[67] As well as the administrative *kleisourai*, it should be noted that the individual passes and defiles through the mountains continued to be referred to by the same term, so that we read, for example, of the *kleisoura* of Podandos, north of the Cilician Gates, as well as those of Seleukeia or Kappadokia.[68]

Over the same period, the naval arrangements of the empire were expanded, so that by about 830 there were three main naval *themata*, of the Aegean, of Samos, and of the Kibyrrhaiotai, in addition to the imperial fleet and the much smaller provincial fleets of Hellas and the Peloponnese. The maritime front was thus covered in the east, and while continued raiding and piracy was not stopped, it was at least checked and occasionally thrown back. The situation in the west was somewhat different. The loss of Carthage in the 690s and of the North African coast had deprived the empire of its naval bases there. Sicily probably continued to function in this way; and there is some slight evidence for imperial naval activity in the Balearics. Sardinia remained an imperial possession throughout. From the late 840s, however, the Balearics too were providing shelter for Muslim pirates and raiders, and by the early ninth century the empire seems to have lost interest in the western Mediterranean. The failure to provide adequate naval support when Sicily and then Crete were invaded in the 820s proved costly, since the latter in particular rapidly became the source of constant maritime raids on the empire's coastal lands.[69]

---

65 See Blysidou et al., Μικρά Ασία των Θεμάτων, 287–97. On the possible earlier date of *Taktikon Uspenskij*, see Živković, "Uspenskij's Taktikon."

66 See Georg. mon. cont., 839.15–16; pseudo-Symeon, 687, 21–22; Symeon *magistros*, 260.12; Leo Grammaticus, 253.14–15. Seal of a *strategos*: ZV 2137A (C.9). Oikonomidès, *Listes*, 349 and nn. 349, 350 prefers an early date—824—for the probable thematic foundation; see Blysidou et al., Μικρά Ασία των Θεμάτων, 287–97. Note also Živković, "Uspenskij's Taktikon," 59–60.

67 See Winkelmann, *Rang- und Ämterstruktur*, 114 for the seals and other textual evidence; Blysidou et al., Μικρά Ασία των Θεμάτων, 321–29; Oikonomidès, *Listes*, 349 and n. 345 notes that Mas'ūdī, writing in the late tenth century, refers to Koloneia as a "district," which may refer to its status as a *kleisourarchia*.

68 See Attaleiates, *History*, 121 (trans. Kaldellis and Krallis, p. 221) and cf. *Skirmishing*, 23.

69 See Eickhoff, *Seekrieg*, 198 and Pryor, *Geography, Technology and War*, 102–11.

The Balkan front presents a somewhat different aspect, and for two basic reasons. First, the absence of any concept such as *jihād* on either side meant that war for purely ideological purposes was largely absent (even if the Byzantines on occasion invoked Christianity or orthodoxy as one of the justifications for their own efforts): ideological differences focused on territory and resources, although war-making to avenge defeats, maintain honor, or respond to internal political pressures certainly was a factor. Second, Bulgar khans and Byzantine emperors recognized, through a series of agreements beginning with that between Constantine IV and Asparuch in 680, that a territorial boundary could be drawn. In organizational terms, however, the same pattern of defensive arrangements was applied as in the east, with armies established to defend imperial territory, evolving gradually a territorial identity, before this military-administrative system was employed by the government to win back lost territories.

There is no evidence here for a "frontier" society such as evolved in the east, although banditry in the hills and forests meant that "peace" was a relative affair for the rural and urban populations of both sides. The isolated Byzantine fortress-settlements were frequently dependent on their own resources and initiative for their survival, until they were, eventually, incorporated into the expanding territory of the empire from the early ninth century. And both Bulgars and Byzantines were active in moving population groups to occupy border districts and to serve as frontier soldiers—in the first case transferring Slav tribes or confederacies to the regions behind the frontier with the Byzantines, and in the second, the immigration of populations from central and eastern Asia Minor to Thrace.[70]

The development of imperial strategy is marked by the incremental evolution of the military provincial commands, both in respect of the distribution of troops and their command structure, as well as in respect of its administrative importance. For the latter, these were at first merely groupings of provinces across which different armies were based. By about 700–730 they had acquired a clear geographical identity (reference is made to "the provinces of such-and-such an army/command," for example); and by the late eighth century some elements of fiscal administration as well as military organization were organized on this

---

70 See Obolensky, *Byzantine Commonwealth*, 63–65 and Ditten and *Ethnische Verschiebungen*, 158–59. For some more general reflections on medieval frontiers and the culture and society they engendered, see the essays in Pohl et al., *Transformation of Frontiers*.

provincial-military basis, although the late Roman provinces continued to subsist. By the middle of the ninth century, and after the fiscal reforms of Nikephoros I, through which the term *thema* seems to have been formally introduced, it is clear that *themata* had both an administrative and military administrative structure, which was rapidly replacing the vestiges of the late Roman arrangements; and that the *strategos* acted effectively as generalissimo in his province, with at the very least a supervisory authority over fiscal and judicial officials.[71]

The number of commands expanded with the stabilization of the empire's political situation (during the reigns of Leo III and Constantine V), initially in terms of internal subdivisions to create politically and logistically more manageable units; then—from the late eighth century—with the reimposition of imperial authority over former imperial lands in the south Balkans. As the empire reasserted its military strength in the east during the ninth century, so the role and the proportion of full-time "tagmatic" units became ever more important. As a reflection of this, there emerged, first, a centralized overall command of the active field forces, the *tagmata* and similar units, under the *domestikos* of the *Scholai*; and then, from the second decade of the tenth century, the subdivision of this command into two spheres, of the west and of the east.

The final stage began in the first half of the tenth century when, following a series of successful campaigns on both the eastern and Balkan frontiers, a whole range of new military districts under independent commanders had to be established. Initially, this involved the upgrading of former *kleisourai* to the status of *themata* as well as the incorporation of new regions as *themata*, with the difference that they were generally quite small, centered on a particular fortress, delineated clearly as geopolitical entities, and placed under officers often referred to as "lesser" *strategoi*. Along the eastern borderlands they were called "frontier" or "Armenian" themes, because Armenians made up a substantial portion of the population, both indigenous and migratory. This nomenclature also served to differentiate them organizationally and culturally from the older "great" or "Roman" *themata*.[72] But this system was further modified with the stationing of

---

71 For the growth of the *themata*, see Brubaker and Haldon, *Byzantium in the Iconoclast Era*, 723–64; Oikonomidès, *Listes*, 348–54; and Winkelmann, *Rang- und Ämterstruktur*, 72–118.

72 During the tenth century a considerable movement of immigrants from Armenia proper into southeastern and southern Asia Minor took place, which led the government at Constantinople to regard these regions as "lesser Armenia." See Dagron and Mihăescu, *Traité sur la guérilla*, 239–45 and Kühn, *Byzantinische Armee*, 61–66.

ever-larger and militarily more effective detachments of the imperial *tagmata* and similarly recruited professional units in a broad band of fortified centers, most of them serving as the headquarters of newly established small *themata*. From the late 960s, these were grouped into a series of larger commands, each under a *doux* or *katepano*, independent of the local thematic administration, and who were given a general authority over the lesser generals across whose regions their military authority was granted.

These new commands generally encompassed a group of new small *themata*, together with major fortified centers located within the *themata* in their rear. Strategically, these new commands formed a screen of buffer provinces protecting what had now become the hinterland of the old *themata*, each covering a segment of the expanded frontier, strategically oriented for offensive operations, and independent of one another's manpower. Similar arrangements were established in the West. One of the results of this development was, inevitably, the increasing irrelevance of the older thematic militias, which gradually lost most of their military potential and capacity. Instead, the field armies both along the frontiers of the empire and within the provinces were composed increasingly of either mercenary, professional troops or forces sent by subordinate and vassal princes and rulers of the various smaller states bordering the empire.

Warfare and Society in the Ninth and Early Tenth Centuries

Given the nature of our evidence, spread as it is across texts and other sources from several centuries, it is difficult to focus on one short period without drawing on this broader range of material. Warfare was part and parcel of the cultural consciousness of all medieval people, even if they were themselves never directly touched by it, and although attitudes and views changed in some respects according to the times, some constant, or relatively constant, reflections of these views can be identified. Imperial ceremonial was employed at Constantinople to encourage popular support and approval for both emperors and military leaders as well as war itself: triumphal processions, organized on an almost liturgical basis and completely integrated into a Christian thought-world, displays of booty and prisoners, or the acclamations reminding the emperors (and the crowd who were in earshot) of their Christian duty to defend Orthodoxy and the empire, all served this end. Court poets were commissioned to write and declaim verse narratives of military achievements, courage, and skills shown by emperors in wartime—among the best-known are George of

Pisidia's poems on the Persian and other wars of the emperor Heraclius or the verses of Theodosios the Deacon on the recovery of Crete by Nikephoros Phokas. Members of the political and ecclesiastical elite composed letters in praise of the emperor's deeds in war, slanting the approach according to whether or not the emperor actually campaigned, or whether he stayed at home. The glorification of military deeds and of individual leaders or rulers formed part of the stock-in-trade of panegyrists and encomiasts, whether on account of an emperor's successes in war, or his pursuit of its antithesis, peace (since in the Byzantine view the former was often necessary for the achievement of the latter). Warfare thus had its positive image, since it was an unfortunate but mostly necessary means to a divinely approved end.[73]

Warfare imposed itself upon the literature and practice of Byzantine culture through many routes. As already noted in chapter 1 above, warfare and the activities of soldiers and generals were a constant element in the historiographical and chronicle literature of the Byzantine world, not just because it was a convenient literary vehicle for the expression of a range of situations and values reflecting the human condition, but also because it was a constant presence on the Byzantine cultural and political horizon. Thus it features also in hagiographies, in funeral addresses, in speeches praising emperors, in sermons and homilies to the congregations of churches, as well as in private letters addressed to individuals in connection with warfare—death, loss of property, and so forth. Two homilies of the Patriarch Photios written in 860 describe graphically the damage caused and the terror inspired by a sudden raid by a large Rus' fleet that year. A number of letters bewail the effects of warfare—the tears of the orphaned children and widowed mothers, the destruction of crops in the fields, of homes, of monastic communities; the enslavement or death of populations; the driving off of livestock; and so forth.[74] In a series of letters to the Bulgarian tsar Symeon written ca. 912–25, Patriarch Nicholas Mystikos describes graphically the devastation, death, enslavement of the population, and other consequences of warfare in Thrace, a picture reinforced by other writers. Other letters of Nicholas, as well as of other writers,

73 See especially McCormick, *Eternal Victory*; *ODB*, "Enkomion," 1:700–701; Kolia-Dermitzaki, "Byzantium at War," esp. 231–34.

74 See Photios, *Homilies*, 74–110. See, for example, a letter of the tenth-century official Symeon *magistros*, in which the disruptive effects of an Arab raid in the 960s are described: Darrouzès, *Épistoliers*, 2: no. 86.

deal with the effects of the state's demands on the provincial populations for extra supplies and provisions and for hospitality, livestock, and equipment. And although there is often a powerful rhetorical element in many of the letters, they illustrate the realities of warfare and fighting. With warfare came insecurity and uncertainty: the eleventh-century *Life* of Lazaros of Galesion mentions a former prisoner of war of the Arabs, who informed the saint that his daughter had had no news of him and thought that he was dead or still a prisoner.[75]

There evolved in addition, and in relation to Islam in particular, an astrological literature, connected also to the apocalyptic tradition, which claimed to foretell the results of wars.[76] Some branches of this tradition presented the wars and the political fortunes of the two contestants, Orthodoxy and Islam, as tied together by divine will in a cyclical relationship, where first one then the other would be victorious. This literature was of such influence that it was believed by some that when, on the basis of the cycle, it was the turn of the other side to be victorious, there was no real point in resisting—fate and divine providence had already determined the outcome.[77] Horoscopes and other forms of predictions were particularly important when fighting and warfare were at issue, since not only the ordinary soldiers but the senior officers were just as interested in trying to predict the outcome of a conflict. Leo's *Taktika* advises generals to beware of the misinterpretation of signs and portents among the soldiery, and to make sure that they spread favorable predictions to avoid demoralizing the soldiers.[78] As we have seen, emperors going on expeditions were advised to take along not just military handbooks and literature relevant to the practice of war, but also astrological and horoscopic books which would assist them in foretelling the outcome.

75 For a useful survey of this material, see Kolia-Dermitzaki, "Byzantium at War," 220–22. For Lazaros and the former prisoner: *Vita S. Lazari*, 529B–D (trans. Greenfield 150–52); on Nicholas Mystikos and Symeon: *PmbZ* 25885 and 27467.

76 See esp. Dagron, "Apprivoiser la guerre."

77 See, for example, the remarks of Liudprand of Cremona on his visit to Constantinople in 968: "The Greeks and Saracens have books . . . where one finds written the number of years of each emperor's life . . . and whether he will have success or failure against the Saracens. . . . for this . . . reason, the Greeks, full of courage, attack, and the Saracens, without hope, offer them no resistance, awaiting the moment when it will be their turn to attack and that of the Greeks to make no resistance." See Liudprand of Cremona, *Legatio* §39 (ed. Becker, 195–96); and discussion with Byzantine and Arab sources in Dagron, "Apprivoiser la guerre," 42–47.

78 See e.g., Const. 14§101.

One or two examples of this tradition occur in the narrative sources, as when the emperor Constantine VI was told by the court astrologer that if he attacked the Bulgar army facing him at Markellai in Thrace he would win, although several of the generals, who considered their tactical situation unsound, advised against this. Constantine duly attacked, and his army suffered a serious (and thus, according to the author of the text, a predictable and deserved) defeat. In spite of the powerful influence of Christian theology and dogma, pre-Christian traditions of this sort continued to have a certain influence. Alexios I reportedly reached a decision by writing down two possible courses of action on separate bits of paper, leaving them on the altar of a church during a night of prayer, and waited to see which one God's will directed the priest to pick up first on entering the building the next morning. The emperor Manuel I reportedly placed considerable emphasis on astrology and its predictive potential in such contexts, with similar unfortunate consequences.[79]

Outside these more literate milieux, however, attitudes varied dramatically by context. At a distance, warfare was something to be anxious about, to hope to avoid, and to pray for success in. A seventh-century fragment of a touching prayer for deliverance from the Arabs and military success for the army survives from Egypt, for example, which testifies eloquently to this; while epigraphic evidence reflects the attitudes of the provincial population in Syria and Palestine during the sixth-century wars with Persia.[80] The population of fortress towns and villages bore the brunt of both enemy and imperial activity. Unfortunately there is little direct information in contemporary written texts about this. They were liable to lose their crops and their livestock, quite apart from their lives or their personal freedom; while they were subject to both fiscal demands and, more immediately, to forced evacuation from their homes when the local military commander decided they were too inviting a target for the enemy raiders. When enemy raids were

79 See *CPTT* C.199–202. The books used by sailors for "reading the weather" were not simply navigational aids. Some were compiled specifically with naval warfare in mind, and included substantial horoscopic elements, and determined in addition on what days of the week naval warfare should be undertaken or avoided. See Dagron, "Firmament." For Constantine VI see *Theophanis Chronographia*, 467–68 (trans. Mango and Scott, 643). On astrology in Byzantium see *ODB*, "Astrology," 1:214–16 and "Horoscope," 2:947–48. For Alexios I: Anna Komnene, *Alexiad*, 15.4 (trans. Sewter, 481–82/Frankopan, 441); for Manuel I: Choniates, *Historia*, 95–96.

80 See Photiades, "Parchment." For the Syrian material, see Trombley, "War and Society."

expected in the ninth and tenth centuries, and probably before, officers referred to as *ekspelatores* were instructed to go from village to village and bring the population and their movables to safety, in the fortresses and refuges in the hills or elsewhere. Indeed, the threat to the population in some areas, such as Cappadocia, was such that they took to living in caves in the hills, where vast underground complexes were established, veritable troglodyte communities whose vestiges are still clearly seen today.[81]

Over the long term, this situation must have affected the nature of social ties as well as the structure of agricultural and pastoral activity, although we have very little evidence for the forms through which these effects were manifested. Some epigraphic and archaeological evidence from the region of medieval Barata (anc. Gaianoupolis, mod. Maden Şehir) in Lykaonia, southeast of Ikonion may be characteristic. Here the easily defended upper part of a much earlier Roman town was reoccupied from the seventh to ninth centuries (dated by several small churches) and suggests the sort of settlement to which much of the area's rural population would have recourse in time of attack. Inscriptions, probably eighth or ninth century, suggest the centrality of warfare in the lives of those who lived there: "Here lies Mousianos, who endured many wounds," for example, or "Here lies Philaretos Akylas, who died in the war on May 30th in the 4th indiction."[82] Soldiers and their families were no more exempt from these effects than the rest of the population, of course. The parents of young men drafted into the army or called up to fulfill their military service wept and lamented as they took leave of their sons; the more privileged were able to deploy powerful contacts to have their sons released from serving in the army, on grounds of economic hardship, for example.[83]

---

81 While the arrangements described in *Skirmishing*, commented on extensively in Dagron and Mihăescu, *Traité sur la guérilla*, 215–37, reflect the situation in the first half of the tenth century, they can be shown to have applied earlier also; see *CPTT* B.35–38 and comm.

82 See Ramsay and Bell, *Thousand and One Churches*, 525–26, and nos. 13, 42, 43. The dating of the inscriptions remains problematic, but the eighth–tenth centuries are generally accepted.

83 For example, the story of the soldier Leo who was forced to send his only son George on an expedition against the Bulgars in the early tenth century because he was himself too old to serve actively: *Miracula S. Georgii*, 21.6–13, 22.7–9; in a slightly different case, a widow in the early tenth century is able to obtain the help of a high-ranking Churchman in pleading for her son to be exempted from the call-up because of her poverty; see Darrouzès, *Épistoliers*, 2:50 (130–31).

Desertion from military service was clearly a problem, and the military treatises frequently refer to the issue. Provincial officials had instructions to arrest and confine all those who failed to turn up in time for the muster held at the beginning of the campaigning season. The regular muster and updating of the military register was essential in minimizing desertion, as the treatise on campaign organization makes clear. The guardsman Ioannikios, a soldier in the elite *Exkoubitoi* regiment, decided in 796 (after some 23 years of service and at the age of 43) to desert his unit after the defeat at Markellai, in which his unit was badly mauled. Although he became a monk, he was forced to flee from former comrades on at least one occasion years later when he was recognized—testifying also to the effectiveness of military discipline. But desertion was not uncommon: some soldiers ran away to become monks, where they would be able to conceal themselves under a new identity; others may have tried simply to return to their homes and families. And it was treated very harshly, as the military codes make clear, and as the occasional testimony of the narrative sources also shows: in southern Greece in 880 some of the crews and soldiers of the Byzantine ships facing an Arab raiding force deserted their posts and fled. They were tracked down, captured, paraded with ignominy, and (ostensibly) impaled. In fact, while the real deserters were captured, those who were executed had originally been enemy prisoners in Constantinople, but were shipped to the scene of the desertion and executed specifically in order to serve as an example of the fate awaiting deserters. The whole event was an artifice designed to frighten the soldiers and sailors back to their duty. According to the chronicler, it worked.[84]

Soldiers, warfare, and the military more broadly were thus an obvious and ever-present part of the Byzantine world, whether in the provinces or, at a greater geographical and cultural distance, the capital city, Constantinople. Soldiers were drawn from every community across the provinces of the empire, they possessed special juridical privileges and associated social status, and they were an important part of rural and provincial power relations, often drawn into networks

---

84 See below, commentary to Const. 8; and see, for example, Maurice, *Strategikon*, 1.6.7; 8.16; 7.B.13; 8.1.28, 35 (repeated in *Taktika*, 8§7, §20 etc.). For the treatment of those who fail to return from leave in time for the muster, or who deliberately fall behind the expeditionary column, see *CPTT* B.42–45 (and cf. Maurice, *Strat.*, 1.6.4; 7.14 = *Taktika* 8§4, §15); and for the registers and the campaign muster: *Campaign Organization*, §29. For Ioannikios: *Vita S. Ioannici*, 337C–338C, 311B. For the events of 880: *Vita Basilii imperatoris*, 62.14–47.

of patronage outside their military capacity, often with the power to intervene vocally and violently in local and even Constantinopolitan politics and affairs. In addressing the issues Leo saw, from the perspective of an emperor, as affecting his empire's military and its leadership, the *Taktika* barely reveals the tip of this institutional, social and political iceberg. Yet it nevertheless offers the historian a point of entry into a set of social institutional practices that were at least in part responsible for giving middle Byzantine society, culture, and politics their particular appearance in the written and material cultural sources.

PART II

Critical Commentary on the *Taktika*

PROLOGUE

While the *Taktika* was certainly intended as a book of advice rather than as a legislative act or document as such, the prooemium is modeled closely on the *prooemia* to imperial legislation, and bears a particular similarity to that of the *Procheiros Nomos*. In this case it belongs to the "tripartite" model as analyzed by Hunger (*Prooimion*, 188–90), and as with all such legislation, follows a specific format. The first part consists usually of three elements: an *invocatio*, generally invoking the Trinity, followed by the *intitulatio*, with the name and titles of the emperor; and then the *inscriptio*, including the name of the person to whom the document is addressed. In the usual legislative instrument, after the prooemium has set out the reasons for issuing the document, the main body of the legislation follows. This again consists of several elements: the *narratio* introduces and describes the problem, the *dispositio* sets out the emperor's command or solution, and the *epilogus* orders the publication or application of the edict or law. In the case of the prooemium to the *Taktika*, the first section (ll. 7–24) represents the *invocatio* and *intitulatio* only (the *inscriptio* is not appropriate so is omitted), setting out the emperor's duty in respect of God and his subjects and introducing the basic theme of the legislation through a series of standard formulas associated with the role of the ruler: the glory of the empire is most enhanced not in earthly displays of power but in maintaining the peace and harmony of his subjects, which in themselves will ensure stability and the restitution of the established state of affairs. The second section (ll. 25–54), representing in effect a *narratio*, sets out the nature of the problem to be addressed and how it came to imperial attention, hence underlining the usefulness or value of the legislation; and the third section (ll. 55–89), equivalent to the *dispositio*, with a bridging section (ll. 90–96), introduces the measures to be taken to deal with the issues thus identified. The final section (ll. 97–113) may be seen as an *epilogus*, in which the agenda or contents of the work that follows is outlined. Given Leo's representation of the treatise as a whole as a handbook, a *procheiron*, the closeness to the form and structure of the prooemium to the *Procheiros Nomos* was probably quite deliberate.

1–2   Λέοντος ἐν Χριστῷ αὐτοκράτορος τῶν ἐν πολέμοις τακτικῶν σύντο-
μος παράδοσις   The use of the title autokrator (also l. 5) was especially favored by Leo from ca. 900; see Schminck, *Studien*, 92–94 and
chapter 2, pp. 59–60 above. The title makes it clear that Leo's treatise was intended to bring together and codify the advice and information offered in older treatises. It is repeated in the acrostic which
makes up the first letter of each paragraph of Const. 20 (see comm.).

3–6   §1   The invocation is the standard formula at the beginning of imperial legislative acts and is close to those used by Leo both in his *novellae constitutiones* (νεαραί) as well as in the acts of earlier rulers.
For discussion see Hunger, *Prooimion* and, more generally, Dagron,
Emperor and Priest. While following the *Strategikon* by invoking
the Trinity, Leo omits the reference to the Theotokos which Maurice
includes in his prologue, replacing it instead with an emphasis on the
homoousian nature and the honor to be accorded to the Trinity. And
whereas Maurice ends his invocation with the "Amen," Leo transfers
this to the very end of the whole treatise, thereby emphasizing both
the prayer-like character of the whole enterprise as well as the divine
support he has received in undertaking it. Apart from the absence of
the epithet εὐμενής the wording is also the same as that making up
the first part of the acrostic in Const. 20 (see comm.), which again
serves to frame the treatise with a clear emphasis on divine support
and approval.

As with the legislation of his father Basil I, and in his own legislative work, the dominant motifs in the introductory material are
the devotion and faith owed to God, not only by the God-appointed
emperor himself but also by the people, his subjects (see p. 121,
above); other motifs are the divinely inspired nature of the imperial legislation and divinely founded nature of imperial authority;
see Simon, "Princeps legibis solutus." For ἀεισέβαστος see Hunger,
Prooimion, 81–82; *Procheiros Nomos*, Intitulatio (Schminck, *Studien*,
56.4). These motifs are emphasized strongly in the "book of advice,"
composed for Leo by Basil I (although mostly by Photios), Κεφάλαια

παραινετικά (ed. Emminger), 50, §§1–2; esp. 59, §27; they recur throughout the twenty constitutions of the *Taktika*; in the final constitution and the epilogue they come to the fore clearly. For a comment on the visual forms of this imperial imagery, see Wood, *Leo VI's Concept* and Gavrilović, "Humiliation of Leo VI."

7–9   cf. *Procheiros Nomos*, pr.18–20. δορυφορία is to be understood as the guardianship with which the emperor is endowed by God, rather than "guard" in the physical sense of soldiers. This section, down to l. 19, bears clear similarities to ll. 1–45 of the prologue to the *Procheiros Nomos*.

11   ἐπανόρθωσις   Cf. Hunger, *Prooimion*, 33, 44, 103–6. A constant feature of imperial rhetoric was the emperor's role as restorer of justice, order, and stability, always with reference to a supposed earlier and idealized state of affairs.

14   προνοίας   Cf. Hunger, *Prooimion*, 84–94. Imperial care and forethought was directly compared with divine providence, the same word applied in both contexts, and implied the emperor's duty of care for his subjects, as underlined by the phrase at ll. 16–17: τῆς ἡμῶν μετὰ Θεὸν . . . προνοίας.

17–18   νύκτωρ μὲν ἐπαγρυπνοῦμεν   See Hunger, *Prooimion*, 94–100 and cf. Leo IV, *Novella* (Zepos, *Jus*, 1:40.6–7) with Dölger, *Regesten*, 338. A standard attribute of the emperor, who should be sleepless in his devotion to his subjects: cf., e.g., *Epanagoge*, 2.2.

23–24   ὅσον ἡ πεῖρα τοῦ νῦν χρόνου. . . . πρόδηλα   A direct reference to recent events and more particularly the defeats which the empire had suffered at the hands of both Bulgars and Saracens. In 896 the general Leo Katakalon was defeated by an inferior Bulgarian army in Thrace at Bulgarophygon; between the early 890s and 904 the empire suffered a number of naval defeats, culminating in the siege and sack of Thessalonike in 904, after which the situation was stabilized. See the summary with literature and sources in Tougher, *Leo VI*, 178–92.

27–31   ἐπειδὴ δὲ . . . . . . καθίστασθαι   Cf. Syrianos, *Strategy*, 4.9–17.

27–28   John 8:44.

32–36   ἀλλὰ ταῖς στρατηγικαῖς μεθόδοις . . . . . . πολιτεύοιτο.   On the issue of just war and holy war see below on 18§105.508–16. Waging war to secure peace was a standard element of Byzantine imperial ideology and represented a neat way of justifying warfare, both defensive and offensive, regardless of circumstances. Further discussion: Taft, "War and Peace," with background and context in Barnes, *Constantine and Eusebius*, 245–60; Baynes, "Eusebius and the Christian Empire," 168–72; and Dvornik, *Political Philosophy*, 2:614–15, 652–53. Soldiers were fully accepted members of the Christian community, who had a recognized and indeed worthy role to play. Liturgical prayers evolved from the fourth and fifth centuries in which the military role of the emperors and the need for soldiers to defend the faith were specifically recognized: "Shelter their (the emperors') heads on the day of battle, strengthen their arm . . . subjugate to them all the barbarian peoples who desire war, confer upon them deep and lasting peace" is an illustrative example from a fifth-century liturgical text (see Trempelas, *Three Liturgies*, 185–86). But neither warfare nor the killing of enemies were ever praised or regarded as in some way deserving of a particular spiritual reward—whether Christians were able to justify warfare from a defensive need (to preserve Orthodoxy, for example) or in a more clearly offensive context (to recover "lost" Roman territory from non-Christian or barbarian or heretic, still judged as defensive action), killing remained a necessary evil, and reservations were still held. In tenth-century Constantinopolitan acclamations, peace was the rule of the Roman emperors and the Christian *oikoumene*, the civilized world. While acclaiming emperors as victorious over their enemies, warfare was acknowledged as a necessarily unpleasant means to a worthwhile end. See Treitinger, *Kaiser- und Reichsideologie*, from 230 and McCormick, *Eternal Victory*, 245–52. For specific studies of the issues, see Shean, *Soldiers for God*; Helgeland, "Christians and the Roman Army"; Swift, "War and the Christian Conscience"; and Viscuso, "Christian Participation in Warfare." For the later Byzantine development, see Stouraitis, *Krieg und Frieden*.

40–43   νῦν δὲ τῆς τακτικῆς . . . . . . συμβαίνοντα.   Cf. Maurice, *Strategikon*, pr.10–13, which Leo copies almost verbatim. Criticism of the current state of affairs and its consequences for military effectiveness is a common topos also in earlier military handbooks, and given the fact that much of Leo's account of the tactical organization of Byzantine provincial armies is already partly out of date (see chap. 3 above), the extent to which anything changed after Leo's issue of the *Taktika* is open to doubt: cf. Vegetius, *Epitoma*, 1.28; Syrianos, *Strategy*, 15.6–14.

48–50   ἀγυμνασίαν... ἀμελοῦμεν   Cf. Const. 18§149.842–43.

54   ὠφέλειαν   Another standard attribute in imperial legislation, which was intended to be of benefit to the subjects of the empire. See Hunger, *Prooimion*, 123–28.

55–58   Ταῖς γὰρ ἀρχαίαις...... ἐρανισάμενοι   Cf. Maurice, *Strategikon*, pr.15 and *CPTT* C.24–39. Following Maurice, with slight emendations, Leo incorporates this topos into his introductory comments; see Maurice, *Strategikon*, pr.14–17 (in turn following Aelian, *Tactica theoria*, 1.2–3; Onasander, *Strategikos*, pr.17–45 (§§4–8); cf. Arrian, *Techne taktike*, 1.1–2). As noted by Dennis (ad loc.), in the margin to lines 56–57 in W (fol. 1r) a scribe, possibly the copyist himself, has listed Arrian, Aelian Pelops, Onasander, Menas, Polyainos, Syrianos, and Plutarch as (some of) the ancient authors in question. See Vári, *Leonis imperatoris tactica*, 1:33, n. 1 and idem, *Bölcs Leó*, 66–67. Menas was one of the seventeen Spartans who concluded the treaty with Athens in 422/421 (Thucydides 5.19.2; 5.24.1); the Pelops in question was the son of Lycurgus and king of Sparta in the late third century BCE: *Der kleine Pauly*, 4:608–9. Why these two in particular should have been included, however, remains unclear.

58–59   ὅσα καὶ διὰ μετρίας πείρας... ἀνεμάθομεν   Leo copies Maurice again, but whereas Maurice (or the general who composed the *Strategikon*) actually did have a certain amount of experience, Leo had none, as far as we know. See Tougher, "Imperial Thought-World."

59   τῷ καθ' ἡμᾶς καιρῷ καὶ τὴν νῦν καταστάσει   Leo again emphasizes the contemporary situation and expands his exemplar to add this comment.

61   πρόχειρον νόμον   See on l. 94 below; and Schminck, *Studien*, 91, who (following Zachariä von Lingenthal, "Zum Militärgesetz des Leo," 606–8) notes that Leo expands his text, which closely follows the prooemium to the *Strategikon*, to describe his treatise in two places as a "*procheiros nomos*," using the opportunity offered by Maurice's original τὰ πρόχειρα ("the rudiments/basics") to substitute his own phrase with its legal and legislative connotations. To what extent this otherwise unusual phrase may well reflect both a date of composition

close to the legal compilation known under the same title, as well as a conscious invocation of Leo's juridical legislation (see discussion in chapter 2 above; Schminck, *Studien*, 62; van Bochove, *To Date*, 29–56 Magdalino, "Non-juridical Legislation," 176) remains unclear.

62   εἰσαγωγήν   Maurice, *Strategikon*, pr. 23–24; Aelian, *Tactica theoria*, 1.6. This is again a topos of the Hellenistic and Roman military treatises (Rance, *Strategikon*, pr. n. 7), but also recalls Leo's own legal codification: Magdalino, "Non-juridical Legislation," 177.

63   ὑποστρατηγοῖς   The term had two meanings: since the emperor was in fact the supreme commander, *strategos*, under God, all the other senior officers were in fact subordinate, *hypostrategoi*, which is how it is used here. Some of the chroniclers use it in this way also: cf. Theophanes continuatus, 368–69, of Eustathios Argyros (see *PmbZ* 21828). In day-to-day reality the term was applied to any senior officer placed *pro tempore* under a senior commander for a specific campaign, whether the emperor or another senior general. See 4§§8.65–11.73.

66–67   Leo, omitting the somewhat overly humble phrase at this point in Maurice's prooemium, that his work is οὐδὲ γὰρ ἔργον ἦν ἱερόν (on which see Rance, *Strategikon*, pr. n. 9), otherwise repeats this topos of ancient technical writing found throughout the Byzantine literary tradition; see on ll. 75–76 below.

68–69   ὅθεν . . . . . . διερμηνεύσαμεν   Cf. Maurice, *Strategikon*, pr. 29–30. Here Leo reverses Maurice's wording. The latter notes that, for clarity's sake, he has also employed Latin and other technical terms in common use. In contrast, Leo remarks that he is clarifying the ancient authors by translating the Latin into Greek (see 7§§16–35, for example), a process termed *exellenismos*. Similar sentiments are expressed at *Procheiros Nomos*, pr.52–53: τῶν δὲ ῥωμαϊκῶν λέξεων τὴν συνθήκην εἰς τὴν Ἑλλάδα γλῶσσαν μετεποιήσαμεν. . . .

75–76   καὶ ἄχρις ἡμῶν   Part of Leo's own addition to Maurice's prooemium. While it continues with the motif of reviving ancient writings and making the obscure more accessible, there is a certain contradiction in the passage, since he says that these precepts have been put into practice by the ancients and onward "up to our own times," a statement which his earlier sentiments would seem flatly to contradict. It suggests in fact that Leo felt unable entirely to dismiss recent generalship and military practice, and that he was torn between grandiose claims for his own "renovatio" in military science, on the one hand, and the fact that things—despite the recent misfortunes so clearly alluded to—were by no means as bad as his rhetorical trope implied. The observation that the writer has rephrased and clarified the text, removing solecisms and other infidelities, is commonly found in treatises such as *CPTT* C.30–39 with commentary, 182; and *Parangelmata poliorcetica*, §3, with commentary, 163–68.

83   τῆς προθυμίας   Perhaps "zeal" rather than "enthusiasm."

85–89   οὐδὲ γάρ, ... ἐπιζήμιον δαπάνην.   Cf., e.g., Const. 12§3; 14§33.214–15; 20§40.211–12; and compare further *Ecloga*, 18: οὐκ ἐν πλήθει γὰρ δυνάμεως νίκη πολέμου, ἀλλ' ἐκ θεοῦ ἡ ἰσχύς. This is part of the standard rhetoric in which faith in divine providence and justice are the key arbiters for success or failure in battle, and tactical and strategical knowledge and sophistication are superior to mere numbers and courage. See Vegetius, *Epitoma*, 1.1; Maurice, *Strategikon*, 2.1.8–11; 7.A.pr.8–12; and cf. *Rhetorica militaris*, 44.9 (where it is not numbers, but bravery and hard work, that bring victory); and Eramo, *Siriano*, 176–78. The sentiment regarding numbers is a topos that appears also in the work of chroniclers and historians. Psellos, for example, is contemptuous of the emperor Romanos III, whom he accuses of thinking that numbers counted for more than tactical skill and discipline: *Chronographia*, 3.8.

90–96   §9 Ὥσπερ γὰρ . . . . . . παρακελευόμεθα.   Cf. 2§24.163–66; and Maurice, *Strategikon*, pr.42–49; 7.A.pr.4–8. The metaphor of the ship and its helmsman occurs in Onasander, *Strategikos*, 4.202–6 (4§5); 32.902–8 (32§10); and 33.919–23 (33§2) and is ancient. Leo once again refers to his treatise as a sort of *"procheiros nomos,"* and urges the reader to follow its precepts attentively. This ends the third section (ll. 58–102) of the prooemium, in which Leo sets out the measures to be taken to address the issues outlined in the second section (ll. 26–57).

97–113   §10 Here Leo sets out the agenda and contents of the treatise, explicitly incorporating the three constitutions on siege warfare, surprise attacks, and naval warfare which had previously circulated or at least been written (probably) as separate documents.

112–13   Proverbs 1:5.

CONSTITUTION 1

As the editor points out (Dennis, *Taktika*, 13 n. 1) most of this chapter is drawn from the *Tactical Theory* of Aelian, although with comments interpolated by Leo to point out the archaisms.

The constitution opens with a series of short definitions, to a degree reminiscent of the opening sections of Syrianos, *Strategy*, 1, 2, and 14. Strategy, *strategia*, is associated with or derived from *stratos*, army, *strategos*, leader of an army, or "general," and the term *strategema*, "stratagem," a trick or technique for outwitting the enemy (on which see Wheeler, *Stratagem*, esp. 1–24), and was a key concept in Graeco-Roman military thinking and approaches to warfare. "Strategy" meant simply "the art of generalship." In contrast to the modern sense of the word—the art and technique of deploying all available resources to gain the objects of war, a meaning which developed only from the nineteenth century—strategy in the Byzantine context was not readily distinguished from tactics (see below), so that the medieval military treatises which provide us with so much of our information on military matters treat the two as part of a continuum, normally using the word *strategy* to refer to the structure and organization of warfare, the art of planning and directing specific campaigns, bearing in mind geographic and climatic factors, communications, and the dispositions and movements of the military forces available to the general. The (probably) ninth-century treatise now ascribed to Syrianos *magistros* on the subject offers this definition:

"Strategy teaches us how to defend what is our own and to threaten what belongs to the enemy. The defensive is the means by which one acts to guard his own people and their property, the offensive is the means by which one retaliates against his opponents." (Syrianos, *Strategy*, 5.1–5; trans. Dennis)

3–13   §§1–6   See also ep.§58 with commentary, below. For ll. 7–8 read: "The science of strategy is the joint practice of good generals, which is to say, study and exercise with stratagems or indeed with the gathering of symbols of victory." This is the first of two sets of definitions—in this case, the aim, purpose, and characteristics of tactics and strategy—which, as Vernadsky, "'Tactics' of Leo the Wise," 333–34 notes, offer a close structural parallel to the definitions of the role and duties of the emperor and the patriarch in the first three titles of the *Eisagoge*.

14–15   The words are from Aelian, *Tactica theoria*, 2.1. But Aelian seems never to have written a naval section, and this sentence does not appear in Arrian's treatise, based on the common text used by both writers (*Techne taktike*, 1.2.1, and see Ch. 2 above). It was a convention in the Hellenistic treatises to divide warfare between land and sea, although the promise to treat naval matters either later or in a separate treatment was hardly ever carried out. Leo had to compile his own naval section from the material at his disposal; see below, on Const. 19.

22–25   ἦν δέ ποτε . . . παρασκευῶν   Leo notes (following the convention) that these tactics and methods are no longer employed; see Rance, "Date," 716–19.

28–43   §§9–12   The second set of definitions, of the aim, purpose, and characteristics or attributes of a general.

30–33   §10   Cf. Syrianos, *Strategy*, 4.18–26. This is the contemporary definition, and is important in respect of the way in which the position of thematic *strategos* evolved. The general is now not just the commander of an army. He is also the most powerful imperial officer in his province, after the emperor; he commands the military *thema* or army, appointed to that position by the emperor, and he has authority over the imperial officials sent out by the emperor, but whom he himself has appointed, as well as over those he appoints directly in the province. This was not always the case, since from the time of their appearance during the later years of the seventh century until the early ninth century at least the provincial *strategoi* remained officers with largely military functions still, even if they wielded very great influence over other aspects of the affairs of their province. For a good survey of the role and position of the thematic *strategoi* in the ninth to eleventh centuries, see Ahrweiler, "Recherches," 36–52; for the seventh and eighth centuries, see Winkelmann, *Rang- und Ämterstruktur*, 138–40, who stresses the almost entirely military character of the functions of these officers, even at the end of the ninth century. For the eighth and ninth centuries, see the analysis in Brubaker and Haldon, *Byzantium in the Iconoclast Era*, 734–39.

36–37   ὅσαι τε στρατιωτικαὶ καὶ ὅσαι ἰδιωτικαὶ καὶ δημόσιοι   The position of the *strategos* is underlined, having authority over military, civil, and fiscal or public affairs. Military officials always had judicial authority over purely military matters, and their courts were accredited with the same jurisdiction in the army as were civil courts over nonmilitary affairs. Conflicts between civil and military jurisdictions were endemic to the system. In the late Roman period these were partially addressed by the instrument known as *prescription of forum*, by which accused persons could refuse to appear before any court but their own, even for criminal offences (see Haldon, *Byzantine Praetorians*, 304–7, with notes 915–26; Dagron and Mihăescu, *Traité sur la guérilla*, 269–72).

39–43   §12   Again a contemporary observation and description: the thematic commander's duties. Note the reference to employing unexpected attacks against the enemy as well as "regular" warfare. The term *stasis* should be understood more generally, as "unrest," rather than more narrowly as "mutiny," since the *thema* in this paragraph is clearly a region rather than its army.

44–49   §13   Note the careful stress on the general's aim to enjoy imperial favor—perhaps a reflection of Leo's reaction to the activities of members of the Doukas clan in the period 904–8, and especially of the refusal of Constantine Doukas to perjure himself for the emperor's favorite Samonas before an inquiry in 905. See *PmbZ* 20405, 23817, and 26973; Grumel, "Notes chronologiques," 202–7; Karlin-Hayter, "Revolt of Andronicus Ducas," 23–25; and Tougher, *Leo VI*, 208–10 with further literature. For a more detailed discussion of some of the issues of the relationship between ruler and members of the elite at this time, see Strano, "Potere imperiale."

CONSTITUTION 2

Virtually the whole of this constitution is drawn, more often than not verbatim, from the similar list in Onasander, *Strategikos*, 1.55–115 and 2.118–53 (1§§1–25); used also by Maurice, *Strategikon*, pr.36–56; and 7.pr.2–7; and probably by Syrianos, *Strategy*, 4.18–26 (and cf. 3.18–29). The section from ll. 124–37 is close to the passage on the qualities of a general in the tenth-century compilation from Polyaenus, the *Strategemata Ambrosiana*, chap. 1 (for cross-references to the original passages from Polyaenus, *Strategika*, see Melber, *Polyaeni strategematon*, 431–504). Lines 163–92 of this constitution follow Maurice, *Strategikon*, pr., 41–69. In some respects the list of virtues is not dissimilar from those outlined in the Κεφάλαια παραινετικά addressed to Leo by his father Basil I. In addition Leo seems to have been aware of the passages in the ninth-century *Rhetorica militaris* relating to the qualities of the general and more particularly to the ways in which the commander should inspire the loyalty and obedience of his troops by stressing his own fatherly concern and his self-sacrifice on their behalf; see esp. *Rhetorica militaris*, 36.2–11 and cf. 45.1–2. The biblical quotations—from both Old and New Testaments—are standard fare in any such list of appropriate or desirable attributes, and need no comment, although they are taken from the *Strategikon*. But the fact that Leo spends some effort in repeating this list is important, because it does not reflect simply a piece of archaism or repetition for its own sake. On the contrary, there are hints elsewhere in the *Taktika* that Leo was actually concerned with the ways in which the military machinery of the empire, especially in the provincial armies, was easily suborned by the leading officers, especially those from families with major interests in the apparatus of the state. Such families included the Doukai, Phokades, and Argyroi, for example (see Cheynet, "Phocas"; Polemis, *Doukai*; and Vannier, *Familles byzantines*), with members of whom Leo himself had often close (and often very strained) relations; see Tougher, *Leo VI*, 164–66 and 194–95; and for most recent bibliography and discussion of many of these families and their connections, see the essays in Cheynet, *Byzantine Aristocracy*. While by no means autonomous in respect of either their economic or political position—the more or less complete dominance of the system of precedence and imperial favors ensured this—they were nevertheless, as numerous examples demonstrated across the ninth century, able to exercise very considerable influence over the soldiers and on occasion employ them to their own advantage.

This is not simply a question of political opposition, which is relatively rare. Rather, it is an issue of the growth of systems of patronage and power, which enabled senior provincial officials, especially military officials who could muster armed support for their activities, to exercise undue influence to their own advantage or to the advantage of their own patrons. It is also an issue of shifts in perception, as noble lineage or birth became a recognized element in determining social status. The evidence from the mid-tenth-century novels of Constantine VII cannot be used to illustrate the point for Leo's reign, but it is clear from them that this was a well-established and certainly a recognized threat to imperial authority and more importantly to imperial control of resources and manpower. Leo's decision to emphasize in this list the appropriate qualities and correct behavioral patterns for a thematic general, based though it is entirely in ancient or earlier models, does therefore have a real contemporary context. It was at least a statement of how the emperor viewed the role of his senior officers, which could be used to some degree as a moral and political yardstick against which to judge those whom he felt to have strayed from the correct path. It is also apparent that Leo himself regarded "noble" birth as at least one important element in the qualities appropriate to generalship, even if he is ambiguous about the exact nature of the relationship between this attribute and talent or merit. See in particular Strano, "Potere imperiale" and the older but still important contribution of Beck, *Byzantinisches Gefolgschaftswesen*. Leo's comments here should be taken in the context of the final sentiment in Const. 1, above.

28–34   §8   Cf. Basil I, *Κεφάλαια παραινετικά* (ed. Emminger), 57, §22; 61–62, §34.

  50   Cf. John 15:13.

51–57   §11   See on 20§214 below.

58–70   §12   Cf. 20§181.

66   στρατοπέδου   τὸ στρατόπεδον, generally understood as "army" but could also refer to an encampment or an expedition: cf. Trapp, *Lexikon*, 7:1621; *CPTT* 175.

71–113   §§13–16   Onasander, *Strategikos*, 1.112–2.153 (1§§17–25)

96–102   Aristotle, *Politics*, 1.6.1255b, but taken from Onasander, *Strategikos*, 2.131–41 (1§§21–22).

114–30   §§17–18   See Onasander, *Strategikos*, 1.55–79 (1§§1–8), 2.151–68 (2§§1–5).

119   Translate σπουδαῖον as "zealous" or "diligent" here, rather than "serious," which makes little sense in this context (as also at 19§82.453 and ep.§29.123).

131–36   §19   Onasander, *Strategikos*, 2.154–57 (2§1–2)

149–58   §22   Cf. Basil I, Κεφάλαια παραινετικά (ed. Emminger), 50, §2.

149–52   See Maurice, *Strategikon*, pr. 36–39.

152   Deut 6:5; Matt. 22:37.

157–58   Ps. 144 (145):19.

159–62   §23   Cf. Basil I, Κεφάλαια παραινετικά (ed. Emminger), 53, §11.

161–62   See Maurice, *Strategikon*, pr., 39–41 (the parallels noted by Dennis for these lines at Matt. 6:26; Luke 12:22–26 are spurious).

163–69   See Maurice, *Strategikon*, pr., 41–45; and cf. *Taktika*, pr. 90–91 (= *Strategikon*, 7.pr.4–5).

172   Matt. 7:24.

178–79   Proverbs, 13:13.

190–92   Maurice, *Strategikon*, pr.65–68.

193–216   §§29–31   Leo draws heavily on Onasander, *Strategikos*, 4, but this is a well-established motif in ancient writings and recurs in a Christian guise here, as also in *Rhetorica militaris*, 6.3; 7.2, 10–14, for example. See on Const. 20§§169–71 below.

217–20   Sentiments clearly expressed in Basil I, Κεφάλαια παραινετικά (ed. Emminger).

220–21   Maurice, *Strategikon*, 8.2.79 (cf. Plutarch, *Moralia* 187D [Chabrias]) and 8.2.93; and see Basil I, Κεφάλαια παραινετικά (ed. Emminger), 67, §46. Dennis, *Taktika*, 37 and n. 15 claims parallels with Theognis, *Elegies*, 1.949 and 2.1278C, but these passages are simply allusions to a lion catching a fawn. The sentiment is repeated at Const. 20§128.633–34. See chapter 2, pp. 50–51 above.

225–26 ἐν τῇ παραλλήλῳ τῶν τακτικῶν μονοβίβλῳ   Dennis, *Taktika*, 37 n. 16 suggests this is "undoubtedly" a reference to the *Sylloge tacticorum*, but this cannot be correct, since the *Sylloge* itself (e.g., in chaps. 35–36, 38–39, and 44) drew on the *Taktika*, so is certainly a later composition. See chap. 2, on sources, Dain's introduction to the text (*Sylloge tacticorum*, 7–10), and Mecella, "Überlieferung der Kestoi," 98–101.

CONSTITUTION 3

In the MSS this constitution and the next are transposed, with the revised order imposed only by Meursius in his edition of 1612, based on the order presented in the prooemium (see PG 107:693, n. 19; Vári, *Tactica*, 42, note to the title of Const. 3, ll. 566–67). The first section of this Constitution, ll. 3–12, is an extended version of Maurice, *Strategikon*, 8.1.5. The main part, "On Counsel," is drawn from an unidentified source, but bears many parallels with Onasander, *Strategikos*, 3 (περὶ τοῦ ἔχειν τὸν στρατηγὸν βουλευτάς); and see also Basil I, Κεφάλαια παραινετικά (ed. Emminger), 55, §18 for similar concepts. Cf. also ep.§§32–33.

8 οἷον τουρμάρχων The term *tourmarches* apears to be a Byzantine development of the sixth or seventh century, since the commander of a *turma* of cavalry, a troop of some 30 soldiers, had been a *decurio*. How the term *turma* evolved thereafter so that it referred to much larger units is obscure. *Tourmarches*, a title that presumably appeared after *tourma* had come to mean a division of cavalry, occurs for the first time in *Theophanis Chronographia*, 391.1–3 (Mango and Scott, 449), a reference to Persian officers killed in battle with the forces under Herakleios in 625/26: οἱ τρεῖς τουρμάρχαι τῶν Περσῶν. Later in the same year Theophanes' account (320.4–6; Mango and Scott, 450) describes how the emperor sent the *tourmarches* George at the head of 1,000 men to seize the bridges across the Lesser Zab; and George is shortly afterward (325.3; Mango and Scott, 453) described as *tourmarches ton Armeniakon* for 627. The significance of this is unclear, since it is generally accepted that Theophanes used anachronistic "thematic" terminology at certain places in his account of Heraclius's wars, although we should not confuse the terminology for officers and units characteristic of the seventh–tenth centuries with "thematic"

arrangements as such; see Brubaker and Haldon, *Byzantium in the Iconoclast Era*, 744–55. The history of the *tourmarches* is, in fact, a good example of the ways in which such titles were evolving over the long term and of the fact that we should avoid assuming a constant or fixed meaning for the term. Thus by the time of the mid-tenth-century *Skirmishing*, *tourmarches* is the only term to survive from the thematic military administration elaborated in the ninth-century *taktika*, apart from *strategos*. For middling, junior, and occasionally senior officers the term *archon*, which had always had a generic meaning of leader or officer, is frequently used instead, as in τουρμάρχης εἴτε ἕτερος ἄρχων; see *Skirmishing*, 6.14–15, 8.51–52, 12.12–13, 14.18. *Tourmarches* itself disappears from Greek histories and chronicles after the first half of the tenth century. The last reference is to Elephantinos, the tourmarches of the Opsikion in the 930s who captured the rebel Basil (Theophanes continuatus, 421.7–11). The office is entirely absent from the *History* of Leo the Deacon and the *Taktikon Escorial*, both contemporary with *Skirmishing*. Tourmarchs are active in northern Syria in 1032 (Yahya, 520, 524), and apart from legal documents, their final written attestation is a letter of Psellos dating to 1057 (Psellos, *Letters*, 165 [Sathas 423–24]). In Leo, the term is used throughout where Maurice uses *merarches* (see *Strategikon*, 1.3 etc.: a generic term used to cover a range of officers who would in reality have had titles such as dux or *magister militum*). The title of *merarches* appears in no other source until the time of the *Kletorologion* of Philotheos (109.19) and then Leo's *Taktika* and afterward. Whether a *merarches* might also be the senior tourmarch in a *thema*, merely the tourmarch who is based in the same fortress or town as the *strategos*, or bear some other significance, remains obscure; see on 4§9.69 below and note Oikonomidès, *Listes*, 108, n. 65. The title *tourmarches* appears on lead seals of the seventh century and onward, sometimes with a territorial attribution (Dazimon, Thessalonike, Adramyttion, etc.), sometimes with a divisional attribution (Anatolikon, Crete, Opsikion, Thrace, etc.) or the name of a unit (e.g., *ZV*, nos. 3148a & b, ca. 800–850, seal of Nasir, imperial *spatharios* and *tourmarches* of the *Phoideratoi*, a late Roman division, the *foederati*, which came to be associated with a particular region within the Anatolikon), and often with no specific location or attribution at all. See Haldon, "Chapters II, 44 and 45," 328 and n. 354; Brubaker and Haldon, *Byzantium in the Iconoclast Era*, 767–68 and n. 143. But the use of the term *tourma* during the eighth and ninth centuries is at the least suggestive of the fact that the provincial armies were regarded primarily as cavalry forces, or that the cavalry forces were recognized as the key elements among them.

According to the ninth-century geographer Ibn Khurradādhbīh, a *tourma* consisted of 5,000 men, but this is certainly a generalization which did not hold in every case: Ibn Khurradādhbīh, 84 and Kudāma ibn Ja'far, 196. It is clear that a *tourma* could vary considerably in strength, as is shown by a report of the territorial transfers of *banda* during the reign of Leo VI preserved in chapter 50 of the *De administrando imperio*. According to this, the newly established *tourma* of the *Kommata* in Kappadokia consisted of seven *banda*; the new *tourma* of Saniana in the *thema* of Charsianon consisted of three *banda*: *DAI* 1: chap. 50.92–100, 101–5. If we accept somewhat earlier evidence, *tourmai* might often have numbered 1,000 soldiers or fewer, suggested by two passages in the *Chronographia* of Theophanes: in his account of the events of the years 715–16, the future emperor Leo is made to inform the Arab commanders that 1,000 soldiers and their *tourmarches* have entered Amorion to defend it against attack; for the year 792/93, he records that 1,000 soldiers of the Armeniakon *thema* were punished for rebellion, along with their *tourmarches*, who was executed: *Theophanis Chronographia*, 389, 469 (Mango and Scott, 538, 644). The account of the great expedition to Syria and Crete in 911 also suggests that a standard *tourma* often numbered about 1,000 soldiers, but could vary considerably above or below this figure; see Haldon, "Chapters II, 44 and 45," 309–29; and chapter 3, above.

26 Εἰ δὲ μὴ   See §§9–11 and cf. esp. 4§4, 13§9, 20§66, and ep.§§32–33.

41 Καὶ βουλεύου μὲν βραδέως   See 20§9, §88 and Maurice, *Strategikon*, 8.1.5, 2.31; and cf. Onasander, *Strategikos*, 3. While the motto comes, via Maurice, from Isocrates, *Ad Demonicum*, 34, there is no evidence that Leo was himself familar with this text; see on Const. 20.25.

43–48 §9 Καιρὸν δέ . . . γένηται   Maurice, *Strategikon*, 8.2.23.

45–48 καὶ τί μὲν δέον σε . . . γένηται   Cf. Const. 20.66.

49–65 §§10–14   See again 20§66, ep.§§32–33; with Onasander, *Strategikos*, 3.

66–69   §15   From here until the end of the constitution Leo draws on or paraphrases a range of precepts from his wider reading and familiarity
with books of advice or ancient military writers such as Onasander.
Much of this advice reappears in one form or another in constitution
20 and in the epilogue; see 20§§121, 125, and 126 and commentary and
Basil I, *Κεφάλαια παραινετικά* (ed. Emminger), 55, §18.

70–73   §16   See Onasander, *Strategikos*, 1.84–86 (1§10), not a direct parallel,
but indicative of the material from which Leo takes inspiration.

CONSTITUTION 4

Parts of this constitution existed as a separate text excerpted from the tradition of MS A at some point during the late tenth century: paragraphs 1–26 of the text generally known under the title Ἐκ τῶν τακτικῶν τοῦ βασιλέως Λέοντος τοῦ σόφου are taken from Const. 4§§6–33; see Dain, *L'"extrait tactique,"* 9–12, and text at 84–88. The French translation which follows at 91–100 is accompanied by useful notes on Leo's terminology.

Although the list of officers and titles, and their various attributes, is based on Maurice, *Strategikon*, 1.3–5, the opening sections of this constitution are Leo's own work and reflect the conditions and military administrative arrangements and assumptions of his own time.

3–14   §1 Κελεύομεν......δουλείᾳ. This whole paragraph includes important indicators of the ways in which provincial recruitment worked. Note the phrase κατὰ τὴν ἄνωθεν καὶ ἐξ ἀρχῆς συνήθειαν, suggesting a longstanding tradition, and recalling the later phrase, from Constantine VII's preamble to his novel "on soldiers" of 949–59, to the effect that the practice of military service based on the possession of land, in which military lands were, on the basis of a similar longstanding tradition, not to be subdivided or sold off (Svoronos, *Novelles* [Zepos, *Jus*, 1:222; trans. McGeer, 71; Dölger, *Regesten*, no. 673]; Haldon, "Military Service"; and idem, "Recruitment and Conscription," 41–65).

5–6   ἐκλέξῃ ... γέροντας According to the *Taktika*, both soldiers and officers were to be selected according to their experience and fitness for service. The only source from which such a selection could plausibly be made is the military register of the *thema*. The provincial officials associated with the registers, variously referred to in the sources as *stratiotikoi kodikes* or *stratiotikoi katalogoi*, were the *epoptai*. There are representative lead seals for these officials for the ninth century and afterward. See, e.g., *ZV* nos. 1920, 2068; *DOSeals* 1:1.22; 71.9; 2:22.4, 6–7, 8 (= *ZV* no. 3155); 43.1; 3:2.9; 5:6.5. In the late Roman period officials called *epoptai* were concerned with fiscal assessments and, with the *exisotai*, were responsible for assessing tax burdens (see, e.g., *CJ* 10.16.13.pr. [a. 496] and the discussion in Dölger, *Beiträge*, 79–81): they were responsible to the general bank of the prefecture in the fifth and sixth centuries (see Brandes, *Finanzverwaltung*, 198–205), and their successors of the same title are still found in the *sekreton* of the *genikon* in the late ninth and tenth centuries—Philotheos, *Kletorologion* 113.30; and comm., 313. The implication from the passage is thus that many more persons were listed in the military registers than were actually called up (and that many of them might in fact not be fit for service). See Haldon, "Chapters II, 44 and 45," 316–22 and idem, "Military Service," 24–25 and n. 64.

7–10   εὐπόρους ... δυναμένους   The emphasis on their economic well-being is important, again underlining the choice that was to be the responsibility of the theme commander, exercised through his officials. There are a number of technical terms here. The term *strateia* has been the subject of an extended scholarly debate, but in this context can clearly be shown to refer to active military service, supported by servants and family members "in their own households"—ἐν τοῖς ἰδίοις οἴκοις. See Haldon, "Recruitment and Conscription," 49–50, 60 with n. 105 and idem, "Military Service," 29–41. Here the phrase εἰς τὴν ἰδίαν στρατείαν ἀσχολουμένους refers explicitly to the soldiers in question being both the registered holder of the military service and the person on whom the obligation to serve in the army fell. While the technical term *stratiotikos oikos* first appears in the tenth-century legislation, from the 930s onward, this technical phrase, "occupied with their own military service," clearly implies that the arrangements described in that legislation were long-standing, even if uncodified, and were certainly in operation before then. The word *strateia* had a double significance, referring both to actual armed service and the fiscal-military obligations attaching to them. Those households which were not registered as "military" but which belonged to the same fiscal district as the soldier's household could, in the event that the soldier's property was insufficient to support his campaign service, also be asked to contribute to his equipment, weaponry, and maintenance. This arrangement can be traced back to the reign of Nikephoros I, and is associated with the establishment of the military provinces as *themata*, that is to say, as regions with a specific military-fiscal aspect. See Brubaker and Haldon, *Byzantium in the Iconoclast Era*, 744–55. It had in addition been a long-standing practice to raise resources from households that did not or could not provide a soldier, who paid instead a certain sum, the proceeds from which were used by the local military establishment or the central *logothesion* to pay for other, less well-equipped soldiers registered on the military codices in the same fiscal districts, or their equipment. By the middle of the tenth century, however, properties registered as "military" were frequently unable to support the cost of a single soldier by themselves, so were grouped together as *syndotai*, responsible jointly for the maintenance of a soldier (an arrangement different from, even if similar in appearance to, the older system of communal contributions formalized in the early ninth century); or for the raising of mercenary troops. It is unclear at this stage—the early tenth century—how far the system of *syndotai*, as it is described in the mid-tenth-century imperial legislation, had already evolved. See Haldon, "Military Service," 30–32; see also Ahrweiler, "Recherches," 14; Dagron and Mihăescu, *Traité sur la guérilla*, 267. For the term *stratiotikos oikos* (in opposition to *politikos oikos*, used of households not registered as having military obligations), see Ahrweiler, "Recherches," 12–14 and Lemerle, *Agrarian History*, from 133.

7   ἐξπεδίτῳ   Lat. *expeditio*. Cf. *CPTT* 71 and C.665, 689.

8   φοσσάτου   Lat. *fossatum*. Cf. *CPTT* 71, 175. The word can refer either to an expeditionary army (see, e.g., at 11.3–4 below) or to the encampment of such a force. The adverb φοσσατικῶς also occurs, used of an army when it is in the field and marching from camp to camp— cf. *Skirmishing*, 13.1.4; 18.1.5); Dagron and Mihăescu, *Traité sur la guérilla*, 179–80. In this sense, a *phossaton* signified a larger expeditionary force, in contrast to a raid, *kourson*: ibid., 177–79 and McGeer, *Sowing the Dragon's Teeth*, 230–48.

10–11 ἐλευθέρους . . . δουλειῶν   See also on 20§71.344–54 below. The soldiers' households and their land were exempt from all state taxes except for the *demosion*, the basic land and hearth taxes—an important advantage soldiers held over ordinary taxpayers in view of the many supplementary taxes as well as extraordinary impositions placed on taxpayers, extracted to support passing military forces or state officials, to defray the costs of fiscal and other officials in carrying out their duties, and so forth (see Harvey, *Economic Expansion*, 104–9). This was an ancient tradition. The difference between "military households" and "civilian households" (*stratiotikoi oikoi* versus *politikoi oikoi*, members of which might also be referred to as ἀστράτευτοι, which is to say, not registered for a *strateia*) was not especially medieval: its origins lie in the standard and entirely normal late Roman distinction drawn between those groups who enjoyed specific immunities in respect of certain state demands and those who did not, such as those owing service in respect of the post (*exkoussatoi tou dromou*), of provisioning military personnel (*prosodiarioi*) and those who worked in imperial armories. On fiscal privileges and their origins, see the texts cited at Haldon, "Recruitment and Conscription," 54 n. 94; 60 n. 104; idem, "Military Service," n. 104; Dagron and Mihăescu, *Traité sur la guérilla*, 264–69; and Patlagean, "L'impôt payé par les soldats" for the late Roman origins of these special categories; also Palme, "Spätrömische Militärgerichtsbarkeit," with extensive literature on the legal situation of the soldier in Roman and late Roman society at 375–77. The *Taktika* (20§71; see comm.) suggests that soldiers drafted for state *aggareiai* (compulsory labor or similar duties) when other nonexempt subjects were not available were to be paid for their labor, implying that they may not always have been appropriately recompensed. Along with the Church and the peasantry, soldiers held a special position in Byzantine concepts of the body politic: Constantine VII noted that "the army is to the state as the head is to the body; neglect it, and the state is in danger" (Svoronos, *Novelles*, 118.1–2; Dölger, *Regesten*, no. 673). Later in the *Taktika* (11§9.45–53 and cf. 9§§1, 16–18; 20§209 with comm.), Leo VI describes peasants and soldiers as the two pillars upon which the polity was founded. The soldier's property was, in theory at least, protected while he was away on state service, again a long-standing legal right: *Digesta* 4.6.45: the soldier was considered, when on active service, to be "rei publicae causa absens"—absent in the service of the state. For the principle of *in integrum restitutio*, see *CJ* 2.50.1, 3, 4, 6, 8; 52.1–7 ("De restitutione militum et eorum qui rei publicae causa afuerunt") and the following sections; and Kaser, *Das römische Zivilprozessrecht*, 330–31.

12 συστρατιώτην The emperors saw themselves symbolically as the father of their soldiers, the soldiers' wives as their daughters-in-law. The reference to soldiers as the *systratiotai* (or *synaspistai*), the comrades-in-arms, of the emperor, and of the general who commanded them (*Rhetorica militaris*, 28.2, 4; 36.2; 55.2) is also ancient (see discussion at *CPTT* C.453–54, and commentary, 242–44): Trajan, for example, referred to "my excellent and most loyal fellow soldiers": *Digesta* 29.1.1; and many other examples can be cited going back through Menander Rhetor to Xenophon and Plato; see Eramo, *Siriano*, 143–44 n. 65. By the same token the general was understood to be the father of his army, to teach, lead, and chastize just as a good father does. Cf. *Rhetorica militaris*, 22.2: ἄνδρες ἀδελφοὶ καὶ συστρατιῶται; 22.4: καὶ γὰρ διὰ τὸν πόθον ὑμῶν φιλῶ καὶ αὐτὸς στρατιώτης καλεῖσθαι, ἵν᾽ ὑμῖν συστρατιώτης κληθῶ, καὶ τὴν τοῦ πατρὸς οὐκ ἀναβάλλομαι προσηγορίαν· εἰ γὰρ καὶ ὡς βασιλεὺς ὑμᾶς προτρέπομαι, ἀλλὰ καὶ ὡς πατὴρ παραινῶ καὶ ὡς ἀδελφὸς διεγείρω....; 53.2: ὅπερ τοίνυν ἐστὶ πατὴρ πρὸς τέκνα, τοῦτο στρατηγὸς πρὸς στρατεύματα....

14 τοῦ δημοσίου τέλους The public burden, i.e., the land tax and its associated surcharges. Since there is no evidence that soldiers were exempt from paying the hearth tax, or *kapnikon*, the term *demosion* should be understood to subsume this also. See Dagron and Mihăescu, *Traité sur la guérilla*, 264–66; Harvey, *Economic Expansion*, 102–4; *ODB* 1:610, 3:2022.

15–16 τὰ λεγόμενα βάνδα The term derives from Lat. *bandum*, in turn borrowed from the Gothic word for a banner or standard, and by extension the unit of soldiers which it identified, see Kramer, "Ein Gräzismus," 197–207; and idem, "Papyrusbelege"; Trapp, *Lexikon*, 2:263. The term was current by the middle of the sixth century and by the time of composition of the *Strategikon* it was well established in this sense (see Rance, *Strategikon*, 22 n. 53), although in this text it is used only of cavalry units. Leo's phrase τὰ λεγόμενα suggests that *bandon* was the actual term commonly used. The immediately foregoing term, *tagma*, is the standard Greek word for any unit of soldiers, and since Leo follows Maurice in many respects he also refers to infantry units generally by the term *tagma*. In the opening sections of this Constitution the term *bandon*, for a unit, is taken to apply to both infantry and cavalry, and its meaning appears no longer to be restricted. The term *arithmos* occurs occasionally, but was also used of one of the imperial elite units, the *Vigla*, or "watch" (just as the substantive "*ta tagmata*" is employed during the ninth and tenth centuries to describe the imperial elite units in and around the palace or Constantinople, in opposition to the provincial armies, *ta themata*).

The size of units varied according to circumstances. Leo follows Maurice, *Strategikon*, 1.4.8–9, 23–25, 32–33 exactly in emphasizing that a unit should be strictly of 200–400 soldiers and that units should in any case be of varying strengths in order to confuse or deceive the enemy. The (probably) mid-tenth-century *Sylloge tacticorum* remarks that the basic cavalry unit, the *bandon*, by this time known also as an *allagion* (see Trapp, *Lexikon*, 1:56), numbered 50, but that anything up to 400 was a possibility, especially for infantry. The writer then notes that the commonplace (i.e., thematic) units—*banda*—number anything from 50 to 150, whereas the "imperial" *allagia* can be as large as 320 or 350 or at most 400 in strength, "such as the present unit of the Thrakesion, numbering 320, or that of Charsianon, numbering from 300–400"; while some of the western *tagmata* number 400 also. See *Sylloge tacticorum* 35.4–5. To judge from the use of the epithet "imperial" here, this latter information probably refers to units raised and paid on a full-time basis, the regional as well as the imperial *tagmata* based in the various provinces of the empire (for such regional *tagmata*, see Kühn, *Byzantinische Armee*, 251–59, with 123–24). As Leo makes clear—following Maurice—the actual size of units was left to the discretion of the commander in the field; see 4.140–41, 183–87, 204–6, 215–20. Note that these *banda* are not necessarily the same as the territorial *banda*, or *topoteresiai*, in each *tourma* within the theme (on which see chap. 3 above, and on Const. 18.599–600 below).

16–17 δεκαρχίας ... τὰ λεγόμενα κοντουβέρνια   See also on ll. 79–81 and
139–44 in this constitution, below; Maurice, *Strategikon*, 1.5; and
Trapp, *Lexikon*, 2:344. The way in which Leo deploys the term again
suggests that *kontoubernion*, Lat. *contubernium*, was still the standard usage. It has several overlapping meanings. Originally referring
to a tent group of 8 (infantry) or 10 (cavalry) soldiers in the Roman
and late Roman army (see Mihăescu, "Éléments latins," 497), it seems
to have remained as part of the vocabulary of the late Roman army
beyond the seventh century, and referred also to the basic file (ἀκία,
στίχος, ὄρδινον) into which such subunits were organized in the battle line. A tent group could consist of as few as 5 cavalry troopers,
two such groups would make up a dekarchy of ten soldiers, but the
normal arrangement seems to have been a group of ten soldiers who
marched, fought, and quartered together. Cf. *Sylloge tacticorum*, 35.12
(Κοντουβέρνιον δὲ καλεῖται τὸ ἑκάστου τάγματος βάθος ἢ πάχος);
45.11; and esp. *Campaign Organization*, 1.35–40, where the equiva-
lence of *dekarchia* with tent group is made explicit. On the ranks and
titles themselves, see below.

19–29 §3 The next sections represent a revised version of Maurice,
Strategikon, 1.3–5, following Onasander, *Strategikos*, 2.157–68 (2§§3–5).

20 δρούγγους   Late Latin *drungus*, from a probably Celtic (Gaulish)
origin; see Rance, "Drungus" and Trapp, *Lexikon*, 2:412. In the late
Roman period the word had two meanings, referring on the one
hand to a specific tactical formation which units of an army could
adopt to deal with a particular situation, and on the other to several units brigading together: cf. Vegetius, *Epitoma*, 3.16.3; Maurice,
Strategikon, 2.1.19 and 2.2–3; 9.3.103–4; and the comments of Rance,
Strategikon, 22 n. 52 and 39 n. 5. The latter use seems rapidly to have
become the standard meaning, and suggests that the tactical brigading of cavalry units in this way in the wars against the Avars in
the last decades of the sixth century had a lasting effect on tactical
organizational structures, with officers entitled *droungarioi* becoming formal elements of the hierarchy of command. In Leo's time, the
droungos was a purely tactical subdivision of the army, although the
adverb *droungisti*—paraphrased generally by expressions such as "in
a group" or "mass"—appears in Leo and in the *Sylloge tacticorum* (see

on Const. 14.318 below). In the text of the *Taktika* it replaces the *moira* of the *Strategikon*, but whereas the latter seems to have been a theoretical descriptive term employed (only) in the *Strategikon* to explicate tactical and organizational principles, *droungos* was in practical everyday use in the period from the seventh century onward (and probably already in the sixth century, although the evidence is slim). At various points in the *Taktika* Leo asserts that there were three *droungoi* in a *tourma* and three *tourmai* in a *thema* (e.g., 4.76–78, 199–203 below), but in practical terms he seems to assume that this hierarchy was a notional structure rather than a general reality. That *droungoi* existed at all at this period as tactical subbdivisions is a moot point. See below, at l. 76, on *droungarios* and chap. 3 above.

20–30   Leo stresses (1) that the thematic commander should choose his leading officers with care, and (2) that their affluence should not be a hindrance to their appointment, a recognition also of the realities of provincial society, in which the wealthier families often supplied officers to the local army, and where relationships of patronage and clientship must have played a significant part in daily affairs (see Strano, "Potere imperiale"). This is especially true since the central administration was relatively apathetic regarding some (although not all, and certainly not fiscal) provincial matters. In the context of the ability of members of provincial elites to exercise influence, this left a good deal of political space within which local affairs could be manipulated (see Neville, *Authority*, 39–40, 99–135). The final sentence of the section might confirm this—Leo clearly felt that the direct investment by a wealthy officer in his soldiers' equipment or supplies would pay off in respect of loyalty in battle—and the implication is, politically also.

30–36   §4 See Onasander, *Strategikos* 3.171–81 (3§§1–3). While drawn from a much older treatise, Leo may well also have had in mind the relationship between the thematic *strategos* and the commanders of the local forces, many of whom may have represented both the provincial gentry as well as the local interests of their soldiers.

37–43   Leo expands on the foregoing paragraph, following his exemplar, Onasander, *Strategikos*, 3.

46–60   §§6–7 Here Leo gives a list of all the senior and other ranks, titles and functions to be detailed in the following sections. Throughout the *Taktika* Leo follows Maurice, *Strategikon*, in taking ἀκία and ὄρδινον as "file," whereas in the *Praecepta* of Nikephoros it generally (although not exclusively) means "row" or "rank"; see Nikephoros, *Praecepta*, 1§10.107; 2§14.142; 3§4.49, 55; 4§§7.82, 20.213; etc. On the confused use of the terms, ll. 140–41, below. In the ancient authors ἀκία (Lat. *acies*) referred to a row or horizontal line, in contrast to *stichos* or file, a vertical line, a distinction which had largely been abandoned by the late sixth century.

50 τοῦ στίχου   Should be translated "of the file," since it refers to the file or ἀκία to which these grades belong.

57 καντάτωρες   This term appears at Maurice, *Strategikon*, 2.19 and again at 7.B.17.28 (although this section is a later recapitulation and not part of the original text; see Rance, *Strategikon*, note to 7.B.17 title); it reappears in *Taktika* 12§§56–57, 98, otherwise found in no other military treatise and, as Leo notes here, an older term no longer in use.

61 Στρατηγὸς   See above, on 1§§10.30–11.35. The rest of this Constitution, up to l. 122 is drawn from Maurice, *Strategikon*, 1.3–5, with modifications of terminology by Leo.

69 ὑποστράτηγος . . . μεράρχης   Cf. 12§61.457–58 below, and comm. on 3§1.8 above. Following his comment that each thematic *strategos* is technically the *hypostrategos* in his province representing the emperor, the supreme commander, Leo notes that the *hypostrategos* of the thematic *strategos* is "nowadays" the so-called *merarches*. While the title is used in Maurice as a generic term for a senior officer of this type (and certainly appears in no other source until the tenth century), Leo's statement, and the evidence of late tenth-century sources, would suggest that the title did come to be employed, even if on a restricted basis, in this sense. The tenth-century evidence would suggest that the term *merarches* was used on occasion of the senior *tourmarches* within a theme, although *merarches* and *tourmarches* are also presented at times as synonymous (e.g., Philotheos, *Kletorologion*, 109.18–19; in *CPTT* C.447 and 504 the merarch is placed after the tourmarchs of a theme in the order of precedence). The *merarches* (sometimes called *meriarches*) of the theme, had his own staff and may have commanded the *tourma* in which the thematic headquarters was established

(for a seal of a *merarches* of Hellas see *DOSeals* 2:8.31; *PmbZ* 27020). Merarchs appear almost entirely in the tenth and eleventh centuries, with the exception of the single reference in the *Kletorologion* of Philotheos and a mention of a certain Machairas, merarch of Charsianon according to Genesios (but kleisourarch according to Theophanes continuatus and Skylitzes), for the year 863; see *PmbZ* 4656 with 31238 and 31244. Other seals of the tenth and eleventh centuries: Niketas, merarch (11th c., Konstantopoulos, "Molybdoboulla," no. 207); Konstantinos, merarch of Knossos (11/12th c., Tsougarakis, "Seals of Crete," 141 and 151); Stephen, merarch (mid-11th. c., Fogg 2059); Michael, meriarch (mid-to-late 11th c., Fogg 1992); Eudokimos . . . , merarch (ca. 980–ca. 1020, Fogg 1822; there is also a seal of probably the same person with the rank of *tourmarches*: Fogg 1322); Niketas Zaras, merarch (10th/11th c., DO 1955.1.3419); and from the Corinth excavations nos. 2719 (3384) (Demetrios, imperial *spatharokandidatos* and merarch, 10th/11th c.) and, uncertainly, 2778 (7863) (Leo N., *protospatharios*, merarch, and patrikios of ?, 10th/11th c.) (my sincere thanks to Lain Wilson for supplying the DO and Corinth sigillographic material and for discussion of the question of the relationship between tourmarch and merarch). There is thus no evidence that *merarches* in fact has an unbroken history from the early seventh into the late ninth century, and it may be a classicizing revival (albeit with specific functional application) of the late ninth century, in the same way that terms such as *archegetes*, *hoplitarches*, and *taxiarches* reappear from the middle of the tenth century; see Oikonomidès, *Listes*, 335–36; McGeer, *Sowing the Dragon's Teeth*, 203. As the second-in-command of the *strategos*, he was based at the theme headquarters with the latter. See *CPTT* 249–50 for further discussion.

70-71   §10   The definition taken from the *Strategikon* is updated. See on 3.8 above.

72-75   §11   The text clarifies again the older language of Maurice, who uses *moirarches* and *moira* rather than *droungarios* and *droungos* (except when referring to the grouping together of a number of units for a particular tactical purpose, see above on l. 22: *droungos*), although the latter term may already have been current then; and *merarches* and *meros* rather than *tourmarches* and *tourma*. There can be no doubt that from the middle of the seventh century at least the title *droungarios* represented a permanent position within the command structure of the empire, and there are numerous lead seals of the period from the seventh into the tenth century to attest to this. The first reference to a *droungarios* is to Theodotos, the *megaloprepestatos droungarios*, who accompanied the *magister militum* Elias on an embassy to the Persian king Siroes in 628 (*Chronicon Paschale*, 731.5; trans. Whitby and Whitby, 186). This is a near-contemporary report and there is no reason to doubt its accuracy. The context suggests that *droungarios* was the title given to a senior officer (holding the dignity of *megaloprepestatos*) under a *magister militum* (and note that in Maurice, *Strategikon*, 1.4.15 the term *merarches* is glossed by *stratelates*, the standard Greek equivalent at this time for *magister militum*). There are many lead seals extant for *droungarioi*, both without any localization as well as for *droungarioi* belonging to a particular *thema* or a specific place or region. In addition, many otherwise unspecific seals of *droungarioi* include a relatively high rank (*apo eparchon*, *hypatos*, etc.), suggesting that these may have been naval commanders rather than simple corps commanders. For a full list see under *droungarios* in *PBE* and *PmbZ*. The vast majority of all seals of *droungarioi* belong to the seventh, eighth, and ninth centuries. The rank and position of such officers was changing rapidly by the end of the ninth century—see discussion on l. 76, κόμης below.

75   τῶν λεγομένων κομήτων   See on l. 76 below.

76   §12   κόμης δέ . . .   By the early seventh century the title *comes* had become the standard late Roman equivalent for *tribunus*, signaling the commander of a unit of soldiers—*tagma*, *arithmos*, *numerus*. It also retained a nonmilitary value and was borne by various civil officials (having been originally a class or grade of imperial office-holder); see Trapp, *Lexikon*, 4:852. There is extant a large number of lead seals for such military officers dating from the seventh century onward. The majority bear no detail other than the invocation and the title, but no rank; some display markedly provincial characteristics. Of these it is likely that those with higher ranks, such as *apo eparchon* or *stratelates*, for example, as in *ZV* 916, 1679, and 2094,

are not *kometes* of *banda*, but more important officials, such as the *kometes* of Abydos, in charge of important customs establishments. That it was a formal command is confirmed by a letter of Theodore of Stoudios, who refers (*Ep.* 160) to a deserving man who had been promoted to the position of *komes*, even though that of a *tourmarches* would not have been too good for him; while there are many references to provincial *kometes* in the hagiographical record; see Haldon, "Chapters II, 44 and 45," 201–352, 322–24. For detailed lists see under *komes* in *PBE* and *PmbZ*.

It is worth pointing out here that while Leo updates Maurice, his version of the structure of the provincial armies is already somewhat outdated. By the late ninth century the positions of *droungarios* and *komes* were often held by a single person, suggestive of important changes in the way provincial forces were organized. Provincial, that is to say thematic, officers referred to as *droungarokometes* now make their appearance. In the *Kletorologion* of Philotheos *droungarioi* and *kometes* are listed together as δρουγγάριοι τῶν βάνδων, κόμητες ὁμοίως (Philotheos, *Kletorologion*, 157.9–11; 109.23–24). It seems that the relatively senior grade of *droungarios* was losing its status, and by the time of the Cretan expedition of 911 the relationship between *droungarioi* and *kometes* was already very variable. Some 42 *droungarioi* and 42 *kometes* made up the officers for the 5,000 Mardaites who sailed with the expedition, for example, and they were each paid a campaign salary of 12 and 6 *nomismata* respectively, suggesting that there existed in this case a clear distinction in rank and status: *De cerimoniis*, 656.10–12; ed. Haldon, "Chapters II, 44 and 45," 209.79–81. It seems in fact likely that the *droungarioi* were actually the commanders of *banda*, with the *kometes* as their second-in-command.

This process, whereby the rank of *droungarios* is increasingly assimilated to that of *komes*, evolved at different rates in different areas. The Armenians of Sebasteia had 960 soldiers, 5 *tourmarchai*, 10 *droungarioi*, and 8 *kometes* in the 911 expedition; and to the expedition of 949 the Thrakesion army contributed a total of 64 *droungarokometes* (*De cerimoniis*, 661, 664.2; the *droungarokometes*: 663.6–8). These *droungarokometes* could each have been set over *banda* of some 50 or so men (Haldon, "Chapters II, 44 and 45," 325–26). A short passage dating originally to the period of Basil I but revised before the middle of the tenth century specifies the hierarchy of command for the *themata* as running from the *strategoi* to their *tourmarchai* and from the latter to their *droungarokometes*, "so that each . . . *bandon*" has the appropriate equipment (*CPTT* C.653–64). For detailed discussion see Haldon, "Chapters II, 44 and 45," 324–28.

77–81   §§13–14   Κένταρχος . . . Δεκάρχης . . . Τετράρχης   Roman and earlier treatises placed great emphasis on the fighting qualities of the leader and secondary ranks in each file. See Aelian, *Tactica theoria*, 5.5; Arrian, *Techne taktike*, 6.5; Syrianos, *Strategy*, 15.86–100; and Maurice, *Strategikon*, 1.5.1. Here Leo omits the *ilarches* or senior "centurion" included by Maurice between the *komes* and the *kentarchos;* see Rance, "*Campidoctores, vicarii*," 400–401. Where Maurice has *hekatontarchos* Leo has sometimes the more usual *kentarchos*, derived from assimilation of the Latin *centurio* with the Greek *hekatontarchos* (see Trapp, *Lexikon*, 4:820). Standard terms like kentarch were beginning to replace traditional Latin military grades by the late sixth century, although the latter continued to be employed as well. In Maurice, *Strategikon*, the lower noncommissioned officers or grades known from the traditional establishment as *semissalis, biarchus,* and *circitor* may be those described by the Greek forms of *tetrarches, pentarches,* and *dekarches*, already common in cavalry units in the middle of the sixth century in the Greek-speaking regions of the empire, although it has been noted that some of the Latin grades are not certainly fixed ranks but probably distinctions of seniority and entitlement with an administrative rather than a tactical function (Rance, "*Campidoctores vicarii*," 397), and the same applies to these Greek equivalents. Thus in various documents from the Nessana archive, dating from the early years of the sixth century to 596 (Kraemer, *Nessana*, 4–6, 20–22): a *primicerius* and *dekarchoi* are mentioned (nos. 35, 37), as well as a *bikarios* (no. 134); together with clearly administrative functionaries such as a *scriniarius* and an *optioprinceps* (nos. 19, 36), illustrative of the mixture of titles in contemporary use. In one document (no. 37), the *dekarchai* appear to be in charge of small groups of between 5 and 10 men, although the nature of the document—a list of detached camel-riders organized in subunits of varying size—may not reflect any standard organization. From the seventh to tenth centuries the titles of the main grades of officer are attested in a wide range of sources. Many of these are topoi, and Philip Rance has pointed out that they reflect a common literary usage derived less from Hellenistic military writings than from Biblical and patristic precedents; see, for example, Exod. 18:21, 25; Deut. 1:15; I Macc. 3:55, and for kentarchs/hekatontarchs, pentekontarchs and dekarchs compare with *Miracula S. Demetrii*, 230.20; *Miracula S. Therapontis*, mir. 16.1; *Vita Theodori Studitae*, chap. 21 (used figuratively); and cf. Wortley, "Military Elements." For more concrete examples of actual ranks, see *Vita S. Philareti*, 127.1–2; *Acts of Xeropotamou*, 1.39 (kentarch, for AD 956); and the well-known descriptions in the Arab geographers' accounts of the Byzantine army, which list the *kometes*, "kontarchs" and dekarchs: Ibn Khurradādhbīh, 84; Kudāma ibn Jaʿfar, 196. See Trapp, *Lexikon*, 2:344 and 6, 1259 for *dekarchos* and *pentarches/os* respectively. Older titles from the pre-Diocletianic establishment also appear; see *Vita S. Stephani Iunioris* (ed. Auzépy), §54 (154.11), where a soldier is promoted to the rank of *kentourion*. It is likely, but not certain, that this is a deliberately archaizing form used by the hagiographer, although the continued existence of units whose origins lay both before and after the early fourth century is not to be doubted.

Rance, "*Campidoctores, vicarii*," 397–98 suggests that in the *Strategikon* the titles of hekatontarch, dekarch, pentarch, and tetrarch were generic terms employed by Maurice to describe tactical functions, and even if they represent the appearance of new terms, they certainly do not reflect any major reform of the army, as has often been assumed. This would appear to be correct for such grades in Leo's *Taktika*, but while very likely, it seems also to be the case that some of them—kentarch and dekarch at least—survived and became part of the regular hierarchy of command thereafter. Pentarchs and tetrarchs never appear in sources other than the tactical treatises, so it would be reasonable to suppose that these terms were merely descriptive and functional terms drawn from such texts, rather than permanent ranks. See also on 4.139–44 below. Further on *campidoctores* and centurions see Janniard, "Centuriones ordinarii"; different interpretation in Wheeler, "Legion as Phalanx, part II."

The best example for continuity of rank titles from later Roman into middle Byzantine times is provided by the group of units known formally as the *arithmos* or *arithmoi*, but more commonly during the ninth and tenth centuries called the *Vigla* or Watch. The singular and plural forms both occur; since the commander was a *droungarios*, and since there is evidence for a duplication of its administrative staff (such as the *chartoularioi*), it seems fairly clear that the regiment was originally made up of two or more distinct sections or *banda* (see Haldon, *Byzantine Praetorians*, 236–41). The *Vigla* was a field unit drafted into service as a palatine guard (*basilikon tagma*) in the late eighth century, almost certainly by the regent Eirene in about 786, and possibly from the army of the *thema* of *Thrakesion*. The internal establishment of the unit is known from the *Kletorologion* of Philotheos of 899 and from incidental information, mostly sigillographic, of the late eighth and ninth centuries. From these sources we learn that the *arithmoi* included *kometes* (in charge of the various *banda* or *arithmoi* making up the brigade), *kentarchoi*, *bandophoroi*, *doukiniatores,* and *mandatores*. See Philotheos, *Kletorologion*, 115.21–32; and refs. in Haldon, *Byzantine Praetorians*, n. 637. The *doukiniatores* may reflect the *ducenarii*, but since their position on the list of 899, which was strictly hierarchical, was lower than the *kentarchoi*, it may be that they represent the guides—*ducatores/doukatores* of the field units (cf., for example, the anonymous tenth-century treatise *Campaign Organization*, index, s.v.).

85 βανδοφόρος   The standard title from the sixth century for the unit standard-bearer, cf. *Sylloge tacticorum*, 35.13; and for the early ninth century, Auzépy, "Vie de Jean de Gothie," §1.7 (although the *Life* is largely a fiction, the use of contemporary technical terms would be quite usual); Trapp, *Lexikon*, 2:263. Note the late seventh-century reference from documents of the exarchate to *draconarii* and *bandofori*, among other ranks or functions of the late Roman establishment; see Brown, *Gentlemen and Officers*, 59 and references. Leo omits the escort of the standard-bearer, the enigmatic "cape-bearer," found in the *Strategikon* at this point, as well as the mention of the *taxiarchai* of the *Optimates* and the *armati* or esquires/attendants of the soldiers of this unit.

86–88 §17 Cf. 12§37.273–84. The original term *deputati* has, according to Leo, now been replaced by the word *skribones*, written also *skribantes* or *kribantes* (cf. *Sylloge tacticorum*, 35.1; 45.3; and Trapp, *Lexikon*, 7:1576), although the origins of this usage remain unclear. The meaning of *deputati* is clear, the term being employed in the fifth and sixth centuries for any soldier deputed to a particular task. The term was also used of medical orderlies in Alexandria, according to a text in the *Codex Theodosianus*, 16.2.42 and 43 for a. 416 and 418 (repeated at *CJ* 1:3.17 and 18). In the *Strategikon* it clearly reflects soldiers deputed to take care of the wounded and render some basic medical assistance, although it is not clear whether this was ever a standard practice in the late Roman army. See also *Strategikon*, 2.9, with commentary (Rance). Whether this tradition was actually maintained in the later Byzantine army is not clear—Leo does not include details about their duties given in the *Strategikon* (2.9) at this point, for example, but expands on their role, following Maurice exactly, at Const. 12§37.273–84, and not employing the term *skribones*. The *Sylloge tacticorum* 45.3, following Maurice, *Strategikon*, 2.9, states that there should be ten *daipotatoi* or *kribantes* per unit, who should be unarmed and follow the unit in action at a distance of some 40 *orguiai*, that is to say, between 74 and 84 m (see Schilbach, *Metrologie*, 22–26—there were several *orguiai* in use; in the *Strategikon*, this distance is given as "about 100 feet," considerably less than the distance given in the *Sylloge tacticorum*), carrying the injured back on their horses through gaps in the line to safety, and where they can be treated by the field doctors. The only *skribones* known from the late Roman and middle Byzantine periods are senior officers, associated with the old *excubitores*, for whom several lead seals exist, who functioned as imperial diplomats, messengers, and enforcers from the late fifth into the late seventh century; and the like-named officers of the reformed successor unit, the *Exkoubitoi*, an imperial *tagma* created from the older unit at some point in the reign of Constantine V. This is the meaning retained by the Suidas, 1.4.696 (p. 388). See Haldon, *Byzantine Praetorians*, 161–64, 292–93. The term appears to derive ultimately from the word for a clerk or record-keeping official, also found as *scribas*. The origins of *skribon* in the sense used by Leo is entirely obscure. But the misspelling or misunderstanding in the variants, such as *kribantes*, may reflect the association with injury and the word *krebation* or bed/stretcher, suggesting that soldiers actually were deputed to such duties at this time.

89 Μανδάτωρες   Cf. *Sylloge tacticorum*, 35.13. The standard term since the late Roman period for a messenger, both in this context as well as in various palatine and judicial departments. The *mandatores* in a military context, as is apparent from both Maurice, *Strategikon*, and the information repeated by Leo, also have the general function of heralds whose duties included also communicating orders and instructions to the soldiers. See Oikonomidès, *Listes*, 298–99, 310 and n. 125.

91–94 Λοχαγὸς . . . Σεκοῦνδος . . . Οὐραγὸς   See on ll. 77–81 above; Trapp, *Lexikon*, 7:1536. Leo copies these terms directly from Maurice, *Strategikon*, 1.3.20–21. Apart from their tactical descriptive value in both Maurice and this text, they have a purely antiquarian value. There is no evidence to suggest whether they remained part of the standard active military vocabulary of the time, and they do not represent ranks or grades, but rather a set of functions. In the translation note that "row" or "line" should be understood as "file."

96–122 §§22–31   Leo repeats exactly the definitions in Maurice, although it is important to note that neither of the terms *cursores* or *defensores* was standard in earlier Roman treatises. The various terms listed are all functional roles assigned to particular soldiers or groups of soldiers in order to achieve specific tasks, whether tactically in battle or in the course of the army's march. By Leo's time the term κούρσωρες (l. 101) is the equivalent of προκουρσάτωρες or κουρσάτωρες found in later treatises, all likewise derived from an earlier Latin form (see, e.g., *Sylloge tacticorum*, 35.15; 40.1–5; Nikephoros, *Praecepta*, 2§3.19–20), and can refer not simply to the screen of light cavalry in the van of the main battle line, but scouts, skirmishers, or raiders as well: cf. Kekaumenos, *Strategikon* (Spadaro), 2.25 (p. 66.12) (Litavrin, 134.22); McGeer, *Sowing the Dragon's Teeth*, 67, 284. The terms *kourson* (e.g., Maurice, *Strategikon*, 1.2.93; 5.2.9; see Mihăescu, "Éléments latins," 497–98) and, by the middle of the tenth century, *monokourson*, referred to different types of enemy raiding activity. See Dagron and Mihăescu, *Traité sur la guérilla*, 178–79; Trapp, *Lexikon*, 4:875–76; and Litavrin, *Kekavmena*, 354–55.

103 Μίνσωρες   Cf. 9§7.33–39; 12§43.323–25; 20§174.866–68. In the late Roman period μήνσορες (*mensores*) described the army's surveyors, responsible both for surveying potential camping locations as well as quartering and billeting troops: cf. *CTh* 7.8.4 (a. 393 = *CJ* 12.12.40.1); Vegetius, *Epitoma*, 2.7.9; and Maurice, *Strategikon*, 2.12, from which Leo takes this passage. There were also groups of *mensores* acting as billeting officials for various senior officers, including the *magister officiorum* and the various *magistri militum*: Jones, *Later Roman Empire*, 582–83; *Der kleine Pauly*, 3:1226. As Leo notes here, the "modern" Byzantine term was *minsorator* (see Trapp, *Lexikon*, 5:1027–28); although Syrianos, *Strategy*, 26, continues to employ μήνσορες, and several texts of the tenth century illustrate their activities: *Skirmishing*, 13.20–22, where they are described as "an advance party of troops, whom the Romans generally call *minsouratores* . . . to get the camp arranged . . ." (trans. Dennis). See also Const. 17.272–74. The late tenth-century anonymous treatise on campaigning describes their work in detail. For imperial camps an experienced senior *minsourator*, accompanied by "the *minsouratores* of all the other officers," equipped with an appropriate measuring cord (in this case of 2,000 m) was sent ahead, appropriately protected by accompanying troops, to determine and measure out the encampment. From this it appears that each senior officer would normally nominate some soldiers under an officer to act as *minsouratores* while on the march; see *Campaign Organization*, 1.41–50; 6.8–14; see also McGeer, *Sowing the Dragon's Teeth*, 348. The *minsouratores* were associated with another group, the *doukatores*, scouts and guides drawn from the army and from local populations. See on Const. 9§7.35 below.

These officers should not be confused with the imperial *minsourator*, an official of the imperial court under the *papias* or concierge of the imperial palace, responsible especially for the imperial tent; see *CPTT* B.48 (and see comm. to B.117); and C.162–77; and cf. Genesios, 4.37 (trans. Kaldellis, 110); *Vita Basilii imperatoris*, 71.16–18.

103 τὰ ἄπλικτα ἤτοι τὰ φοσσάτα   Two standard words for encampment. See *CPTT* 155, and above, on 4.8; Trapp, *Lexikon*, 1:159–60. In the

course of the second half of the eleventh century the term κατοῦνα also appears, from medieval Latin (*canto*) and Italian (*cantone*): Trapp, *Lexikon*, 4:812; Litavrin, *Soveti i rasskazi*, 356–57, probably the result of the presence of Frankish and other western mercenary troops as well as the earlier peresence of Byzantine troops in southern Italy and Sicily before the 1070s.

105 Ἀντικένσωρας   In the late Roman period these were responsible for establishing the route of march and ensuring the army arrived safely at the encampment, as well as for ensuring adequate access to water and other resources: Maurice, *Strategikon*, 1.3.32; 1.9.22; 37; 2.12, etc. They do not appear in Vegetius, nor in other earlier Roman military contexts (although they appear in John Lydus, *De magistratibus*, 3.27.4 [115.9 ed. Wünsch]) as writers of rescripts. As Leo notes again, they were by his time no longer a separate specialist role, but assigned to the *minsoratores*, and he omits them from his later references to the activities of the latter. Cf. Mihăescu, "Éléments latins," 155–56; Trapp, *Lexikon*, 1:128.

110 Σκουλκάτωρες   Another term Leo simply copies from Maurice where it appears in the original, but which he often replaces with a generic word such as κατάσκοποι (the term employed throughout in Syrianos, *Strategy*) or a more contemporary term such as βιγλάτω-ρες or similar. See Const. 17.448–512 and cf. *Sylloge tacticorum*, 7.2, 25.1–3 (= Syrianos, *Strategy*, 42); *Skirmishing*, 6.12, 10.101, 14.82–84 (*viglatores*); and *Campaign Organization*, 18.22–33. See Dagron and Mihăescu, *Traité sur la guérilla*, 216; and Trapp, *Lexikon*, 7:1575. These are to be distinguished functionally from the screen of "scouts" or *prokoursatores* already mentioned above (see comm. at 96–122), although their duties often overlapped. The term is rare outside of the military treatises, but also appears in the *DAI* 1: chap. 53.57.

116–19 §§29–30   The enumeration of functions and titles at *Strategikon*, 1.3 does not list the *enedroi* and *notophylakes* at this point. Leo inserts them, since he is here making a complete list of the range of functions and definitions found in his sources. Cf. Maurice, *Strategikon*, 1.4.29 and 2.5. For their functions in the fourth–fifth centuries, see Wheeler, *Stratagem*, 83–84, 86, 98. During the tenth century, and probably after Leo was writing, the traditional term for rearguard, *notophylakes*, was replaced by the term *saka* (e.g., *Sylloge tacticorum*, 23.7–8, 29.1; 46.17–21; *Skirmishing*, 16.17, 45; *Campaign Organization*, 31.3–4), probably from Arabic *saqat* (see Dagron and Mihăescu, *Traité sur la guérilla*, 59 n. 4 and McGeer, *Sowing the Dragon's Teeth*, 283 and n. 31), although Moravscik, *Byzantinoturcica*, 2:263–64 sees a Turkic origin. The task of the rearguard was not simply to protect the army but also to pursue or turn back deserters, stragglers, and the like, cf. Const. 17§§56–57.

120   Τοῦλδον   Lat. *tultum* (*fero*); cf. Dain, "'Touldos' et 'touldon'." For further discussion see on Const. 10 below.

123–24   ἡ τοῦ στρατηγοῦ προέλευσις ... δομέστικον   The *proeleusis* of the thematic commander was in effect his immediate staff and retinue, and included the key officials or officers of his household military administration: cf. *Sylloge tacticorum*, 49.4 and see *CPTT* 168–69; Trapp, *Lexikon*, 6:1388–89. Here the *komes tes kortes* and the *domestikos* of the *thema* are named; but the staff also included several other officers as well as ordinary soldiers, who were to be armed and equipped well, and presumably served as the general's personal guard (cf. Const. 6.8). In the lists of officers and soldiers dealing with the Cretan expedition of 949 the following is entered regarding the *strategos* of the Thrakesion theme: ἀπὸ τῆς προελεύσεως τοῦ στρατηγοῦ ὁ πρωτομανδάτωρ, ὁ πρωτοκαγκελλάριος, ὁ πρωτοβανδοφόρος, πρωτοδομέστικοι ϛ', πρωτοκένταρχοι ϛ', προελευσιμαῖοι πεζοὶ ρ'. (Haldon, "Chapters II, 44 and 45," 217.205–6)

From other evidence it is clear that there had taken place some evolution and change in much of the military administration of the provinces between the time of Leo's writing and the middle of the tenth century. Nevertheless this list gives some idea of the thematic general's staff. It probably also included the so-called *merarches* or *meriarches*, the senior tourmarch of a theme and (probably) in charge of the *tourma* in which the thematic *strategos* was based; see *CPTT* 249–50. Each of these named officers was at the head of a particular group of subordinates who staffed the administration of the *strategos*. The precise function of each remains unclear. Each senior officer below the *strategos* also had his own staff or *proeleusis*: in 949, for example, the tourmarch of the Theodosiakoi, one of the divisions of the Thrakesion theme, had a *proeleusis* which included 4 staff officers (*proagetai*), a *protomondator* and a *domestikos* (Haldon, "Chapters II, 44 and 45," 217.210–11 [*De cerimoniis*, 663]).

The *domestikos* of the theme was probably an adjutant to the *strategos*. He is not to be confused with senior or junior tagmatic officers of the same name, although his role was similar to that of junior *domestikoi* in the *tagmata*. See *CPTT* comm. to C.505 (p. 250).

The *komes tes kortes* was in charge of the tent and security arrangements for the *strategos*. Each *komes* appears to have had a number of attendants called *kortinarioi*. See *CPTT* C.570–73; Oikonomidès, *Listes*, 341. As well as meaning "tent" or "pavilion," the term *korte* could refer also to the general's baggage and equipment, and the personnel to manage it: cf. Trapp, *Lexikon*, 4:865.

127–33   §33   An important passage in which Leo sets out the roles of three key imperial officials responsible for important functions in the provinces, and which emphasizes the degree of central control wielded at this time by the government at Constantinople. Each *thema* had one *protonotarios*, an official appointed from the department of the *sakellion*, the general fiscal inspectorate, in the palace at Constantinople. As Leo notes here, his role was the civil administration of the theme, and among his duties was that of liaising between central fiscal departments and with local officials in respect of thematic finance, military supplies, and so forth. For the origins and functions of the *protonotarios*, see Brubaker and Haldon, *Byzantium in the Iconoclast Era*, 679–82. See *CPTT* 167, 236; Oikonomidès, *Listes*, 315; Haldon, "Chapters II, 44 and 45," 289–90 with sources and literature; and Haldon, *Warfare, State and Society*, 143–45. The role of the *protonotarios* is comparable to that of the earlier imperial *procuratores* vis-à-vis senatorial governors; see *Der kleine Pauly*, 4:1151.

The *chartoularios* of the theme was appointed from the military *logothesion*, the fiscal department of the central government responsible chiefly for military recruitment and supply. He liaised with the provincial *chartoularioi* of the general *logothesion*, through the *protonotarios*, and his main duty was maintaining the military registers. See Oikonomidès, *Listes*, 314; Haldon, *Byzantium in the Seventh Century*, 212.

The *praitor* or judge—*dikastes*—of the theme was responsible for judicial matters, and was appointed, in theory at least, from the bureau of the Quaestor at Constantinople. See Oikonomidès, *Listes*, 322–23. His position had evolved by the first half of the ninth century out of the original civil governors of the late Roman provinces. See Haldon, *Byzantium in the Seventh Century*, 266–76.

134-36 §34 Cf. Maurice, *Strategikon*, 1.4.1; and see on ll. 157–61 below. The relationship between the following advice, and the reality of unit structures which any commander would be faced with, is unclear. Following earlier precedents that go back into the classical period, Leo later advises that relatives and friends or neighbors should be banded together in the same squads. No doubt this was practical when recruits or registered soldiers were called up from their village communities to go on campaign (see, for example, *Vita Lazari*, 529C [trans. Greenfield p. 151], for the middle of the eleventh century, where explicit mention is made of the people from the same village community who went off to war together), but it cannot have left the general much leeway to restructure his forces in the way the treatises recommend. Maintaining an esprit de corps without familial and communal identities and loyalties requires regular contact and training as well as continuity among junior officers, such as was the case with Roman legionary units of the late republic and empire and as is the case in most modern armies. Since there seems to have been no regular structure of grades below that of kentarch, it is likely that the appointment of these junior grades was regularly made on an ad hoc basis, and reflected the needs of the moment. This may well be one factor in the very variable morale of Byzantine as well as other medieval armies, and would also suggest the compensatory importance of religious motivation.

137-39 See on 77–84, above; cf. Maurice, *Strategikon*, 1.5.

139 εἰ δυνατὸν δὲ καὶ τοξεύειν εἰδότας The question of archery remains problematic, but Leo's concern is a real one. Leo largely copies Maurice, and while the *Strategikon* is concerned that as many soldiers as possible should be trained in archery, it nowhere suggests that absence or incompetence in archery is a problem. There are few direct and contemporary references to ninth- and tenth-century Byzantine archery. In a passage taken from the *Strategikon* to which Leo makes an important addition (τῆς γὰρ τοξείας παντελῶς ἀμεληθείσης καὶ διαπεσούσης ἐν τοῖς Ῥωμαίοις τὰ πολλὰ νῦν εἴωθε σφάλματα γίνεσθαι) he comments strongly on the damage caused by the lack of competent archery: Const. 6§5.32–33 = *Strategikon*, 1.2.28–30). See also Const. 11§41.235–38 for a very similar sentiment. The general recommendation ("all Roman recruits up to the age of 40 to be ordered to carry bows and quivers, regardless of the level of skill they have attained") is taken verbatim from *Strategikon*, 1.2.28–30, but Leo's specific point about the decline in archery and the evils which have resulted from it are contemporary, and can be assumed to reflect the actual situation. His description of the mounted lancer/archer: Const. 6§§1–13 (with certain changes = *Strategikon*, 1.2.3–58).

That archery was an integral element in Byzantine weaponry and tactical dispositions is not to be doubted (indeed, at Const. 18§125.612, Leo implies that the Romans can rely on their excellent supply of bows and arrows in fighting the Saracens of Tarsos; while at the battle of Versinikia in 813 Byzantine archery caused considerable dislocation among the Bulgarian formations: Skylitzes, *Synopsis historiarum*, 6.82–90 (trans. Wortley, 5), but there is only limited information on the type of bow employed, the draw used by the troops, and their effectiveness in battle. Roman archers of the sixth century were taught the Hunnic, or steppe release, using the thumb together with the index and middle fingers, rather than the Mediterranean release, employing the first three fingers, although both releases were clearly known: the *Strategikon* states that archers could shoot using either the Roman (i.e., the Hunnic) or the Persian method: *Strategikon*, 1.1–2; and cf. Syrianos, *Strategy*, 44–47. This is not mentioned specifically in the tenth-century treatises, and the *Taktika* of Leo, in paraphrasing this section from the *Strategikon*, omits this detail; but Leo the Deacon remarks that Nikephoros Phokas taught his soldiers "to draw the arrow to the chest," which seems to suggest the Mediterranean release (Leo the Deacon, *Historia*, 50.23; Talbot and Sullivan 100). Leo, however, frequently employs phrases and expressions from classical sources, so that his descriptions are not necessarily to be taken at face value. His account of Nikephoros's siege of the Muslim stronghold at Chandax (Candia) on Crete, for example, is taken directly from an account in Agathias of the siege of Cumae by Narses (see examples cited in Sullivan, "Instructional prescriptions," 181–82). In general on Roman archery, see James, "Dura-Europos" and Kolias, *Byzantinische Waffen*, 214–38, with literature. For the late Roman background, see Amatuccio, *Peri Toxeias*; and for a skeptical view of Procopius's account of the late Roman mounted archer, see Kaldellis, "Classicism," 187–201.

139–44   See also on ll. 77–81 above. The functions of these different tactical grades is repeated from older treatises, in particular Maurice, *Strategikon*, 1.3.18–21; 3.5.17–19, etc., and while Leo does add some clarification, suggesting that they were still relevant and understood, at least in terms of tactical maneuvering, there is no evidence to corroborate this. The basic file (for cavalry) was understood as consisting of ten soldiers, headed by a dekarch, with the pentarch (in charge of four other men) as his second, and the tetrarch or "file-closer" at the rear of the file. These were not "officers" in the sense of permanent positions, but functional, tactical positions, and—following Maurice—they were to be selected from the strongest and most experienced, with the rest in between them. When the file was extended laterally, so that it was five rather than ten deep, then the dekarch, three soldiers, and a tetrarch made up one file, the pentarch with three soldiers and the other tetrarch made up the other (cf. ll. 84–89 above). Note that Leo places the pentarch "in the middle of the file" (4§14.80; and cf. *Sylloge tacticorum*, 35.1.8; 46.4–5, where the pentarch is simply the fifth man in a file of ten), whereas in the *Strategikon* he is second, behind the dekarch; by the same token, at Const. 18§143.772–73 Leo makes the pentarch the equivalent of the *ouragos*, the last man in a file of ten (omitting mention of the tetrarch, who normally fulfilled this role), and conflating pentarch with tetrarch again at 18.799–800 (800 pentarchs for 400 dekarchs, where one might expect 400 of each). Whether this reflects a real tactical organizational change, or Leo's own confusion or misunderstanding, is not clear.

140–41   τῶν ... ὀρδίνων "Files," late Lat. *ordo, -inis*, Hellenized by the late sixth century as ὄρδινον, the equivalent of Greek ἀκία. See on ll. 40–52, 53–60 above and Trapp, *Lexikon*, 5:1145. In Leo's time and later *ordinos* was still in general use in military contexts, although there is frequent confusion as to its precise value—following Maurice, Leo uses it generally of files, as opposed to ranks or rows, the usage attested in, e.g., the *Praecepta militaria* of Nikephoros (1§10.107; 2§14.142; 3§§1.4, 6.49, etc.) and the anonymous *Campaign Organization* (10.58), as well as in texts based more closely on ancient antecedents such as the *Sylloge tacticorum* (32.1; 45.11–16 etc.—see under ὄρδινον in the index). Following Maurice, Leo also used the verb ὀρδινεύειν, from Lat. *ordinare* (e.g., l. 149), although the word does not appear in the other military treatises and may therefore be assumed to have been no longer in common use. See Mihăescu, "Éléments latins," 497 and idem, "Littérature byzantine," 359–61.

145   ἐπιλέκτους   Since Leo is copying directly from Maurice at this point, the term "select" refers simply, as in the *Strategikon*, to the dekarch, pentarch, and tetrarchs. Elsewhere, however, the term refers more particularly to the permanent or professional soldiers who constituted the standing element of the thematic armies; see, e.g., Const. 18§26.144; §§136, 143, 146, and commentary.

151   παῖδες   The question of servants and other support for soldiers in the Byzantine armies is under-researched, but see the useful discussion in Kolias, "Ein zu wenig bekannter Faktor." For the ancient and later Roman periods, see Welwei, *Unfreie*, esp. 167–80.

152   τὰ σαγμάρια   Maurice, *Strategikon*, 1.5.4. Lat. *sagmarius* (e.g., Vegetius, *Epitoma*, 2.10.4), derived from *sagma*, a pack-saddle. See *CPTT* C.202 and comm., 185, with sources; Trapp, *Lexikon*, 7:1519. The term was the generic word for any pack animal, whether a mule or a horse. A number of other terms were also in use, such as *parippion*, packhorse, or *molarion*, mule, for example. Several of these had a specific technical value. See *CPTT* 184–86, Leone, *Animal*.

157–61   §40   This appears to be Leo's own additional remark, and emphasizes the varied size of different units as well as the flexibility of structure of the Byzantine provincial armies, although the numbers for the different tent groups he gives all reflect organizational patterns he could find in the treatises he consulted. In fact, the usual number for a squad or tent group seems to have been ten men in the cavalry; see above, on ll. 16–17. The comment on soldiers grouped together in permanent squads, however, also reinforces the notion which he expresses in the next paragraph (ll. 162–67), drawn from Onasander (*Strategikos*, 24), whereby relatives and friends should be grouped together to stiffen morale and fighting spirit. This was a standard recommendation of the tenth-century manuals and reflected the realities of the recruitment and mustering arrangements; see Nikephoros, *Praecepta*, 1§2.10–13; 3§10.73–80; and cf. *Campaign Organization*, 2.4–12.

162–67 §41   Cf. Onasander, *Strategikos*, 24.

168–72 §42   Maurice, *Strategikon*, 1.5.1, 2.7.

173–209 §§43–50   Maurice, *Strategikon*, 1.4. Leo changes or qualifies the unit names in the *Strategikon* (*moira, meros, chiliarchia*) with their ninth- and tenth-century equivalents *droungos* and *tourma*. The numbers involved remain unchanged, however. As noted already, numbers clearly varied more widely than this, since the near-contemporary *Sylloge tacticorum* refers to cavalry *banda* of as few as 50 and as many as 400, and infantry units of between 200 and 400 (*Sylloge tacticorum*, 35.4–5).

205–9 §50   Cf. Const. 12§34.

210–12 §51   See the longer passage about doubling the number of standards shown to confuse an enemy at *Strategikon*, 2.20.

213–330   Having taken most of his material thus far from the first two books of the *Strategikon*, Leo now begins to excerpt from book twelve (specifically 12.B.7–8), which is devoted to the infantry. This seems to be part of his scheme to reorganize the material in a way he considers more useful for the commanders who, he hopes, will be using the *Taktika*.

214–16   It is interesting that Leo omits from this short list (although he retains it in other places, see on 7§38.266 below) those categories which he either did not understand or which no longer appeared relevant. The *campidoctor* or "senior drill-sergeant" in late Roman infantry units (cf. Vegetius, *Epitoma*, 3.6.23) is omitted, since his role probably no longer existed as a result of the way in which Byzantine infantry fighting and organization had evolved in the interim, although some insignia referred to as *kampidiktouria* and associated with the imperial palace were kept in the church of the Lord in tenth-century Constantinople (see *CPTT* 272–73 with references). The *bandophoroi* are not listed, presumably because he has recently dealt with them;

he omits the *drakonarioi*, standard-bearers who carried dragon-like devices which filled with wind when held aloft, adopted originally in the second century from either Germanic or Sarmatian armies: Coulston, "Draco Standard." The rank of *draconarius* survived, but whether such officers actually carried the *draco* is open to question. See Brown, *Gentlemen and Officers*, 275 for an eighth-century Ravenna *draconarius*. Imperial ensigns or standards referred to as *drakontia* were listed as being stored in the church of the Lord in the tenth century: *CPTT* 272.

Also omitted here are the *armatourai*, originally weapons drill specialists under the *campidoctor;* see Rance, "Simulacra pugnae." The term may later have signified "armorers," although this is problematic; see Mihăescu, "Éléments latins," 1:491; 2:157; idem, "Littérature byzantine," 56–57; and Trapp, *Lexikon*, 2:199. Leo does include the essential *samiatores*, "polishers," who presumably worked on maintaining weapons and armor, along with bowyers and fletchers.

215   διαφόραις γλώσσαις   This is also a sentiment borrowed from Maurice (cf. 12.B.7.4, where "Roman," i.e., Latin, Greek, and Persian are listed), but would naturally be of value. Leo names no particular language.

220–23   §54   As noted above, this passage is taken from the *Strategikon*, 12.B.7, dealing with the infantry; in fact he has not yet dealt with the cavalry baggage train explicitly (see l. 120 above).

222–23   τοῦ λεγομένου καραγοῦ   Cf. 11§39.220–21; late Lat. *carrago*, the late Roman term for wagon train, from a probably Germanic source—according to Ammianus, it was from the Gothic, and generally referred to the wagon-circle or laager; see 31.7.5 and 7 (Goths), 31.2.18 (Alans) and 31.2.10 (Huns); cf. also Vegetius, *Epitoma*, 3.10.16 (trans. Milner 89). See Mihăescu, "Éléments latins," 498 and Trapp, *Lexikon*, 4:764. The word does not appear in the late tenth-century treatises which describe contemporary reality (such as the *De velitatione bellica*, the *Praecepta* of Nikephoros Phokas, or the anonymous *De re militari*), but Leo uses it here in the sense of a laager or wagon-circle rather than, as in the *Strategikon* (e.g., 12.B.7.10; 18.2; 22.99) the train of wagons which may be formed in such a defensive arrangement: Κάραγος γὰρ λέγεται ὁ διὰ τῶν ἁμαξῶν . . . περιορισμὸς εἰς ἀσφάλειαν τοῦ στρατοῦ (ll. 224–25).

224–25 §55 For the *karagos* see below on 11.40–42.

234–87 §§58–69 This section follows Maurice, *Strategikon*, 12.B.8.3–36, who in turn uses older material: cf. Aelian, *Tactica theoria*, 8.3–9.10, 15.2–16.3; also Arrian, *Techne taktike*, 9–10 and note Syrianos, *Strategy*, esp. 15 (trans. Dennis, 47–53). This is the ancient Hellenistic infantry formation.

241–46 §59 Cf. Asklepiodotus, *Techne taktike* 2.7, 9; 4.4; Aelian, *Tactica theoria*, 9.6, 10; 11.6; 16.6; 20.2; and Arrian, *Techne taktike*, 14.2 and 18.4.

244 ὁπλιτῶν ἤτοι σκουτάτων Where Maurice simply writes *skoutatoi*, Leo demonstrates his knowledge by referring also to the *hoplitai*. See also on 14§69.493–94 and Janniard, "Armati, scutati."

269–71 Slightly emending Maurice (*Strategikon*, 12.B.8.18–20) Leo omits the *vicarius* and the *campidoctor* from the reference to the commanding officers of the individual *banda*.

272–73 Τὰς μέντοι . . . δεκαὲξ ἀνδρῶν Leo follows the *Strategikon* in its assumption that the infantry line will be in files of 16 men, who also make up two tent groups or *kontoubernia* (see, e.g., *Strategikon*, 12.B.9.20–30) although at ll. 157–58 he also notes that the general should decide on the strength of the tent groups and files according to circumstances.

273 τὸν τῆς μάχης κάματον *Strategikon*, 12.B.8.22 has simply τὸν κάματον. Cf. Nikephoros, *Praecepta*, 1§8.87: οἵ τε τῷ καμάτῳ ἀτονοῦντες.

275–87 §§67–69 *Strategikon*, 12.B.8.23–36.

288–330 *Strategikon*, 12.B.9.3–41.

330–32 Note Leo's own concluding statement, ending with the verb διωρισάμεθα, emphasizing once more the official nature of these instructions and recalling the language of imperial legislation.

CONSTITUTION 5

The source of Leo's fifth constitution has not been traced, although it includes elements that appear in Aelian, for example. From the vocabulary employed and the way in which the various items of armor and weaponry are described, it would appear that this Constitution is Leo's own creation (or that of one of his excerptors), drawn from Constitution 6, which he compiled in turn from Maurice, *Strategikon*, 1 and 12. As usual, however, he has brought his descriptions up to date where terms have either gone out of use or changed their meaning; and given the presence in the imperial palace of soldiers of various categories wearing a range of different costumes and bearing different types of arms, it is a reasonable supposition that he knew what he was talking about, even if his familiarity with the military on campaign was somewhat limited, as is recognized: see Parani, "Dressed to Kill." The archaeology of Byzantine weapons is still in its infancy, and literary accounts or (usually highly stylized) pictorial representations provide the greater part of the available evidence. For some recent assessments of the material evidence, see the detailed review of the textual, archaeological, and pictorial evidence in Grotowski, *Arms and Armour*, which includes a complete catalogue and extensive bibliography on all the technical terms found in Leo's *Taktika*. See also Haldon, "Some Aspects"; Nicolle, "Byzantine and Islamic Arms"; Schwarzer, "An Eleventh-Century Shipwreck"; Dawson, "'Fit for the Task'"; the material cited in Kolias, *Byzantinische Waffen*; and the literature cited in the commentary which follows. The detailed work of Grotowski, *Arms and Armour*, represents the latest term-by-term analysis and discussion of the technical vocabulary that follows, and should be used in conjunction with this commentary to Const. 5.

3–8   §1   Cf. 6§1, 11§41. Leo's general instruction to the thematic commander to ensure that military equipment is maintained in sufficient quantity and in good order to meet the exigencies of warfare.

Apart from the overall direction of the *strategos*, the officers under his authority are also to be responsible—the *tourmarchai* in particular will be intended. Note that this equipment is εἰς ὅπλισιν καὶ ὑπηρεσίαν τοῦ ὑπὸ τὸ σὸν θέμα μαχίμου στρατοῦ, which is to say, for the standing element in the theme which formed the core of the provincial armies. The term *to makhimon* is frequently employed by Leo as well as in later treatises of the period (e.g., *Skirmishing*, 19.2 [214, 17 Dennis], 21.5 [226.58 Dennis]; *Campaign Organization*, 8.4, 9.11, 15.15–16), in contrast to terms such as "conscript army" (ἀνδραποδώδης στρατός, cf. Const. 18§142.767) or service corps (τὸ ὑπουργικόν), to denote the difference between the élite of the *thema*, the fighting force made up of those who could properly maintain their equipment, or who were maintained from the fiscal resources of their community, and as more-or-less full-time soldiers; and the less able and less well-equipped soldiers conscripted through the *strateia* but of little or limited military value. Note also Nikephoros, *Praecepta*, 4§14.165–66: here the term οἱ χυδαῖοι probably refers not just to noncombatants but to poor thematic conscripts also, in contrast to the full-time units, referred to simply as οἱ σχολάριοι. By the middle of the tenth century if not already in Leo's time, this word was often used of *tagmata* in general, whether Constantinopolitan (the four "imperial" *tagmata*) or of any other "professional" unit; see Kühn, *Byzantinische Armee*, 73–74. Leo also employs the term *epilektoi*, referring not, as in Maurice, to the selected officers of a unit, but rather to the fighting core of each thematic army (cf. Const. 18.696–97, 765–66, 798–807). See for further comment Dagron and Mihăescu, *Traité sur la guérilla*, 184–86.

The arrangements for procuring equipment and weaponry remain relatively obscure. Until the period of the early Islamic conquests the state exercised a quasi-monopoly over the production and distribution of weapons, and also managed a series of *fabricae* or imperial workshops which produced other equipment such as shields and helmets. After the 640s few of these workshops remained within the imperial frontier, and of those that remained—at Sardis, Nicomedia, Adrianople, Caesarea, and Thessalonike—nothing is known (on Caesarea see Haldon, *Byzantine Praetorians*, 318–20). There was an arms workshop or workshops in Constantinople, but the relationship between the official in charge of these—the *archon tou armamentou*—and provincial production is obscure (ibid., 319–21 with nn. 972–77). By the early eighth century, some evidence shows that provincial soldiers could be responsible for obtaining and providing their own weapons and armor, hence marking the abandonment of the state monopoly which had been introduced by Justinian (Haldon, "Military Service," 21–23; although it should be added that we have no way of knowing to what extent this monopoly was either observed or enforced). There is some evidence in inscriptions as well as texts for provincial armorers and weapon makers. By the ninth century, the provincial military officers, through their own officials, were commissioned with raising the necessary extra weapons and equipment, which was done by applying compulsory levies on provincial craftsmen and artisans; see Haldon, "Chapters II, 44 and 45," 291–94.

9 τοξάρια ... κουκούρων   As noted above, there is very little archaeological evidence for items such as the Byzantine bow and its accessories. The *Sylloge tacticorum* gives some details on the dimensions of the bow used by Byzantine soldiers, which together with descriptions and illustrations of the curved Byzantine bows suggest that the basic model remained that of the Hunnic bow, adopted in the fifth and sixth century, measuring 1.17–1.25 m (45–48 in.) in length, with arrows of 0.7 m (27 in.). This composite reflex or recurve weapon was constructed of several elements, chiefly wood, horn, and animal sinew. One of the lists associated with the Cretan expedition of 949 includes "Roman bows"—τοξαρέας Ῥωμαίας—among other weapons for the fighting crew of a warship, suggesting that a specific type of "Roman" bow had evolved (or had at least been in use), different in some respects from the bows of the mounted nomads whom the Byzantines employed in their armies; see Haldon, "Chapters II, 44 and 45," §45.104 (= *De cerimoniis*, 669.21). See Haldon, "Byzantine Military Technology," 39 and Kolias, *Byzantinische Waffen*, 214–38. For related bows of Khazar type, see Gorelik, "Arms and Armour," 127–47, at 133 (with pl. 9, figs. 1–18).

The bowcase, *thekarion*, was a case shaped to fit the bow when strung, and with a flap or cover to protect the weapon in inclement weather (the alternative was a long leathern sheath used to cover the bow when it was unstrung, associated archaeologically with the steppe nomad cultures). There are several types of bowcase known from both archaeological and representational contexts, but since it is this type which is described in Maurice, *Strategikon*, 1.1.14–17, it is unlikely that any major differences had evolved since the sixth century. See Haldon, "Byzantine Military Technology," 21 and n. 52. The quiver, *koukouron*, was of a type common to the whole region from the southern steppe and eastern Europe to the Iranian plateau and beyond, and consisted of an elongated trapezoidal box that could hold anything from 30 to 60 arrows. The word itself—cf. mod. German *Köcher*, French *carquois*— is considered to be of Hun or Avar derivation (Kolias, *Byzantinische Waffen*, 228) rather than Germanic (*pace* Mihăescu, "Littérature byzantine," 59). See Haldon, "Byzantine Military Technology," 22 n. 52 and Kolias, *Byzantinische Waffen*, 227–29 with sources and literature. For illustrations, see Gorelik, "Arms and Armour," pl. 9 and Nicolle, "Arms of the Umayyad Era," nos. 151–57.

As far as the arrows are concerned, there is little material evidence. Some archaeological data exist for late Roman and early Byzantine arrowheads from sites in Italy and the Balkans (see Haldon, "Some Aspects," 74–75), and a number of iron arrowheads (leaf-shaped), for example, have been recovered from Byzantine Amorion, the military headquarters of an important command. These were excavated from probably eighth- or ninth-century contexts; see Lightfoot, "Amorium Project," 328. Central Asian evidence may offer useful parallels. Khazar tumuli or *kurgans* from the northwestern Caspian region (near Verkhnechiryurt) have revealed several items of military equipment, including three-flanged and leaf-shaped arrowheads, sabers, *pallashes* (a straight-bladed, or occasionally very slightly curved, type of saber), pieces of riveted lamellar, mail, and straight swords. Since Khazar soldiers certainly served as mercenaries in the Byzantine armies from the ninth century on, this sort of equipment may well have been commonplace in Byzantine territory too. The flanges on these arrowheads were extremely pronounced, and the heads were tanged, unlike the earlier Avar, Lombard, and Byzantine arrowheads, which were socketed. For a summary of the material with further literature see Magomedov, *Obrazhovanie Khazarskogo Kaganata*, figs. 18, 19, 22.

9–30 §§2–3   Cf. Maurice, *Strategikon*, 1.2.

9 σπαθία   Byzantine swords were on average some 3 ft./1 m in length, but this varied considerably. On swords, blades, and hilts see Grotowski, *Arms and Armour*, 342–57; Kolias, *Byzantinische Waffen*, 136–47; Dawson, "'Fit for the Task'," 6; and below, on ll. 17–18. The archaeology of Byzantine swords barely exists; see Kiss, "Frühmittelalterliche byzantinische Schwerter" and La Salvia, "Fabbricazione delle spade." The study of late Roman swords has important implications for the period after the late sixth century; see Kazanski et al., "Byzance et les royaumes barbares." That for neighboring cultures, in particular for the Balkans and medieval west, but also for the steppes to the north of the Black Sea as well as, to a lesser extent, for the Islamic world, offers more information. See, e.g., Nicolle, "Arms of the Umayyad Era"; al-Sarraf, "Close Combat Weapons," esp. 167–73 and 176; Gorelik, "Oriental Armour"; Kennedy, *Armies of the Caliphs*, 173–75; and Gorelik, "Arms and Armour," 127–47, and figs.

10–12 σκουτάρια . . . σεσαμιωμένα   See the discussion at Const. 6§21.114–15, §22.133 below. The different types of shield had distinct functions, although it is by no means clear what these were, except when the treatises make this explicit. For shield types, construction, and use, see the useful survey of the textual evidence in Kolias, *Byzantinische Waffen*, 88–131. The *Sylloge tacticorum* gives dimensions for shields of the period: shields for the infantry, round, or four- or three-cornered, were up to 1.4 m in diameter; for light-armed troops circular, about 0.8 m in diameter; for the cavalry, circular, about 0.7 m in diameter for the light cavalry, and up to 1 m in diameter for the heavier troops (38.1; 39.1, 8). It also describes large, kite-shaped shields for heavy infantry, and it is entirely possible that it is from their contacts with Byzantine troops in Italy, or as mercenaries in the Byzantine armies elsewhere, that western cavalry and infantry began to adopt this type of shield, usually considered an entirely western European development, especially associated with the Normans. See Grotowski, *Arms and Armour*, 208–36; Haldon, "Byzantine Military Technology," 33–34; Kolias, *Byzantinische Waffen*, 105–8, both with further references; and Dawson, "'Fit for the Task'," 2–10.

12–15 Syrianos, *Strategy*, 16.31–39.

12 κοντάρια μικρά, ὀκτάπηχα   The length given of 8 cubits, or 12 feet (3.65 m), is found also in the *Sylloge tacticorum*, 38.3; 39.1 as the minimum length for an infantry spear. Since Leo discusses spears in more detail when dealing with the armament of different types of soldiers, they will be dealt with in the commentary to Const. 6, below.

13–15   Leo's brief comment on the classical Macedonian pike, employed in the phalanx, and described in the Hellenistic and Roman treatises from which he drew some of his information: cf. Aelian, *Tactia theoria*, 14.2 and Polyaenus, *Strategika*, 2.29.2.

15–18   Various missile or cutting weapons. *Riktaria* (also spelled *riptaria*) were also referred to as *akontia*, and were effectively throwing spears or javelins, some 2.7 m min. length with the head; see on Const. 6§22.131–32 below; *Sylloge tacticorum*, 38.6; 39.8. See Kolias, *Byzantinische Waffen*, 185–213 for a complete discussion of spears and lances. For the single- and two-bladed axes, see ibid., 162–72, esp. 167–70; and more generally on medieval war-axes, different types of which were used by soldiers of all the empire's neighbors, see Gorelik, "Arms and Armour," 131, 134, 137 (Khazars and Magyars); Pedersen, "Scandinavian Weaponry," 29–31 (both with archaeological evidence); al-Sarraf, "Close Combat Weapons," 162–64 (textual and visual evidence); and Grotowski, *Arms and Armour*, 318–20.

The term *bastagia* derives from βασταγή, which meant originally "a (means of) transport" and was applied to the wagons and other forms of carriage of the public transport system in the late Roman period (see Jones, *Later Roman Empire*, 834; and cf. Lydus, *De magistratibus*, 1.13; Trapp, *Lexikon*, 2:269). In the *Strategikon* it applies likewise to transport—wagons and ox-teams—hired or conscripted by the government to move supplies and matériel for the army; see 12.B.22.117: διὰ βασταγῆς δημοσίας. By the tenth century this meaning appears to have been forgotten—the state transport service is referred to as the *dromos*, and its slow division (wagons and carts) seems in any case to have been abolished by the administrative reforms under Justinian in the middle of the sixth century (see Dunn, "Kommerkiarios," 17–19 with older literature) and the word *bastagion* seems to mean generically something by which an object is carried or transported; see Trapp, *Lexikon*, 2:269. That the fast postal/transport system did continue into the middle Byzantine period is suggested by the reference to δημόσια ὀχήματα, used by the general Manuel when he fled from Pylai, opposite Constantinople, to the Syrian frontier in 879; see Leo grammaticus, 218.15–19; pseudo-Symeon *magister*, 632.11–13, although it has been pointed out that while the term *ochema* was usually employed to describe lighter, two-wheeled vehicles, it could also refer to the postal animals themselves. In this case, when Manuel had reached his destination, he hamstrung the horses so they could not be returned, suggesting the latter usage of the word on this occasion. See Belke, "Verkehrsmittel," 49. On occasion the term could also refer to ordinary horses: in the eleventh-century *Life* of Philaretos the Younger of Calabria Roman cavalry horses are referred to to as τὰ τῶν Ῥωμαίων ὀχήματα; see *Vita S. Philareti iunioris* (ed. Caruso), 96, n. 20.

17–18   παραμήρια, μαχαίρας μεγάλας μονοστόμους   Read as "*parameria, large single-edged blades/swords*"; and cf. pseudo-Symeon, 697.9: μάχαιραν μεγάλην, τὸ δὴ λεγόμενον παραμήριον (of the weapon given to the young Leo VI by Santabarenos in order to provoke the suspicion of the emperor Basil I); a ceremonial *paramerion* was also carried with the emperor on expeditions: *CPTT* C.219 (cf. Trapp, *Lexikon*, 6:1217). The word meant originally a dagger or knife carried "next to the thigh," but it appears to have an additional and alternative meaning in Leo and in the tenth-century treatises. In Leo, and in the *Sylloge tacticorum*, the *paramerion* appears beside the regular *spathion* or straight-bladed two-edged sword (described as of the same length—just under 1 m or 36 in.), but as a single-edged weapon, which the soldiers wore on their waist belts—διεζωσμένους/ παραζωννύσθωσαν—(in contrast to the *spathion*, worn from a baldric across the shoulder "in the Roman manner," although the *spathion zostikion* was likewise worn from a waist belt); see *Sylloge tacticorum*, 39.2; Const. 6.19–20. Yet the same word is also used of other edged weapons slung from a belt, indeed the *Sylloge tacticorum*, 38.5 refers to infantry *parameria* as two-edged; see Kolias, *Byzantinische Waffen*, 137–38. The word thus appears to refer in general to any edged weapon slung from the waistband (and "next to the thigh"), but can refer to the single-edged weapon specifically, since the *Sylloge tacticorum* (39.2) states καὶ ἕτερα δὲ μονόστομα ξίφη τοῖς διστόμοις ἰσομήκη . . . ἃ δὴ καὶ παραμήρια λέγονται. Long, single-edged knives are known from archaeological contexts of the late Roman period, more especially from the eastern empire and Sasanian Iran or neighboring territories, and their use appears to have been widespread. They are relatively short (no more than 70 cm in length); see Kazanski et al., "Byzance et les royaumes barbares d'Occident," 172–76. The weapon described here is considerably longer than these, however, and seems most probably to refer to a pallash, also slung from the waist, and first introduced by the Avars; see Haldon, "Military Technology," 31; Fettich, "Das Kunstgewerke," 14, fig. 12; and Laszlo, "Études archéologiques," 228–29, 232–33 with pl. 46, 51–53. For a careful assessment of the evidence, see Grotowski, *Arms and Armour*, 357–60. Such pallashes are well attested archaeologically from the Balkans and steppe regions from the late sixth and seventh century onward, associated with the Sabirian Huns (in the Caucasus), with the Bulgars south of the Danube in the late seventh century, and with the Khazars in the southern steppe from the eighth century on. They are generally about 1 m in length, sometimes slightly longer, and fit the description of the middle Byzantine *paramerion* exactly. See Gorelik, "Arms and Armour," 129–30, 133. Whether Byzantine cavalrymen, or foreign mercenary or allied troops, also used the shorter, more strongly curved Turkic saber, remains unknown. It appears from the late seventh to the ninth centuries in archaeological contexts neighboring Byzantine territory. See esp. Zakharov and Arendt, "Studia Levedica, 2," esp. pls. 3, 6, 7; and cf. Fülöp, "Awarenzeitliche Fürstenfunde," esp. 183, and figs. 14.1–6. For Pecheneg sabers, see Pletneva, "Pečenegi, torki i polovtsi," esp. 159–60, 168. For a possible parallel, see Nicolle, "Arms of the Umayyad Era," no. 120 (a single-edged proto-saber from the Altai, 6th–10th c.). The archaeological data is slender, however. One pallash, from a Bulgarian site, probably of the late seventh century, and several swords from excavated burials in the regions north of the central Danube, dating from the seventh to the tenth century, may be of Byzantine manufacture. The pallash or saber measures approximately the same in length as the single-edged *paramerion* of the *Sylloge tacticorum*: Werner, *Malaja Pereščepina*, 25–27 and

Kazanski and Sodini, "Byzance et l'art 'nomade'." In the case of the swords the weapons were characterized by decorated bronze pommels and cross-guards, which distinguished them from the types usually associated with the regions where they were found: Kiss, "Frühmittelalterliche byzantinische Schwerter," 193–95 (and catalogue at 199–207). Although their identity as Byzantine is not absolutely certain, the similarities between these weapons and similar swords, dated to the sixth-seventh centuries, found at Corinth and Pergamon, is suggestive, and it may be possible to begin establishing a concrete typology.

19–20 λωρίκια The word derives from Lat. *lorica*, and from the sixth century at least generally referred to a shirt or long coat of mail, or simply any chain mail. Here Leo notes that such coats can also be constructed of either horn or boiled leather, presumably of scale, or in lamellar style, that is to say, of plates connected by thongs, a style of defense that can be traced in particular to the steppe nomads.

21–22 ἐπιλωρικὰ ἱμάτια Padded or quilted jackets worn over the armor, long for the cavalry trooper, knee-length for infantry.

22 κλιβάνια A Byzantine development influenced by the steppe warriors with whom the empire was in constant contact, consisting of a sleeveless jacket of lamellar, with splint arm guards attached at the shoulders, and constructed of metal or leather—quite likely from the Khazars, who in the course of the late seventh and eighth centuries developed a new type of flexible upper-body armor consisting of iron plates riveted together; see esp. Tsurtsumia, "Evolution of Splint Armour," with literature; Grotowski, *Arms and Armour*, 125–62; Gorelik, "Arms and Armour," 134–35 with illustrations; Haldon, "Military Technology," 20–21, 34–37; Kolias, *Byzantinische Waffen*, 37–60; Dawson, "Kremasmata, kabadion, klibanion"; idem, "Suntagma Hoplon," 84–85; and see on Const. 6§2.9 and §4.27 below. This long coat of mail or quilting is copied by Leo directly from the *Strategikon*, but although it is not mentioned at all in the more realistic and independent mid-tenth-century sources, he brings his account up to date by noting that it can be of lamellar; and the pictorial evidence suggests that it had not fallen out of use, for several representations appear to depict such knee-length coats, made of what may be

either scale or lamellar. Indeed, the pictorial detail—with the "skirt" of the garment divided from the waist for riding—corroborates the written accounts to an extent. Such armors appear in an eleventh-century Byzantine manuscript illumination, on frescoes or bas-reliefs in provincial chapels and churches of the tenth and eleventh centuries, and in minor arts—e.g., on caskets, items of metalwork, and icons, as shown by Grotowski, *Arms and Armour.* For central Asian parallels of the same period, see Khudyanov, *Yužnoi Sibiri i Tsentralnoi Asii,* with illustrations.

κασσίδας   Helmets are once again a subject for which there is little material evidence apart from representations in manuscripts of frescoes and the very general descriptions in the military treatises; see Kolias, *Byzantinische Waffen,* 75–85; for the background and context of helmet evolution: Böhnen, "Zur Herkunft"; Bishop and Coulston, *Roman Military Equipment,* 167–72; James, "Evidence from Dura Europos"; Coulston, "Arms and Armour," 21, 23–24; and Nicolle, "Arms of the Umayyad Era," nos. 164–78. A helmet of the tenth century, probably Byzantine or Bulgar, with the bowl from a single sheet, and with riveted strengthening bands and holes for the attachment of a ventail, has been excavated from a site in central Bulgaria. Archaeological evidence from the south Russian region and the northern Caucasus has revealed helmets of a slightly different style but similar construction, wrought from a single sheet, with a spike or fitting for a plume, similar to those described in the Byzantine treatises; See Nicolle, *Medieval Warfare Sourcebook,* 2:76–77; and Kirpichnikov, *Drevnierusskoe Oruzhie,* esp. fig. 10. Helmets could be open (cf. *Sylloge tacticorum,* 38.7; 39.9), or have ventails and face-coverings of mail attached (ibid., 38 5, 39.3; Nikephoros, *Praecepta,* 3§4.35–37).

23   ποδόψελλα . . . ὕλης   Leg guards and arm guards were made of various materials—leather, metal, bone, or even (for infantry) of wooden (6§21.124, but copying Maurice, *Strategikon,* 12.B.4.6–7) splints on a cloth or leather backing, construction techniques popular and usual on the Eurasian steppe, and certainly familiar to the Byzantines of the ninth and tenth centuries from Khazar or Magyar warriors; see Grotowski, *Arms and Armour,* 183–91; Haldon, "Military Technology," 28, 37 (and on splint armor in general see 16); Gorelik,

"Arms and Armour," pl. 11.5, figs. 15–19; pl. 11.12, figs. 7–10; and Kolias, *Byzantinische Waffen*, 65–74.

24–26   περιτραχήλια . . . σιδηρᾶ   The translation and text should be emended here; see the critical apparatus to ll. 24 and 25 and Vári, ed., 92.1176–77: the readings ἐρίου (in M and W) rather than κεντούκλου (in A and two later dependent MSS) and λίνου (in all MSS) rather than ῥίνου (Dennis's own emendation, which is not justified by the manuscript tradition), should be preferred, to read, "Let those who do not have neck guards of mail have iron ones, lined with wool on the inside and with linen outside." (Although since *kentouklon* refers to quilting or padding [see on l. 24 below] the latter might be an acceptable reading and would have made sense, in the context, to the tenth-century scribe or redactor of A.) This neck protector or gorget is taken from the *Strategikon*, 1.2.2, where it is stated to be a borrowing from the Avars. It occurs in none of the late tenth-century treatises and is most likely no longer a contemporary item of equipment: in the Ambrosian paraphrase of the *Strategikon* (in A) the term περιτραχήλια is glossed as μανίκια (ed. Leoni, *La parafrasi Ambrosiana*, 33.12). This word usually refers in tenth-century sources to a neckband or gorget, chiefly of a decorative or maybe hierarchical value, see Kolias, *Byzantinische Waffen*, 80–81. Leo's account here suggests he may have confused the *peritrachelion* with the ventail or curtain of mail which was attached to the helmets of heavy cavalry troopers according to the *Praecepta* of Nikephoros: 3§4.35–37; and cf. Kolias, *Byzantinische Waffen*, 37–44 on *zaba* (and see below on 6.9).

24   κεντούκλου   From Lat. *centuculus*, "patchwork," thus quilted/padded, rather than felt, for which cf. Lat. *coactilia*, Gr. πιλωτό; see Trapp, *Lexikon*, 4:820; Kolias, *Byzantinische Waffen*, 54–55; Mihăescu, "Éléments latins," 1:487; and idem, "Littérature byzantine," 1:204–5; 2:55–56.

25   νευρικὰ   from the substantive νεῦρον, boiled leather or dried meat, although also used of a type of shellfish: cf. *CPTT* C.147–48 and 203; Trapp, *Lexikon*, 5:1076; connected also with νοβερονίκιον (?); see idem, 5:1082. Cf. Nikephoros, *Praecepta*, 3§5.39: horse-armor made ἀπὸ κεντούκλων καὶ νεύρων κεκολλημένων—"of padding and leather glued together." The phrase **νευρικὰ τὰ ἀπὸ κεντούκλων διπλῶν**

γινόμενα is very similar to that at Const. 19.91: τὰ λεγόμενα νευρικὰ ἅπερ ἀπὸ διπλῶν κενδούκλων γίνεται, where *neurika* is a substantive, meaning something like "leather-and-felt-armor." Here it has the sense of padded or quilted leather protection made up of squares or patches of leather sewn onto a backing or stitched together.

27   σωληνάρια   Arrow guides, a slender channeled board or plate of wood used to shoot small darts from an ordinary bow; see Amatuccio, *Peri Toxeias*, 98–100 and esp. Nishimura, "Crossbows" and Pétrin, "Philological Notes," 270–76. This device appears only in Leo, derived from Maurice, *Strategikon*, 12.B.5.4, and in the *Sylloge tacticorum*, 38.8–9, derived from Leo and Maurice. They are not mentioned in the more contemporary *Praecepta*, nor in the later treatises, and it may be doubted that they were still employed.

28   σελλοπούγγια   A saddlebag, Lat. *sella + punga*, a pouch or bag/sack, perhaps from a Germanic word. Cf. Mihăescu, "Éléments latins," 1:494; "Littérature byzantine," 53; Trapp, *Lexikon*, 7:538.

28–29   λωρόσοκκα, πέδηκλα   The lasso and the hobble were employed for roping in horses from open pasture and restraining them on the horse lines: cf. *CPTT* C.134–35: σωκάρια to be provided for the horses of the imperial cortège. The lasso also had a military application, for capturing or disabling an opponent, as several accounts reveal, and was associated in particular with steppe nomadic mounted warfare; see Moravcsik, "A hunok taktikájához"; Sullivan, "Constantine VI"; the detailed comment in Rance, commentary to *Strategikon*, 1.2.42; and cf. Suidas, 1.4.278 (p. 346); Trapp, *Lexikon*, 5:597; 7:1583. For πέδη-κλα, see ibid., 6:1250.

29   σεληναῖα σιδηρᾶ μετὰ καρφίων αὐτῶν   This seems to be the first literary reference to the horseshoe (see Trapp, *Lexikon*, 7:1537). There seems little doubt that the term σεληναῖα refers to horseshoes, since the same term is used in a less military context in one of the treatises on imperial military expeditions ascribed to Constantine VII (see *CPTT* C.82–84: τὰ σ′ σαγμάρια . . . ἐπιφερόμενα καὶ σελιναῖα; 131–32: ἐκ δὲ τοῦ βασιλικοῦ βεστιαρίου ὀφείλει λαμβάνει σίδηρον ν′ λίτρας λόγῳ σελιναίων. . . . Note Nikephoros Ouranos, *Taktika*, 174 (Dain, *La "Tactique,"* 84), where the verb καλιγωθῆναι—to shoe—is associated explicitly with the σεληναία, as at *CPTT* C.82–84, 632. In the

Byzantine archaeological context they are attested from several sites, albeit sparsely: at the rural farmstead site at Çadır Höyük in north central Anatolia, for example, one half-moon shaped example was uncovered, dated to the middle Byzantine period (9th–late 11th century), with a span of about 11 cm (see Cassis et al., "Çadır Höyük"). A ninth-century Arab source gives details of methods of shoeing the hoof, and there is little reason to doubt their widespread use by this time; see the extracts from Ibn Abu Hizām in the twelfth-century *Book of Agriculture* of Ibn al 'Awwām, *Kitāb al-filāha*, 2.2:100–101; and Hyland, *Medieval Warhorse*, 86. Evidence for the use of horseshoes in the late Roman army is both sparse and ambiguous, although there is some evidence that their use increased in the fifth century and afterward, and we must assume that their more widespread use is a Byzantine or at least medieval innovation. See Hyland, *Equus*, 123–24; and Dixon and Southern, *Roman Cavalry*, 229–33. Archaeologically they are attested increasingly from the ninth and tenth centuries in central Europe, and their spread has been connected with the breeding of a more specialized and heavier western and central European warhorse; see Kasparek, "Stand der Forschung," 42. The Byzantine veterinary treatises dealing with horses make no mention of horseshoes, primarily a reflection of their antiquarian tradition; see McCabe, *Byzantine Encyclopaedia*, esp. 299–300; with Doyen-Higuet, "Contribution." Horseshoes are not mentioned in Maurice, and their presence in Leo emphasizes the fact that the provincial cavalry trooper was to a large extent responsible for all the basic needs for both himself and his mount. Farriers were clearly present with a cavalry force, but it was up to the individual soldier to make sure he had spare horseshoes and nails. The archaeological data thus show that the term *selenaia*, "moon-shaped," accurately describes a horseshoe like a crescent-shaped plate, not very different from the modern and western medieval horseshoe (cf. the Norman and western medieval horseshoes of the twelfth century and later in Hyland, *Medieval Warhorse*, 75).

Leo makes no reference at all to protective hoofplates for the cavalry, to protect them against such devices as caltrops, although such equipment was certainly employed—Syrianos, *Strategy*, 17.16–19 refers

to such iron plates, and the late eleventh-century *Life* of the Calabrian
St. Philaretos the Younger describes the Byzantine cavalry as employing them at the battle of Traina/Dragina (mod. Troina) in Sicily in
1040, thus saving their horses from the caltrops scattered around the
enemy camp: *Vita S. Philareti iunioris*, 96 n. 20 (ed. Henschenius,
608: the Latin translation of the original Greek text, on which see
Beck, *Kirche und theologische Literatur*, 582). For the battle, with further sources and literature, see Felix, *Byzanz und die islamische Welt*,
210; note also Rance, "Date," 729–31 (with English trans. at 731). An
addition (probably of the tenth century) to the collection of military
maxims known under the title of *Leonis imperatoris strategemata*
(see chapter 2 n. 16 above) incorporated into the *Sylloge tacticorum*,
refers to a certain Onias who, in wishing to avoid being pursued by
the enemy, reversed his mount's horse shoes (τὰ σωληναῖα) so that it
appeared he had gone in the opposite direction: *Sylloge tacticorum*,
99.5; see Krentz and Wheeler, *Polyaenus*, 1:33; 2:1075. See also comen-
tary to Const. 14§41.277–92 and 20§147 below.

29–30   For these elements of cavalry mount armor see below on Const. 6.

32   βούκινα μικρὰ καὶ μεγάλα   Leo follows Maurice in his use of the
term *boukinon*, except on the one occasion when Maurice also
employs the word *taurea* (*Strategikon*, 12.B.16.4 = Const. 7§50.345);
while he adds the definition of the tuba: ὅ ἐστι μικρὸν βούκινον, to
Maurice's original wording (Const. 7§50.346). In the *Strategikon*,
therefore, at least three distinct instruments are in use for different types of signal: the trumpet/bugle; the tuba, a small bugle, and
the *taurea*, or horn. This agrees with Vegetius, *Epitoma*, 3.6: *Tuba
quae directa est appellatur, bucina quae in semet aereo circulo flecti-
tur, cornu quod ex uris agrestibus, argento nexum, temperato arte spi-
ritu quem canentis flatus emittit auditur* ("The trumpet [*tuba*] is the
name for the straight instrument. The bugle [*bucina*] is that which is
bent back on itself in a bronze circle. The horn [*cornu*] is that which
is made from the wild aurochs, bound with silver, and when mod-
ulated with a skillful breath emits a note of singing wind" (trans.
Milner, 70, with n. 9). Trumpet or horn blasts signaled a range of
activities, in both drill and battlefield signals; for when animals were
to be watered or returned to the encampment (see on Consts. 7§18.128

and §50.346; 9§75.370; 11§19.105–7 and §23.145; 13§11.54–55 below); and for signaling the daily routine within the camp (see on 17§89.520). See for detailed discussion Maliaras, "Musikinstrumente."

33 τριβόλους . . . τελείους Read: "caltrops tied together *and attached to large pegs*," not "hardened into very sharp points." Caltrops were employed in large numbers by the Byzantines and their enemies. They could be of various sizes, but always consisted of four spikes of iron or wood, so that whichever way they fell one spike always pointed upright (see descriptions at Vegetius, *Epitoma*, 3.24.3–4 and Procopius, *Wars* 7.24.15–18; with Rance, *Strategikon*, 4.3, note 16). For the Cretan expedition of 949, for example, 500,000 caltrops were ordered for 20 warships, to be hurled into enemy vessels when at close range, to cause maximum disruption (see Haldon, "Chapters II, 44 and 45," chap. 45, 130–31 [*De cerimoniis*, 671], and 278). They were often used joined together by a cord for easy retrieval and with a peg to hold the line in place. Cf. *Campaign Organization*, 2.17 (262 Dennis) (= *Sylloge tacticorum*, 22.5; 6): each soldier is to carry a string of 8 caltrops, each *kontoubernion* or dekarchy is to have a small iron stake to which they can be attached. Cf. Syrianos, *Strategy*, 29.25–32. For examples from archaeological contexts, see Bishop and Coulston, *Roman Military Equipment*, 155 and fig. 111, nos. 8–11; and cf. Gilliver, "Hedgehogs," and Tsurtsumia, "Τρίβολo⊠." For an example of their use against the Byzantines by Arab forces in eleventh-century Sicily, see *Vita S. Philareti iunioris*, 96 n. 20 (ed. Henschenius, 608; and see on l. 29, above); for earlier late Roman/Sasanian examples, Rance, "Date," 732 with nn. 90 and 91.

36 κιλίκια See Maurice, *Strategikon*, 10.3.9 and Rance, *Strategikon*, commentary: goat-skin/leather: the word can refer both to the skins as well as to covers or finished items made from goatskin. Cf. *CPTT* comm. to C.176, at p. 208, and cf. Vegetius, *Epitoma* 4.6.2; Trapp, *Lexikon*, 4:831–32; Mihăescu, "Éléments latins," 1:487–88.

38–39 τοξοβολίστρας, μαγγανικὰ λακάτια ἑκατέρωθεν στρεφόμενα The first term refers to large bow-ballistae, almost certainly tension rather than torsion weapons. From the late fifth century tension-spanned weapons appear to have gained in popularity over the more complex torsion artillery. The vocabulary employed in the Byzantine military treatises, where it sheds any light on the matter at all, reinforces this probability: the term *cheirotoxobolistra*, with the use of the term *toxon*, bow, for example (Roman torsion-powered machines were frequently differentiated from tension machines by the presence or absence of this term). The evidence for the continued use of simple torsion-powered machines, such as the onager-type weapon (with a single arm set vertically through a horizontally mounted spring), is often ambiguous, but there is little reason to doubt their continued use into the seventh century and beyond; see Maurice, *Strategikon*, 10.3.8–12, 17; and for the fourth century, for example, Ammianus Marcellinus, 19.7.6–8; 20.7.10 (where they are referred to as *scorpiones*); 23.4.4–7 etc. See also Vegetius, *Epitoma*, 2.10.25; 3.3.14; 4.8–9. But such machines were both more difficult to construct and much heavier than the equivalent tension- or traction-powered engines. See Chevedden, "Artillery in Late Antiquity," 148 and 160–63; and Huuri, "Zur Geschichte," 51–63, 212–14. The main difference between a *toxobolistra* and a *cheirotoxobolistra* appears to have been one of size and the fact that the former was spanned mechanically by a windlass at the end of the stock: note the contrast between *cheirotoxobolistrai* at *De cerimoniis*, 669.21–670.1 (Haldon, "Chapters II, 44 and 45," 225.105) and the τοξοβολίστραι μεγάλαι μετὰ τροχιλίων (large bow-ballistae with windlasses/pulleys) at 670.10–11 (Haldon, "Chapters II, 44 and 45," 225.116). Both types of weapon are referred to, sometimes indirectly, as *cheiromaggana* (e.g., Nikephoros, *Praecepta*, 1§15.150–55) in the technical military treatises, and *toxobolistrai* appear in some of the narrative chronicles also. See esp. the naval treatises of the tenth century: ed. Dain, *Naumachica*, 1.60 (= 6.57); 7.122.3, 10, 11. For examples from historiographical works: *Theophanis Chronographia*, 384 (for 713–14 Mango and Scott, 534); *Vita Basilii imperatoris*, 59.16 (212).

μαγγανικὰ λακάτια appears to be another type of ballista: cf. 6.148: καὶ βαλίστρας ἤτοι μαγγανικά, τὰ λεγομένα ἀλακάτια, στρεφομένα κυκλόθεν (and 14§74.519–20) suggesting that the term *elakation* or *alakation* was the demotic word for these weapons; see, e.g., *De cerimoniis*, 671.2 (Haldon, "Chapters II, 44 and 45," 227.123), where it occurs as *eilaktion*. Originally the word referred to a distaff or pole, but was used in later Greek of a winch for drawing nets out of the water: Du Cange, *Glossarium Graecitatis*, 474–75; *Suidae Lexicon*, 2.559 no. 190.

For its application to a winch or windlass, see Demetrakos, *Μέγα λεξικὸν*, s.v. ἠλακάτη (2). In the tenth-century treatise *Parangelmata poliorcetica* it meant a windlass, and this was presumably its nickname (compare the late Roman onager, or "mule," a torsion-powered vertically mounted single-arm stone-thrower: Chevedden, "Artillery in Late Antiquity," 137–39; idem, "Hybrid Trebuchet"; and Sullivan, "Tenth-Century," 199 with references). These weapons are referred to as *ta magganika alakatia*, "the windlass artillery/machines," in the same contexts as the *toxobolistra* at 14§74.519–20 (taken from Maurice, *Strategikon*, 12.B.18.9, but with Leo's detail added—the *Stategikon* refers blandly to "stone-throwers"); and importantly at 15§26.156–57 as "stone-throwing machines called *alakatia* or *tetrareai*" (but which could also shoot incendiaries; see 15§26.157–58). The two are not the same but complementary, as is made clear in the mid-tenth-century anonymous treatise on siege warfare: τετραρέας, μαγγανικὰ καὶ τὰς λεγομένας ἠλακάτας καὶ χειρομάγγανα (*De obsidione toleranda*, §14 van den Berg; 156 Sullivan). *Tetrareai* are probably traction-powered trebuchets; cf. Haldon, "Chapters II, 44 and 45," 273–74; and Rance, *Strategikon*, 10.1, note 9 for discussion of late sixth-century artillery and Maurice's text.

As we have seen (see also 6.150–51, 14.519–20), these *alakatia* are described as mounted on carts and swiveling from side to side (i.e., like the late Roman *carroballista,* as described in both the anonymous *De rebus bellicis* and Procopius; see Chevedden, "Artillery in Late Antiquity," 154–63 for the sources and their interpretation; and Procopius, *Wars,* 5.21.14–18; *Anon. de rebus bellicis*, §7), following Maurice, *Strategikon*, 12.B.6.8–9, where wagons with *ballistae* swiveling to both sides are listed. Since Leo takes such trouble to update his information, and since the *alakation* appears in several other tenth-century sources, all this would support the notion that it was a windlass-spanned, swivel-mounted bow-ballista, different from the *toxobolistra*, that could sling stones and that could be mounted on a carriage (which would exclude the possibility that the term referred to a traction trebuchet-type weapon). There are no explicit accounts of such weapons being employed in battles of the Byzantine period, however. Most sources refer simply to "stone-throwers." It must be admitted that it is not at all clear that Leo actually understood the differences between these various devices. On the other hand, the somewhat later *De obsidione toleranda* §66 van den Berg; 172 Sullivan: διὰ τῶν τετραρίων καὶ τῶν μαγγανικῶν καὶ τῶν λεκατῶν (*sic*) is no more specific.

40 λόγῳ ἀρμαμέντου   See Mihăescu, "Éléments latins," 491 (Lat. *armamentum*, "arsenal"). While taken from Maurice, *Strategikon*, 12.B.6.16, there is no reason to doubt that ninth- and tenth-century armies were usually accompanied by wagons with spare weapons and raw materials for the armorers: cf. Nikephoros, *Praecepta*, 1§14; *Campaign Organization*, 31.14–19, as well as for the siege-train, if present; see, for example, Attaleiates, *Historia*, 151.13–15 (trans. Kaldellis and Krallis, 277) (with discussion in Belke, "Verkehrsmittel," 55) for the substantial number of wagons reported to have accompanied the imperial army between Theodosioupolis and Mantzikert in 1071; and for a similar siege-train in the 1176 campaign ending at Myriokephalon, see Choniates, *Historia* (ed. van Dieten), 179.

46 ναυκέλλια ἤγουν πλοῖα μικρά   *Naukellia* (also found as *naukla*), from Lat. *naucella* (see, e.g., *Dig.* 33.7.17.1; Du Cange, *Glossarium Graecitatis*, 987; Trapp, *Lexikon*, 5:1067); Mihăescu, "Éléments latins," 274–75; Cosentino, "Per una nuova edizione," 82–83.

48 ἀτεγίας   A shelter or hut. Cf. Trapp, *Lexikon*, 2:226.

53–72 §§10–13 Leo's final paragraphs sum up the purpose of the foregoing constitution, which is to list the range of equipment and supplies which the general and his subodinates should bear in readiness for war. As noted above, his material seems to be drawn largely from books 1, 2, and 12 of the *Strategikon*, but he has attempted to modernize the language or clarify certain key points (for example in respect of the technical words for items of artillery). In fact Leo notes expressly that he has presented the information, in effect, "as far as his memory serves"—presumably from his reading of the relevant books of Maurice. In the following constitution he repeats the information, but following Maurice more closely, and in specific relation to cavalry and infantry.

58 λαμπρὰ καὶ τεθηγμένα   For bright and polished weapons, see on Const. 14§§33–34, 98; 6§4.28; 20§188.953–56. Cf. Onasander, *Strategikos*, 28–29; Maurice, *Strategikon*, 7.B.15.

CONSTITUTION 6

The first part of this constitution, on cavalry armament (ll. 3–106), is taken directly from Maurice, *Strategikon*, 1.2; the second part, on the infantry (ll. 107–58), represents *Strategikon*, 12.B.1, 4, 5, and 6 (in the order 4, 5, 1, and 6, however). Leo takes the last section (ll. 164–228) from Aelian, *Tactica theoria*, 2.11–13 (for cavalry) and 2.1, 3, 7, 8, and 14 (for infantry). In the first two sections on cavalry and infantry Leo copies Maurice almost word for word for the most part, changing key terms as usual to reflect current usage (thus, as elsewhere, substituting *tourmarches* and *droungarios* for *merarches* and *moirarches*) as noted in the commentary to earlier constitutions.

3-7 §1 Δεῖ τοίνυν ὁπλισθῆναι τοὺς στρατιώτας . . . πρὸς τὴν ἑκάστου ποιότητά τε καὶ δύναμιν Although taken from the *Strategikon*, Leo alters the original wording to reflect the conditions of his own time, in which the ability of the individual registered *stratiotes* to adequately arm and equip himself was a concern of the local military command, many of whom belonged to a local aristocracy or social elite. See on 11.41 below.

9-20 §2 See above, comm. to 5.9–30, for the technical terminology.

9 The word *zaba* is probably of Turkic/Altaic origin, appears in Latin and Greek in the sixth century, and is often an equivalent for *lorikion*: Mihăescu, "Éléments latins," 486; idem, "La littérature byzantine," 49; Haldon, "Military Technology," 19–21, 37 n. 126; Kolias, *Byzantinische Waffen*, 37–43; idem, "Ζάβα, ζαβαρεῖον, Ζαβαρειώτης." This long, ankle-length armor is specific to cavalry, and first appears in the *Strategikon*. Fitted with rings and laces, it was adjustable in length and rolled up to fit in weatherproof carrying cases. Here Leo simply copies Maurice, but probably by this time, and certainly by

the time the *Praecepta* of Nikephoros were compiled, the word simply means strips or pieces of mail attached to other items of armor or
to helmets: cf. Kolias, *Byzantinische Waffen*, 65–67; McGeer, *Sowing
the Dragon's Teeth*, 70. Note that at 14.549 (= Maurice, *Strategikon*,
12.B.23.16), Leo replaces Maurice's term *zabatos* with his own *lorikatos*, and at 18.250 *zabai* is replaced by *lorikia*. Such armor was both
expensive and time-consuming to fabricate, and it is clear from later
texts, especially the *Praecepta*, that coats of lamellar or dense quilting, worn over a *klibanion* of lamellar, were far more usual. Leo has
already noted (5.20–21) that if this armor cannot be of mail, then it
should be of leather lamellar.

16–17   ἐν δὲ τοῖς τοξοζωνίοις ῥίνια καὶ σουβλία   Archers were meant to
carry basic tools for maintaining their weapons, and a file and awl
were essential for keeping arrows and strings in good condition.
These were presumably kept in a small pouch attached to the sling or
belt for the bowcase. Cf. *Sylloge tacticorum*, 39.5, where these items,
along with a knife, glue, and "other tools" are kept "on the strap of
the quiver," perhaps in loops or a pouch. *Soublia* is from Lat. *subla*:
cf. Banfi, "Problemi di lessico balcanico," 18–19.

17   κοντάρια καβαλλαρικὰ μικρά   The lances have a cord or sling
attached so that they can be slung over the shoulder. This is again
directly from Maurice, who adds that it is a borrowing from the
Avars. It is unlikely, although not impossible, that it was an actual
feature of contemporary cavalry fighting in the ninth century and
after, since the later, nonderivative treatises make no mention of it,
although steppe cavalry such as that from the Khazars may have continued to employ it; see Gorelik, "Arms and Armour," pls. 11.1.1–3,
5–7; 11.4. Leo gives no dimensions for the lance, but the *Sylloge tacticorum*, 39.1 gives a measurement of 8 cubits (3.7 m) plus a lance head
of greater than one span (i.e., of 23.4 cm); why Leo refers to them as
"small" is not clear.

22–23   χειρομάνικα . . . χρήσιμον   From χείρ and Lat. *manica*. Although in
the *Strategikon* this probably refers to mailed or otherwise protective
gauntlets, by Leo's time the word appears to have been understood to
mean a lower arm guard. The terms *manikellion* and *cheiromanikon*
were often equivalent, see below on l. 123 and Kolias, *Byzantinische*

Waffen, 64 and n. 3. Leo alters Maurice, *Strategikon*, 1.2.3 to note that these should be worn "by those who may possess them." According to the *Praecepta*, 3§4.28–31, such arm guards were to be of thickly stitched silk and cotton, with mail attached, presumably wrapped around, creating a defense which would absorb shock and resist cutting and stabbing also.

27–29   §4   Not from the *Strategikon*, but Leo's own addition.

27   κλιβάνια   The *klibanion* was, as noted at Const. 5.19–22, an upper-body armor of metal or hide and of lamellar construction. According to the *Praecepta*, 3§4.27, they could be fitted with sleeves down to the elbows, probably of similar construction, or of splint-armor attached to a backing. From the lower edge of the *klibanion* a skirt or pleated arrangement of *kremasmata* could be hung, made of thickly woven coarse silk and cotton strips, and with strips of mail—*zabai*—attached, similar to the *manikellia*. Cf. McGeer, *Sowing the Dragon's Teeth*, 69–70; Kolias, *Byzantinische Waffen*, 47–48; on its construction and materials, see Dawson, "Suntagma Hoplôn"; and Tsurtsumia, "Evolution of Splint Armour." Richly decorated *klibania* were also worn by the emperor for ceremonial; see *CPTT* C.753 and 278–79.

28   περικνημῖδας . . . ποδόψελλα   See above, on Const. 5§3.23. Leg guards or greaves, also called χαλκότουβα (below, l. 123; Nikephoros, *Praecepta*, 3§4.37). In the latter case they might originally, as the name suggests, have been of bronze or metal; leg guards of wood or leather were also usual at this period. Tubular leg guards or greaves, of iron, have been excavated from a Khazar grave at Borisovskiy, dated to the eighth or ninth century (see Gorelik, "Arms and Armour," 135 and pl. 11.5.25, and cf. nos. 23, 24 for splinted greaves). The assumption is that such guards were constructed of splints attached to a backing of leather or cloth. See Kolias, *Byzantinische Waffen*, 70–74. For decorated greaves in imperial ceremonial contexts, see *CPTT* C.753 and 279.

29   πτερνιστῆρας   This is an odd detail in the context and the term occurs in no other treatise of the period, nor in Maurice; see Trapp, *Lexikon*, 7:1477; although reference to spurs and spurring horses on do appear in the oldest version of the epic *Digenes Akritas*, 1.128 (καταπτέρνισμα); 4.1178 (πτερνιστήρια). Spurs had been in use since Roman times, although there is some debate as to whether they were

widely employed before the eleventh century. See Vigneron, *Cheval dans l'antiquité*, 61 and Grotowski, *Arms and Armour*, 392–94. See Dixon and Southern, *Roman Cavalry*, 58–59 with further literature.

ἐπιλώρικα Discussed at 5.19–22, but it can be noted here that it was sleeveless and woven from coarse silk and cotton; see Nikephoros, *Praecepta*, 3§4.31, and Kolias, *Byzantinische Waffen*, 58–61. Ceremonial *epilorika* could be richly decorated: *CPTT* C.749–50 and 277.

30-34   §5   While Leo takes the first part of this paragraph directly from Maurice, the second sentiment, regarding the harm resulting from a lack of archery in the Roman army, is his own, and we may reasonably assume it reflects a genuine contemporary concern; see commentary to 4.139 above. In the *Strategikon*, however, the term *neoteros* meant a recruit (see Pertusi, "Ordinamenti militari," 664 and Haldon, *Recruitment and Conscription*, 24), whereas it is unlikely that it retained this particular meaning in Leo's day, given the established mode of conscription within the thematic provinces.

31   τοξοφάρετρα This is probably not a specific item of equipment but rather a collective term, and it is unclear whether Leo has actually understood the *Strategikon* here. It is used by Maurice to mean "bow and arrows" or "archery kit"; see *Strategikon*, 12.B.5.2 and 20.9, for example, where this meaning is apparent; and see on Const. 5.8 above.

37-39   §7   This additional weaponry for mounted soldiers is Leo's own addition; the *Strategikon* restricts javelins to the infantry. Given the lance, bow, and edged weapons carried by Leo's version of the cavalry trooper, two additional javelins might seem an unnecessary encumbrance. Later treatises describe the heavy or medium cavalry weaponry as consisting of all or some of lance, bow and sword, axe or mace.

40-43   Τοὺς δὲ . . . σέλλας Horse armor for the animal's head and breast, constructed from quilting or leather, probably lamellar. See Grotowski, *Arms and Armour*, 395–96. Contemporary steppe horse-armor offers good parallels, see Gorelik, "Arms and Armour," 135, 141, with pls. 11.3 and 4. Leo adds to the definition in *Strategikon*, 1.2.35–39 the detail about the protective hangings attached to the saddle, presumably reflecting his personal observations of the more heavily armed cavalry (of the *tagmata*?). The *Sylloge tacticorum*, 39.6 gives more detail: this armor should be of mail, or of iron or horn lamellar (or scale), and can also protect the horse's rump. From a somewhat later date, the *Praecepta* (3§5.38–45) describe two types of horse armor for the *kataphraktoi*: first, complete protection for the whole upper body of the mount (presumably in separate sections)—head, breast, flanks and rump—constructed of quilting and boiled leather, so that only its eyes and nostrils are revealed; second, an ox-hide *klibanion* to protect the animal's breast and forequarters, split to allow free movement of the legs.

47 αἱ σέλλαι   Cf. Trapp, *Lexikon*, 7:1538. There is no archaeological evidence for Byzantine military saddles, but pictorial evidence combined with written evidence has been used to arrive at a reasonably clear account of their appearance. Unlike the late Roman cavalry saddle, the Byzantine saddle had a developed—higher and broader—cantle and pommel, a result chiefly of the need to accommodate a different riding position with the advent of the stirrup (see on l. 49, below). The Byzantine saddle almost certainly was modeled originally on that of the Avars and, later, of the Khazars; see László, "Grabfund von Koroncó"; Hyland, *Medieval Warhorse*, 5–8; Dixon and Southern, *The Roman Cavalry*, 70–74; Grotowski, *Arms and Armour*, 383–86; and Vanderheyde, "Monture." For Khazar and Magyar saddle forms and decoration, with which the Byzantines would have been quite familiar, see also Gorelik, "Arms and Armour," pls. 11.9.44 and 45; 11.13.35, 36.

49 τὰς δύο σιδηρᾶς σκάλας   For the term *skala*, see Kraft, "Lat.-griech. *Scala*"; cf. Mihăescu, "Éléments latins," 1:494 and idem, "Littérature byzantine," 53; Trapp, *Lexikon*, 7:1558–59. Although mentioned in the *Strategikon*, and repeated by Leo, the stirrup does not appear in any Byzantine written source until the middle of the tenth century, where it occurs in a story related to a hunting accident in the 860s: Symeon *magistros*, 132.2 (p. 260.10). Archaeologically a few examples are known from middle Byzantine (9th–12th century) sites, such as at the rural site of Çadır Höyük in north central Anatolia: Cassis, "Rural Settlement," 5. In pictorial representation of military saints they appear on an eighth-century icon of St. Merkourios from Mt. Sinai: Grotowski, *Arms and Armour*, 380–82 (esp. 381, n. 10). Since the reference takes stirrups entirely for granted, it is reasonable to

suppose that their use was by this time widespread and unremarkable. Stirrups appear to have been introduced to Europe through the Avars in the last quarter of the sixth century, although there is some slight evidence to suggest that they arrived shortly before then: Ivanišević and Bugarski, "Étriers byzantins." The archaeological evidence is from Avar graves dating before ca. 600; see Garam, "Bemerkungen," 253, 258–61 and fig. 1. There is still some debate on the issue, but a late sixth-century date is now generally accepted, and there is an extensive archaeological literature; see Curta, "Earliest Avar-age Stirrups" and Werner, "Ein byzantinischer 'Steigbügel'." For a recent survey and discussion, see Genito, "Archaeology of the Early Medieval Nomads"; older literature: Bivar, "Cavalry Equipment," 286–88; Haldon, "Byzantine Military Technology," 22–23 n. 57; and cf. Halsall, *Warfare and Society*, 173–74. By the late seventh century stirrups had appeared in the Islamic world, but there is no evidence that they were adopted from the Sasanians; rather they appear to have moved eastward from the Roman empire into the Caliphate; see the summary of the literary and archaeological evidence in Kennedy, *Armies of the Caliphs*, 171–73. Stirrups have been found in substantial quantities from Khazar graves from the seventh to the tenth century. See Gorelik, "Arms and Armour," pl. 11.9.53–58; 11.13.37–46. Stirrups certainly enhanced riding and mounted combat skills, but they are no longer thought to have had a revolutionary impact. Indeed their original function was to assist in mounting; see Bachrach, "Animals and Warfare," esp. 737–48; Genito, "Archaeology of the Early Medieval Nomads," 235; and esp. Gillmor, "Stirrups and Stirrup Thesis," with further literature. The stirrup may have promoted the introduction or development of a new form of saddle, in turn likely originating with the Avars; see Grotowski, *Arms and Armour*, 383–84; Haldon, "Byzantine Military Technology," 23 and n. 57; and Garam, "Sepolture de Cavalli."

49–50   For the lasso, hobble, and saddlebag, see on 5§3.28–29.

50–51   τριῶν ἢ τεσσάρων ἡμέρων δαπάνην Logistical arrangements rarely receive any mention in medieval written sources, and this information, even though repeated directly from the *Strategikon*, is important. For the Byzantine military diet, see Kolias, "Eßgewohnheiten und Verpflegung," 197–99. The provisions carried in saddlebags were intended usually to obviate difficulties if the troops were unable to return to a camp or supply train during an action, and would include a kilo or so of twice-baked bread or *paximation* (*bucellatum* in the later Roman period), dried meat, and a canteen of water. See Haldon, "Chapters II, 44 and 45," 294–98 for grain and other provisions.

53–54 τζικούριν δίστομον   For types of battle-axe, see Kolias, *Byzantinische Waffen*, 162–72; and above, on Const. 5§2.15–18.

56–66 §§12–13 Leo changes the wording slightly from the text of the *Strategikon*, mentioning only linen and woolen clothing (Maurice also includes a type of sackcloth in his list), and omitting the late Latin terms γουννία ἤγουν νοβερονίκια (a type of large hooded cape; see Rance, comm. to *Strategikon*, 1.2.50) and describing the protective cape in question simply as a large and roomy *kentoukla* with wide sleeves, suggesting that the term *kentouklon* by itself can refer to an item of quilted clothing.

68 διὰ τὸ ἀναγκαῖον τῆς χρείας   Not simply to cover any contingency— this is also a reference to the need to dig latrines; see *CPTT* C.214–16, 308, and p. 213 and cf. *Hypotheseis*, 57.3.

69–77 Τοὺς δὲ . . . ἐπινοεῖν   Maurice, Strategikon, 1.2.62–63 has: Χρὴ ἀναγκάζεσθαι τοὺς στρατιώτας, καὶ μάλιστα τοὺς τὰ φαμιλιαρικὰ λαμβάνοντας, πάντως παῖδας ἑαυτοῖς ἐπινοεῖν ἢ δούλους ἢ ἐλευθέρους. Leo changes the terminology to accord with his own times (the *familiaricum* appears to be an allowance issued to assist soldiers maintain servants and family members; see Rance, comm. to *Strategikon*, 1.2), but it is not clear that by *tagmata* he means anything more than ordinary provincial units (which is the way the word is used throughout the rest of the treatise). Whether he means to differentiate here between the officers and soldiers of the *tagmata*, and those of the provincial units—*ta thematika banda*—remains uncertain, but unlikely, since Leo makes no other reference in the treatise to the palatine *tagmata* (although this does not necessarily mean he did not also understand them to be included in his general account). Where Maurice refers to the annual payment of the soldiers' *roga*, Leo makes reference to the time of both the *roga* and of the *adnoumion* or muster, at which the soldiers' equipment and their servants were registered or checked by the permanent officers of the *thema* or the appropriate officials in the *tagmata*. Leo's account fits with what little is known of the procedures by which the provincial soldiers were selected for duty and mustered from the eighth or early ninth century until the eleventh. See in general Haldon, *Recruitment and Conscription* and idem, "Military Service," 20–29.

Whether soldiers could furnish their own pack animals, as the final sentence suggests (following the *Strategikon* which, however, implies merely that they should have a certain number of pack animals per group or per individual), to be checked at the muster also, is unknown. All the (limited) evidence we possess suggests that the state supplied pack animals for imperial expeditions only (and then partly through impositions on individual officers and departments of the court); otherwise animals were conscripted or levied from the tax-paying population of the empire according to need. See *CPTT* C.59–135, with 160–62, 184–98. Given the emphasis in both Leo and later writers on selecting the better-off registered soldiers from the thematic registers for active service, it is not impossible that they also supplied their own pack animal. Later treatises certainly warn against marching with too many pack animals and servants, which both overstretches the resources available and can cause confusion and disorder in the army (e.g., *Campaign Organization*, §15; §17; Nikephoros, *Praecepta*, 1§17.166–71).

79–86   §16   Leo follows the *Strategikon* again in these details relating to pennants and standards. To what extent his information reflects the realities of the army of his own time is difficult to say, however, although it is corroborated by the *Praecepta* (4§7.76–84): each *bandon* should have a pennant, marked with its own recognizable symbol, and its spare horses held in reserve should also have one. The senior officers and commander also have their own banners (5.2.15–19). See McGeer, *Sowing the Dragon's Teeth*, 73; Dennis, "Byzantine Battle Flags"; Dufrenne, "Aux sources des gonfanons"; McGeer and Cutler, "Battle Standards and Flags"; Babuin, "Standards"; and *CPTT* 270–74 for unit and imperial banners and flags.

87–89   §17   The need to carry spare weaponry and especially arrows is obvious. Arrows were supplied both through impositions on local populations and from imperial storehouses: cf. Nikephoros, *Praecepta*, 1§14.137–43 in reference to "imperial" arrows; and cf. on Const. 5.3–9.

92   ὀπισθοκουρβίων   The cantle of the saddle, see Mihăescu, "Éléments latins," 1:493. On saddles and related equipment see Grotowski, *Arms and Armour*, 383–92.

96–103   Οὐδὲ τοῦτο δὲ ... ἀπορῶσιν   See also on 9§3.14–16 below. A frequently repeated concern is that the presence of soldiers should not harm the local population. Leo repeats, with minor changes (the introduction of merchants, for example) the same paragraph from the *Strategikon*, 1.2.86–91. It is not quite clear whether the implication is that the prices of the requisite commodities are too high for the troops to afford, or not available. In either case the *strategos* has somehow or other to provide for his soldiers by purchases from merchants. Accounts of the campaigns of Romanos IV in the late 1060s and early 1070s report the detachment of small units or troops of soldiers sent off to purchase corn, and of the rear guard lagging behind the main body of the army to protect those sent to purchase supplies: Attaleiates, *Historia*, 107.23–108.1, 126.14–15 (trans. Kaldellis and Krallis, 197, 231). In fact Leo seems to ignore the existing arrangements, managed by the *protonotarios* of each *thema*, whereby supplies and provisions were raised from the local population and (in theory) written off against the taxes owed from the following year, or were simply taken by compulsory purchase; see Haldon, *Warfare, State and Society*, 143–48 with sources and literature.

107–33   See Maurice, *Strategikon*, 12.B.4–5.

114–40   §§21–22   While Leo's account in the *Taktika* of the armament of the infantry is drawn from that of the *Strategikon* of Maurice, certain details not given in the older treatise suggest an attempt to describe contemporary arms and the sort of equipment which the average peasant conscript in the provincial armies might have afforded: the large, round shields, for example, as well as the battle-axes, which do not appear in the *Strategikon*. Leo lists only mail or lamellar armor for the picked men of each file (ll. 121–22), while the mid-tenth-century *Sylloge tacticorum* prescribes either mail or lamellar armor (the lamellar of either iron or horn), but adds that if this is not possible, then quilted garments of cotton or coarse silk should be worn. Further on in this constitution (ll. 136) Leo, like the later *Praecepta militaria*, lists only *epilorika* as defensive armor for infantry, the equivalent of the *kabadion* of the *Praecepta* (a quilted, knee-length tunic of coarse silk and cotton; see Kolias, *Byzantinische Waffen*, 55–57 and Dawson, "Suntagma Hoplôn," 81–82), representing a still more realistic appreciation of the situation; see McGeer, *Sowing the Dragon's Teeth*, 184–85, 204–5. By the same token, the *Praecepta* ordains that infantry be equipped with thickly quilted caps—*kamelaukia*—rather than iron helmets, in contrast to the helmets recommended by Leo. Leo retains the well-established and functional division between light and heavy infantry, while noting (ll. 107–10) that the ancients classed infantry in three groups: heavy, light, and "peltasts" (best rendered as "skirmishers"), the last of which was no longer to be differentiated from the second group.

114   On swords see on 5§2.9–10 above.

115   κοντάριν   The spear was the key weapon wielded by infantry. Leo gives no dimensions here, but at 5.12 (above) he gives a figure of 8 cubits, or 3.74 m (12 feet). The *Sylloge tacticorum*, 38.3 states likewise that spears should be 8–10 cubits long (3.7–4.7 m), with heads or blades of 1.5 spans (*spithamai*) (0.47 m or 13.5 in.), cavalry lances were 8 cubits in length with blades of over a span. The *Praecepta* of Nikephoros (1§3.29–31) and the *Taktika* of Nikephoros Ouranos (56.3.33–35) both give a longer dimension for infantry spears: they were to be 25–30 *spithamai* long (5.85–7 m). While such long spears have been doubted as impractical (see Kolias, *Byzantinische Waffen*, 192 and n. 42 and McGeer, *Sowing the Dragon's Teeth*, esp. 206), Leo the Deacon does note that Byzantine cavalry at this period did indeed use very long spears (see 143.19–21; trans. Talbot and Sullivan, 188; cf. 132.16–17 for the "very long" lance used by John Tzimiskes), and this may reflect a particular development of the period. McGeer suggests that Phokas is using—uniquely at this point in the text—a "small" *spithame* in this context. But this seems less than likely in view of Leo the Deacon's evidence. Note that at 5§2.12, Leo refers to "small spears, 8 cubits in length": the distinction is implicit, between these "small" spears, and longer ones about which he offers no further detail, but which clearly existed. On spears and spearheads, see Kolias, *Byzantinische Waffen*, 191–98. Although archaeological evidence for Byantine spearheads remains to be identified, material from neighboring cultures in the Balkans and on the steppe provides useful comparative material; see, for illustrative examples, Gorelik, "Arms and Armour," pl. 11.8.1–22. See also Dawson, "'Fit for the Task'," 7–10 and Grotowski, *Arms and Armour*, 318–29.

115–16   σκουτάριν . . . τέλειον.   The translation here is not exact, and should read "a shield of the kind called thyreos, oblong and large, but in any case (πάντως), a perfectly round (one)," taking πάντως in this sense, rather than as "altogether." Kolias, *Byzantinische Waffen*, 91 and n. 27, noted that there was a lack of clarity here, and that this was recognized by the author of the parallel passage in the *Taktika* of Nikephoros Ouranos (see Vári, *Tactica*, 1:115.286–87), who changed the wording from πάντως δὲ στρογγύλον τέλειον to εἰ δὲ μή, στρογγύλον τέλειον ("a shield of the kind called thyreos, oblong and large, but if not [then] a perfectly round [one]"). But this is merely a certain opacity of expression on Leo's part rather than either a contradiction or corruption in the text.

The fact that Leo uses the classical term *thyreos* (see also 14.625) may at first suggest no more than a simple classicizing tendency, since this is the term used by his exemplars Aelian, Arrian, and Onasander (e.g., Onasander, *Strategikos*, 20.667–68 (20§1) and Arrian, *Techne taktike*, 11.5–6); it may equally suggest the appearance of an item not described in his late Roman sources, for these large shields are somewhat problematic. The probably ninth-century Syrianos, *Strategy*, 16.5–9, lists similar shields, however, specifying a large shield of some 7 *spithamai* across (1.6 m or 63 in.), and, as we have seen, so do the later tenth-century treatises (see Rance, "Date," 723–29). Such a large shield would have been an unwieldy piece of equipment. It is not mentioned in late Roman historical sources (see Southern and Dixon, *Late Roman Army*, 99–101; and cf. Bishop and Coulston, *Roman Military Equipment*, 172–73), and must have been introduced after the period of the *Strategikon*, which contains no reference to such an item. The *Sylloge tacticorum*, 38.1 gives a clearer picture: Τῶν μὲν ὁπλιτῶν καλουμένων ἢ τετράγωσι ἔστωσαν αἱ ἀσπίδες εἰς στενὸν ἀποτελυτῶσαι κάτωθεν σπιθαμῶν ἓξ ἔγγιστα, ἢ καὶ τρίγωνοι πλὴν ἀνδρομήκεις σχεδὸν κατὰ τὰς τετραγώνους, ἢ καὶ στρογγύλαι σπιθαμῶν τριῶν πρὸς τῇ ἡμισείᾳ ("Let the shields of the so-called "hoplites" be either four-sided, narrowing at the base to six spans at the narrowest, or three-sided but of the height of a man, like the four-sided shields, or oval and (with a radius of) three and a half spans" [author's trans.]), and cf. *Sylloge tacticorum*, 45.33 for such tall shields "which some also call *thyreoi*." A width of six spans for infantry shields is corroborated by the *Praecepta*, 1§3.28–29, which adds that they should be even larger if possible. A "span" measured 9 inches, so that these shields would be some 54 inches (1.40 m) across, or even slightly larger; see Schilbach, *Metrologie*, 16–18. For shield types and shapes at this time see Kolias, *Byzantinische Waffen*, 105–14, and on shields in general 88–131; also Bruhn-Hoffmeyer, "Military Equipment," 84–90. Leo gives no dimensions, but at different points mentions their characteristic features: they are large, called *thyreoi*, heavy; they cover the whole of the body, can be four-sided, and are the height of a man (cf. 6§31.204, 14§91.624–26, 20§183.919–20). In the passage already cited from the *Sylloge tacticorum* (45.33), the description of their deployment in a defensive formation is based on a similar passage in the *Strategikon* (12.B.16.33–38), but whereas in the latter the shield wall is made up from two ranks, with the second rank leaning forward and resting their shields on the shield bosses of the front rank to create a two-tiered wall of shields, in the *Sylloge* the front rank alone closes up with overlapping *thyreoi*.

The origin of such large shields is uncertain. The empire's main enemies carried relatively small or standard-sized shields—see Kennedy, *Armies of the Caliphs*, 171—although Fatimid soldiers carried shields the height of a man, according to the tenth-century *De administrando imperio*, 1: chap. 15.13; the *Strategikon*, 11.4.45 reports that the Slavs employed very strong shields, "difficult to carry." It is quite possible that such large shields were adopted by Byzantine infantry, largely recruited from the rural peasantry of the empire during the eighth and ninth centuries, as a means of offering additional protection and support for what were regarded as second-rate soldiers with very limited equipment.

That the shields were to be painted uniformly by *bandon* or *tagma* is confirmed in the *Sylloge tacticorum*, 38.2, and a great deal of other evidence shows that shields or the leather covers stretched across them were decorated both by coloured fields and by specific symbols, both religious and secular. See Kolias, *Byzantinische Waffen*, 124–31, and on shield construction, development, and decoration, see Grotowski, *Arms and Armour*, 208–53, and Babuin, "Standards."

117–18 κασσίδα τούφιν μικρὸν ἔχουσαν Leo copies Maurice, *Strategikon*, 1.2.24–25; 12.B.4.3–4 for both cavalry (see above, ll. 11–12) and infantry in respect of the various decorative tufts or plumes attached to the soldiers' armor, repeated also in the *Sylloge tacticorum*, 38.5; 39.1, 3. Apart from tufts or plumes attached to helmets (corroborated by archaeological evidence for the plume-holders on the crown of helmets; see on Const. 5§3.22 above) such decorations, which may also have served to differentiate one unit from another, at least where permanent divisions existed, were also attached to the shoulder-pieces of body-armor. See Kolias, *Byzantinische Waffen*, 61–64. Leo's prescription is probably optimistic: *Praecepta*, 1§3.23–24 makes no mention of iron helmets for infantry at all, assuming they will have a protective helmet of quilted silk and cotton, called a *kamelaukion*, held in place with a binding wrapped around the skull. Given the social and economic status of foot soldiers already alluded to above, this is a more likely form of head cover at this time than Leo's helmets, except perhaps for tagmatic units funded and equipped by the state directly. Such an item is reflected in a similar padded cap of the eighth or ninth century from the northern Caucasus region, covered in decorated silken fabric, with a neck guard such as is described in the tenth-century treatises for the majority of infantry, although it may well also have served to cover a helmet; see Kirpichnikov, *Drevnierusskoe Oruzhie*. On *kamelaukion* (which had a variety of meanings), see Kolias, *Byzantinische Waffen*, 85–87 with sources and literature.

118 σφενδόβολα Slings were a universal infantry weapon, for both light and heavy infantry, although the treatises differ in ascribing them to one or the other or both. The *Sylloge tacticorum*, 38.10 remarks that the slings of the light infantry should be no shorter than 6 spans long, that is to say, 1.40 m or 54 in. See Kolias, *Byzantinische Waffen*, 254–59.

123   κατὰ δὲ . . . φλαμουλίσκια μικρά   The term *mela* refers to shoulder pieces, possibly with a defensive/protective function. See Kolias, *Byzantinische Waffen*, 41–42.

124-25   χειρομάνικα . . . χαλκότουβα   See on 1.28 above.

129   κουκούρα μεγάλα   See on 5§2.9 above. The *Sylloge tacticorum*, 38.8 agrees with Leo on the number of arrows to be carried (30–40), whereas the *Praecepta*, 1§4.32–39, ordains that archers should carry two bows, two quivers of 40 and 60 arrows each, as well as their small shield and other weapons.

129-30   σωληνάρια ξύλινα   See on 5.27 above. Their arrows or darts were useless to the enemy because they were too short to be loosed from an ordinary bow without the wooden arrow guide, so could not be shot back again.

131-32   βηρύττας, ἃ λέγεται ῥικτάρια   See on 14§60.437. Leo copies Maurice's term, derived from the Latin *verutum*, but by the late ninth century it was certainly obsolete, and replaced by the generic "*riktarion/ riptarion*" for any javelin or spearlike missile. The *Sylloge tacticorum*, 38.6; 39.8, gives the length of the infantry and light cavalry *akontion* as ca. 2.7 m. The *verutum* appears to have been originally a short heavy javelin of 3.5 feet in length used by heavy infantry (see Vegetius, *Epitoma*, 2.15.5–7, although there remain some unresolved problems in respect of his terminology), but the term clearly had lost this technical meaning by Leo's time. See Bishop and Coulston, *Roman Military Equipment*, 79, 202. Leo (Const. 7§3.24–25) uses this term as an equivalent also for the *martzubarboulon* or *matzobarboulon* (= Lat. *mattiobarbulus*), originally a short, lead-weighted javelin or *plumbatum;* see Vegetius, *Epitoma*, 1.17, 2.15.4, 3.14.10. For the etymology of the term see Mihăescu, "Littérature byzantine," 59–60 (and cf. Trapp, *Lexikon*, 2:274, 277); Buora, "Nuovi studi"; and Grotowski, *Arms and Armour*, 318–19. For the use of the weapon, see Bishop and Coulston, *Roman Military Equipment*, 200–202; and Rance, *Strategikon*, notes to Maurice (*Strategikon*, 12.B.22–25). Both terms are taken directly from the *Strategikon*, and since neither word occurs in other tenth-century treatises, we may assume that Leo is indulging his antiquarian interest at this point.

133-39   Maurice, *Strategikon*, 12.B.1.

133   No dimensions for the light infantry shields are given by Leo, nor in the *Sylloge tacticorum*, 38.10. At *Sylloge tacticorum*, 39.8, however, the light cavalry shields are given as either oblong or circular and of 3 or 4 spans—so from 0.7–0.94 m (27–36 in.)—in width/length. The light infantry shields are likely to have been of similar dimensions. The *Praecepta*, 1§4.36–37 refers to the small shields carried by the archers as *cheiroskoutara*, since they can be strapped and held to one arm while still allowing the soldier to use his bow. See Kolias, *Byzantinische Waffen*, 109–10.

134–39   Here Leo inserts the opening section from the *Strategikon* on infantry (12.B.1) and changes the subject back from light infantry to all infantry. Apart from the better soldiers who were to stand in the front two ranks (see on ll. 114–39 above), infantry soldiers were to be equipped with *epilorika* (only) as defensive equipment, apart from their shields and helmets. Their knee-length tunics are common to all the treatises (*Strategikon*, 12.B.1.2–3; *Sylloge tacticorum*, 38.11; Nikephoros, *Praecepta*, 1§3.14–16)

136–40   τὰ δὲ ὑποδήματα ... χρήσιμόν ἐστιν   Taken with some changes in vocabulary from Maurice, *Strategikon*, 12.B.1.3–8. The emphasis is on practicality—they should be studded with hobnails and square-toed (or at least not pointed or long-toed). The *Sylloge tacticorum*, 38.11 notes for the light infantry, in a comment clearly taken from this passage of Leo, that their footwear should be μετρίως καθηλωμένα, "lightly nailed," implying that the boots or footwear of the heavy or regular infantry was more heavily hobnailed. The *Praecepta* describe boots for the infantry with protective leggings which could be folded or rolled up to the thigh or doubled around the shin and calves to afford protection. See McGeer, *Sowing the Dragon's Teeth*, 62. Attaleiates describes Byzantine cavalry soldiers wearing high boots or leggings of leather which had to be cut down when they were forced to march or fight on foot; see Attaleiates, *Historia*, 41.4–8 (trans. Kaldellis and Krallis, 73). In general on military footwear, see Grotowski, *Arms and Armour*, 191–203.

It is impossible to know whether hair regulations were actually enforced: short hair was in ninth-century hagiography regarded as a sign of iconoclast leanings, or at least loyalty to the iconoclast

emperors among members of the "older generation" of palatine officials (who could be identified by the fact that they shaved themselves
very closely) as well as of the population at large; see Speck, *Kaiser
Konstantin VI*, 603–4 and n. 87a; Ševčenko, "Hagiography," 116 and
n. 18. See, also *Vita S. Stephani iunioris* (ed. Auzépy), 137.21–138.9 and
commentary, 233, according to which Constantine V had ordered
court officers to shave closely. That this was a style through which
supporters of imperial iconoclasm could be identified is borne out by
a later hagiography, in which the imperial officer Kallistos is disciplined, while at his post in the palace, by the emperor Theophilos for
his unkempt appearance and unshaven beard: *De XLII martyribus
Amoriensibus*, no. 2 (de Callisto), 22–36; see 24.30–25.2. Adult emperors remained bearded, of course, on their coinage.

142–49   Maurice, *Strategikon*, 12.B.6.

142–44   ἀμάξας ... ἀργῶσιν   While Leo (and *Sylloge tacticorum*, 38.12), following the *Strategikon*, 12.B.6, prescribes carts for each tent group of
eight infantry, the late tenth-century treatises suggest that mules and
pack animals were the norm for transporting the unit baggage and
supplies. See for example Nikephoros, *Praecepta*, 1§§14.137–39, 17.166–
67; 2§1.5–9; 5§5.47–52; and *Campaign Organization*, 9.35–36; 10.15–
16; 31.14–23. See also *CPTT* C.332–70 and commentary, 235–37. The
Praecepta (2§1) stipulates one mule for two infantry soldiers, which
would suffice to carry all their equipment and provisions for up to
24 days at least. Wagons or carts were certainly employed at times,
especially when a siege-train was required to move heavy equipment,
but for the most part pack animals were the norm by Leo's time. His
reference to between ten and twenty carts carrying extra arrows along
with hardtack and millet may be realistic, but would have slowed an
army down greatly unless, as discussed below (ll. 155–59) and later
in the *Taktika*, the fighting troops or some of them were sent on at
their own faster pace. The logistical problems associated with large
numbers of carts would have been considerable. For detailed discussion see Haldon, *Warfare, State and Society*, 158–66; and below on
Const. 10 on the baggage train or *touldon*.

145–54   χειρομύλιν ... περισσοῦ   The list of equipment and supplies is taken
directly from Maurice, although it receives some corroboration from

contemporary sources. The treatise on imperial military expeditions compiled at the order of Constantine VII in the 950s includes a short note to the effect that thematic commanders must issue orders to their *tourmarchai* and on to their *droungarokometes* to ensure that every *bandon* had its bootmakers and smithy as well as its "set of tools," which included an axe, an adze, and a chisel and mallet. The same order was issued to the commanders of the *tagmata* to be passed on down to the respective units: *CPTT* C.653–64.

The hand mill was an essential item, since the soldiers had to produce their own flour for bread and hardtack (see also *Sylloge tacticorum*, 38.12; and Maurice, *Strategikon*, 12.B.6.5): Hand mills can be operated with considerable efficiency: experiments with hand mills from the Roman fort at Saalburg demonstrated that 4–6 men could mill up to 220 pounds of grain into flour in one hour: *Saalburg-Jarhbücher* 3 (1912): 75–95 and Erdkamp, *Hunger and the Sword*, 35 with n. 37.

153–54   πίστον καὶ παξαμάτιν   Cf. below on 13§12.58–66. Soldiers were issued with two main varieties of bread or flour: wheat flour for simple baked loaves or double-baked hardtack, referred to in late Roman times as *bucellatum* and by the Byzantines as *paximadion* or *paximation* (see Trapp, *Lexikon*, 6:1203). In campaign conditions, it was normally the soldiers themselves who milled and baked this. The hardtack was more easily preserved over a longer period, was easy to produce, and demanded fairly simple milling and baking skills. In addition πίστον (cf. Lat. *pistum*, "pounded"), which sometimes occurs simply to refer to grains in general, usually meant millet: Trapp, *Lexikon*, 6:1306; Du Cange, *Glossarium Graecitatis*, 1173–74; Sophocles, *Greek Lexicon*, 892, πίστον, "coarsely ground millet"; Mihǎescu, "Éléments latins," 1:489–90; Kolias, "Eßgewohnheiten," 198–99. The version in the *Sylloge tacticorum*, 38.12, 39.3, which is taken from Leo, lists *paxamata* together with κέγχρος ἀληλεσμένη, ground millet, which was used to make a porridge-like meal; see Kolias, "Eßgewohnheiten," and more generally on the Byzantine military diet; see also Davies, "Roman Military Diet." Note that Nikephoros Ouranos (*Taktika*, 10.13, following Leo more or less exactly), explains the term as: πίστου, τουτέστι κέγχρου, ἐκλελεπισμένου ἐλαφροῦ, "*pistum*, that is millet, light from being husked." Hardtack could be baked in field ovens—*klibanoi*—or simply laid in the ashes of campfires, an advantage when speed was essential. The *Sylloge tacticorum*, 57.2, notes that the best such bread was baked in thin oval loaves cooked in a field oven, and then dried in the sun (although the passage derives ultimately, via the probably tenth-century compilation known as the *Apparatus bellicus*, from the *Cesti* of the second–third century CE writer Julius Africanus).

155–59   §24 Various ratios of mules or pack animals to soldiers are provided in the sources (and see above on ll. 142–44). A rate of 1 mule or pack animal for 10 men appears to have been standard when troops needed to move ahead of the main train quickly. The average standard supplies for 10 men for 8–10 days (of ca. 1.3 kg per man per diem) implies a maximum required load of some 130 kg, excluding fodder for the pack animal. This is a full load for a medieval mule, whose average capacity ranged around the 95–114 kg mark, and we can reasonably assume that in such circumstances the soldiers would not be issued a full maximum daily ration. For detailed analysis of pack animal loads and food and fodder requirements, see Haldon, *Warfare, State and Society*, 166–70, 281–92; with discussion in Erdkamp, *Hunger and the Sword*, 27–83; Kroll, *Tiere*, 171–72.

160–89   Here Leo copies or paraphrases Aelian, *Tactica theoria*, 2.11–13; cf. also for ll. 160–68 and 178–89 and Arrian, *Techne taktike*, 4, although Leo did not use the latter; see chap. 2, pp. 48–50. See also Const. 7§§67–69.

171   τὰ λεγόμενα νῦν μέναυλα The exact meaning of the term has caused some discussion, primarily because it appears to have changed meaning at some point after the *Taktika* was composed. Leo uses it to mean simply a lance or spear, which could be used in hand-to-hand combat (11§22.133, 135; 19§14.87; §15.100; §16.107; §69.385), although at 11§22.133 and 135 there is an implication that *menaulon* can refer to a stouter spear of some sort; or that it can be thrown, like a javelin (9§71.358). Here it is clearly a javelin or throwing-spear of some sort that can be hurled from the saddle, so certainly not any sort of heavy pike. By the 930s, in contrast, it meant a somewhat more specialized weapon, longer and sturdier, a type of pike up to 3.5 m in length, employed by heavy infantry as a defense against cavalry (*Sylloge tacticorum*, 38.3; 47.16, 22), and this is the meaning it retains in the *Praecepta* of

Nikephoros (1§8.83). See McGeer, *Sowing the Dragon's Teeth*, 209–12 for discussion; cf. Trapp, *Lexikon*, 5:998. Such a weapon clearly could not be thrown very effectively, and Leo may, at Const. 9.358, be echoing the original meaning of the term, which derives from Lat. *venabulum*, a type of heavy hunting javelin: Kolias, *Byzantinische Waffen*, 194–95; Grotowski, *Arms and Armour*, 320–23.

192–215   See Aelian, *Tactica theoria*, 2.7–9; cf. Arrian, *Techne taktike*, 3, 11.5–6; Onasander, *Strategikos*, 20 (20). Leo is hopelessly confused with regard to Macedonian armament, blending classical hoplite with Macedonian phalanx equipment.

218–29   Aelian, *Tactica theoria*, 12, 14.1–3.

CONSTITUTION 7

This constitution is drawn largely from the *Strategikon*, with some passages from Aelian's *Tactical Theory*, from Onasander, *Strategikos*, and possibly also Syrianos, *Strategy*. Leo rephrases much of what he takes from these authorities, but only very lightly, occasionally glossing an older or obsolete term with the contemporary word. He may also have been influenced by the section on tactical order and discipline in the *Rhetorica militaris*, 41.1–4; and note 40.2: καλὸν γὰρ πρότερον προπαρασκευασαμένους τὰ ὅπλα, ὧν χωρὶς ἀδύνατον πολεμεῖν.

Order and discipline, even if frequently compromised by poor leadership or inadequately trained troops, remained throughout the empire's history part of the Byzantine self-image and an important aspect of military organization. Tactical cohesion and order were among the foremost concerns of the commander. The treatises repeat the same injunctions and instructions because they remained entirely appropriate (e.g., Nikephoros, *Praecepta*, 2§§4–12; 4§§11–12). The real issue was the extent to which individual commanders were able to apply them effectively and to train their soldiers properly. The *Taktika*, as we have seen, is clearly an attempt to establish a widely applicable system that all the emperor's senior commanders would apply in practice. The tenth-century historians' accounts corroborate the treatises of the period in stressing these aspects. The order and cohesion of Roman forces is mentioned by several writers, as in the battle before Tarsos in 965 (Leo the Deacon, *Historia*, 58–59; Talbot and Sullivan, 106–8), at the battles of Dorostolon (Leo the Deacon, *Historia*, 139–53; Talbot and Sullivan, 184–96), between the well-ordered battle line of the Roman forces under Isaac I Komnenos in 1059 and the Pechenegs (Psellos remarks on the Pechenegs' dismay at the unbroken line of Roman shields facing their assault: Psellos, *Chronographia*, 7.70), or the battles fought by Romanos IV against the Turks in 1070 (where, in spite of the criticisms of poor generalship and lack of discipline among the troops made by the contemporary eyewitness reporter of the events, Michael Attaleiates,

Roman units seem to have still fought and marched with order and cohesion: Attaleiates, *Historia*, 114, 126, 160; trans. Kaldellis and Krallis, 209, 229-231, 293). Michael Psellos remarks on the superb discipline of Bardas Skleros's cavalry, who were able even under heavy attack and while withdrawing to wheel about, countercharge, and drive their enemies from the field: Psellos, *Chronographia*, 1.13. The harsh tactical discipline imposed on his forces by Basil II was singled out for praise by Psellos, but was exceptional only in its rigor (*Chronographia*, 1.33 [trans. Sewter, 26]); and he notes with some contempt that the emperor Romanos III thought numbers counted for more than skill and discipline (*Chronographia*, 3.8 [trans. Sewter, 43]); Choniates and Kinnamos refer on several occasions, as do the historians of the tenth and eleventh centuries, to the training and exercising of the troops in field tactics and fighting skills, which may have given them a slight advantage, when well-led, over their enemies: Choniates, *Historia*, 12, 29–30, 77; Kinnamos, *Epitome*, 126. Alexios I imposed a similar order on his battle line, in which no one was to advance in front of the line as it moved, and where cohesion and solidarity were the key elements (Anna Komnene, *Alexiad*, 7.3 [trans. Sewter, 224 / Frankopan, 193]). Whether or not Byzantine commanders did indeed impose order and discipline, it was certainly seen as a key to success and as something the good commander should strive to achieve. Note the remarks of the author of the *Vita Basilii imperatoris*, 36.10–32 (132–34), reflecting the views of a mid-tenth-century writer, and especially 36.18–25: "Were it possible for everyone to learn military science or art without study and considerable practice, authors of works on tactics who devote so much labor to this topic would be merely ranting senselessly, as would the greatest among emperors and generals with many triumphs over many enemies to their credit: for none of them ever dared attack the enemy ranks with an undrilled and untrained army."

3–7   §1   Leo's own cautionary words about the need for training and exercise. For a Russian translation with some notes: Kučma, "Metodika."

8–18   Onasander, *Strategikos*, 9.324–33 (9§§2–3).

18–20   Onasander, *Strategikos*, 10.337–40 (10§1).

21–28   §3   Maurice, *Strategikon*, 12.B.2–3, except that Leo adds the phrase about the rough and even ground. Cf. Onasander, *Strategikos*, 10.362–79 (10§§4–5).

25   σαλίβα   Possibly of Arabic origin: *saliba* = "cross." A general term for a mace or club; see Leo VI, Ἐκ τῶν τακτικῶν τοῦ βασιλέως Λέοντος

τοῦ σόφου, §38: ὅτι μαρτζοβάρβουλον ἐλέγετο ἡ νῦν σαλίβα. And Kolias, *Byzantinische Waffen*, 176–78. At Const. 6§22.133–34 (= Maurice, *Strategikon*, 12.B.5.8); 6§23.146 (= *Strategikon*, 12.B.6.6); 7§41.290 (= *Strategikon*, 12.B.12.12); 7§55.384 (= *Strategikon*, 12.B.16.45); 9§56.282 (= *Strategikon*, 12.B.20.10); 14§75.522 (= *Strategikon*, 12.B.18.12), Leo regularly replaces *martzobarboulon* in the Strategikon with the word *tzikourin*, an axe. Kolias, loc. cit., therefore suggested that the term might also be employed to mean a type of axe. It seems more probable that Leo was not in fact sure what precisely a *martzobarboulon* was, and it occurs only once, at this point, in the *Taktika*. See Trapp, *Lexikon*, 7:1524.

29–40   §4   Maurice, *Strategikon*, 1.1.5–18.

41–45   §5   Almost certainly παρακοντακίου should be read as παρακονταρίου and related to the exercise of the same name. See Dennis, *Taktika*, 107 n. 3.

46–53   §6   Cf. *Sylloge tacticorum*, 9.1–3; *Strategikon*, 1.9.3–5, 47–54; and Const. 9§§2–4.

51   κενά τινα   Read καινά τινα ("novel . . . thoughts"). Cf. 9§4.20.

54–97   §§7–14   Onasander, *Strategikos*, 10.342–79 (10§§2–5)

76–83   §10   The difference between mock battles and large-scale training exercises is not always clear, and both were a regular aspect of ancient and medieval military culture; see for a general survey and discussion of the relevant sources, with extensive literature: Rance, "Simulacra pugnae," for the tradition and the ways in which it was incorporated into Maurice in particular; and Wheeler, "Legion as Phalanx, Part I," 314. For later Byzantine examples, cf. the battle staged by Heraclius in 620/21, before his first campaign against the Persians: cf. George of Pisidia, *Exp. Pers.* 2.120–62 (repeated by *Theophanis Chronographia* 304.3–11, trans. Mango and Scott, 436); or the mock battles and training that Nikephoros II put his troops through in 963 before his campaign against Tarsos (Leo the Deacon, *Historia*, 35–36; Talbot and Sullivan, 87–88); references to training exercises under Basil I: *Vita Basilii imperatoris*, 36 (132–34). The anonymous late tenth-century treatise *Campaign Organization*, 28 and 30 likewise emphasizes the critical importance of regular training, both in combat and tactical skills as well as in activities such as setting up and deploying out of an encampment, as does the treatise on *Skirmishing*, 19.23–28. For the tenth century in particular, see McGeer, *Sowing the Dragon's Teeth*, 217–21. Mock battles using weapons with their blades or heads removed are also a part of this tradition: cf., for example, Arrian, *Techne taktike*, 34.8 and 40.4; Vegetius, *Epitoma*, 1.15.1; with Rance, "Simulacra pugnae," 262–64.

80   τὰ λεγόμενα χαρζάνια   Cf. Onasander, *Strategikos*, 10.362–75 (10§4) and see Du Cange, *Glossarium Graecitatis*, 1733. In Onasander (l. 364) the equivalent phrase is "leather straps" (ἱμάντων ταυρείων), in turn derived from Xenophon, *Cyropaedia*, 2.3.17–18; see Schellenburg, "Einige Bemerkungen," 189–90. *Charzania* appear also at *De ceri-moniis*, 623.12 and 624.6, where the word appears to refer to an item of headwear or something placed over the head and onto the shoul-ders, thus perhaps a similar meaning. In S. Italian and Sicilian Greek dialect the word Χαρζανίτης appears as a family name, derived from *Charsianites*, i.e., from the region of *Charsianon*, although this is not secure and may refer to a trade—a *charzanites* would (prob-ably) be a leather-worker or strap-maker. Cf. Caracausi, *Lessico greco della Sicilia*, 617. Note the ninth-century Constantinopolitan family of *Monocharzanites*: Skylitzes, 84.93; trans. Wortley, 85 (or *Morocharzanoi* in Theophanes continuatus, 154.16).

88–92   §12   Onasander, *Strategikos*, 10.369–75 (10§5)

93–95   §13   Cf. Onasander, *Strategikos*, 10.376–79 (10§6)

98–108   §15   Maurice, *Strategikon*, 7.B.17.48–59 (and cf. 3.5.123–26); but see also *Rhetorica militaris*, 41.3–4 on the importance of not getting ahead of the line or breaking ranks to pursue the enemy.

109–16   §16   This paragraph is Leo's own, albeit echoing the opening words of Maurice, *Strategikon*, 1.2. Cf. *Taktika*, pr.65–71 and Const. 7§46.

117–248   §§17–35   Maurice, *Strategikon*, 3.5 (and probably based on a now-lost late Roman drill-book; see Rance, *Strategikon*, notes to bk. 3). These paragraphs (§§17–26) represent the clearest example of Leo's promise at pr.§6.68–70, to translate obsolete or obscure Latin terms into com-prehensible Greek. At l. 130 ἐν ἀραιοτέροις ... διαστήμασι should read: "in more open intervals"; and at 1.131 ἐξ ἴσου περιπατεῖτε should read: "At an even pace, proceed." Again Leo changes the wording in places—at the beginning of a new set of exercises or commands, for example; or in terms of replacing the older terminology (e.g., ll. 206–7, 224, where words such as *moira* and *meros* are replaced by *droungos* and *tourma*, etc.). His translation of the Latin depends on the paraphrases in the *Strategikon* of the commands to be issued and carried out, as well as on the fact that the MS of the *Strategikon* which Leo had at his disposal was often garbled or faulty. For a detailed discussion and analysis of these commands and the movements described, see Rance, *Strategikon*, commentary to *Strategikon*, 3.5; and Rance, "De Militari Scientia." On exercising and training cavalry mounts, see Hyland, *Medieval Warhorse*, 28.

141 ὀπίσω τοῦ ὀρδίνου   Translate "at the rear of the file."

148 ὅτ' ἂν ἡ τοξεία ἄρχεται γίνεσθαι   *Pace* Dennis, this refers to the archery of the enemy: "when the (enemy) archery commences," i.e., when the unit comes within their range.

153 τριπόδῳ μόνῳ ἤγουν κινήματι συμμέτρῳ   Maurice, *Strategikon*, 3.5.33 simply has τριπόδῳ. The word is best rendered as "canter," the only gait with three footfalls: τρίποδον and κάλπη are equated in the hippiatrical tradition; see Adams, *Pelagonius*, 4–5, 60, 209–38, and esp. 598–602. There is considerable lack of clarity on these terms, so that Du Cange takes the word to mean "trot" (*Glossarium Graecitatis* 1608), as does *LSJ* s.vv. κάλπη, καλπάζω: "trot"), followed by most modern authorities (e.g., Maurice, *Strategikon*, German trans. at 157; McGeer, *Sowing the Dragon's Teeth*, 305–6). Standard late Roman practice, as described in the *Strategikon*, was to attack over a short distance and at a canter; see Rance, "Battle," 368. The *Praecepta*, 4§13.142 likewise order that the heavy cavalry wedge should attack at a canter, not at a trot, using the phrase βῆμα τριπόδος.

157 προκλάστας   Cf. Const. 12.204 (= *Strategikon*, 2.5.23), an alternative for the *koursores* of the *Strategikon*; or the term Leo refers to as the "modern" word *koursatores*, although of course it occurs in ancient treatises also—see on 6.96–106.

163 σαγιττοβόλον   The term "bowshot" was in common use as a unit of (approximate) measurement, although of course bows of different strength, or differently strung, would achieve different distances. The bowshot has been explained as the flight distance, within which an arrow might have varying penetrative effects depending on a range of factors; see McLeod, "Range of the Ancient Bow," where the average distance of a bowshot for a composite reflex bow (the type of bow used by east Roman armies and their adversaries on the eastern frontier) is calculated as 290–335 m. An explicit definition is found in the tenth-century *Sylloge tacticorum*, 43.11, where an "average bowshot" (ἡ συμμέτρος τόξου βολή) is 156 *orguiai*, or just under 330 m (see Schilbach, *Metrologie*, 42). See McGeer, *Sowing the Dragon's Teeth*, 68; and Rance, *Strategikon*, notes to 2.4.9.

192   πουβλικίζεσθαι   See Maurice, *Strategikon*, 3.5.78. Leo simply copies Maurice's term (see also 7§29.204 = *Strategikon*, 3.5.83: προπουβλικίζεσθαι), but while it was not in common use, he or his redactors may have known it from the legislative language of the novels and the legal handbooks with which he was certainly familiar. See Trapp, *Lexikon*, 6:1359.

211–12   ὥστε τὴν συμβολὴν πρὸς αὐτοὺς εἰκάζεσθαι   For an alternative interpretation based on the way in which the author of the *Strategikon* described these exercises, see Rance, *Strategikon*, note to 3.5.92–93, who translates it as "so that they can use them to estimate the course of their attack." The ten cavalry troopers positioned ahead of the main body of the *bandon* act, in effect, as a target for the charge of the bulk of the unit.

253   διωρίσασθαι   Again recalls the legislative aspect of his *diataxeis* or constitutions.

254–461   Maurice, *Strategikon*, 12.B.11–16.

266   ὁ καμπιδούκτωρ ἤγουν ὁ ὁδηγὸς τῶν τόπων   An indication of the obsolescence of the function and title in Leo's time is that he explains it in this way, perhaps misled by the term *ducator/doukator*, referring to those sent ahead of an army to scout out routes and possible encampments. In fact, the *campidoctor* had been up to the late sixth century at least the senior drillmaster (see on 4§52.214–16 above). Leo adheres to the traditional file-structure of the *Strategikon*, with files of 16 infantry = two infantry *kontoubernia* (although he also accepts that such groups can be of five or ten cavalry, or four or eight or sixteen infantry, according to the situation); see 4§40.157–58 and on 4§66.272–73. Such a 16-deep infantry formation was almost certainly obsolete by Leo's time, and even if the reformation of infantry tactics

carried through in the middle decades of the tenth century had not yet been initiated (see McGeer, *Sowing the Dragon's Teeth*, 202–10; Haldon, *Warfare, State and Society*, 217–22), a 16-deep formation for formed troops seems unlikely, although on occasion such deep formations may have been necessary to combat enemy cavalry. It is clear, however, that Leo repeats the information here, as elsewhere, because the principles around which these maneuvers are based will nevertheless apply in practical military tactical terms (even if the actual formations derive from Hellenistic models). In Leo's time it was already becoming evident that well-trained heavy infantry were a key component of a successful offensive force, hence the general interest in and emphasis upon ancient, Hellenistic, and/or Roman infantry tactics. Note that the *Praecepta* state that "such formations are no longer employed and this type of phalanx is impractical." Instead, the heavy infantry line is to be 7 or 8 deep, with two lines each of two heavy infantrymen separated by 3 ranks of archers (Nikephoros, *Praecepta*, 1§§7.62–65, 8.78–79) or a similar arrangement together with a line of light troops with javelins (1.9.94–95), but the lines can be doubled up for specific tactical purposes, as the same treaise goes on to explain.

269–75   §39   See 12§§51–52.

276   Μὴ συμπλέκεσθαι δὲ αὐτὸν (sc. τὸν ἄρχοντα) τοῖς ἐναντίοις   See 14§3 and commentary on 14.690–702 (§§99–100); and cf. 20§§153, 159, 193.

290   βαρδούκια   Another, and by the tenth century standard, term for maces, also referred to as σιδηροραύδιον or σιδηροράβδιον. See Kolias, *Byzantinische Waffen*, 176–77, 181; Trapp, *Lexikon*, 2:265.

304–12   §43   A major concern was to prevent over-hasty and careless pursuit of fleeing enemy troops, both because of the dangers of an ambush or false retreat, but also because of the potential for indiscipline and the breakup of the army. See also 20§82.

322–24   §46   Cf. Prolog. 6; 7§§17–35.

325–461   §§47–65   Although derived largely from the *Strategikon*, and hence also Aelian, it should be recalled that the late ninth and early tenth centuries seem to have seen a real revival of interest in infantry formations, so that the information from older treatises, whether Maurice or earlier *exempla*, is not merely antiquarian. As the evolution of infantry tactics demonstrates—exemplified in the somewhat later *Sylloge tacticorum* and in the *Syntaxis armatorum quadrata*—in this respect Leo's *Taktika* was already indicative of a significant shift in Byzantine approaches to the ways in which cavalry and infantry were to operate together.

335   φούλκῳ περιπατεῖν   See comm. to l. 370 below.

358   Περιπατεῖν ἴσως   Refers to the pace of march, so "march evenly" has the sense of "march at an even pace" (see Rance, *Strategikon*, 12.B.16.17).

343–49   §50   Cf. 11§19, 12§53.

345–46   ἢ βουκίνῳ ἢ τῇ ταυρέᾳ ... ἢ τῇ τούβᾳ, ὅ ἐστι μικρόν βούκινον   Leo copies Maurice (12.B.16.4–5), but adds the qualifying "which is a small *boukinon*" for the tuba. What exactly these different instruments represent remains a little vague; see Maliaras, "Musikinstrumente" and on Const. 5§4.32 above.

365   εἰς τὰ ἄρματα ἀλλήλοις ἐγγίζωσιν   Leo misunderstands the original in the *Strategikon*, 12.B.16.24, which has εἰς τὰ βούκουλα, that is to say, "to the shield bosses": the formation was in close order so that the shields overlapped shield boss to shield boss, affording maximum solidity and protection to the line. The term *boukoula* for shield boss seems to have dropped out of use, a process paralleled by the disappearance of shield bosses themselves, as methods of carrying shields (through straps) made them unnecessary (and through which the term *boukoula* came to be applied to the shield itself); see Trapp, *Lexikon*, 2:290; Kolias, *Byzantinische Waffen*, 98–102. The term *arma*, which in general meant "arms," could also mean "shield": cf. *De cerimoniis*, 302.6; and Trapp, *Lexikon*, 2:199.

370   Φούλκῳ   In the late Roman army the *fulcum* was a dense formation formed by locking or overlapping the shields of the front rank and by the second rank reaching forward and resting their shields on the shield bosses of those in front, thus creating a complete wall to protect against missiles. See Wheeler, "Legion as Phalanx, Part I," 351–53; also Janniard, "Végèce"; and esp. Rance, "*Fulcum*," who argues that this was not a Germanic but an entirely Roman development, and that it is this that Maurice's text describes here. But either Leo misunderstands his original or a common ancestor of MW has introduced an error, by having the second rank rest their shields "on those in front" (if one translates literally εἰς τοὺς ἔμπροσθεν) rather than the better reading εἰς αὐτὰ τὰ τῶν ἔμπροσθεν σκουτάρια, of VE or the similar readings of B or A (as noted in the critical apparatus), to refer to the shields of those in front, which is what the translation here reasonably assumes; see Rance, "*Fulcum*," 317–18. Note the gloss to the term *phoulkon* by a later redactor, in AVBE, who clearly understood the formation correctly, at l. 335: ἤγουν τοὺς ὀπίσω σκέποντας τὰς τῶν ἔμπροσθεν κεφαλὰς τοῖς σκουταρίοις καὶ οἱονεὶ κεραμωθέντας περιπατεῖν. Note that κεραμωθέντας is the term employed in describing the same arrangement at Onasander, *Strategikos*, 20.671 (20§1)

By the tenth century the word *phoulkon* seems to have lost its technical value and refers rather to a specific ordered body of troops (designated to defend a foraging party, for example; see on 17§36.191–97 below) or any dense or compact body of troops; see Nikephoros, *Praecepta*, 4§2.19 and the editor's commentary at 71–72 with references; the word appears with the same meaning in the treatises *Skirmishing* and *Campaign Organization;* see Dagron and Mihăescu, *Traité sur la guérilla*, 224 and n. 18. Whether such a tactic for infantry was still employed, therefore, is hard to say, although the *Sylloge tacticorum*, 43.7, in mentioning the spacing of the infantry soldiers when in the *testudo* formation, refers to it as: ὃ δὴ καὶ σύσκουτον ἡ δημώδης ὀνομάζει φωνή, suggesting that the formation had a common or demotic name and thus that it was still employed; and ibid., 45.33, which reproduces the formation from *Strategikon*, 12.B.16.33–38 and Const. 7§54, but brings the passage up-to-date with contemporary body-length shields (*thyreoi*) in the first rank, rather than the two layers of smaller shields in the original account. Yet judging from the general situation of infantry, as discussed above, and from the sort of maneuvers or field-discipline assumed by later treatises such as the *Praecepta* of Nikephoros, a considerable degree of skepticism must remain.

381–82 Leo simply replaces Maurice's "Adiuta.... Deus" (*Strategikon*, 12.B.16.42–43) with "βοήθει.... ὁ Θεός," but whether this is an accurate reflection of the battle cries of his own time is unknown, although it does reflect actual late Roman practice under Heraclius; see Rance, *Strategikon*, 12.B.16, note 136 with literature.

383–86   οἱ δὲ σκουτάτοι... αὐτῶν   It is not clear whether Leo misunderstood or emended the text of the *Strategikon* here. The original in the *Strategikon* (12.B.16.43–48) reads οἱ δὲ σκουτάτοι, οἱ εἰς τὸ μέτωπον τεταγμένοι.... εἰ μὲν ἔχουσι μαρτζοβάρβουλα ἢ ῥιπτάρια, ἀναπαύοντες τὰ κοντάρια εἰς τὸ χαμαὶ ῥίπτουσιν ἐκεῖνα. Εἰ δὲ μήγε, ἀναμένοντες μέχρις οὗ ἐγγὺς ἔλθωσιν, τότε ἀκοντίζοντες τὰ κοντάρια αὐτῶν ἐπιλαμβάνονται τῶν σπαθίων αὐτῶν... ("in respect of the heavy infantry... if those deployed at the front have *mattiobarbuli* or missiles, then, fixing their spears into the ground, they throw these. But if not, waiting for the enemy to come close, they then hurl their spears and draw their swords....."). In M and W Leo omits the reference to the spears so that his text reads, "As for the heavy infantry, if those who are drawn up at the front have maces or axes or missile weapons, they throw them down(ward), but if not, waiting for the enemy to close, they then hurl their spears or missiles and, drawing their swords...." Dennis has emended the text to render εἰς τὸ χαμαὶ as εἰς τὸ ἄμα, which makes sense, but is not justified by the manuscript tradition (note that the later MSS AVBE all omit the phrase). It seems better to leave the text as it is in M and W, and assume that Leo did not fully understand the movements described. In the original the expression ἀναπαύοντες τὰ κοντάρια εἰς τὸ χαμαὶ means "resting/fixing the spears in the ground" (Rance, *Strategikon*, note to 12.A.7.54–55; B.16.43–48). While Leo's emendation or omission (unless it is the mistake of the copyist of the exemplar from which W and M were copied) makes nonsense of the maneuver, it is possible that he or his copyist was influenced by a similar passage in Syrianos, *Strategy*, 36.14–20 where the first three ranks of an infantry formation attacked by cavalry are to lay their spears on the ground, shoot at the approaching enemy with their bows, then pick up their spears once the enemy are closing with them.

462–67   §66   Leo's own final comment on the importance of training exercises and mock fighting.

468–70   Cf. Aelian, *Tactica theoria*, 24.4; Arrian, *Techne taktike*, 20 (and see on 6§§25.160–26.168).

471–73   See below on ll. 486–92. Leo's admission that he does not always comprehend his ancient authors points both to substantial changes in technical terminology and to his own lack of understanding, as well as corruptions in the transmission and recopying of his authors. See Dain, *L'histoire*, 107–15, 134–47.

477–85 §68   A summary of Aelian, *Tactica theoria*, 25.2–29.9, 30.1–31.4.

486–94 §69   See Aelian, *Tactica theoria*, 42.1; and cf. Asklepiodotus 10.14; Aelian, *Tactica theoria*, 27.3, 28.2; Arrian, *Techne taktike*, 23.3, 24.2–4; and Syrianos, *Strategy*, 24.38–57. Leo's final remark (ll. 493–94) shows that he does not in fact understand these commands, which—as Dain showed—suffered considerably in the process of their transmission through various later recensions of Aelian (Dain, *L'histoire*, 146–47; and n. 145, n. 2). Some of these commands are missing from the two surviving recensions through which Aelian has come down to us, and Dain suggested that Leo supplied them by reference to a version of Arrian's *Techne taktike*. But as noted in chapter 2 above (pp. 48–50), it is more likely that, since the emperor or his redactor was working from a third version of Aelian's text, which has not survived but which Dain has shown must have existed, the commands in question were present in this third recension.

CONSTITUTION 8

This constitution copies the *Strategikon*, 1.6–8 more or less exactly, but Leo changes certain key terms or phrases to bring the text up to date, and adds at the end a final additional paragraph relevant to the social conditions of his own time. The regulations were originally divided into three groups, those to be read out to the soldiers (1.6), those to be read out to the officers (1.7), and the list of military punishments (1.8). Leo ignores these subdivisions and simply lists all the regulations, albeit in the same order.

The list of regulations and punishments in both the *Strategikon* and the *Taktika* is closely related to a second set of legal or para-legal prescriptions, namely the so-called *Nomos stratiotikos*. Referred to also as the military code or "mutiny act," this code survives in two versions, an earlier—ascribed to the seventh or eighth century (Ashburner, "Byzantine Mutiny Act," repr. in Zepos, *Jus*, 2:75–79)—and a later—likely from the tenth century (Korzensky, "Leges poenales militares," repr. in Zepos, *Jus*, 2:80–89). It is based on prescriptions from the *Digest* and the *Codex Iustinianus*, as well as on the *Strategikon* of Maurice. The second version is clearly also connected with writings such as the *Taktika* of Leo VI and the *Basilika*. Its use as more than a very general guide to the theory of military discipline for the period is limited, in view of the problem of dating. See Köpstein, "Profane Gesetzgebung," 145; Kučma, "Νόμος στρατιωτικός"; also Karayannopoulos and Weiss, *Quellenkunde*, 395–96. For the long-standing Roman tradition regarding discipline, see Giuffrida, "Disciplina Romanorum"; and for the later history of the text, see Verri, *Leggi penali militari*.

The so-called "military code" was for many years closely associated with the *Ecloga* of Leo III and Constantine V, and with other texts purportedly dating to the period from the late seventh into the first half of the ninth century, texts such as the *Appendix Eclogae*, a privately commissioned collection, widely employed (judging from the rich manuscript tradition) throughout the following centuries as a reference work and companion to the *Ecloga*; see Burgmann and Troianos, "Appendix Eclogae," 90–93. Similarly taken to be related was the *Nomos Mosaikos*, or Mosaic Law, a selection of some 70 extracts from the Greek Pentateuch, arranged in 50 chapters, again supposed to have been produced in the first half of the eighth century: Burgmann and Troianos, "Nomos Mosaikos." The *Ecloga* and the *Appendix Eclogae*, together with the *Nomos Mosaikos*, the *Farmer's Law* (Νόμος γεωργικός), the *Rhodian Sea Law* (Νόμος Ῥοδίων ναυτικός) and the *Military Code* (Νόμος στρατιωτικός) were thus supposed to represent the basis for secular law throughout the period in question.

But some significant objections to these dates have been raised, as well as against the assumption of their association with the *Ecloga*, although the consensus among legal historians remains in favor of a seventh- or eighth-century date. Thus the supposedly Justinianic content of these texts finds little basis in reality. Indeed, there are a number of important divergences, both in legal terms as well as in the ethic underlying them. Similarities in style, language, and moral tone have been noted between some elements in these texts and the juristic texts inspired or commissioned by the patriarch Photios; see Schminck, "Probleme" and idem, "Bemerkungen." Schminck concludes that the *Farmer's Law* is, in fact, in part at least the responsibility of Photios; that the Rhodian Sea Law was compiled as part of the *Basilika*, and indeed that part of it was compiled by Leo VI himself; and that the "military code" was compiled or at least commissioned by Leo VI. In addition, Schminck argued that the (probably) earlier of these two collections, that edited by Ashburner, already reinterprets the vocabulary taken from the older Justinianic legislation in a sense more typical of the second half of the ninth or first half of the tenth century than of an earlier period, most particularly with respect to the technical use of *strateia* and related words, so that it is possible that this collection, in its currently transmitted form, was in fact a product of the reign of Leo VI. It would certainly fit with his interests and activities in the codification of military and legal material.

As noted, however, these conclusions have not met with general acceptance, and the origins of the *Military Code* and *Farmer's Law* remain associated with the seventh and eighth centuries. Indeed, and as noted in chapter 2 above, the *Military Code* was in existence and associated with the manuscript tradition of the *Ecloga* of Leo III and Constantine V by ca. 886. By its nature it must have had an independent existence, similar to some of the smaller treatises on military matters (that on encampments, for example, or that on certain infantry formations: cf. the text discussed in McGeer, "*Syntaxis Armatorum Quadrata*"); and while the second collection can certainly be placed well into the tenth century, both by contents as well as by its more technical presentation and language, and appears to draw in several instances on the *Taktika* itself, this does not invalidate a much earlier date for the original compilation. See for a detailed critique of Schminck's hypotheses Burgmann, "Nomoi."

The extent to which proper military discipline, as reflected in Leo's Const. 8, was actually imposed is not known. Discipline and obedience to commanding officers was certainly understood as a key element of military order and necessary to success; see *Rhetorica militaris*, 15.1; and esp. 43.1–4 and 44.1–10. On the basis of the sources for the period, the most able commanders and leaders were the most likely to effectively apply military discipline, something clearly recognized in the military treatises. Financial generosity, either on the part of individual commanders or officers, or the government, was a crucial ingredient in encouraging soldiers to follow orders and accept the discipline necessary for effective fighting (on largesse before battle or campaign, see *CPTT* C.261–62, and pp. 225–26). Success generated good morale and self-confidence (as often in the tenth and early eleventh centuries); failure and defeat produced low morale and defeatism (as the military events of the late seventh century might suggest). See Troianos, "Ποινές," esp. 33–34; Kolias, "Στρατιωτικὰ ἐγκλήματα"; and for the late Roman background see esp. Palme, "Spätrömische Militärgerichtsbarkeit," with extensive literature.

A strict code certainly prevailed in elite units such as the imperial *tagmata* and in units that had a particular loyalty to their commanding officer. Although hagiographical in nature, and written in a mode hostile to the iconoclast emperor in question, a mid-ninth-century source includes a plausible account of an officer who was upbraided for his unkempt appearance while at his post in the palace (see *De XLII martyribus Amoriensibus*, no. 2.22–36: de Callisto, 24.30–25.2). Discipline was probably least effective in the militia-like thematic forces (Kaegi, *Byzantine Military Unrest*, 293–324). In contrast, Attaleiates records an attack by a group of drunken Varangians on the emperor Nikephoros III Botaneiates (Attaleiates, *Historia*, 294–96; trans. Kaldellis and Krallis, 537–39).

In Const. 13§6 and 20§18 (following Maurice, *Strategikon*, 7.A.6; 8.1.15) Leo advises that acts of insubordination or indiscipline immediately preceding a battle should be ignored by the officers, in case the troops should be demoralized or alienated by the usual punishment. This tells us both that discipline was indeed enforced by punishment, although whether of the severity or consistency described in the various versions of the so-called "*Military Code*" is unclear, but also that the morale of the armies could be fairly fragile. A good example comes from the late ninth century, when the commander of a naval force sent to confront a Saracen fleet in the Adriatic found that some of his crews had deserted before he could join battle. Although he found and arrested the deserters, he decided rather than punishing them to have some Saracen prisoners executed in their stead, disguising them beforehand as Byzantines and thus avoiding shedding the blood of his own men, as the text says. This quickly reestablished discipline, improved morale, and led to the consequent defeat of the enemy fleet. See *Vita Basilii imperatoris*, 62.14–47 (222–24), a story repeated by Skylitzes (*Synopsis historiarum*, 154–55; trans. Wortley, 149–50), and in a somewhat different version by Genesios (4.34; trans. Kaldellis, 105–6), who implies that punishment for deserters was both usual and severe. Tight discipline under a good commander in whom the army had faith clearly seems to have gone together with good morale and success, as the case of the strict discipline enforced by Nikephoros Phokas seems to demonstrate; see Leo the Deacon, *Historia*, 57–58; Talbot and Sullivan, 105–6. See further on Const. 13§6.26–35 and 20§4, below. Conversely it could affect morale very badly, as when Romanos IV had a man punished during the ill-fated Mantzikert campaign against the wishes of his fellow soldiers and in spite of his protestations in the name of the icon of the Virgin which accompanied the army; see Attaleiates, *Historia*, 152.21–153.14 (trans. Kaldellis and Krallis, 279). For the ways in which deserters were recovered and dealt with in the late Roman period, see Kaiser, "Fahndung nach Deserteuren" and below on Const. 8.56–61.

7–10   §2   Maurice, *Strategikon*, 1.6§§1–2; Ashburner, "Mutiny Act," §22–23; Korzensky, "Leges," §6.

11–12   §3   Maurice, *Strategikon*, 1.6§3; Ashburner, "Mutiny Act," §45.

13–15   §4   Maurice, *Strategikon*, 1.6§4; Ashburner, "Mutiny Act," §46; Korzensky, "Leges," §11.

13   βαγεῦσαι   The verb evolved from Lat. *vagari*, "to wander," and is used also in the sense of an absence without permission or leave: cf. *CTh* 7.1.12 (384), 16 (398); *Dig.*, 49.16.3.2; cf. Trapp, *Lexikon*, 2:257.

κομεάτος   Likewise derives from the Latin *commeatus*, meaning a leave of absence: cf. Trapp, *Lexikon*, 4:852. This regulation has again been altered by Leo to make it relevant. The original merely stipulates that if a soldier goes absent without leave after his furlough is ended he will be stripped of his military status and left to the authority of the civil authorities (since military status brought with it a number of fiscal and juridical privileges, this was a serious punishment; see Haldon, "Military Service," 42 and n. 104 with literature). Leo updates the clause to qualify *kommeaton* as "the release of the soldiers to their own households," reflecting the nature of provincial military service in his own time.

14   ταξατίωνα   Refers to a garrison (cf. e.g., *Theophanis Chronographia*, 370.7; trans. Mango and Scott, 516; *DAI* 1: chap. 45.69) and by association in this context, to garrison duty or servile duties: the soldier remains registered as a *stratiotes*, thus can still be called upon for personal service, but suffers a substantial reduction in status.

16–18   §5 Maurice, *Strategikon*, 1.6§5; Ashburner, "Mutiny Act," §41; Korzensky, "Leges," §5.

19–21   §6 Maurice, *Strategikon*, 1.6§6; Ashburner, "Mutiny Act," §13; Korzensky, "Leges," §12.

22–24   §7 Maurice, *Strategikon*, 1.6§7; Ashburner, "Mutiny Act," §48; Korzensky, "Leges," §9.

25–27   §8 Maurice, *Strategikon*, 1.6§8; Ashburner, "Mutiny Act," §44; Korzensky, "Leges," §7. Officers were responsible for the conduct of the men under their immediate command. Although later than Leo by half a century, a good example is reported by Leo the Deacon in his account of a campaign under Nikephoros Phokas against North Syria: an infantry soldier, becoming weary, discarded his shield. Having been found out by the emperor, he was ordered to be punished by his officer (*lochagos*): he was to be lashed, have his nose cut, and be paraded around the camp as a warning to others. The junior officer failed to carry this out, however, and when this was discovered by the emperor, the latter suffered the same punishment: Leo the Deacon, *Historia*, 57.19–58.8; Talbot and Sullivan, 105–6.

28–30   §9 Maurice, *Strategikon*, 1.6§9; Ashburner, "Mutiny Act," §50; Korzensky, "Leges," §13.

31–32   §10 Maurice, *Strategikon*, 1.6§10; Ashburner, "Mutiny Act," §49; Korzensky, "Leges," §10.

33–36   §11 Maurice, *Strategikon*, 1.6§11. The term ἀπόλυσιν clearly means release from service, in this case when the opportunity arises; the original text in the *Strategikon* read ῥεπαρατίων, Latin *reparatio*, which has been interpreted either as a leave of absence or as an additional cash sum for the replacement of equipment. For junior officers' duties in respect of unit discipline, see also on ll. 25–26 above and the example reported by Leo the Deacon.

37 Maurice, *Strategikon*, 1.7§12; Ashburner, "Mutiny Act," §22–23; Korzensky, "Leges," §6.

38–42 §§13–14 Maurice, *Strategikon*, 1.7§13; Ashburner, "Mutiny Act," §49; Korzensky, "Leges," §10 (cf. *Strategikon*, 1.6.10 and above, ll. 30–31)

40 παραχειμαδίῳ Winter quarters. Troops are described in the treatises and narrative histories as being εἰς παραχειμασίαν: cf. *Sylloge tacticorum*, 22.1–2; 56.2; Skylitzes, 394.70–71 (trans. Wortley, 372); Cedrenus, 508.19–20; 608.18–19; and cf. the verb παραχειμάζειν (*Syll. tact.*, 87.5).

43–46 §15 Maurice, *Strategikon*, 1.7§14; Ashburner, "Mutiny Act," §47; Korzensky, "Leges," §8. This is a regulation which in a late ninth- or tenth-century context makes sense only in relation to full-time or "professional"/mercenary units. This was, of course, also the case with the soldiers to whom it applied in the late sixth-century *Strategikon*, but would make little sense in relation to the provincial soldiery of the *themata*, who generally served on a seasonal basis.

47–49 §16 Maurice, *Strategikon*, 1.7§15; Ashburner, "Mutiny Act," §13; Korzensky, "Leges," §12.

50–55 §§17–19 Maurice, *Strategikon*, 1.7.17–1.8.3.

53–54 §18 The "other punishments" are presumably those included in the *leges militares* not repeated here. Note also, e.g., *Procheiros Nomos*, 39.53; *Epanagoge*, 40.71 (on those who steal weapons or pack animals while on campaign, although the regulation was intended to cover nonmilitary personnel)

56–61 §20 Maurice, *Strategikon*, 1.8§16; Ashburner, "Mutiny Act," §8; Korzensky, "Leges," §14. For the situation in the Roman army, see Seston, "Fahnenflucht"; Wesch-Klein, "Hochkonjunktur für Deserteure"; and Wierschowski, "Kriegsdienstverweigerung."

62–69 §§21–22 Maurice, *Strategikon*, 1.8§17; Ashburner, "Mutiny Act," §9; Korzensky, "Leges," §15.

70–74 §23 Maurice, *Strategikon*, 1.8 §18; Ashburner, "Mutiny Act," §10; Korzensky, "Leges," §16.

75–79 §24 Maurice, *Strategikon*, 1.8 §19; Ashburner, "Mutiny Act," §12; Korzensky, "Leges," §17.

80–81   §25   Maurice, *Strategikon*, 1.8 §20; Ashburner, "Mutiny Act," §11; Korzensky, "Leges," §18.

82–87   §26   This is Leo's own addition (see also on 19.119–23 below), and is an important indicator of the ways in which members of the military establishment in the provinces were able to exploit their positions of authority through the relations of patronage which had evolved between them and the provincial soldiery. Similar problems were the subject of imperial legislation in the fifth and sixth century; see *CJ* 9:12.10 (a. 468); Just., *Nov.* 30 (a. 536); 116 (a. 542). The tension between the interests of the state and those of local landlords and powerful men became increasingly apparent through the tenth century. Leo's remarks prefigure later, mid-tenth-century legislation as reflected in a novel of Constantine VII of perhaps 947, which condemns those who exempt soldiers from military service in return for gifts, and which also alludes to the misconduct, incompetence, and venality of provincial officials and military officers: Svoronos, *Novelles*, esp. C1 (124.121–25–129) and C3 (126.149–57); Engl. trans. in McGeer, *Land Legislation*, 75–76, with commentary at 68–70; and discussion in Lemerle, *Agrarian History*, 122–23. A document of the Athonite monastery of Iviron for the year 975 refers to *stratiotai* who have fled to the estates of powerful persons and the Church; see *Acts of Iviron*, 1: no. 2.3–4. The presence of soldiers in the provinces in itself affected local power relationships, and the social environment in which they were rooted. In the first case, local military commanders might use their troops to improve their personal social and economic situation, by employing them to coerce the population in one form or another. Provincial military commanders were endowed with authority over considerable resources in manpower and the coercive power vested in their office. This became especially problematic after the ninth century, when military officers in the provinces frequently came from landed families—ninth- and tenth-century legislation explicitly mentions the fact that soldiers are attracted into the "private" service of "powerful" persons, and that officers take soldiers away from their duties and employ them in their own service. The growth of military retinues at this time and thereafter is a basic feature of Byzantine social development. See Magdalino, "Byzantine Aristocratic *Oikos*."

CONSTITUTION 9

See on Const. 14.141–45 and 18.550–56 below. This constitution is compiled from elements of Onasander, *Strategikos* and Maurice's *Strategikon*. Leo's main concern here is to provide guidance for commanders who are entering territory in which there is an active enemy to deal with but in which the terrain or landscape may also pose as great a threat to the army. He would certainly have been familiar with some of the results of carelessness on the part of Byzantine commanders, both in the rugged wooded terrain of the Balkans, in fighting against the Bulgars, as well as on the eastern front, in the Taurus and Anti-Taurus mountain passes. Byzantine forces had suffered several defeats as a result of being trapped on unfavorable ground, most catastrophically, of course, the destruction of the army under the emperor Nikephoros I in Bulgaria in 811. More recently Leo will have been very aware of the nature of the warfare along the eastern frontier, and he must also have had at his disposal reports of some of the experiences which his own father, Basil I, an extremely effective field commander, had made in the wars against the Paulicians and the frontier territories of the Caliphate in the 860s and 870s, and he duly refers to this in the course of this section. For detailed discussion see Haldon, *Warfare, State and Society*, 149–62.

Leo's advice and that of the older authorities upon whom he depends for most of his material is remarkably similar to that found in much more recent treatises on the subject; see, for example, the excellent volume by Col. G. A. Furse, *Art of Marching*. The evidence for the period from the ninth through to the twelfth centuries suggests that these regulations were generally followed, even if not in every detail; see, e.g., *CPTT* B.107–15, 134–50; C.474–96; *Campaign Organization*, 10.

There were changes over time, some of them evidenced by shifts in terminology. The technical names applied to the various elements of the column evolved, for example, with the introduction of the Arabic word *saqat*, Hellenized as *saka*, for the rearguard (see on 4§§29.116–30.119 above). At 9§§45–47 Leo describes a defensive marching formation in which divisions or brigades were drawn up in double columns, the better to resist attacks from the flank while on the march, perhaps as the result of the order adopted for a particular campaign or campaigns, possibly those of John Kourkouas in the 920s to 940s—cf. *Sylloge tacticorum*, 47. This may lie behind the appearance during the first half of the tenth century of a defensive formation where the various divisions of the army—posted on the flanks, in the van, and in the rear—formed in effect a loosely articulated square formation, so that in the event of an attack they could close up to form a solid bulwark from which to resist the enemy; see on 18§§113.550–114.556.

With minor variations, noted in texts deriving from the time of Basil I, for example, as well as of the period from the late tenth to late twelfth centuries, the basic pattern for marching columns was followed by all major expeditionary forces (*CPTT* B.107–50; C.479–96, 561–69, information taken from the campaigns of Basil I and the generals of Leo VI; *Campaign Organization*, 10, 12–15; Nikephoros Ouranos, *Taktika*, 63–64; and the discussion in McGeer, *Sowing the Dragon's Teeth*, 332–41). Anna Komnene, writing of Alexios's campaigns in the early twelfth century, shows that fifes or pipes were used to maintain the tempo of the march (*Alexiad*, 15.7, trans. Sewter, 491/Frankopan, 451).

In general on Byzantine land routes, transport, and travel see Belke, "Verkehrsmittel"; Dimitroukas, *Reisen und Verkehr*; and the essays in Macrides, *Travel in the Byzantine World*. On roads and routes, see also Belke, "Pflasterstrasse"; idem, "Communications: Roads and Bridges"; and Kučma, "Organizacii."

> **2** ὁδοιπορίας The standard term for a march: cf. *Campaign Organization*, 10. Whether Leo is aware of the technical value, in Maurice and the older military treatises (see, e.g., *Strategikon*, 9.1.25 = Const. 17.5.34; Aelian, *Tactica theoria*, 36–37; Arrian, *Techne taktike*, 28), of the term πορεία to refer to a specific marching formation, an order of march (τὴν ἐκδεδομένην μετὰ τάξεως ὁδοιπορίαν: *Campaign Organization*, 30.8–9), as opposed to a march, is not clear.
>
> **3–8** §1 Onasander, *Strategikos*, 6.269–73 (6§10)
>
> **9–13** §2 Onasander, *Strategikos*, 6.283–89 (6§13)
>
> **14–16** §3 Maurice, *Strategikon*, 1.9.3–5. See also on Const. 6§19.96–103 above. One of the major concerns of all the military handbooks, Roman, late Roman, and Byzantine, is the impact of an army on the producing population and on the countryside through which it

passes; see Haldon, *Warfare, State and Society*, 145–47, 234–47 and Erdkamp, *Hunger and the Sword*, 84–140. The need to avoid harming the provincials by permitting the army to forage and extract supplies without proper administrative controls is often repeated—although even where such controls were established, the presence of a large force of soldiers, their animals, and their followers will rarely have been welcome: Maurice, *Strategikon*, 1.6.19, 13; 1.9 = Const. 8§§10, 14.

17-20   §4   Cf. Const. 7§2.

21-96   §§5-21   This passage (except ll. 60–66) = *Strategikon* 1.9.6–63.

26   ἀποσκευὴν   Cf. Maurice, *Strategikon*, 1.9.10. Here Leo uses the term *aposkeue*—"baggage"—to differentiate the unit pack animals and carts from the main train or *touldon* (on which see Const. 10 and commentary) of the army; see Const. 5§5, and 6§23.142–43 and commentary; and cf. below, l. 50.

29-30   μὴ συνάγειν ... λοιμώττειν   The ability of an area to supply a substantial army for more than a very short period was of great concern: cf. a novel of Tiberius Constantine remitting taxes for certain eastern provinces, but alluding to the hardship caused by the presence of soldiers and warfare: Just., *Nov.* 163 (also in Zepos, *Jus*, 1: coll. 1, nov. 12). The monks of the island of Gymnopelagesion in the Aegean were forced to abandon their home in the tenth century as a result of the constant requisitioning of their livestock and other produce by passing vessels of the imperial fleet: *Act of Lavra*, 1: no. 10, ll. 15–18. The tenth-century *Treatises on Imperial Expeditions* (*CPTT*), and the list of major base camps or *aplekta*, describe the process by which the different corps met up, according to the direction and target of the campaign; see *CPTT* A; B.97–100. The point was to minimize the impact of an army on Roman territory: each division or subdivision is to march independently, and with its own baggage, until an agreed rendezvous or until scouts or other lookouts report the enemy, at which point, as the text goes on to say, the whole force should reunite. The Syrian campaign of Romanos III in 1031 and the Mantzikert campaign in 1071 illustrate some of the problems: conflicts with local populations within imperial territory over the seizure of livestock or produce; the problem of ensuring that the areas to be passed through could support the army; and the need to secure supplies in advance when entering areas which had already been devastated or where the population had fled. See Skylitzes, *Synopsis historiarum,* 379–80 (trans. Wortley, 358–59); Zonaras, 4.130–31; for 1071: Attaleiates, *Historia,* 148.14–17 (trans. Kaldellis and Krallis, 271).

33 ὡς πρὸ ἓξ ἢ ἑπτὰ ἡμερῶν ἢ καὶ δέκα   The distance of a day's march varied, of course. In contrast to the armies of the period up to the third and fourth centuries, the medieval East Roman army did not have a system of paved military roads at its disposal. The rapid strategic dispersal or concentration of troops was thus greatly hindered and, although small forces were able to move to intercept invading troops fairly quickly, the assembly and movement of larger armies was a slow and cumbersome process. For discussion and statistics see Haldon, "Roads and Communications"; Belke, "Verkehrsmittel"; and with further examples, Haldon, *Warfare, State and Society,* 164–66 and Erdkamp, *Hunger and the Sword,* 62–83. For more general discussion, see Bachrach, "Animals and Warfare"; with Furse, *Art of Marching*; Van Crefeld, *Supplying War,* 28–29; Engels, *Alexander the Great,* 154–56; Elton, *Warfare in Roman Europe,* 244–45 for further discussion and evidence; and Roth, *Logistics of the Roman Army.*

35 δουκάτωρες   Lat. *ducator,* a term not used in the *Strategikon.* The term applies primarily to local guides, people from the regions through which the army is marching and more particularly from the areas bordering enemy lands, who are familiar with the tracks and roads and paths of the backcountry as well as with the major routes, and who also know something of the enemy's land so that they can guide the army to water and areas which can provide fodder and supplies. See the detailed description at *Campaign Organization,* 18.3–15, which differentiates between local peasants, who know their own area, and the *doukatores* who have a wider knowledge. Cf. *CPTT* B.116–19 and 171. For the *minsoratores,* whose main responsibility was measuring out and locating the encampment, see on Const. 4§24.103 above.

40 Leo has already noted (4§25.105) that the *antikensores* are no longer a recognized body, and that the duties they formerly exercised are now carried out by the *minsoratores.*

42–46 §9 The method for dealing with difficult roads or tracks is illustrated by an account of Basil I's clearing and leveling a rough mountain road in the Taurus range near Koukousos during the campaign of 877–78 against Germanikeia: *Vita Basilii imperatoris* 48.3–5 (168). Cf. Const. 17§58.337–40.

45–46 τοὺς δὲ. . . . δουλεία   Read: "The men detailed . . . should not be subject to scouting or any other duty." Such troops were relieved of other duties because of the burdensome nature of these tasks. See Rance, "Noumera or Mounera"; and cf. *Vita Basilii imperatoris*, 48.3–8 (168), describing the rigors of cutting a road through rough territory during one of the campaigns of how Basil I; and cf. Anderson, "Campaign of Basil I."

46 βίγλᾳ   Lat. *vigilium*, see Du Cange, *Glossarium Graecitatis*, 199–200; Trapp, *Lexikon*, 2:278. Its general meaning was "watch," "watchpost," "lookout," "reconnaissance," but there were a range of specialist terms derived from it, such as ἐξώβιγλα, ἐσώβιγλα, καμινόβιγλα, etc.: cf. *Skirmishing*, 1; *Campaign Organization*, 4.

48–49 διὰ τὸν φόβον τῶν ἐπιτιμίων   Leo appears to have misunderstood the original text of the *Strategikon* (1.9.29–30) here, where the term οἱ ἐπιτίμιοι referred to the ceremonial guard or retinue of the commander of the army which, together with his personal guards and attendants—ἰδικοὶ αὐτοῦ ἄνθρωποι—should precede him in the column. Instead Leo takes the word to be ἐπιτίμια—punishments or penalties (perhaps, therefore, referring to the military regulations in the preceding constitution). The *Sylloge tacticorum*, 49.4, avoids the ambiguity by rewording the sentence completely. The *banda* in question in the *Strategikon* are standards.

60–66 §14   A rare passage where Leo makes reference to an event about which he was himself directly informed. For Basil I's campaign against Germanikeia in 877–78, see *Vita Basilii imperatoris* 48 (168–74); *TIB* 2:82–83; Vasiliev, *Byzance et les Arabes*, 2.1:82–94. Note also the account of the ensuing triumph celebrated on his return to Constantinople, in *CPTT* C.742–807 and 268–69.

73–85 §§16–18   Repeated emphasis on safeguarding the local rural population from the effects of the passing army. See on ll. 14–16 above.

92–96 §21   Emphasis on keeping the army as far away from prying eyes as possible is a theme common to all the treatises, and was certainly practiced as far as was feasible. Basil's campaign of 877–78, already referred to, was a case in point, since he chose to follow the minor roads and tracks through the mountains, both to surprise his enemy as well as to minimize their numbers. See Anderson, "Campaign of Basil I"; Lemerle, "L'histoire des Pauliciens"; and McCormick, *Eternal Victory*, 154 and n. 84.

97–103 §§22–23 Onasander, *Strategikos*, 6.274–82 (6§§11–12)

104–11 §24 Onasander, *Strategikos*, 6.283–89 (6§13) but somewhat reworded and expanded; repeats the sentiments expressed at ll. 9–13 above, also drawn from the same chapter of Onasander.

112–14 §25 Onasander, *Strategikos*, 10.381–83 (10§7), very abbreviated paraphrase.

115–18 §26 Onasander, *Strategikos*, 6.290–92 (6§14); but the reference to protecting merchants is repeated elsewhere, since the army's equipment and supplies were clearly at least in part dependent on such sources. See below on Const. 11§41.

119–25 §27 Onasander, *Strategikos*, 7.295–303 (7). The treatise written for Nikephoros II Phokas on skirmishing or guerilla warfare in the Taurus and Anti-Taurus mountains deals with the issue of holding defiles and controlling passes in extensive detail. See in particular §§3, 11; and the comments in Dagron and Mihăescu, *Traité sur la guérilla*, 219–20. But the precautions set out here were standard, cf. *Sylloge tacticorum*, 49.8; *Campaign Organization*, 14 and 19.

122 τὰς λεγομένας κλεισούρας Cf. Lat. *clausura*. See Castellvi, "*Clausurae*"; Napoli and Rebuffat, "*Clausurae*"; and Trapp, *Lexikon*, 4:837. The word had two complementary meanings. From the late eighth and early ninth century it was applied to frontier districts, with an administrative identity, built around a pass or defile through the Taurus or Anti-Taurus mountains, and commanded by an officer with the title *kleisourarches*. See Ahrweiler, "Recherches," 81–82; Oikonomidès, *Listes*, 342; and *CPTT* 248. These kleisourarchies (*kleisourarchiai*), created from subdivisions of the *themata* from which they were detached, may represent the crystallization of a new defensive policy: a locally focused defense, involving a "guerrilla" strategy of harassing, ambushing, and dogging invading raiders, designed to stymie all but the largest forces and to prevent both the pillaging of the countryside and the economic dislocation which followed. Leo's expression here suggests that he means to refer to this administrative unit. As well as these administrative *kleisourai*, however, any individual pass or defile through the mountains could be referred to by the same term, so that we read, for example, of the *kleisoura* of Podandos, north of the Cilician Gates, as well as those of Seleukeia or Kappadokia: e.g., Attaleiates, *Historia*, 121 (trans. Kaldellis and Krallis, 221); *Skirmishing*, 23; Dagron and Mihăescu, *Traité sur la guérilla*, 219. See also on Const. 18§120.584–87 below.

126–29   §28   Onasander, *Strategikos*, 7.299–303 (7§2). Seizing the passes and defiles in advance of a possible enemy attack was a well-established strategy. Cf. e.g., *Skirmishing*, 3.

130–87   §§29–41   Onasander, *Strategikos*, 6.218–68 (6§§1–9).

188–254   §§42–51   Maurice, *Strategikon*, 9.4.

207–34   §§45–47   See also on 18§§113.550–114.556.

208–9   See Maurice, *Strategikon*, 9.4, and Rance, ibid., notes 74 and 75 for the classical terms for various types of marching order employed by Maurice and copied by Leo, although it is not clear that Leo understood their technical value.

217   ἐξπλήκτους   Lat. *explicitus*, which Leo copies from Maurice. In the *Strategikon* it refers to lightly armed troops, those deputed to particular tasks without armor or baggage: e.g., 2.9.4; 5.5.6; 11.4.172, 183 etc. Leo clearly understood this, since at 17§50.279–80 he mentions οἱ δὲ ἔξπληκτοι, ἤγουν οἱ εὐπλήκτως ἄνευ βάρους, that is to say, "the *explektoi*, or those lightly protected and without burdens..." (Dennis trans., emended slightly). See also 10§15.70, 13§13.69, 17§17.96, §39.215, §40.226, etc. See Trapp, *Lexikon*, 3:544, where the entry should be emended accordingly.

235–40   §48   A stratagem occasionally documented in narrative accounts and present in older handbooks, e.g., Polyaenus, *Strategika*, 2.1.30; 3.9.62 (= *Hypotheseis*, 54.23); Maurice, *Strategikon*, 9.4.45–51.

246–49   §50   See Rance, *Strategikon*, 9.4, note 83.

255–74   §§52–55   Maurice, *Strategikon*, 12.B.19.

275–371   §§56–75   Maurice, *Strategikon*, 12.B.20.

278–80   λωρίκια ... τζικούρια   Leo "modernizes" this list of equipment and weapons, replacing terms such as *zaba, berutta, akontia mikra Mauriskia,* and *martzoubarbouloi* with words more familiar to the tenth-century reader. For comment on the terminology in the *Strategikon*, see Rance, *Strategikon*, 12.B.20, notes 186–88.

292–93   μέχρι ἑνὸς σημείου   As well as referring to a sign more generally (including a standard, pennon, or a signal; see, e.g., 4§16, §38.154, etc.; 7§4.33; §18.125; Nikephoros, *Praecepta*, 1§10.108; *Skirmishing*, 10.17; *Sylloge tacticorum*, 20.5; 95.17) the term was also used of a milestone and by extension referred also to a (Roman/Byzantine) mile, or approx. 1480 m (Schilbach, *Metrologie*, 32–34). In Byzantine texts after the seventh century it occurs largely when copied from older sources, as here (cf. Maurice, *Strategikon*, 7.B.7.7, 15.11; 9.2.29–30; 12.B.20.21–22).

300   ἀποσοβεῖν   Cf. Nikephoros, *Praecepta*, 4§8.85 for the term ἀπο-σοβηταί or "those who repulse," and McGeer, *Sowing the Dragon's Teeth*, 74.

305   <δρούγγους, τοῦτ' ἔστι>   Supplied by the editor following the MS tradition of the *Strategikon*, but in all the MSS of Leo (MWA, VBE) the phrase, with its original sense of "in irregular groups" is omitted, suggesting that, at this juncture at least, either something had dropped out of the version of Maurice used by Leo, or that it made no sense to a tenth-century redactor, where a *droungos* was a larger body of troops (although at l. 309 the phrase is retained and seems correctly understood).

331–34   εἰ δὲ ἔμπροσθεν … μέτωπον   The translation should be emended to read, "If they appear in front of either one or both of the center divisions, having turned toward the position to the right and having resumed their frontal position, the other two divisions move up and likewise deploy alongside them forming the battle line with their flank as their front." However, if we assume that Leo understood the technical value of the terms Maurice uses, the translation should be emended to read: "… the other two divisions move up and in the same manner deploy beside them in the battle line and thus adopt a formation extended along a front." πλαγία, in the military treatises, had the meaning of "extended" (as "in line," while ὄρθιος meant "narrow," as "in column"), as at *Strategikon*, 12.B.20.58–62. See Rance, *Strategikon*, 12.B.20, note 203. But as the next comment suggests, Leo probably did not understand this technical meaning.

355   ἐπὶ μέτωπον... πλάτος   See preceding entry: Leo's explanatory εἰς πλάτος, suggests that he is unaware of the technical value of πλαγία, understanding it literally, as the translation indicates.

358   Cf. Maurice, *Strategikon*, 12.B.20.84–98 (with Kolias, *Byzantinische Waffen*, 194–95). Note the contradiction between this translation of *menaula* as javelins and that at 6.171 (and see commentary, above), where they appear as (heavy) spears or lances.

361   ὑποτασσόμενοι   Translate as "deploy behind." See Rance, *Strategikon*, 12.B.20, note 208.

369–73   §§75–76   See in greater detail the commentary to Const. 11.

CONSTITUTION 10

The whole of this short constitution is based on Maurice, *Strategikon*, 5, apart from the final paragraph, which summarizes points already made very briefly in Constitutions 4 and 9. In *Sylloge tacticorum*, 23, the same points are made, but with greater emphasis on the details incorporated in *Taktika*, 9, on moving through defiles and difficult country, either into or out of enemy territory, while in the *Praecepta* of Nikephoros (e.g., 4§§6–7) the emphasis is on protecting the baggage train by placing it in the center of the column or at a safe distance from the scene of any engagement. Cf. also *Skirmishing*, 16.3–13. For detailed discussion of the Byzantine baggage train—*touldon* or *touldos*—see Haldon, *Warfare, State and Society*, 158–66. Leo repeats Maurice in noting that on occasion infantry units accompanied by baggage animals with provisions for 8–10 days should be sent on ahead. He also states that infantry units were usually accompanied by light carts, probably the sort of 2–wheeled vehicle that could be drawn by a single mule for which the Edict of Diocletian lays down regulations (*CTh* 8.5.48, giving maximum loads for 2-wheeled and 4-wheeled vehicles); whereas cavalry units would have a train of pack animals that could keep up with them (see on Const. 10§17.77–79 below). The *Praecepta* suggest that pack animals alone were employed for most purposes; but both forms of transport were employed according to circumstances, even if pack animals seem to have been the preferred and more usual mode by the tenth century.

The ratio of pack animals or carts to men is not particularly clear. The *Praecepta* of Nikephoros (2§1) stipulates one mule for two infantry soldiers, which would suffice to carry all their equipment and provisions for up to 24 days at least. For infantry units Leo, following Maurice, states that there should be one cart for each dekarchy to carry the supplies, tents, etc. for the soldiers, with an additional number of carts or wagons—ten or twenty—for the whole *bandon* to transport spare weapons and additional provisions; and when they move up to the front away from the train or camp there should be one mule for each dekarchy carrying provisions for up to 10 days (Const. 6§23.142–43, 154–58). He also notes (10§§11–12) that when (cavalry) units move up to the front they should each take their spare horse, or a pack animal (for infantry, see below on ll. 48–49), loaded with some 20–30 lbs. of hardtack and millet (i.e., enough for a standard ration per diem for 10 days), along with temporary shelters or a small tent (although this passage is from the *Strategikon* and is intended for cavalry, it is not clear that Leo retains this distinction, and may believe that this can apply to infantry also).

5   παλλικάρια   In the *Strategikon* the word used is παλλίκας (s. πάλλιξ), with the meaning of either "boys" or "grooms/servants." It evolves later into the meaning of an esquire and eventually a warrior or fighter: cf. Trapp, *Lexikon*, 6:1181; and above, on Const. 4.151–52.

6   τέκνα ἢ συγγενεῖς   It is generally accepted that soldiers at all periods were often able to take wives and children with them, especially when still in friendly territory, who would then contribute to both managing the baggage as well as the range of menial and other tasks required by an army on campaign. They were both an advantage and a liability, of course, since they expanded the requirement in food and to some extent transport, could encumber rapid movement, and were a risk in security and camp discipline. Byzantine sources virtually never mention them explicitly, except in the military treatises, and even then only obliquely—as in this instance, for example (and cf. ll. 12–16 below), and *Campaign Organization*, 15, which warns against taking more noncombatants than is absolutely essential into hostile territory. Cf. Erdkamp, *Hunger and the Sword*, 41–42.

17–20   §4   The number of mules or horses to be looked after by a single man varies from 4 to 10. In Const. 4§38.152–53 (= Maurice, *Strategikon*, 1.5.4.18–20), Leo stipulates one man for every three or four beasts; the "professional" unit of muleteers and baggage train attendants, the *Optimatoi*, were in the ninth and tenth centuries allocated one mule or packhorse per soldier, with an assistant from the imperial stables for every team of 10 mules or 20 packhorses to aid in maintaining the animals' loads and harness (*CPTT* C.332–46, and 235). For the imperial household baggage, special arrangements were made, according to which animals were recorded and given a seal or identity

marker, so that theft and fraud could be minimized. See *CPTT* C.338.
It is likely that these "seals" (σφραγῖδες) were very like those which
survive from the late sixth-century, bronze marker-plaques for animals of the public post, bearing the inscription: "animal belonging to
the sacred *armamenton*, by imperial decree not to be conscripted for
aggareia": *Année épigraphique* (1992), nos. 1825, 1945; and cf. Bendall
and Morrisson, "Protecting Horses in Byzantium."

25   σαγμάρια   The standard word for pack animal (see 4§38; 5§6.42;
9§37.168; §56.283; 10§17.77; 11§38.215), so that Leo seems to misun-
derstand the term *adestraton* completely, which he employs only
here. In Maurice (e.g., *Strategikon*, 5.2) it refers to remounts or spare
cavalry horses, not pack animals (cf. *Strategikon* 1.2.72–73, 5.18–22;
12.B.6.10–11). It is possible, however, that the implication is that the
reserve horses or remounts could also be employed to carry supplies
or equipment under certain circumstances.

25–26   ἀδέστρατα   = Lat. *ad* + *dextratus* = "at/by the right hand," suggestive
of the animal as led by a servant; see Trapp, *Lexikon*, 1:19; Stephanus,
Thesaurus, 1:637; Mihăescu, "Éléments latins," 1:494–95; idem, "Littérature byzantine," 201–3. Leo takes the word from Maurice. It appears
in no other Byzantine source for this period (although it clearly continued in use in the Latin west; cf. medieval Latin *dextrarius* [war-
horse] and Old French *destrier*: Du Cange, *Glossarium Latinitatis*, 3:92;
Niermeier, *Lexicon*, 328 s.v.). It refers to the spare or reserve animals,
those called in contemporary texts συρτά and sometimes παρίππια.
The rate of remounts remains unclear. In this constitution Leo, following the *Strategikon*, notes that remounts should be held back
at the base camp, with the rest of the baggage train; but that otherwise when soldiers moved off to fight they would each take one spare
horse or remount (l. 48), so that the standard ratio was probably 1:1
(Rance, *Strategikon*, 5.2, note 6). Similar provisions are mentioned in
the *Praecepta* of Nikephoros. The tenth-century treatise on imperial
military expeditions suggests a reserve stock of about 20%: 100 animals
for 482 (*CPTT* C.389–91, but these are pack animals and the rates may
have been reckoned differently). Two of the late tenth-century treatises
also imply a remount rate for advance units and the main lancer division (as opposed to mounted archer units) of 1:1: *Skirmishing*, 14.35–36

(193 Dennis); Nikephoros, *Praecepta*, 1§17, 4§1 (and on not having excessive spare horses on raids: 4§77; Gyftopolou, "Riding.").

36–40 τότε ... πολέμου Advice repeated in *Skirmishing*, 16.3–13.

48–49 τὰ ἀδέστρατα ἤγουν σαγμάρια The *Strategikon* mentions only the
adestrata here; Leo has added the word *sagmaria* again to clarify, either
as a mistaken gloss or perhaps with the implication that remounts
could carry provisions or equipment at times. See above on ll. 25–26.

49 σαγία διπλᾶ By the tenth century the *sagion* (Lat. *sagum*) was an
alternative or equivalent for *mantion*, both meaning a short cloak or
tunic. Originally just a military garment, by the sixth century it had
acquired a broader range of meanings: cf. *CPTT* 260, 280. Following
Maurice here, Leo uses it to refer to a large outer cloak, in this case
double-layered, to serve as a cover and protection in the field.

50 τὸ λεγόμενον καμάρδιν Again copied from the *Strategikon*. The term
comes from Lat. *cameratus* (*camera*, Gr. καμάρα), meaning arched or
vaulted, describing a small tent or canopy. This word appears only
in the *Taktika* and in no other middle Byzantine source, see Trapp,
Lexikon, 4:753; Du Cange, *Glossarium Graecitatis*, 559; Sophocles,
Greek Lexicon, 624; Mihăescu, "Éléments latins," 1:487.

51 δαπάνην ... ἐλαφροῦ See on Const. 6§23.153–54 above. According
to *Skirmishing*, 8.13–14, basic rations for soldiers consisted of bread
and either cheese or dried and salted meat. In some cases we know
that herds of animals—sheep and cattle—were driven along with the
army in order to provide a supply of fresh meat: the treatise on imperial military expeditions is explicit in this, as is Michael Attaleiates
in his account of the Mantzikert expedition: *CPTT* text C.146–47 and
comm. at p. 202, where sheep, lambs, cattle and calves are listed; and
Attaleiates, *Historia*, 151.15–17 (trans. Kaldellis and Krallis, 277). See
Foxhall and Forbes, "Sitometreia." Soldiers' rations from the Roman
period onward and into the early modern era averaged about 1.3 kg
per diem, mostly made up of bread with a small proportion of
dried meat or fish, so that 30 pounds (1 Byzantine pound = ca 320 g:
Schilbach, "Metrologie," 174) = 9.6 kg would be sufficient for 10–15 days
on a short ration. See Haldon, *Warfare, State and Society*, 166–68,
and 281–92; Erdkamp, *Hunger and the Sword*, 27–45; general considerations: Garnsey, *Food and Society*.

56 χόρτον ἢ ἄχυρον   We possess very little information on Byzantine military mounts; see extended discussion on 17§§68.392–72.427 below. Leo (17§70.403–5) repeats Maurice's description (*Strategikon*, 9.5.10–12) of the space occupied by a cavalry mount, 8 feet × 3 feet (0.9 m × 2.4 m). If this is an accurate reflection, then it has been calculated that it would represent a horse of between 14.3 and 15.2 hands in height (1.5–1.57 m), and based on zooarchaeological evidence from both the Roman and medieval periods, this would appear to be a reasonable estimate of the average Byzantine cavalry mount. See Hyland, *Equus*, 67–70; eadem, *Medieval Warhorse*, 33; and, more generally on horses in Byzantium, Kolias, "Horse" and other contributions in the same volume. Roman cavalry horses required something in the order of (in modern weights) 20 lbs (9 kg) of fodder per day in rest conditions: 5–6 lbs (2.2–2.7 kg) barley and a further 10–15 lbs (4.5–6.8 kg) hay or grazing: Hyland, *Equus*, 90; Dixon and Southern, *Roman Cavalry*, 208–17, esp. 210–11. The area required for grazing depended on several factors—quality of pasturage, seasonal variations, and so forth. The basic requirements per horse amount to four to five hours' grazing per day, so that in that time twenty horses would graze one acre of medium-quality pasture. On campaign, they were probably fed less. In addition, the amount of required pasturage increased considerably where barley feed was not available. Horses also consume an average of 22.75–36.4 liters of water per day—the amount varies according to breed, temperature, nature of work, and so on: Hyland, *Equus*, 96; see also Gladitz, *Horse Breeding*, 127–28 and further literature. In modern weights, one day's fodder for a cavalry *bandon* of 200 troopers will thus have amounted to some 1400 kg, without barley (or 10 acres/2.5 hectares of pasture), and not taking into account the needs of any remounts or spare horses, which would effectively double the requirement. As well as the horses, the demands of mules and—depending on circumstances—oxen also had to be accounted for. A mule requires some 2 kg of hard fodder plus up to 6 kg of dry or green fodder (or 12 kg of pasturage) per day, in addition to the 20 liters of water it will drink; an ox needs 7 kg of hard fodder and 11 kg of dry or green fodder (or as much as 22 kg of pasturage) and 30 liters of water, so that, again depending on the terrain and other circumstances, this is a substantial requirement to be taken into account. See Haldon, "Roads and Communications," 144–46 and Erdkamp, *Hunger and the Sword*, 37–41, 70–83.

62   ἐκλυθῆναι   Read "become enfeebled" not "come unstrung"

66–70   §15   Cf. *Sylloge tacticorum*, 49.12; *Praecepta*, 4§§6–7; *Campaign Organization*, 10.

77–78   εἴτε ἐν ἁμάξαις . . . συμμίκτου   This is Leo's own final comment, and would seem to support the idea that carts or wagons would be present usually only when there were also infantry present, whereas cavalry forces used pack animals for their baggage.

CONSTITUTION 11

This constitution is drawn from Onasander and, for the most part, 7.B.13 and 12.B.20 and 22 of the *Strategikon*. The advice reflected traditional as well as contemporary practice insofar as commanders of field forces were generally expected to set up encampments and marching camps while engaged in active operations, although there are several examples of generals who failed to do this and whose troops suffered the consequences. In the *Strategikon* of Maurice about a third of 12.B.22 has been seen as derived from earlier compilations, in particular a treatise on military encampments, no longer extant; see Zuckerman, "Chapitres peu connus," esp. 373–85. This (probably) late Roman treatise, drawing in its turn on earlier sources, was written in Greek, and evolved into two distinct traditions. The longer tradition was employed in *Strategikon*, 12.B.22; by Syrianos *magistros*, writing in the ninth century; by the anonymous tenth-century author of *De re militari* (*Campaign Organization*); and by the anonymous author of the (probably) ninth-century *Apparatus bellicus* (see above, chapter 1, n. 35). The shorter tradition survives in a brief section on encampments in the *Taktika* of Nikephoros Ouranos, and in *Strategikon*, 12.C. Leo's sections 13–40 (ll. 73–230, with the exception of ll. 116–42, on the tactics employed in Bulgaria and Syria by the general Nikephoros Phokas) follow *Strategikon*, 12.B.22, and thus likewise derive ultimately from the same common source.

On Byzantine military encampments in general, see Haldon, *Warfare, State and Society*, 152–54; the discussion in McGeer, *Sowing the Dragon's Teeth*, 347–59; and the older but still valuable articles of Kolias, "Περὶ ἀπλήκτου"; Grosse, "Römisch-byzantinische Marschlager"; and Kulakovskij, "Vizantijskij lager." For the established Roman tradition of the Principate, and the sources from which the ancient and later treatises on castrametation were based, see Baatz, "Quellen zur Bauplanung," and the summary in Hanel, "Military Camps." But, as noted below, Leo's account of the methods that might be employed to defend an encampment contains some important and significant differences from the earlier tradition.

Failing properly to entrench a camp or to maintain proper watches and pickets was frequently a reason for the defeat of an army. There are many examples,
but a clear case occurred in 883, when Kestas Styppeiotes (*PmbZ* 23699), the commander of a column dispatched to attack the city of Tarsos, failed properly to site,
entrench, and defend his camp or set up a proper watch. The result was a substantial defeat of the otherwise superior Roman forces; see *Vita Basilii imperatoris*, 51
(184–88). The early tenth-century Arab historian Tabari's version of events is more
detailed and shows up the commanding officer's carelessness and incompetence.
See Vasiliev, *Byzance et les Arabes*, 2.2:9. For further examples, see McGeer, *Sowing
the Dragon's Teeth*, 354–59. During his disastrous campaign into Bulgaria in 811
it would appear that, after the sack of Pliska, the different divisions of the imperial force were left to set up their own camps, and that no proper entrenchment
was ordered—the consequences are well known; see Theophanes, *Chronographia*,
490–91 (trans. Mango and Scott, 672–73) and *Scriptor incertus* (ed. Iadevaia),
30.82–89. Although the text was compiled in the second half of the ninth century, it was based on earlier and probably eyewitness accounts. See Markopoulos,
"Chronique de l'an 811"; Dujcev, "Chronique byzantine de l'an 811"; discussion in
Brubaker and Haldon, *Sources: An Annotated Survey*, 179–80.

Marching camps were intended to protect an army in hostile territory. Scouts
and guides responsible for locating suitable sites and for laying out the camp went
ahead of the main force (see on *minsoratores* and *doukatores* in Consts. 4 and 9
above). As this constitution makes clear, the chief requirements were a defensi-
ble situation, a good supply of water and forage for the horses and pack animals,
and adequate space for the different contingents which had to be accommodated.
Details also survive from the late tenth-century texts of the order in which the
tents of the different units were to be laid out, the distances between them, the
system employed for establishing watches and picket lines, passwords and camp
security, and associated matters. Great stress was laid on camp security: passwords were issued for each watch, and watch-commanders were enjoined to allow
no one past without the correct password (usually the name of a saint or similar
symbol of Orthodoxy). Elaborate arrangements were in place regarding the cir-
cuits and patrols made by the watch at regular intervals; and there was often more
than one perimeter, an inner and an outer, particularly if the emperor was present; see below, and e.g., *CPTT* C.420ff.; *Campaign Organization*, 1–5. As described
in the various *Taktika*, camps were protected by ditches and palisades, sometimes
cut locally, sometimes made from the spears of the infantry; entrances were placed

so that they could be covered by archers and not easily rushed; and a well-drilled system of maneuvers was practiced to enable a force under attack to march out against the enemy, to retreat into the camp, or to set up and entrench a camp while under attack. The historians' accounts corroborate the information offered by the treatises, often in some detail, as we have seen.

The Byzantines were not alone in defending and entrenching their marching camps, of course. It is clear from the treatise on *Skirmishing* that it was standard practice in the armies of the Caliphate and the N. Syrian emirates. At a later date it was clearly the usual practice when in enemy territory for both Anatolian Turks and German soldiers of the Second Crusade: Choniates, *Historia*, 21, 64 (the German troops did not entrench their camp in 1147 because they were on Roman territory). Psellos notes that the Pechenegs did not entrench or fortify their camps with a ditch (implicitly contrasting them to the Romans), and seems to think this an indication of their barbarity: *Chronographia*, 7.68 (trans. Sewter, 242).

3–9 §1 Maurice, *Strategikon*, 12.B.20.91–94; and cf. Onasander, *Strategikos*, 8.306–16 [8§1–2]). See on 4.8 above. The *phossaton* referred by extension to the expeditionary army as well. These marching camps or *aplekta* are not to be confused with the more permanent establishments, referred to by the same term, which served both as mustering points for imperial armies gathering for extensive campaigns as well as centers of supply and logistics and in some cases as sources of livestock. The origins of these are obscure, and the question of when they first appeared remains unanswered. Malagina, the base nearest to Constantinople, is first mentioned when the *tagmata* were ordered to assemble there in 786/87 by Eirene; it was seized and sacked by an Arab force in 798/99 (there is some disagreement over the chronology): *Theophanis Chronographia*, 462, 473 (trans. Mango and Scott, 636, 651), and appears to have been a base for the imperial baggage train and a stock-raising center. Malagina was on the Sangarios, in Bithynia; see Foss, "Byzantine Malagina," following Sahin, *Katalog der antiken Inschriften*, 22–23, and 150. Although Malagina is mentioned in the late seventh-century pseudo-Methodius Apocalypse among the places at which an Arab force would winter in preparation for the attacks on Constantinople in the period 674–78, this cannot be used as *terminus a quo*, since this section of the text has been shown to be a much later interpolation (late eighth or early ninth century); see Aerts, "Zu einer neuen Ausgabe," esp. 129–30. The other bases were all strategically located for expeditions directed at the eastern or southern frontier, and again, may have been established as part of a systematic strategy under Constantine V—there is no evidence for their existence before this time. See discussion in *CPTT* A and commentary, pp. 155–57.

10–22   Onasander, *Strategikos*, 8.306–16 (8§§1–2). See also *Sylloge tacticorum*, 22.1 and on 159–64 below. Concern over the location of camps is frequently expressed in the treatises; see ll. 155–74, below. Reference to "rising vapors and foul smell" which "bring pestilence and deadly diseases" (ll. 18–19) reflects a standard premodern assumption about the causes of disease; see Borca, "Towns and Marshes," and especially Nutton, "Medical Thoughts"; Nutton, *Ancient Medicine*, 25–27; and more generally Traina, *Paludi e bonifiche*. In fact, it is foul, brackish, or otherwise polluted water that brings disease, in particular dysentery, the major scourge of armies everywhere before the arrival of cholera from the Indian subcontinent in the middle of the nineteenth century; see Biraben, "Diseases in Europe." Marshy and low-lying, sodden terrain is also home to malaria, of course, a further threat to the well-being and effectiveness of the soldiers. The major exacerbating factors in the spread and intensity of disease were length of stay and density of numbers, combined with situation. We have little direct evidence about the incidence of disease in Byzantine military camps, nor about mortality rates more generally. Leo, following the *Strategikon*, based in turn on Onasander and a long tradition of Roman thinking in this respect (see, e.g., Vegetius, *Epitoma*, 1.21–22, who also denounces the loss of skill in entrenching and defending a camp among the armies of his own day), advises that marching camps be established in healthy sites, and that they be of short duration. Sieges were an especially dangerous situation, since the soldiers were generally resident outside their objective for weeks or months at a time, and the possibilities for disposing of human waste diminished fairly rapidly after the first weeks. Disease was far more deadly than battle—in the Boer War, the Crimean War, and the American Civil War, for example, death from disease ran at over twice the rate of attrition from battle; see in general the older but still useful Prinzing, *Epidemics Resulting from Wars*. While conditions were different in many respects, mortality from disease, especially dysentery and related enteric infections, often reached as much as 30% of a force in sixteenth- and seventeenth-century European wars. Work on records for the Crusades suggests an average mortality rate from noncombat causes, including disease, of some 15–20%; see Mitchell, *Medicine in the Crusades*. Precise Roman figures are few, but what evidence there is suggests that similar numbers might be unfit for duty due to illness: Nutton, *Ancient Medicine*, 25; Rosenstein, *Rome at War*, 130–31. Two recognized contributory factors in this were poor nutrition of the soldiers, on the one hand, and poor footwear and clothing, on the other; see Tallett, *War and Society*, 105–7. Climate and other demographic factors obviously affected both, but the emphasis in the Roman and Byzantine tradition on adequate supplies and appropriate equipment—including footwear—might suggest that these factors were generally less influential, even if it is also true that we have no evidence other than the recommendations in the treatises that proper footwear, for example, was actually worn.

On the other hand, the hazards attending any encampment with respect to disposal of waste, both human and animal, were considerable. Permanent Roman military camps were generally well-equipped with latrine and watering facilities—filtration tanks for maintaining a reasonably pure water supply and latrines and cesspits established well away from barracks; see Davies, *Service in the Roman Army*, 211; Davidson, *Barracks of the Roman Army*, 1:233–36. While there is little doubt later Roman and Byzantine soldiers did not benefit from the level of sophistication achieved by the Roman army medical services in the third century, either with respect to sanitation or to surgical services (see Nutton, *Ancient Medicine*, 171–86; von Petrikovits, *Die Innenbauten römischer Legionslager*, 105–6; and Johnson, *Roman Forts*, 202–10), nevertheless the regulations and recommendations in the *Taktika* suggest at least an awareness of the most important issues of military health; and the evidence of the chronicles (for example Leo the Deacon or Attaleiates) also suggests that good commanders tried to observe them as closely as possible. It has also been suggested that the Roman military tradition of food preparation in small tent groups or *contubernia* of 8 men or so minimized the possibilities for enteric infections to spread across a whole army through poor mass food preparation: Jackson, *Doctors and Diseases*, 133. Assuming that Maurice, Leo, and the other Byzantine treatises, which refer to squads or mess groups of eight or ten, reflect actual practice (and there is no reason to doubt this in the current state of our knowledge), then this consideration may also have helped minimize Byzantine illness rates in this context.

The emphasis on entrenching even if for only one night, and the need to set pickets and watches carefully, reflects not just the sources Leo employs but the realities of warfare both in the Balkans and on the eastern front. Imperial forces had suffered several defeats as a result of laxness in these respects (see below), and Leo was certainly only too aware of them.

23–27   §4   Onasander, *Strategikos*, 9.319–23 (9§1); cf. *Sylloge tacticorum*, 22.2.

29–30   Onasander, *Strategikos*, 9.324–33 (9§§2–3); cf. on Const. 7§2.8–20 above.

31–36   §§6–7   The concern for ensuring adequate supplies and pasture is repeated throughout the *Taktika* and is a major theme in all the treatises. The welfare of merchants is likewise an important theme in the *Taktika*, and reflects the twofold nature of Byzantine logistical arrangements, depending partly on markets and supplies provided through merchants and traders, and partly on state requisitions and taxation in kind.

37–44   §8   There is no similarity here to Maurice, *Strategikon*, 7.B.13 (see Dennis, *Taktika*, 196, ad ll. 39–66), which concerns scouts. No similar passage is found in the *Strategikon*, but Leo seems to be basing his warning on Onasander, *Strategikos*, 8.306–310 (8§1), as well as drawing on advice in later sections of this constitution that are drawn from Maurice.

39   ἔχεις δὲ καὶ βίγλας ἔξωθεν   There were a number of terms employed to describe guards or scouts and their duties. The general term φυλακή referred to any guard or watch, usually fixed. The slightly more specialist *vigla* (Lat. *vigilia*) could refer to a fixed or mobile watch, scout, or picket, and those who patrolled the roads or perimeter of a camp might be referred to as βιγλάτορες. In specific contexts, however, the watches or pickets might be qualified, as ἐξώβιγλα, for example, the outer cordon of pickets or outposts, as opposed to ἐσώβιγλα, the inner cordon (e.g., *Skirmishing*, 15.23–42). For a detailed discussion, see Dagron and Mihăescu, *Traité sur la guérilla*, 215–18. The sentries who patrolled the inner perimeter of an encampment or a fortress belonged to the κέρκετον (Lat. *circa/circare* > *circitum*, cf. the verb κερκετεύω, to patrol/make rounds); see, e.g., Nikephoros, *Praecepta*, 5§7.64–68; and esp. *CPTT* C.310, 420 and commentary with further literature and references, 232, 240.

40–42   ἢ χάρακα πήξεις, ὃ λέγεται σταβαρῶσαι, εἴτε ἀραιῶς εἴτε πυκνότερον, ὡς ἡ δύναμις ἔχει, ἤτε διὰ ξύλων τελείων ἢ δένδρων κοπέντων.   See Trapp, *Lexikon*, 7:1597. Leo's description suggests a barricade of stakes or cut branches, an interpretation confirmed by the paraphrase of Nikephoros Ouranos in his version of the same passage in his own *Taktika*: ἢ κόψον δένδρα καὶ σταβάρωσον αὐτὸ ἢ μετὰ τῶν τοιούτων δένδρων ἢ μετὰ ξύλων μεγάλων (Vári, *Taktika*, 286.1002). In the Ambrosian redaction (A = Ambrosianus B 119 sup., written between 959–63, and probably by a single scribe/redactor; see Leoni, *La parafrasi Ambrosiana*, 18–28; Cosentino, "Syrianos's 'Strategikon',' 243–46) of excerpts from the *Strategika* of Polyaenus, the so-called *Strategemata Ambrosiana* (10.2; see Dain and de Foucault, "Stratégistes byzantins," 364–65) the verb σtυβαροῦν appears. In the same manuscript and in the paraphrase of Onasander's *Strategikos*, 52.589 (Korzensky and Vári, 72), the term recurs as στάβαρα, and with the same meaning (... καὶ διὰ σούδας καὶ διὰ χάρακος, ἤτοι σταβάρων ...). In the late tenth-century anonymous *Parekbolai* (9.2, also based on Polyaenus via several intermediaries; see Dain and de Foucault, "Stratégistes byzantins," 368–69) the word appears as στάβαρον. The term σταβαρῶσαι and its derivatives seem to be contemporary—late ninth century at least—and occur thereafter in several texts, including Eustathios of Thessalonike's commentaries on Homer; see Eustathios of Thessalonike, 2:492.18 (cited by Du Cange, *Glossarium Graecitatis*, 1424–25 s.v. Στάβαρα); and in some monastic archival documents; see *Acts of Lavra*, 298–99, l. 15 (a. 1108); *Acts of Pantokrator*, 165.6 (a. 1400).

It appears as a synonym for the ancient term σταύρωμα, a palisade (cf. *LSJ*, s.v., and, e.g., Polyaenus, *Strategika*, 8.23.7 = *Hypotheseis*, 16.2). Different versions occur in the tenth-century texts, as noted. Additionally, in the *Apparatus bellicus* (date uncertain, possibly pre-Leo: ed. Zuckerman, 368.95) we find the term στάβαρα, in a passage certainly drawn from an earlier late Roman treatise, but into which this more "modern" term has been introduced by the copyist or redactor. It indicates a type of *cheval de frise* (in its simplest form a horizontal wooden axis with a series of crossed sharpened stakes, cut branches or spears attached along its length), equivalent to the classical term σταύρωμα. A similar list in the early or mid-tenth-century *Sylloge tacticorum* (22.4) has: Σκληρῷ δὲ τῶν τόπων ὄντων ἔσθ' ὅτε καὶ ῥαδίως ὀρύσσεσθαι μὴ παρεχομένων ἢ τειχῶν ἐκ πλίνθων ἢ λίθων ἢ κορμοῦ δένδρων, ἢ σταυρώμασιν ἢ ἀμαξῶν τῷ πλήθει τὴν παρεμβολὴν πᾶσι τρόποις ἀσφαλιζέσθω, thus using the classical term rather than the newer word. The standard usage of σταύρωμα occurs in Skylitzes, *Synopsis historiarum*, 39.45; trans. Wortley, 42 (τὴν Ἀδριανούπολιν καὶ χάρακι καὶ σταυρώμασι... ἐπολιόρκει) when Michael II cuts the city off with palisaded entrenchments. The word στάβαρα/στάβαρον/ σταβαρῶσαι and its derivatives seem to reflect the conflation of the older term with a word of Slavic origin, cf. mod. Russian ставитъ, "to place/put/fix" and ставка, "a placing/setting." A likely derivation is from the noun *stobor*, an old Balto-Slav word present in several south Slavonic languages such as Bulgarian, Slovenian, and Serbo-Croatian, with meanings such as "pillar, post, stake." For detailed discussion of the etymological possibilities and derivations, see Vasmer, *Slaven*, 237–38 and 247; idem, *Russisches etymologisches Wörterbuch*, 3, s.v. *stoborije*. I am especially grateful to Professor Mary MacRobert and to Dr. Jonathan Shepard for their advice in this matter.

In contrast to words such as τάφρος or φόσσα, which can mean only a ditch or trench, the term χάραξ (and related terms such as χαρακώσεις), while it also referred to a ditch (cf. *Campaign Organization*, 1.88: Ὁ δὲ χάραξ βάθος μὲν ἐχέτω ποδῶν ἑπτὰ....) had a broader meaning, and could refer also to a defensive barricade or palisade, and by extension was equivalent to φοσσάτον: cf. Maurice, *Strategikon*, 12.C.1 (Καταγραφῇ χάρακος ἤτοι φοσσάτου) and *Sylloge tacticorum*, 21.4 (τῇ τοῦ χάρακος τάφρῳ) and the treatise of Syrianos, *Strategy*, 29, which may be of ninth-century date. It can take one of several forms, either wagons and carts set up as a laager (a defensive perimeter also referred to as a *karagos*; see below); or a barricade of wood or stone, or a barrier of stakes or cut branches, set densely or more openly and referred to by the term *stabarosai/stabara/stauroma* (see above). Alternative terms for ditch were χάνδαξ and σοῦδα. The former appears from the ninth century on (e.g., Nikephoros, *Praecepta*, 5§6.54–55: τάφρον . . . τὸν παρ᾽ ἡμῖν λεγόμενον χάνδακα), derived from Arabic *kndq*, a ditch or trench: cf. Sophocles, *Greek lexikon*, s.v. For the latter see e.g., Nikephoros, *Praecepta*, 5§5.49; Nikephoros Ouranos, *Taktika*, 176, ed. and trans. Zuckerman (in "Chapitres peu connus," 381–82), ll. 12–13 (τὸ ὄρυγμα τῆς σούδας), and discussion in Dain, "Σοῦδα," 233–41. The term *karagos* is late Latin *carrago*, perhaps of Gothic origin (cf. Ammianus Marcellinus, 31.7.7), although the first element may be Latin (cf. *carrus*); see Mihăescu, "Éléments latins," 1:498; idem, "Littérature byzantine," 380. Whereas terms such as *charax, taphros*, or *touldon* went into standard medieval Greek usage (see Sophocles, *Lexicon*, s.v. and Trapp, *Lexikon*, s.v.), *karagos* appears only in Maurice and Leo (Trapp, *Lexikon*, 4:764), and although he states that it is "the term we use" (11.220; cf. 4.224), it seems most probable that this does not reflect actual contemporary usage at all, but is adopted simply from Maurice, his source.

To what extent the instructions and advice on castrametation which Leo repeats here were put into practice clearly varied according to circumstances, but all the evidence suggests that the standard precautions were observed when in enemy territory, the more so in view of the continuously updated textual tradition discussed above. The treatise of Leo Katakylas drew on the standard campaign practice of the ninth century, and refers as a matter of course to the army establishing a defensive perimeter every evening and to the security measures that were usually put in place: *CPTT* B.148–50. Leo the Deacon, a chronicler of the wars of the emperors Nikephoros II Phokas and John I Tzimiskes, notes that the Romans customarily fortified their camps with a ditch and bank surmounted by spears, a technique described exactly in contemporary military treatises.

Note that this practice is not found in the late Roman period, and appears to be a Byzantine, or at least post-Roman, development, perhaps under the influence of neighboring peoples (see below). He often mentions generals throwing up earthworks around besieged cities or fortresses, and his testimony together with that of other chroniclers of the period suggest that Roman forces were thoroughly accustomed to such undertakings. Other accounts report the establishment of marching camps along the advance, to which Roman forces were able to withdraw on their return march: Leo the Deacon, *Historia*, 8.16, 16.18–20, 58.11–12, 70.5, 72.21, 142.1–143.5, 171.11 (Talbot and Sullivan, 62, 68, 106, 119, 123, 187, 213); Attaleiates, *Historia*, 109.5–7, 117.11–12, 118.11–13, 119.12–15, 120.9–10, 151.8–10 (trans. Kaldellis and Krallis, 199, 213–15, 217, 219, 275); cf. McGeer, *Sowing the Dragon's Teeth*, 347–57. Theophylact Simocatta (*Historia*, 6.6.5) describes the distance covered by the army on certain occasions as representing a "march of so many camps," indicative of the standard practice—the distance from Heraclea in Thrace to Drizipera was 4 camps; that from Drizipera to Dorostolon on the Danube was 20 camps. The tenth-century treatise *Skirmishing* likewise speaks of marches in terms of a day or the distance between two camps: e.g., 22.15; 23.2. The passage as a whole is interesting, insofar as, when Leo relies on Maurice, there is no question but that a trench or defensive barrier should be established when the army camps for the night; whereas Leo implies here (ll. 42–44) that commanders really only had to do this when in enemy territory and when they were aware of an enemy force nearby. When on home ground or on exercise, or when there was no threat from the enemy, it was optional—or was generally seen as such. This view is reinforced in the *Praecepta* of Nikephoros, 5§6.55–58 and repeated in the *Taktika* of Nikephoros Ouranos: 62.6.

In the sources which give details of marching camps, such as in Leo the Deacon's account of a camp established by John Tzimiskes during his Balkan campaign of 971, the perimeter is defended by the serried shields of the infantry units encamped around the cavalry and baggage, the spears stuck in the ground and pointing out through the shields—see Nikephoros, *Praecepta*, 5§§3, 5 (= Nikephoros Ouranos, *Taktika*, 62.3–5); Leo the Deacon, *Historia*, 142.1–143.5 (Talbot and

Sullivan, 187). This arrangement is referred to as the *skoutaroma*, and may be a post-sixth-century development, since it is not mentioned in the *Strategikon* or earlier Roman treatises such as Vegetius; but see Wheeler, review of Charles, *Vegetius*, n. 12. See McGeer, *Sowing the Dragon's Teeth*, 347–59 for other accounts, including a description in Yahya of Antioch relating to a similarly defended encampment set up by Romanos III during his campaign in Syria in 1030. See also Syrianos, *Strategy*, 28.28–29, which describes the same arrangement as τὸν κοινὸν χάρακα. In both Leo the Deacon and Yahya, this practice is described as "usual" or "standard" for the Byzantines; and a perimeter made up of the shields of soldiers of the *Vigla* is explicitly referred to as surrounding the imperial encampment on an expedition when the emperor is present: cf. *CPTT* C.429–31. If the treatise on *Strategy* ascribed to Syrianos is indeed of ninth-century rather than sixth-century date, the defensive systems here described may well reflect standard middle Byzantine practice from the time of Basil I at least, if not earlier; see Rance, "Date," 719–23. It is worth noting that archaeological evidence from Khazar graves of the ninth century includes flanged spearheads pierced with a hole close to the socket. These have been interpreted as a means by which the spears can be linked to form part of a spear-and-shield fence around an encampment (see Gorelik, "Arms and Armour," pl. 11.8.14–18), a defensive system ascribed by the eleventh-century Persian writer Abu Sa'id 'Ab' al-Hajj b. ad-Dahhak b. Mahmud Gardīzī in his *Chronicle* as typical of the steppe people. See Martinez, "Gardīzī on the Turks," 154–55 ("Their leader orders every man in the army to make a peg sharpened at one end the length of three cubits, [so that] when the army dismounts those pegs are implanted [in a circle] round about the army. [Then] a shield is hung from each peg until the encampment becomes like a walled town. [Thus] if an enemy were to attempt a surprise attack by night and elude the sentries he would be unable to do anything, for the encampment is like a fortress by reason of those pegs") (although "stakes" is a better translation for these 3–cubit-long [1.4 m] "pegs").

45–53 §9 Leo's concern for the peasantry and farmers—the tax base of the state, in effect—is well known and occurs elsewhere in the *Taktika*. See on 4§1.11–12 above and cf. 20§§193 and 209 below.

54-66 §§10-11 Maurice, *Strategikon*, 7.A.13; cf. 5.4. Leo omits the reference to the possibility of an attack by "the Scythian peoples." The emphasis is on forethought and advance planning, and must have been especially important in the more arid regions or toward the end of the summer, when both pasture and fodder were difficult to come by.

57 ὡς ἤδη σοι καὶ ἐν τῷ περὶ ὁδοιπορίας κεφαλαίῳ προδιεθέμεθα But there is no such advice in Const. 9, on marches; a similar reference, although for a slightly different situation, does occur at 10§13.

67-72 §12 Maurice, *Strategikon*, 7.A.7.

67 εἰς δευτέραν τύχην The expression δευτέρα τύχη, referring to a sudden reversal, can be traced at least as far back as the late Hellenistic *Aesopica*, or *Lives of Aesop*, where it refers to "an unexpected turn for the worse": see Perry, *Aesopica*, Vita G, sections 57–59.70sqq. While compiled probably during the Second Sophistic, elements may well be several centuries older. The expression seems to be proverbial, although there is no trace of it in the corpus of Greek proverbs; and while it is unusual for what may have been a standard usage to disappear for some centuries before reappearing, it probably had a vernacular life for centuries before resurfacing in the late sixth century. Theophylact Simocatta (*Historia*, 5.4.5) uses it in the same way on one occasion: ἐς πολεμίαν γὰρ γῆν ἀναστρέφεσθε, ἐν ᾗ τὸ νικᾶν ὑπερένδοξον καὶ τὸ τῆς δευτέρας τύχης βαρὺ καὶ λεγόμενον. . . . See Const. 13§7.37–37; 18§66.319; 20§125.623.

73-104 §§13-18 Maurice, *Strategikon*, 12.B.22.2–32, with Rance, *Strategikon*, 12.B.22, notes 241–43; and cf. *Sylloge tacticorum*, 22.4–10. The arrangement repeated from Maurice by Leo was thus to have the light troops, those with javelins and bows, encamped close to the defensive perimeter and the wagons, and behind them a substantial space (which in earlier times would have been referred to as an *intervallum*)—300–400 Roman feet (approx. 94–125 m)—before the main body of soldiers and the horses, to protect the latter from enemy archery or missiles. Maurice also includes a section (12.C), however, which offers different measurements, similar to those given in chap. 176 of the late tenth-century *Taktika* of Nikephoros Ouranos (381.4–10 ed. Zuckerman): only 200 feet between the wagons and the tents of the army. As noted already, the discrepancy reflects the existence of two traditions deriving from the same late Roman source or sources, in turn representing texts which described actual practice under different circumstances (Zuckerman, "Chapitres peu connus," 384–85). The (possibly) ninth-century anonymous Syrianos, *Strategy* (28) and the section on camps in the late tenth-century *Apparatus bellicus* both reflect the tradition upon which Maurice's section 12.22 is based, describing the infantry (heavy infantry outermost; as noted, however, Leo—following Maurice—omits these and refers merely to the light infantry) as quartered close to and around the inner perimeter of the camp. Between them and the rest of the army a path or space of some 300–400 feet was left, between the infantry tents and the cavalry, who were quartered in the center, as far from the danger posed by enemy arrows as possible (although Syrianos, *Strategy* gives no actual measurements); this was also intended to permit inspection of the defenses and the guards. The *Praecepta* of Nikephoros ordains a similar arrangement, with infantry around the perimeter, cavalry within, and at the center the pack animals, a bowshot from the perimeter: *Praecepta*, 5§5.48–49 (a bowshot, according to the *Sylloge tacticorum*, 43.11, was 156 *orguiai*, or just under 330 m; see on Const. 7.165 above); the late tenth-century anonymous treatise *Campaign Organization*, 1 gives even more precise detail: there should be a space of 22 *orguiai*, which is ca. 44 m, between the perimeter and the infantry tents; a width of a further 44 m for the latter; and then an additional twelve meters between the infantry tents and the cavalry, making a total of 100 m between perimeter and cavalry. The *Sylloge tacticorum* (22.7) largely follows Leo, but does not distinguish between light and heavy infantry camped within the perimeter. Like Leo (and Maurice) it stipulates a cleared area around the outside of the ditch, to be scattered with caltrops and small pits against enemy cavalry, of some 8–10 *orguiai* (16–20 m); and another clear area between the inner defenses (wagons and infantry), and the tents of the officers, their units and the horses, of 300 Roman feet, or about 94 m. All agree that this arrangement is to protect the latter from enemy archery.

Although its origins lie in the format and layout of the classical legionary and auxiliary camps of the empire, the Byzantine marching camp had evolved away from this early pattern, a reflection of shifts in both tactical organization as well as in strategic requirements. Unfortunately there are no excavated or even surveyed examples of Byzantine camps against which to compare the treatises, largely because they were very temporary establishments which can have left little trace and so they have never been the object of any extensive archaeological surveys. The general assumption is that the details given in the *Strategikon* reflect actual practice in the later Roman period; see Southern and Dixon, *Late Roman Army*, 132–33, but even they make little reference to surveyed or excavated examples. Yet while for both the later Roman as well as the Byzantine period there is no archaeological or topographical evidence, the corroboration from historians' accounts would suggest that there is little reason to doubt them. For further detailed discussion and literature, see McGeer, *Sowing the Dragon's Teeth*, 348–59.

The basic plan of the early and middle Byzantine camp was rectangular, divided internally into four quadrants separated by centrally crossing paths or roads leading to entrances in the middle of each side. This is the pattern described in the anonymous (possibly ninth-century) Syrianos, *Strategy*, 28 and in the *Strategikon* of Maurice, 12.B.22, and it is repeated in this constitution of the *Taktika*. While Leo's account reflects his sources, he does also incorporate contemporary terms. By the time the *Praecepta* of Nikephoros and the anonymous treatise on *Campaign Organization* were compiled, two or even three sets of intersecting paths were ordained, further subdividing the internal space of the camp (Nikephoros, *Praecepta*, 5; *Campaign Organization*, 1). But in both cases, the encampment is intended for an imperial field army, reflecting differences in status and function between various units; see McGeer, *Sowing the Dragon's Teeth*, 349–52.

The exception to these arrangements is preserved in chap. 176 of the *Taktika* of Nikephoros Ouranos, which reflects a (probably) ninth-century text, and differs in both its lack of detail, on the one hand, and in the different arrangements for a camp, on the other, from chap. 62 of the same treatise, which follows closely chap. 5 of the *Praecepta* of Nikephoros. This short chapter is important, because it is the only text to mention the four imperial *tagmata* (*Scholai, Exkoubitoi,*

Vigla/Arithmos, and *Hikanatoi*) in the context of a field army on campaign. The *tagmata* continued to campaign throughout the tenth century (cf. *Campaign Organization*, 1.157–62, where the *Scholai* as well as "the other *tagmata*" are mentioned, with the *droungarios* of the *Vigla*, along with troops of the *Hetaireia* and the new unit of the *Athanatoi* or "Immortals"). Soldiers of all four imperial *tagmata* were frequently referred to collectively as οἱ σχολάριοι rather than by their separate unit names—see, e.g., the documents for the 949 Cretan expedition, ed. Haldon, "Chapters II, 44 and 45," chap. 45.37; 40; 42–45 (Ἀπὸ τῶν περατικῶν ταγμάτων ὁ ἐξκουβίτωρ μετὰ τοῦ τοποτηρητοῦ αὐτοῦ καὶ παντὸς τοῦ τάγματος αὐτοῦ, ἀρχόντων καὶ σχολαρίων. . . . ὁ ἱκανᾶτος μετὰ τοῦ τοποτηρητοῦ αὐτοῦ καὶ παντὸς τοῦ τάγματος αὐτοῦ, ἀρχόντων καὶ σχολαρίων. . . .). The reference to the four "original" *tagmata* in this text preserved by Nikephoros Ouranos might refer either to a regular nonimperial expeditionary encampment of the first half of the tenth century (the *Hetaireia*, established in the ninth century, campaigned only when the emperor was in command: cf. *CPTT* C.421–24; the *Athanatoi* were a unit established by John I Tzimiskes), or to an earlier period, as described in the *Three Treatises* ascribed to Constantine VII, and probably reflects conditions in the ninth or early tenth century (although the text does refer to sixteen taxiarchies of infantry, following its late ancient model). See *CPTT* B.107–15; 148–50; C.474–78.

78 λάκκους μικρούς, ἔχοντες πάλους ἐντὸς πεπηγμένους Follows Maurice, *Strategikon*, 12.B.22.118–19, but without the small pits, which Leo adds and which, at Const. 14§42, he takes from Maurice, *Strategikon*, 4.3.52–65. See below on Const. 14§42.293–305; and Gilliver, "Hedgehogs."

81 παραπόρτια The Ambrosian paraphrase of Onasander has τὰς πυλίδας, ἤτοι τὰ παραπόρτια (see, e.g., Korzensky and Vári, 73.596–97, 79.693); and Maurice, *Strategikon*, 10.3.21, 24; 12.B.22.9 uses both παραπόρτια and παραπύλιον. Leo uses παραπόρτια here and at Const. 15§4.22 and §45.280, and παραπύλη at 15§45.282; see below and Trapp, *Lexikon*, 6:1221. While Leo copies Maurice, late tenth-century authorities describe a somewhat different layout for the gates, with from eight to twelve gates evenly spaced around the perimeter:

Nikephoros, *Praecepta*, 5§4 (and McGeer, *Sowing the Dragon's Teeth*, 77 and 352; *Campaign Organization*, 1.71–87; and see the diagrams in Dennis, *Three Treatises*, 329).

91–92   Καὶ ἕκαστον τουρμάρχην ... πλατείας   Locating the commanding officer to one side or in one sector of the central section of the camp had been standard Roman practice, and following Maurice (12.B.22.15–21; with Rance, *Strategikon*, 12.B.22, note 247) is repeated by Leo here; see Hanel, "Military Camps," 401–3 and the older but still valuable account in Webster, *Roman Imperial Army*, 167–70. Whether this arrangement had fallen into desuetude or not is unclear, since the late tenth-century treatise *Campaign Organization*, 1.107–8 merely notes that the imperial tent or headquarters is to be in the center of the encampment, as do the *Praecepta* of Nikephoros, 5§1 and, in its turn, the *Taktika* of Nikephoros Ouranos, 62.1. The *Sylloge tacticorum*, 22.7 notes simply that the subordinate commanders are to pitch their tents in the center of their divisions, which occupy the four quadrants of the camp, and although there is no specific stipulation about the general's tent, all four divisions are to ensure that they leave room around the general's tent to avoid congestion in the event of panic or an attack.

The arrangements for setting watches and patrolling the perimeters described in the latter treatises are very close, both in the units involved as well as in the system of watches and pickets, to those implied in the *CPTT* C.474–76, 570–79, the information for which was drawn from the reigns of Basil I (867–86) and Leo himself. There were a number of technical terms associated with camps and watches, some of which appear in the treatises, some of which are absent. Leo uses neither the word κέρκετον for the watch patrol (see on ll. 37–44 above) nor the term φῖνα for perimeter, for example (see *CPTT* 226), whereas he does use πεδατοῦρα, which in Maurice and earlier late Roman sources refers to sections of defended walls (see Mihăescu, "Éléments latins," 497) or even stretches of defended *limes* (Dietz, "*Cohortes, ripae, pedaturae*"; Minkova, "Some More Information"), but which by the ninth and tenth century referred to the soldiers who guarded a perimeter, or a wall, also; see Haldon, *Byzantine Praetorians*, 541–42; idem, "Strategies of Defence," 145, and cf. *CPTT* C. 438, 575. Cf. also on Const. 15.299 below.

97 τὰς ἑσπερινὰς μίσσας "Dismissal," derives from Latin *missa*. Cf. *Theophanis Chronographia*, 237.20: πρὸ τοῦ μισσεῦσσαι, and for the development of the term and its meanings in the Byzantine period: Mihăescu, "Littérature," 373, with Trapp, *Lexikon*, 5:1031.

101–4 §18   Here, some of the duties of the unit *mandatores* are outlined, illustrative of the relative efficiency—in theory at least—of the tactical arrangements of the Byzantine armies of the period. See above on 4§18.89.

105–7 §19   For prayers and hymns in the army on campaign, see Pertusi, "Acolouthia militare"; Dennis, "Religious Services"; and Caseau and Cheynet, "Communion du soldat." Daily prayers, and prayers especially before combat, were a standard feature of Byzantine armies on campaign (cf., for example, Nikephoros, *Praecepta*, 6§3.32–46 on fasting before battle, followed by the soldiers and officers taking communion). Priests regularly accompanied the army; and on an imperial expedition, liturgical vessels and a small imperial chapel (presumably a tent), as well as four priests in the emperor's personal service, accompanied the imperial cortège; see *CPTT* C.183–84; 389. Note the difference between "large" and "small" trumpets or bugles, and see above on 5§4.32; below, on l. 145 and on Const. 15§20.109–21; and cf. Trapp, *Lexikon*, 2:289.

106–7 τοῦ εἰθισμένου βουκίνου . . . τρισάγιον ὕμνον   Maurice, *Strategikon*, 12.B.22.33–35.

108–15 §20   Maurice, *Strategikon*, 12.B.22.35–43. See on ll. 37–44 above. The somewhat later *Sylloge tacticorum* (22.9) states that during the morning and evening hymn the so-called "extended litany" (τὴν λεγομένην ἐκτενῆ) should take place (see *ODB*, 2:1234; Trapp, *Lexikon*, 3:475 with references and literature) and that singing and dancing after the evening meal should be discouraged (but see also Dennis, *Taktika*, 203, note 10 with literature) in order to prevent any sort of disorder; the *Praecepta* of Nikephoros, 6§2 stipulates something similar, although with greater emphasis on the soldiers' participation in the service. For the *Trisagion* see *ODB*, 3:2121.

116–21 See 20§21 (and see Maurice, *Strategikon*, 8.1.27). The opening clause in the first sentence of this section follows the opening sentence of Maurice, *Strategikon*, 12.B.22.4; the next sentence, on lighting numerous campfires to make the enemy think the army is still in the camp, is from Onasander: *Strategikos*, 10.413–18 (10§13); and cf. Polyaenus, *Strategika*, 1.40.8; 3.9.50; cf. also 3.11.15 (*Hypotheseis*, 46.6); 4.18.2; 7.11.4.

121–27   This well-known story is repeated in greater detail in the later *Skirmishing*, 20 and mentioned again in the *Taktika*, at 17§65.373–76 (and see commentary).

121–22   Νικηφόρον ἴσμεν, τὸν ἡμέτερον στρατηγόν   Nikephoros Phokas, the elder, the grandfather of the emperor Nikephoros II and the father of Bardas Phokas. See Cheynet, "Phocas," 291–96; Honigmann, *Ostgrenze*, 82–83 (although misdated to the year 900); Grégoire, "Carrière;" and *PmbZ* 25545, 20769.

122   ὅτε κατὰ Συρίας ἀπεστάλη παρ' ἡμῶν   The campaign in question is not mentioned in the Arabic chronicles for the period and seems to have taken place at some time between 886, when Nikephoros returned from a command in Italy, and 895/96, in which year he probably died; see Cheynet, "Phocas," 293–95. Leo gives no details beyond this, but we must assume that Nikephoros was able to slip away unseen at some point, using this particular stratagem, before the enemy forces could regroup after their defeat outside Adana. The identity of Apoulpher, the eunuch, "the commander of the Saracens" (*PmbZ* 20548), remains uncertain, although he is probably the same as Yāzamān, the emir of Tarsos who died in 891/92 (*PmbZ* 28463).

128–42   §22   This expedition is probably that of 895, when Nikephoros was recalled from the eastern command, but the truce which was confirmed in 896 meant that he saw no fighting.

The three-legged device "invented" by Nikephoros functioned as a large caltrop or anticavalry device, and was intended as an effective perimeter defense where no entrenchment was dug around the camp. Whether Nikephoros really invented these *triskellia* fitted with blades, or merely made use of an existing device is not clear—the *Taktika* of Nikephoros Ouranos refers to τρισκέλλια μετὰ τζιπάτων which should be thrown, together with caltrops, around the outside of the trench surrounding the camp: Nikephoros Ouranos, *Taktika*, 65.11. The phrase τάξιν μεναύλου refers to the stoutness of the third and longer "leg," to which the blade was to be attached, and suggests that the term *menaulon* was already beginning to be employed to mean a pike rather than a regular spear or lance; see on Const. 6§27.171 above. McGeer, *Sowing the Dragon's Teeth*, 166, and idem, "Tradition and Reality," 134–35 suggests that the three-legged device involved a full-length spear with a long blade, but the dimensions Leo gives, and the functions ascribed in both Leo's *Taktika* as well as that of Nikephoros Ouranos make it clear that it was much smaller than this—the longer leg was to be of 5 or 6 spans, that is to say 117–40.4 cm (45–54 in.), whereas spears and lances ranged from 8 to 10 cubits long, or 3.7 –4.7 m (146–85 in.), with heads or blades of 1.5 spans (47 cm or 13.5 in.) (see above on Const. 6.17; 114). The device was thus small enough to be dismantled and carried by the soldiers without too great an additional burden, effective enough to repel or hinder cavalry, and big enough to be held and used as a weapon, all the characteristics described here by Leo. The term *labdaraia* or *labdarea* was used to describe various devices. In this case, it refers to a fairly small-bladed tripod, as we have seen. Elsewhere it can be used of larger artillery devices, probably traction-powered trebuchets (see Haldon, "Chapters II, 44 and 45," 274–75); or of stouter, larger tripods used to prevent an enemy rolling rocks and logs down against an attacking force during a siege; see *Parangelmata poliorcetica*, §6.2, and commentary, 172. For the defensive arrangements for marching camps see on ll. 40–42 above.

143–230 §§23–40 Maurice, *Strategikon*, 12.B.22.33–131.

145 τὰ βούκινα τρισσάκις σημαίνειν Rance, *Strategikon*, commentary to 12.B.22.1, notes 253, 255 for the earlier Roman trumpet signals; and see below on Const. 13§11.54–57 (with commentary to ll. 106–7 above and to 15§20.109–21 below). Leo copies Maurice's abbreviated version, which compresses three separate signals for three separate actions into one "triple blast," except that whereas Maurice distinguishes between the *tuba* and the *bucina*, the former to signal the beginning or end of activities within the encampment, the latter for military activities such as striking camp or battlefield signals, Leo notes only that there should be "larger" and "smaller" *boukina* (see on ll. 106–7 above). On trumpets and other instruments, see above, on Const. 5.32. The late tenth-century treatise *Campaign Organization*, 9, prescribes

three separate trumpet blasts. A little before dawn a first blast signals that the troops should prepare for departure. A second blast signals that a small advance force should exit the camp and prepare to guard the army's departure, that the imperial tent should be taken down, and that the troops and their pack animals should station themselves in their marching positions. At the third blast, the imperial column and the rest of the army marches out in order. None of the other treatises confirms this procedure, although there is no reason to doubt it, in view of its simple practicality. That the practice represents a standard and unbroken tradition, reflected only obliquely in Leo and Maurice, is likely, but cannot be proven. Trumpets of one sort or another were certainly employed for signals during battle, as attested by several sources—see, for example, Leo the Deacon, *Historia*, 8.5, 14.8–9, 15.17, 22.18, 24.16–17, etc. (Talbot and Sullivan, 61, 67–68, 75, 76, 108); and Maliaras, "Musikinstrumente," 84. A remarkably similar, disciplined, and impressively organized procedure employed by Mongol armies is described by the Dominican David of Ashby in the 1270s, where a large kettledrum replaces the Roman bugle; see Brunel, "David d'Ashby," 42–43; Engl. trans. in Sinor, "Inner Asian Warriors," 135–36.

148–51 §24 See on Const. 5§4.33 above and *Sylloge tacticorum*, 22.5, 6; 38.12.

152–54 Οἴδαμεν δὲ ... καὶ ἀναγκαία As Rance, *Strategikon*, 12.B.22.6, note 257, points out, the use of the archaic term στρατοπεδεία reflects Maurice's use of one of a number of earlier texts on encampments, better preserved in *Campaign Organization*, 1.

155–56 See Rance, *Strategikon*, 12.B.22, note 258 for a variant translation (where πλάγια is translated as "slopes," and ἐγκειμένους τόπους as "low-lying places").

162–64 §28 Digging latrines outside the perimeter was a standard recommendation, even if the cause of infection was itself misunderstood. See on ll. 10–22 and 25–27, above; Onasander, *Strategikos*, 8–9.310–23 (8§2–9§1); Vegetius, *Epitoma*, 3.2.12; and see *CPTT* C.308.

168–72 ποταμοῦ δὲ ... εἰς τὸ ταράσσειν αὐτόν. The danger of horses or pack animals polluting sources of drinking water was considerable. Cf. *Campaign Organization*, 1.56–59; and note older prescriptions on the issue in *CJ* 12.35.12 (= *CTh* 8.1.13 [a. 391]) regarding military pollution of civilian water sources. The treatise *Skirmishing* (5.6–8) notes that the commander should make sure his soldiers do not muddy the water supply and hence endanger their health; while the *Sylloge tacticorum*, 61—following the earlier anonymous *Apparatus bellicus*—devotes a section to the dangers of poisoned water. A horse drinks—depending on context, climate, and a range of other factors—up to 6 gallons of water per day. A large body of horses will thus generate very substantial amounts of urine, and keeping this clear of any sources of drinking water was essential. For comment on the original passages from the *Strategikon*, see Rance, *Strategikon*, 12.B.22, notes 260, 262.

175   See the comments with references in Rance, *Strategikon*, 12.B.22, note 263.

180–86   §32   Regulations and advice on keeping cavalry and infantry in separate encampments and locations until the enemy posed a clear threat are repeated from Maurice and derive in turn from older treatises. See, e.g., Vegetius, *Epitoma*, 3.8.19.

§37   For earlier historical examples of the danger described here see Rance, *Strategikon*, 12.B.22, note 270.

215–16   ἢ διὰ σαγμαρίων ἢ ἐπὶ τούτῳ ἀφωρισμένων ἵππων   Although the translation reads "requisitioned" for the verb ἀφωρισμένων the term rather implies "allocated" (from among the horses already with the army). Leo here updates his text by omitting the reference to a baggage train of camels in the equivalent paragraph of Maurice, and replacing it with the neutral *sagmaria*, "pack animals." The extent to which camels were used in military contexts at this period remains unclear. For the later Roman context, see Bartosiewicz and Dirjec, "Camels in Antiquity"; Dąbrowa, "Dromedarii"; and Graf, "Camels, Roads and Wheels." Note also Kroll, *Tiere*, 172–74. See also on Const. 18§106.517–22 below.

220   καραγὸν   For the *karagos* see above on 11.40–42.

231–38   Τοὺς δὲ ἄρχοντας . . . στρατεύματι   The instruction to officers to employ the winter period by soliciting through the chain of command information regarding the soldiers' requirements for mounts and equipment is repeated elsewhere and seems to represent the standard way through which the provincial forces were supplied; see

CPTT C.653–59 with commentary, 256, 259; and on Consts. 5§1.3–8 and 6§1; and 6§19.96–102 above. The text at this point does not specify how this was achieved, but earlier Leo notes that merchants are to be one source of such requirements.

For the serious concern with archery, see on Const. 4§35.139 above; also 6§5, 20§81.

236 τοὺς ἀστρατεύτους   That is to say, those not registered in the military *kodikes* and subject to a *strateia*, or obligation to furnish a soldier. See the detailed comments to 4§1.3–11 above and 20§§71 and 81 below.

CONSTITUTION 12

This Constitution is drawn entirely from Maurice, *Strategikon*, bks. 2, 3, and 7, with some changes in vocabulary and phrasing both to bring the text up to date as well as to impress it with Leo's own style. Entitled "On Advance Preparation for Battle," it concerns chiefly the training and drilling of the troops for battle, discussion of the use of various commands and signals, tactical maneuvering and the different formations of the (primarily) cavalry forces which it considers.

Leo's emphasis on order, on the tactical division of a field army into at least two lines, all comes straight from Maurice, of course, but it also reflects the practice of his own day. Order, discipline, the coherence these generated, and reliance on collective effect rather than on individual prowess were the characteristics which Byzantines considered differentiated themselves and their methods of waging war from their enemies. The tactical substructures described by Leo—tent group or squad, *bandon*, *droungos*, and *tourma*—were an essential element in this: right up until the end of the empire units were organized into subdivisions placed under junior officers, in a chain of command which facilitated the coherent management of often very disparate forces. Such qualities as discipline and order are frequently repeated in all the military treatises, and they are alluded to also implicitly as well as explicitly in some of the narrative histories. The differences between Byzantine order and discipline and Frankish haste and indiscipline, for example, described in the *Strategikon* of Maurice (Maurice, *Strategikon*, 11.3) and repeated by both Leo and, independently, by Anna Komnene in the early twelfth century, typifies this perspective. In reality Carolingian armies appear often to have been well-disciplined and ordered (but so were their adversaries, Saxons and Bretons, for example, who frequently defeated Frankish armies of infantry and cavalry; see Bachrach, "Charlemagne's Cavalry"; and esp. Reuter, "Carolingian and Ottonian Warfare"; Halsall, *Warfare and Society*, 90, 119–33, 215–27); the Frankish leaders of the twelfth century were often able tacticians who outwitted the supposedly more subtle Byzantines (e.g., *Alexiad*, 5.4; trans. Sewter, 162–66 / Frankopan, 135–38), and it should be stressed that the mere existence of a military disciplinary code and the assumption of Roman discipline is no proof that such discipline was always enforced, or indeed enforceable: context and the quality of leadership were crucially important pre-requisites for effective discipline. But while the Byzantine view of themselves and their enemies was laden with value judgments, it nevertheless indicates the centrality of military discipline and Roman tradition in Byzantine military thinking.

9–118   §§2–16   Maurice, *Strategikon*, 2.1.5–94; and cf. Const. 18§§143.770–149.846. See also on Const. 18§136.696–723, below. The insistence on at least two lines of battle reflects both the prescriptions of the *Strategikon* as well as experience. The *Strategikon* reflects already a shift in Roman battle tactics as the empire responded to the influence of the Avars and other nomad peoples it had to face during the late sixth century, noting that the greater the degree of subdivision of the various units, the more flexible the battle formation. The *Strategikon* notes that the "older military writers" emphasized this, and goes on to point out that the Avars and Turks "do not draw themselves in one battle line only, as do the Romans and Persians, staking the fate of tens of thousands of horsemen on a single throw. But they form two, sometimes even three lines, distributing the units in depth" (*Strategikon*, 2.1; trans. Dennis, 23). Followed faithfully by Leo's text, it goes on to prescribe a variety of basic formations, depending upon numbers, designed to meet various eventualities in the field. In each case, two battle lines are ordained, the first line with outflankers on the right and flank guards on the left, and with a third line made up of the baggage train, reserve horses, and two bodies of rearguards behind the flanks (*Strategikon*, 3.8–10). The extent to which this description marks a real change in late sixth-century tactics is difficult to assess. Evidence for the double battle line of Byzantine armies is sparse, and the degree to which these tactical distinctions were maintained after the middle of the seventh century is impossible to say. The merits of having more than one battle line continued to be recognized, and the fear that the front line might turn and run was taken into account by the treatises, including here in the *Taktika*, when describing the various formations a commander might employ. But it is equally apparent from the *Taktika*, as well as from later writers, that the Byzantine battle order for cavalry, consisting of two distinct lines which could strike the enemy's front in succession, was regarded as an essential element in the Roman potential for victory, and clearly differentiated the imperial forces from their opponents. For further discussion, see Haldon, *Warfare, State and Society*, 205–23.

14–17   §3   Repeats pr.§8.

14–23   §§3–4   Maurice, *Strategikon*, 2.1.8–18.

21   δημοσίου πολέμου   In the generic sense employed throughout the *Taktika* by Leo this term simply means "pitched battle," but *demosios polemos*, in the sense of a publicly proclaimed or declared war, was the opposite of an undeclared or covert war or encounter, ἄσπονδος or ἀκήρυκτος πόλεμος, see Chrysos, "Νόμος πολέμου," 203–4.

31–39   §6   Maurice, *Strategikon*, 2.1.82–94.

119–212   §§17–28   Maurice, *Strategikon*, 2.2–5 (although Leo expands on *Strat*, 2.5.21–24 at ll. 201–12). This presentation of the divisions of the battle line represents the basic formation for Byzantine armies from the late sixth century onward. According to the military handbooks from Maurice onward, it consisted in essence of a tripartite line—left, center, and right—with flank guards and outflanking units on the left and right wings respectively, and with a second line and a third, reserve line, behind the front line. The general himself should have a small reserve attached to his person, which could be dispatched as appropriate to strengthen the attack or the defense. Units could also be concealed behind the flank of the first or second line, both to cover these from an outflanking move or an ambush, as well as to sweep around the enemy's line to take them in the rear. There is little reason to doubt that the tripartite battle line, along with the order and discipline assumed to accompany Byzantine forces on campaign, were real enough, as in accounts of various battles, such as Versinikia in 813 (*Theophanis Chronographia*, 500; trans. Mango and Scott, 684; *Scriptor incertus* (ed. Iadevaia), 41–43; Theophanes continuatus, 13–15). At the battle of Anzen, near Dazimon, in 838, the imperial forces fought in three divisions, the center under the command of the emperor Theophilos himself, with the *tagmata*: Leo Grammaticus,

222, 224; Symeon *magistros*, 130§§27, 32; Genesios, 3.14 (48–49; trans. Kaldellis, 63–64); Theophanes continuatus, 127–29. Discussions of other battles, but which refer actually to the battle of 838 can also be found in the same sources at different locations: Genesios, 3.9 (43–44; trans. Kaldellis, 57–58) and Theophanes continuatus, 113–14 and 116–18, for a battle supposedly earlier than Anzen, but actually describing aspects of that fought in 838; and Genesios, 4.14 (65–66; trans. Kaldellis, 82–83) and Theophanes continuatus, 177 for a battle of Anzen in 858, but again describing that of 838. Arabic accounts: Vasiliev, *Byzance et les Arabes*, 1:331–32, 333–34 (Mas'ūdī); 299–301, 309 (Tabarī); 275 (Ya'qubi). In battles at Lalakaon, in the Armenian-Paphlagonia border region in 863 (Theophanes continuatus, 179–83; Genesios, 4.15 (68–69; trans. Kaldellis, 85); Skylitzes, *Synopsis historiarum*, 99–101 (trans. Wortley, 100–101); Arabic sources in Vasiliev, *Byzance et les Arabes*, 1:277 (Ya'qubi); 325 (Tabarī); for discussion: Huxley, "Bishop's meadow," including all the previous secondary literature and analysis of the sources), and at Bathys Ryax in 872/73, imperial forces were again organized by thematic division: *Vita Basilii imperatoris*, 42 (150–54); Genesius, 4.36 (86–88; trans. Kaldellis, 108–9); and Skylitzes, *Synopsis historiarum*, 138 (trans. Wortley, 136); cf. the discussion in Kaegi, *Byzantine Military Unrest*, 281–82. Before the battle of Acheloos in 917, the field army of *tagmata* and *themata* was drawn up in order, unit by unit, prior to advancing against the Bulgar forces: Skylitzes, *Synopsis historiarum*, 203.81–96 (trans. Wortley, 197). An orderly line of battle and the clear division of the army into independent corps consisting of several smaller divisions are attested throughout, and clearly continued to be the basis upon which Byzantine armies were disposed for battle. The general order of battle, as far as it can be inferred from the historical accounts, therefore, consisted of divisions organized by provincial command/*thema*, and from the time of Constantine V onward with the imperial *tagmata* in the center, both as a reserve behind the main line when the emperor was present as well as the mounted core of the line of battle.

127–41   §§19–20   Maurice, *Strategikon*, 2.2.

136–41   §20   Maurice, *Strategikon*, 2.3, 12.

142–47   §21   Maurice, *Strategikon*, 2.4.

179–84 §25 Maurice, *Strategikon*, 2.5.

194–200 §27 Maurice, *Strategikon*, 2.6.1–16.

213–17 §29 This comment seems to be Leo's own addition, added perhaps to reinforce the previous points.

218–60 §§30–34 Maurice, *Strategikon*, 2.6. At 237–42 and 245–49 Leo alters considerably the wording of 2.6.20–35, however, which deals with a group of specifically late Roman formations: the *phoideratika, vixellationes arithmoi, Illyrikianoi, Optimatoi,* and *ethnikoi,* or allies. On these, see Haldon, "Administrative Continuities," 10–14; Schmitt, "From the Late Roman"; and Rance, *Strategikon,* chapter 7. For Leo's earlier summary of these sections of the *Strategikon,* see Leo VI, *Problemata,* 2.28–30. Leo may have intended the different depths to apply to different types of unit (for example, thematic as opposed to tagmatic, and so forth), although there is no explicit evidence for this. It is in fact most likely that the sort of details here described were left entirely to the discretion of the local officers or commander, and that local tradition about who fought alongside or with whom guided the actual formations in combat.

261–63 §36 Maurice, *Strategikon,* 2.7. Although there is no other corroborating evidence for actual mess or tent groups from this period or earlier, the late tenth-century treatises, such as the *Praecepta* and *Campaign Organization,* take them for granted, and there is no reason to doubt that they were part of the normal structure of every Byzantine unit of infantry or cavalry. Cf. Nikephoros, *Praecepta,* 1§2, 3§10, 5§1, 6§1; *Campaign Organization,* 1.37–40, 2.17–18. The latter text equates the tent group implicitly with the *dekarchia* for both infantry and cavalry.

264–72 §§37–39 Maurice, *Strategikon,* 2.8. Much of this might be seen as common sense, even if entirely drawn from the *Strategikon,* but since we have virtually no evidence for actual combat situations from this period it is impossible to say whether there was ever any likelihood of different tactical weaponry and armament for soldiers in a Byzantine line of battle after the middle of the seventh century. While tactical functional distinctions do appear to have been a reality by the second half of the tenth century (see McGeer, "Infantry versus Cavalry"), Leo's treatise was compiled at the very beginnings of such changes, and it is far more likely that, with the exception of the tagmatic soldiers, the regular thematic cavalrymen were equipped more or less as they could afford to be, and that if there were different tactical roles ascribed to specific groups within the line of battle, there was little if any possibility of ensuring that they had different specialist armament.

273–302 §§37–39 Maurice, *Strategikon*, 2.9, and above on 4.86–88 (where Leo glosses the late Roman term with what he claims is the contemporary word, *skribones*). Cf. 12§96.

297–301 On stirrups, see 6§10.49 above.

303–14 §§40–41 Maurice, *Strategikon*, 2.10. Pennons and banners are frequently mentioned in the treatises, both those following earlier texts, such as Leo, as well as the more contemporary writings of the middle and late tenth century; see above on Const. 6§16.79–86.

273–302 §§37–39 Maurice, *Strategikon*, 2.9, 12.

303–14 §§40–41 Maurice, *Strategikon*, 2.10 and 7.B.17.14–16.

315–22 §42 Maurice, *Strategikon*, 2.11.

323–25 §43 Maurice, *Strategikon*, 2.12. On surveyors see above on 4.103.

326–50 §§44–47 Maurice, *Strategikon*, 2.13.

351–59 §48 Maurice, *Strategikon*, 2.14.

363–79 §§50–52 Maurice, *Strategikon*, 2.16.

380–94 §§53–54 Maurice, *Strategikon*, 2.17, but with additional reinforcing remarks from Leo on the need for silence (391–94). See also Const. 20§204.

395–400 §55 Maurice, *Strategikon*, 2.18.21–25; and cf. Const. 20§204.1052–1055. For prayers and the presence of clergy with the army, see on 11§19.105–7 above. On blessing the standards and other religious observances, and the role of military "chaplains," see on 13§1.3–5 and 18§127.620–25, below. In the *Strategikon* the troops are to shout out the *nobiscum deus* three times by division only as they marched out of the camp, and should thereafter remain silent, the war cry being reserved until the last minute, when hand-to-hand combat is reached (cf. also *Strategikon*, 12.B.16.42–43: "adiuta. . . . deus" [= Const. 7§55.381–82], at the point at which the lightly armed archers loose their arrows). Leo here omits *Strategikon*, 2.18.1–20 (and cf. Vegetius, *Epitoma*, 3.5.4–5), which recommends that soldiers must not shout out the "nobiscum (deus)" as they move into the attack, since this will both frighten the inexperienced soldiers and encourage the more experienced or braver to charge forward in a disorderly manner and break the formation. Leo's text advises instead that the troops remain silent as they march out of their camp, shout out the victory cry of the cross (the νικητήριον τοῦ σταυροῦ) as they marched into the attack, and then—once again following Maurice—cheer and shout other battle cries as hand-to-hand fighting began. This detail of Leo's text seems to reflect actual practice: at the battle of Bathys Ryax, in 872/73, the (admittedly interdependent) chronicles of Genesios and the *Vita Basilii imperatoris* report that the Byzantine forces shouted: "σταυρὸς νενίκηκε!" as they charged down onto the Paulician army under the command of Chrysocheir: *Vita Basilii imperatoris*, 42.31–32 (154); Genesios, 4.36 (87; trans. Kaldellis, 109). While these texts were a product of the tenth century, there seems no reason to doubt that they are based on earlier accounts nearer to the events described, even if exactitude may be partial. According to Nikephoros Ouranos, who paraphrased much of the *Taktika* in his compilation at the end of the tenth century, the *niketerion* was an antiphony: the commander shouted out "ὁ σταυρός" and the soldiers responded with "νικᾷ!" (Vári ed., 58, 68). The notion that, when battle was imminent, the troops should remain as silent as possible while drawn up in their ranks, has a long pedigree (note also Leo's comment in Const. 20§204); see Goldsworthy, *Roman Army at War*, 196–97, and with some reservations regarding actual practice, Cowan, "Clashing of Weapons."

The aim was to unnerve the enemy by the utter silence and discipline of the Roman lines. That this was indeed practiced on occasion in the late Roman period is clear from a description given by Theophylact Simocatta of a battle fought between Romans, with Persian allies, against Persian rebel forces in 591. They had already had their standards blessed by the clergy accompanying the army, and may also have participated in holy liturgy to purify their souls and to pray for victory. The war cry had been shouted on leaving camp: Th. Sim., *Historia*, 5.9.5–7.

Once drawn up in their positions no unnecessary movements were to be undertaken. Soldiers and subordinate officers were to await the orders to advance, which was also to take place with minimum noise and commotion according to the *Praecepta* of Nikephoros, 4§11.110–12. Only when they were on the point of clashing with enemy soldiers was the battle cry to be shouted again, in an attempt to unnerve the enemy: cf. Maurice, *Strategikon*, 8.2.46. In practice, of course, the use of a battle cry or war chant depended very much on the circumstances—there are several examples of soldiers who, having secretly been able to surround an enemy force, were then encouraged before the attack to make the maximum noise in terms of regular war cries, trumpet blasts, and drum beats, as well as blood-curdling yells to terrify the enemy, especially if the Romans were fewer in number, and this is something which the treatises also recommend. As noted already, the smaller Roman force terrified the Paulician army encamped at Bathys Ryax before charging down from the surrounding hills to annihilate it in 872/73; while as his troops marched in a dense line-of-battle order against the enemy, John Tzimiskes ordered a similar effect to cow the Rus' forces at Preslav in 970; and for the use of drums to disturb or cow the enemy, cf. Leo the Deacon, *Historia*, 24.16–17; 110.12; 133.6–8 (Talbot and Sullivan, 76, 160, 179). Note that the *Taktika* warns of the frightening effect the drums and cymbals used by the Tarsiots can have on Byzantine forces, 18.518–20 (and commentary, below). See also on 18.517–22 below.

By the time of Leo's writing, therefore, the victory cry of the cross seems to have become standard. By the middle of the tenth century, the troops were instructed to utter a slightly different variant: "Lord Jesus Christ, our God, have mercy on us. Amen": Nikephoros, *Praecepta*, 4§11. The extent to which these battle cries were employed across the whole army is difficult to ascertain, particularly where non-Christian allies or mercenaries were involved. Or even more clearly in the case of Muslim auxiliaries or mercenaries (as in the late eleventh and twelfth centuries, when the empire employed considerable numbers of Turk soldiers), it is unlikely that the obviously Christian war cries were demanded from any but the indigenous Byzantine soldiers.

401–8   §56   Maurice, *Strategikon*, 2.19, but Leo expands the passage, and follows it with an explicit call for officers and soldiers to stress the fact that the soldiers fight for their faith and will be rewarded accordingly.

409–20 §57 See for a detailed discussion and analysis of Byzantine military harangues and rhetoric Karapli, Κατευόδωσις στρατού; also Koutrakou, *Propagande*, 350–86. The ninth-century *Rhetorica militaris* sets out in great detail the various tropes and techniques, topics and themes, which a military harangue might employ, and some of this is reflected quite evidently in this paragraph of the *Taktika* as well as elsewhere in Leo's treatise. See, for example, the very similar language and motifs in *Rhetorica militaris*, 9.2: λαμβάνεται δὲ τὸ δίκαιον ἀπὸ τοῦ ζήλου τῆς πίστεως; ἀπὸ τῆς πατρίδος, ἀπὸ τῆς πρὸς τοὺς ὁμοφύλους ἀγάπης, ἀπὸ τῆς τῶν ἀδικησάντων τιμωρίας· δίκαιον γὰρ τούτων ἕκαστον; each of these is further elaborated at 10.1–14.9: Leo follows exactly the same order in this paragraph. He seems to have drawn on these passages and motifs in several places in the *Taktika*. See as well Dagron's discussion, "Byzance et le modèle islamique," 228 and chapter 2, above; and concerning exhortatory speeches and calls that the *kantatores* or heralds were to give to the soldiers see commentary to 14§31.204–8. For μισθός as spiritual reward see on Const. 18.592–98, with Dagron, "Byzance et le modèle islamique," 221 and n. 14. The strongly Christian content is important, because it foreshadows Leo's comments on the best way for Christians to combat Islam, and in the context of warfare specifically motivated by religious antipathy, outlined in Const. 18 (see on 18§125.612–15, §127.620–25 below); the emphasis on fighting for God, spiritual salvation, kindred, and nation (ll. 412–17) is repeated exactly in Const. 18§127.622–23, but this is not a formulation introduced for the first time, since the *Rhetorica militaris* deploys similar language (10.1–2; 37.7[6]), even if it does not specifically mention Islam (which in the context of the war in the east would have been unnecessary). Old Testament parallels and language were a prominent feature of the military harangues that have survived, as well as of other texts associated with fighting for the faith and the sacrifices to be made on behalf of the Chosen People; see Markopoulos, "Ideology of War," 53–54; Pertusi, "Acolouthia militare," and Detorakis and Mossay, "Office byzantin." Whether Leo also had in mind particular speeches delivered by military leaders of his own or recent times cannot be known. The themes he mentions here follow those in the *Rhetorica* *militaris* as well as those found in known military harangues—see the speech which an emperor (probably Basil I) reportedly delivered to his troops on the march into Syria, recorded by Constantine VII, *CPTT* C.466–73, where the soldiers are "soldiers of Christ," and where the troops are encouraged to show their bravery for both God and emperor. See also Vári, "Exzerptenwerke"; Ahrweiler, "Discours inédit," 399; esp. McGeer, "Two Military Orations"; literature and sources from the ninth–tenth centuries in *CPTT* 243–44; and Stouraitis, "Methodologische Überlegungen," 279–80. That these speeches were actually composed by Constantine VII is, of course, doubtful; see Ševčenko, "Re-reading Constantine Porphyrogenitus," esp. 186–87 and n. 49. Whether any of these speeches were actually delivered before battle remains problematic; see Stouraitis, "Just War," 238–39, 243–45.

Military speeches are at the same time one aspect of a literary tradition focusing on the glorification of military deeds, or of individual leaders or rulers (e.g., *Rhetorica militaris*, 29.2–5; 30–33), which formed part of the stock-in-trade of panegyrists and encomiasts, whether on account of an emperor's successes in war, or his pursuit of its antithesis, peace, or indeed both (since in the Byzantine view the former was often necessary for the achievement of the latter); see especially, and in general, McCormick, *Eternal Victory*; more specifically: *ODB*, "Enkomion," 1:700–701; and Kolia-Dermitzaki, "Byzantium at War," 231–34. Such a tradition is not confined to the Hellenic/Roman world and its tradition, of course; see Karapli, "Speeches of Arab Leaders," and more specifically Karapli, Κατευόδωσις στρατοῦ.

420   μᾶλλον ἢ χρημάτων πλῆθος   Leo may have had in mind the donatives and rewards regularly handed out to senior officers as well as, on occasion, to regular soldiers, in the course of campaigns, described in detail in the three treatises ascribed to Constantine VII; see *CPTT* C.501–11, 536–47, for example. In spite of Leo's insistence on spiritual rewards, however, he also places considerable emphasis on this aspect, a point in which the *Rhetorica militaris* is also insistent; see on Const. 16§§2.6–5.32 and 20§192 below.

421–30   §58 Maurice, *Strategikon*, 2.20.1–13, omitting the last 7 lines. The advice is repeated at Const. 12§99.

431–51 §59 Maurice, *Strategikon*, 3.11.1–19, but again with somewhat modified introductory and concluding sections and an up-to-date vocabulary for the titles of officers. Leo omits the emphasis in the *Strategikon*
on written orders—here he underscores the verbal exhortation and
delivery of instructions from officers to soldiers.

452–55 §60 Maurice, *Strategikon*, 3.11.19–21.

456–546 §§61–74 Maurice, *Strategikon*, 3.12–15.

457–58 τὸν ὑποστρατηγὸν. . . . τὸν λεγόμενον νῦν τοῦ θέματος μεράρχην
For the *merarches* see on 4§9.69 above.

547–57 §§75–76 These ten lines appear to be Leo's own advice, and reflect
probably contemporary practice, especially as regards the order not
to pass the second line. It also appears to reflect Leo's (or his informant's) own knowledge of the thematic soldiery, many of whom were
under arms only irregularly and certainly lacked the sort of discipline and cohesion to be expected of regular or professional soldiers.

558–75 §§77–79 Maurice, *Strategikon*, 3.16; see on Const. 14§37.240–44
below; and see Syrianos, *Strategy*, 40.

576–79 Cf. Maurice, *Strategikon*, 3.11.3–4 on written orders.

579–636 Maurice, *Strategikon*, 7.B.16.

602 μανδατώρων The word *mandatores* here is a mistake in the version
of Maurice, *Strategikon*, 7.16.20, which Leo was using. The heralds or
messengers would not be carrying out such duties, normally assigned
to scouts or *skoulkatores*. Cf. Consts. 4§26.110; 12§42.317; §97.660;
17§78.466; §80.471–72, §85.496. See Rance, *Strategikon*, 7.B.16, note 104.

637–41 τὸν τρισάγιον ὕμνον . . . τὸ τρισάγιον Cf. Const. 11§19 for the evening *trisagion*.

637–90 §§92–102 Maurice, *Strategikon*, 7.B.17.4–45.

669–73 §99 Maurice, *Strategikon*, 2.20.

688–90 Leo adds his own coda to the advice given in this passage, stressing
that each subordinate senior officer—*komes*, *droungarios* and tourmarch in ascending order of rank and responsibility—must be thoroughly familiar with it.

691–92 Paraphrased from Maurice, *Strategikon*, 7.A.pr. At ll. 669, 674, and
684 Leo retains the form of his earlier compilation, the *Problemata*
(cf. *Prob.* 2.56, 7.15, 5.4–5), although the sections from *Strategikon*,
7.B.16–17 which he uses here are not included in the earlier text. It is

possible that they were prepared for the *Problemata* but, not being included at that time, retained their earlier form when incorporated into this Constitution of the *Taktika*.

693–733   Maurice, *Strategikon*, 7.A.pr.13–53, but note that at l. 718 he omits the advice in Maurice, *Strategikon*, 7.A.pr.34–37, that if the enemy is a Scythian or Hunnic people, then the Romans should attack in February or March, when their horses are exhausted from the hardships of the winter (lack of pasturage); see on Const. 18§60.299–300 below. For §108, and an ancient precedent: Appian, *Spanish Wars*, 87. 378–79 (Scipio Aemilianus's advice on fighting the Numantines).

734–36   §109   Leo's formulaic final paragraph.

CONSTITUTION 13

The constitution is based entirely on Maurice, *Strategikon*, 7.A.1–14, although Leo omits the whole of the prologue, which he excerpts instead in his own preface to the *Taktika*; see above, on *Taktika*, pr.90–94.

3–5 §1 Maurice, *Strategikon*, 7.A.1; cf. Pertusi, "Acolouthia militare"; Haldon, *Byzantine Praetorians*, 568; and McCormick, *Eternal Victory*, 245–52; Dennis, "Religious Services." For prayers and the presence of clergy with the army, see on Const. 11.105–7 above.

12–17 §4 Maurice, *Strategikon*, 7.A.4. See on Const. 12§57.409–20 above, on speeches and harangues to embolden and fire up the troops (and cf. 14§101.703–7); and compare with the points to be employed in haranguing the soldiers set out in the ninth-century *Rhetorica militaris*, 9–13 and 22.1–7. Note that Leo omits the reference in the *Strategikon* (7.A.4.6–8) to written orders, but refers simply to "commands and other orders." For rewards for bravery and discipline, see *Rhetorica militaris*, 45.5–6; 55.4–5; and on Const. 20§§191–92 below.

17–19 ἔτι δὲ . . . τάγμα This presumably is intended to refer, as in Maurice, to the disciplinary ordinances enumerated in Const. 8. The point is repeated at 14§84.570–76.

20–25 §5 See 17§91 and cf. Onasander, *Strategikos*, 14.592–606 (14§§3–4) (cf. Polyaenus, *Strategika*, 2.1.6) and Ammianus Marcellinus, *History*, 24.8.1 for a similar tactic to boost morale employed by Julian during his Persian campaign in 363, on using prisoners to exemplify the enemy's weakness.

26–35 §6 Maurice, *Strategikon*, 7.A.6. Cf. 20§§4–5, commentary to Const. 8 above, and Haldon, *Warfare, State and Society*, 231–32. Military punishments were to be exercised with discretion, and indeed the evidence suggests that in practice the advice given here and repeated from the *Strategikon* was accepted as common sense. For foreign allied troops: 12§90.

42–46 §8 Based on Maurice, *Strategikon*, 7.A.7, but expanded to include the advice that troops may take an early meal if no battle is expected. Resting the army after a march and ensuring that the soldiers and animals were properly fed and watered were crucial to success. A good example is provided by the account in both Attaleiates and Kekaumenos of the inexperienced commander of an expedition under Constantine IX against the Pechenegs in 1049 who failed to encamp or rest his troops before giving battle, with disastrous results: Attaleiates, *Historia*, 32–33 (trans. Kaldellis and Krallis, 57); Kekaumenos, *Strategikon* (Spadaro), 2.64 (Litavrin, 162.12–18), although Skylitzes (*Synopsis historiarum*, 468; trans. Wortley, 437) ascribes the defeat to the commander's failure to listen to the counsel of his fellow officers.

46 ἐκλυθῶσιν Read "become weakened" *vel sim.* rather than "fall apart" (cf. on 10§13.62 and 15§16.87)

47–51 §9 Maurice, *Strategikon*, 7.A.8.1–4, again expanded slightly and updated in respect of technical terms (tourmarchs rather than merarchs)

52–53 §10 Maurice, *Strategikon*, 7.A.8.4–5.

54–57 §11 Maurice, *Strategikon*, 7.A.9. See above on Const. 11§23.145. Although the text here is not especially clear, Leo's "on the night before battle" probably refers to the trumpet blast at the commencement of the first watch of the night, thus at 1800 of the previous day, at which time the officers should make arrangements to have the horses watered, which would take some considerable time and needed to be organized well ahead of the start of any action. Trumpet blasts also signaled when the animals were to be brought back into the camp (see *CPTT* C.145). On the seasonally varying length of the Byzantine day and night see on Const. 17§89.519 below and Grumel, *Chronologie*, 163–64.

55–56 τοὺς ἵππους ἐπὶ τὸν ποτὸν παρασκευάσωσιν ἐξαγαγεῖν Read "they should prepare to lead the horses out to water. . . ." rather than "they should make sure to lead the horses out. . . ."

58–66 §12 Maurice, *Strategikon*, 7.A.10, except that where Maurice has ἄρτου ἢ ἀλφίτου ἢ πίστου ἑψητοῦ ἢ κρέους (bread/wheat or barley or boiled millet or meat) Leo has ἄρτου ἢ ἀλεύρου ἢ πίστου ἑψητοῦ ἢ παξαμάδας ἢ κρέας (bread or flour or boiled millet or biscuit or meat). For *piston* see above on Const. 6§23.153–54; and cf. Nikephoros Ouranos, *Taktika*, 13.12: κέγχρου ἑψητοῦ, boiled millet meal. The derivation of the term ἀργαβία, which Leo copies from Maurice, is unclear, although it appears to refer to a sack or pouch of some description (see Trapp, *Lexikon*, 1: 191), clearly differentiated from the *sellopouggion* or saddlebag. Leo repeats the earlier instruction from the *Strategikon*, in turn following older and traditional Roman regulations, regarding wine in battle or on active service. See also *Strategikon*, 12.B.23.33–34; and cf. earlier regulations enjoining the use of sour wine or winevinegar when on campaign, both in Vegetius, *Epitoma*, 3.3.10; and in imperial law: *CTh* 8.4.4 (361), 6 (360); *CTh* Val. III. *Nov.* 13§4 (445). Cf. also 12§100.

67–74 §13 Maurice, *Strategikon*, 7.A.11 (cf. Vegetius, *Epitoma*, 3.10.7). See on Const. 9.217 above: the term ἔξπληκτος, which Leo copies from Maurice, derives from Lat. *explicitus*, "disentangled/unhindered" and hence also "lightly armed" or "lightly equipped." See Zilliacus, *Zum Kampf*, 224; Mihăescu, "Éléments latins," 158; idem, "Littérature byzantine," 366; Trapp, *Lexikon*, 3:544.

75–81 §14 Maurice, *Strategikon*, 7.A.12.

82–88 §15 Maurice, *Strategikon*, 7.A.14 but with the introductory comment on divine favor. The issue of indiscipline and plundering of enemy camps or wounded and dead before the battle is finished was common and longstanding, as this passage from Maurice, and many examples recorded in narrative sources attest. During a campaign against the Bulgars in 707/8, for example, the cavalry *themata* suffered a defeat because they failed to set pickets and guard the camp carefully, and were caught by surprise while scattered to collect forage for the horses (*Theophanis Chronographia*, 376; trans. Mango and Scott, 525); after the successful landing on Crete in 961 Nikephoros Phokas sent the trusted general Nikephoros Pastilas inland with a small force to scout and report back on enemy dispositions. In spite of his experience, he allowed his column to disperse, attracted by the richness of the plunder and the absence of the enemy forces, and was subsequently ambushed and heavily defeated (Leo the Deacon,

Historia, 9.9–10.17; Talbot and Sullivan, 62–63). See for further discussion Dennis, "Byzantines in Battle," 177–78 and McGeer, *Sowing the Dragon's Teeth*, 320–23.

89–102 §§16–17   Maurice, *Strategikon*, 7.A.13. The reference to what has been previously written about is to 11§10 (see commentary on 11§10.57) and possibly to Const. 10§13. Leo removes the reference to the possibility of a battle with "Scythian" peoples.

103–7 §18   Leo's concluding comment and introduction to the next Constitution.

CONSTITUTION 14

As usual much of this section is drawn from the *Strategikon*, notably books 7.B and 12.B, together with extracts from Aelian and Onasander.

3-7 §1 See on 13§1.3–5 and 11§19.105–7, above, and Maurice, *Strategikon*, 2.18 (ll. 13–17) and 2.19. For the so-called "extended litany" (τὴν λεγομένην ἐκτενῆ) see on Const. 11§20.108–15 above.

8-196 §§2–29 Maurice, *Strategikon*, 7.B.1–14.

12-13 μὴ συμπλέκεσθαι τοῖς πολεμίοις διὰ χειρός Cf. 7§40.276; 12§105; 14§§99–100; 20§§153, 159, 193; and the long passage on this in Onasander, *Strategikos*, 33.914–42 (33§§1–4). The idea has a long history and is a topos of ancient military theory; see Wheeler, "General as Hoplite," 125, n. 17, 152–54 and Beston, "Hellenistic Military Leadership." This advice was taken seriously by Byzantine commanders and soldiers, although there are many examples of leaders who ignored it: the *patrikios* Leo, commander of Adrianople when it was besieged by the Bulgars in 921, was nicknamed Μωρολέων, "stupid Leo," for his frequent personal attacks on the enemy forces: Skylitzes, *Synopsis historiarum*, 218.84–88 (trans. Wortley, 211), although some clearly admired his bravery, calling him instead Θυμολέων, "brave Leo": Theophanes continuatus, 404.18–21; see *PmbZ* 25420. While the personal involvement of the commander was not necessarily either sensible or laudable, even emperors were known to ignore this advice: John Tzimiskes joined his cavalry in a charge against the Rus' at Dorostolon, for example (Leo the Deacon, *Historia*, 153.12–19; Talbot and Sullivan, 196–97); Alexios I fought a single combat with a Cuman horseman before both armies, and won (Anna Komnene, *Alexiad*, 10.4; trans. Sewter, 305/Frankopan, 271); and the young

Manuel Komnenos rashly attacked the enemy, fortunately with success (but was later punished by his father John II. See Choniates, *Historia*, 35); other examples are discussed by McGeer, *Sowing the Dragon's Teeth*, 299 and Dennis, "Byzantines in Battle," 174–75. Of course the heroism of the protagonists is proclaimed by such action, and it is not always clear to what extent such accounts were exaggerated. From this perspective, the story of "stupid Leo" may well be a more plausible account. The reason for this caution is simply the fact that Byzantine armies, like all medieval armies, depended very heavily for their morale and their discipline on the presence of a recognized leader. When the leader fell or disappeared, morale could collapse and the army break up. The pages of Byzantine history are littered with such stories. In 678, for example, when Constantine IV's standard was observed leaving the field (he was suffering an attack of gout), panic and flight ensued (*Theophanis Chronographia*, 359; trans. Mango and Scott, 499); in 917, when the imperial forces under Leo Phokas (*PmbZ* 24408) had driven back the Bulgars in the first phase of the battle, Leo dismounted to refresh himself. But his horse bolted and, seeing the riderless animal and thinking he had been killed, the cavalry began to panic and turn back from the pursuit, enabling Tsar Symeon to reassemble his forces and counterattack (Skylitzes, *Synopsis Historiarum*, 203–5; trans. Wortley, 197–98; for the consequences of the reversal of the imperial banner at Mantzikert see Attaleiates, *Historia*, 161, trans. Kaldellis and Krallis, 293). And in 998 the Byzantine forces had begun the pursuit of the defeated Fatimid forces near Apamaea in Syria when the commander, Damianos Dalassenos, *doux* of Antioch (*PmbZ* 21379), was killed. Demoralization set in, the Fatimid troops rallied, and victory tuned into rout: Amedroz and Margoliouth, *'Abbasid Caliphate*, 6:240–41; also Canard, "Sources," 299–300.

18–24   §4   The treatise on skirmishing or guerrilla warfare is very clear that high ground should be occupied by the Roman forces. If they could not do so, then enemy forces occupying such positions should be avoided, since their advantage was too great; see *Skirmishing*, 3.

34   τὸ εἰρημένον διάστημα   "The prescribed distance," that is to say, four bowshots, described in 12§46.335–47.

37–39   §7 Cf. also 14§§37.240–38.252 on laying ambushes and feigning retreat; *Sylloge tacticorum*, 51.1; and Nikephoros, *Praecepta*, 2§§3–5, 7. Feigned retreat was an ancient tactic, and the concern about troops breaking formation to pursue the enemy was well-grounded in experience, particularly when dealing with mounted enemies who employed the feigned retreat or flight, a tactic familiar to the Romans since the third century and generally associated with barbarian and nomadic peoples, but which soon became part of the standard tactical repertoire of Roman commanders; see Rance, *Strategikon*, 4.3, note 7. Byzantine commanders certainly employed it. At the battle of Versinikia in 813, for example, the retreat of one of the Byzantine wings was at first thought by the Bulgars to be a feigned retreat to draw them into an exposed position, although in the event this proved not to be the case, and the Byzantine withdrawal turned into a rout: Theophanes, *Chronographia*, 500; trans. Mango and Scott, 684; *Scriptor incertus* (ed. Iadevaia), 41–43; Theophanes continuatus, 13–15. It remained a constant danger: in 1070, less disciplined units under Manuel Komnenos fighting the Seljuk Turks rashly pursued the apparently retreating enemy, only to fall into an ambush and be cut to pieces: Zonaras, *Epitomae historiarum* 3 (694–95). See the discussion in McGeer, *Sowing the Dragon's Teeth*, 300–301, 320–23, with further examples.

40–47   §8 Such maneuvers, by which troops were redeployed on the battlefield, required good order and discipline, but were clearly standard in the Byzantine army of the time. At the battle of Anzen, for example, in 838, Theophilos reinforced one of his wings with units under his own command and from the reserve line, a movement which enabled the Romans to push the enemy back (although the final outcome of the battle was a defeat for the Byzantines); see pseudo-Symeon, 636–37, 638; and Theophanes continuatus, 127–29 (and Symeon *magistros* 224, 227). The source tradition on this campaign and battle is extremely confused; see on Const. 12§§17.119–28.212 above.

48–57   §9 Maurice, *Strategikon*, 7.B.7. *Pace* Dennis, *Taktika*, 295 n. 3, there seems to be no connection between this passage and the Scholia to Euripides, *Phoinissai*, 600.

58–59   §10 Cf. Maurice, Strategikon, 8.2.40 and 9.1. Leo inserts this into the texts he has copied or paraphrased from the *Strategikon*.

66 τὴν ἔσω τράφον   Properly τάφρον, although the alternative form is common in the MS tradition. The sections on encampments mention only a single ditch or entrenchment around the camp, however, so this phrase refers more probably to the inner edge or side of the defensive ditch, rather than to an otherwise unattested second, inner ditch. See 11§8.37–44, §13.75–78 and the commentary, above.

73–75 τότε φέρε … ἀφόριζε   Cf. Const. 9§69. While Leo is copying Maurice here, the maneuver described does seem to have been practiced, at least at the time the *Strategikon* was compiled, since Theophylact Simocatta (7.14.3) notes that in a skirmish with the Avars in 598 the Roman right wing was given responsibility for guarding the baggage. Later Byzantine accounts of battles make no specific mention of this measure.

76–96 §14   See 10§13, 11§10.

101–13 See Maurice, *Strategikon*, 7.B.11.1–9; 11.2.103–5. There are somewhat distant parallels between the advice given by Maurice and repeated by Leo, and examples in Thucydides, *Peloponnesian War*, 7.60.5, 61.2–3, 66.3; but no textual dependency is apparent, *pace* Dennis, *Taktika*, 299 n. 4.

106–7 ὡς γὰρ ἐκ θείας ψήφου οὕτως τὸ ἀποτέλεσμα δεχόμενον ἐν πάσῃ δειλίᾳ γίνεται   This advice comes from Maurice and earlier strategists, and in spite of Leo's ascription of a simple fatalism to the Islamic armies at 18§112, it was no doubt a problem for all medieval armies. See also on ll. 703–7 below and esp. with regard to the ways in which scriptural examples could be deployed to point to coming victory to encourage the soldiers (but with possible negative results, as Dagron notes: Dagron, "Byzance et le modèle islamique," 226 and n. 33). For the various recommended rhetorical ploys designed to bolster an army's morale after a defeat, see esp. *Rhetorica militaris*, 54.2 and 56.1–57.9.

110 τὸ δὴ λεγόμενον φυγομαχήσῃς   In Maurice, *Strategikon*, 7.B.11.16, this is a substantive: τὸ λεγόμενον φυγομαχεῖν. Leo's sentence is all in the second person singular, so a slightly more accurate translation would be: "Instead, employ tricks and deception, such that you might attack by surprise and so to speak fight while fleeing. . . ." τὸ δὴ λεγόμενον thus functions here parenthetically as an adverbial qualifier, "so to speak." The verb φυγομαχεῖν recurs at Consts. 18§37.200 and 20§201.1040, and is a term found in the Hellenistic treatises (e.g.,

Onasander, *Strategikos*, 11.525 [11§3]; 31.859 [31§1]), generally referring to avoidance of battle, or delaying and hit-and-run tactics designed to avoid a full-scale confrontation; but it may also have carried the implication of feigned flight. See the comments in Wheeler, *Stratagem*, 45.

122   δι' ὀψίδων "Hostages," from Lat. *obsides*, cf. *Theophanis Chronographia*, 393.6, 394.18 (Mango and Scott, 543, 544) and Trapp, *Lexikon*, 5:1168.

124–25   τῇ ἀπαγορεύσει τῶν προτεινομένων   Read: "in suppressing/restricting the actual proposals" rather than "rejecting. . . ."

129–33   §18   See esp. *Rhetorica militaris*, 54.2, 56.1–5 for advice on the rhetoric to be deployed to hearten and encourage a defeated force.

135   τῇ προειρημένῃ μεθόδῳ τῆς τάξεως   See ll. 114–17. That is to say, the commander should adopt the formation whereby the former first line is placed second, in the rear of the new first line.

141–45   εἰς δύο φάλαγγας. . . . ἀσφαλῶς   A defensive square marching formation, with cavalry dismounted and the horses secured in the middle, was standard procedure when marching under attack, and is described in both ancient and medieval treatises. See Onasander, *Strategikos*, 6.244–55 (6§§5–6); Vegetius, *Epitoma*, 3.6.13–14 describes a column with flanking forces around the center and baggage, serving a similar function. See Elton, *Warfare*, 244–45. A similar formation, more loosely articulated, was employed when marching in hostile territory: a hollow column drawn up with the infantry divisions forming its outer walls, shielding the cavalry, baggage and spare horses, and permitting the cavalry to move out to attack the enemy when appropriate. See the introductory remarks to Const. 9, above, and on 18§§113.550–114.556 below.

148–49   νίκα καὶ μὴ ὑπερνίκα   Cf. Menander, *Sententiae*, 419 (in Maurice, *Strategikon*, 7.B.12); Attaleiates, *Historia*, 26.17 (trans. Kaldellis and Krallis, 45); and Polyaenus, *Strategika*, 8.23.29 (*Hypotheseis*, 4.14). The saying goes back at least as far as the classical dramatist Menander and is probably older. See literature and sources cited by Rance, *Strategikon*, 7.B.12, note 80; and idem, "'Win but Do Not Overwin'." Similar sentiments, especially in respect of false pride and arrogance after victory, are expressed in Basil I, Κεφάλαια παραινετικά (ed. Emminger), 73, §65; and see on Const. 17.117–19 below.

157–58 ἐπεὶ καὶ ... ἔσται See ll. 469–70 below; the text follows Maurice, *Strategikon*, 7.B.12. Contra Dennis, *Taktika*, 303 n. 6 there seems in fact to be no connection with Aristotle, *Physics*, 197a30. See Rance, *Strategikon*, 7.B.12, note 82.

161–62 οὐκ ἀρκεῖ ... ἐχθρῶν Leo changes Maurice's text here: whereas the latter notes that tactics alone are insufficient to guarantee the safety of the army or harm to the enemy, Leo substitutes "strength alone" for "tactics alone."

162–63 ἡ ... διοίκησις Cf. l. 215. Translate as "direction" or "management" rather than leadership, since as the paragraph as a whole (and at 14§33.215) makes clear, Leo wishes to emphasize here the general's organizational skills. Cf. διοίκημα, "matter/affair/administrative act" (Trapp, *Lexikon*, 2:390).

169 λάκκοι Read "pits" rather than "pools of water" (cf. *Sylloge tacticorum*, 94.1; 96.2; *Campaign Organization*, 2.21; Trapp, *Lexikon*, 5:912)

171 Leo often replaces the older term *skoulka*, which is the term in Maurice's original for a party or troop of scouts, with the contemporary *vigla*, which in the *Strategikon* tends to refer to stationary guard posts or sentries. See *Strategikon*, 10.3.47, 53; 12.B.22.28, 37. Other than in the *Taktika*, *skoulka* does not appear in any other tenth-century treatises, although the word *skoulkatores* does appear in the *DAI* 1: chap. 53.57, referring to scouts.

173 ἐν διπλαῖς βίγλαις With the sense of twice as many patrols, frequently passing over the same ground—see ll. 86–88.

181–85 Maurice, *Strategikon*, 7.B.13.11–15 and cf. 7.B.16.20–22.

197–203 §30 This passage is taken from Onasander, *Strategikos*, 10.402–8 (10§§10–11); and cf. *Sylloge tacticorum*, 21.1; 2.

204–8 §31 Cf. 20§72.355–62 and *Rhetorica militaris*, 9.3, 10.1–2, 11.1–2, with commentary to Const. 12§57.409–20, above. Leo strengthens the original passage (Maurice, *Strategikon*, 7.B.6, based loosely on Onasander, *Strategikos*, 36.998–1004 [36§§1–2]), following the model of the ninth-century *Rhetorica militaris*, by adding that the soldiers fight for their faith and their brethren, a sentiment repeated and stressed at Const. 18.620–25. On concern for the souls of the fallen, see Detorakis and Mossay, "Office byzantin" and Pertusi, "Acolouthia militare." This is consonant with Leo's broader intent in compiling the *Taktika*, to emphasize the God-guarded status of the Roman polity, and the Christian piety and sacrifice of its soldiers, and to contrast the Christian with the Islamic state. The point is made especially clear in Const. 18 as has been shown: Dagron and Mihăescu, *Traité sur la guérilla*, 145–49 and Dagron, "Byzance et le modèle islamique."

209–10 Εἰ δὲ καὶ ... προσηκούσης Care of soldiers' families and children was a concern of the late Roman state, although in fact the *Strategikon* itself makes only indirect reference to the issue, since it was enshrined in both imperial legislation and, as far as the non-legal sources can be relied upon, in practice. In about 594 Maurice had reorganized the arrangements for discharged veterans, who were to be paid a regular state allowance or pension, among other privileges. And although the widows and offspring of deceased soldiers also received limited state support (e.g., *CJ* 12.37.16.8 [491–518]) the arrangements by which the sons of soldiers killed in action were permitted to inherit the status and remuneration of their father (presumably as a measure also to maintain recruitment: Th. Sim., 7.1.7; cf. *CJ* 12.47.3; and Jones, *Later Roman Empire*, 675; Haldon, *Recruitment and Conscription*, 21, 23–24; and Whitby, "Recruitment," 80–81) appear to have been extended by Maurice to apply to all children (presumably in a commutable form), not just sons: Scharf, *Foederati*, 96–97.

By the time of the composition of the *Taktika* a very different situation prevailed. The notion that military service transmitted certain privileges may have survived—see Bartusis, *Late Byzantine Army*, 203–4. In the later Byzantine period, there is evidence to suggest that a portion of the revenue granted as *pronoia* to soldiers could on occasion be granted to the widow of the *pronoiarios*; see *Acts of Vatopedi*, 1: no. 52 (I am grateful to Mark Bartusis for bringing this text to my attention). But this seems to have been the result of a specific petition by the original *pronoia* holders, and there is no evidence for any annuities or payments made by the state to soldiers' widows before this. According to a letter of Theodore of Stoudios the widows of soldiers registered for military service had to pay the state in lieu of their husbands' service: Theodore the Studite, *Epistulae*, 7.61–63. This change reflected changes in the way soldiers were recruited, on the one hand, and the hereditary nature of some provincial military service, on the other (for discussion see Haldon, "Military Service," 23–24), for since most provincial soldiers were recruited from peasant farmers whose continued support from their lands and through their families was assumed, a system of pensions was largely irrelevant. For widows, however, the situation was different, yet little if anything was done to assist: the empress Eirene is credited with removing the burden on widows attached to the *strateia*, but this appears to have been short-lived: a century later a letter from the patriarch Nicholas I, for example, asks for release from military obligations on behalf of a widow who cannot afford to equip her son: Darrouzès, *Épistoliers byzantins*, 2:50.130–31. A similar case appears in *Vita Euthymii Iunioris*, 172.19–21. Euthymios's father was registered as holder of a *strateia*, the obligations attached to which fell upon his family after his decease. The only way the widow could support the burden was by registering her son, even though an infant, and thus—the implication—paying a commutation in lieu of actual service. Euthymios was born in the 820s, the *Life* was written ca. 900.

In Leo's time (and probably since the late seventh century) there were thus no effective measures for the relief or support of soldiers' families, nor is there any evidence for any arrangements for state pensions or the like, for the reasons noted already, and as the *Taktika* itself makes clear—see on 4.7–11 above. Soldiers' families continued to benefit from the particular juridical status of those in military service in respect of certain extraordinary taxes and rights of inheritance, but there is no evidence that they received any other support. The situation with the professional and full-time tagmatic troops was similar, certainly in the early ninth century, when the patriarch Nikephoros noted that the disbanded or retired soldiers of the *tagmata* in Constantinople were in extreme poverty because they were no longer issued with the pay or supplies from which they had formerly supported themselves: Nikephoros Patr., *Apologeticus maior*, 556B; the same sentiments are expressed in Nikephoros Patr., *Antirrheticus* 3.492B: ἀπορίᾳ πολλῇ πιέζονται, μάλιστα τῶν βασιλικῶν σιτηρεσίων αὐτοῖς ἐπιλελοιπότων, ἀφ' ὧν τὰ πρὸς τὸ ζῆν μετὰ τῶν ὅπλων αὐτοῖς περιεγίνετο. Leo's remark in the *Taktika* makes it clear that it was largely up to the officers and the provincial commander to assist those in need, although this was a standard sentiment (20.37 = Maurice, *Strategikon*, 8.1.3).

That the state did make some attempt to support widows, however, is clear, although how effective the measures were is unknown. *De cerimoniis*, 694–95 refers to a series of provisions under which "Saracen" prisoners of war (on which see below on 15§39.238–49), described as *gambroi*, that is to say, bridegrooms, were to be settled within the empire and given lands that might then support them, lands which were exempt from basic fiscal dues for a specific period. The possibility exists that these were men who agreed to marry a widow of a soldier, thus maintaining both the widow (and family) as well as the fiscal revenues of the state: cf. *De cerimoniis*, 695.3–5. If this supposition is correct—and this is a big supposition—it would be an example (the only one) of an attempt to cater for the needs of the widows of soldiers who had died on campaign by facilitating a remarriage, with fiscal benefits attached. But apart from this ambiguous text, there is no source which suggests that soldiers, of whatever category, received anything from the state after their service was over, unless it be voluntary on the part of the general; see Const. 20§72.357–59: ἀλλὰ καὶ τὰ τέκνα τούτων καὶ τὰς γαμετὰς καὶ τοὺς ὅλους οἴκους αὐτῶν δέον τῆς παρὰ σοῦ προνοίας ἀπολαύειν, ὦ στρατηγέ, καὶ ἐπιμελείας καὶ ἀντιλήψεως. Hence the stress laid by Leo on the euergetism of the general and his officers in respect of their soldiers' equipment (e.g., Const. 4§3.26–29 = Onasander, *Strategikos*, 2.163–66 [2§5]; 5§1.6–8; 6§1.3–5; cf. 11§41.231–35; ep.§56), and on the voluntary contributions toward the equipment and maintenance of the army which wealthy but militarily inactive persons might be expected to make (see Const. 20§205). The fiscal exemptions and special juridical situation of soldiers was assumed to be sufficient compensation for their efforts. See on Const. 4§1.7–14, and cf. 20§71; *Sylloge tacticorum*, 36.1–2 (repeated from *Taktika*, Const. 4). The treatise *Skirmishing* notes that soldiers were harassed and unfairly punished by civil officials, and that their privileges were ignored. This situation was probably not new in the middle of the tenth century; see *Skirmishing*, 19.28–66.

211–28 §§33–34 Maurice, *Strategikon*, 7.B.15. See Const. 5§11.57–58; 6§4.28; 20§188.953–56 and commentary. Leo follows Maurice here in repeating that the common view—held by both Romans and foreigners—that a dull battle line is more likely to reflect the probable victor than a gleaming one is wrong. In fact, neither view is correct, since the outcome will be determined by God, generalship, and discipline. At Const. 14§98 he explicitly contrasts Onasander's recommendation (*Strategikos*, 28.826–31 [28] and 29.837–39 [29§§1–2]) of a gleaming battle line with what "more recent authorities"—by which he means Maurice—have said; and while at other points in the *Taktika* (see above) he adheres to the traditional view that gleaming and burnished equipment and weaponry is helpful in unnerving the foe, this is to be taken in the context of the passage noted above (14§98), where the advice is to maintain the dullness of the line until the last minute, before revealing the gleaming brightness of the weaponry immediately before contact with the enemy, and thus throwing them into confusion. Mutanabbi, the tenth-century court poet to Sayf ad-Daula, the Hamdanid emir of Aleppo, remarked that the dazzling light reflected from the brilliant armor of the Byzantine forces was such that the onlooker confused the men with their weapons (Vasiliev, *Byzance et les Arabes*, 2.2:347, verse 44; cf. 333 verses 16–17, and 420). The extent to which this advice to keep shining weaponry and armor covered up was actually followed is unclear. Leo the Deacon's account of the wars of Nikephoros II and John I Tzimiskes frequently mentions the soldiers' shining armor: e.g., *Historia*, 8.4. See Kolias, *Byzantinische Waffen*, 61, 83 for further references.

226–27 ὅπερ σημειοῦνται οἱ ἐνάντιοι   Cf. the alternative (and somewhat clearer) reading in ABE: ὅπερ οἱ ἐχθροὶ παρατηροῦνται καλόν ἐστι.

229–411 §§35–58   These sections are all taken by Leo from Maurice, *Strategikon*, 4.1–5.

229–31 Represents Leo's own opening comment to the material taken from the *Strategikon*. The potential for small forces to defeat much larger armies is a common motif, cf. e.g., Syrianos, *Strategy*, 33.13–14; 39–41.

234–39 §36 Maurice, *Strategikon*, 4.1.4–12; and see Const. 17.96–123. A classic example of this sort of ambush or surprise attack took place at Bathys Ryax in 872/73 in a campaign against the Paulicians under their commander Chrysocheir (see *PmbZ* 1153 and 21340). Toward the end of the campaigning season the Paulician leader encamped at Agranes (near mod. Muşalem Kale) in the Charsianon region, with the *domestikos* of the *Scholai* and his force a short distance away at Siboron (mod. Karamağara). From there Chrysocheir marched northeast toward Bathys Ryax, the mod. Kalınırmak Gap on the northeastern edge of the Ak Dağ, close to the point at which the road from Sebasteia (Sivas) divides some 28 km to the northwest, one branch going west, the other north. It was an important strategic location and an established meeting place for the Byzantine forces of eastern Anatolia when campaigning in the region or further to the east. Two thematic contingents under their *strategoi*, those of Charsianon and Armeniakon, were detached to follow him as far as Bathys Ryax. Chrysocheir seems to have been unaware that he was being shadowed, and after a short march pitched camp toward evening in the plain at the foot of the mountains at the head of the valley of Bathys Ryax. The pursuing forces took up their position behind the enemy host and on the saddle between the two hills above them. Eventually it was agreed that a selected force of some 600 men from both divisions would mount a dawn attack, supported by the remaining forces who would create a great clamor with trumpets and beating of drums to panic the enemy soldiers into believing that the whole combined thematic army under the *domestikos* had fallen upon them. The plan worked perfectly. The enemy host fell into a complete panic, some leaping on their horses and fleeing, others cutting the loads off the pack animals and trying to escape on them. Their baggage train and all the booty they had been able to collect was taken. Chrysocheir himself fled with a few bodyguards but was killed during the flight, and the pursuit of the broken army, many of whom then ran into the much larger force under the *domestikos*, stretched across some 30 miles of the province. See *Vita Basilii imperatoris*, 42 (150–54); Genesios, 4.36 (86–88; trans. Kaldellis, 108–9).

240–52 §§37–38 Maurice, *Strategikon*, 4.1.12–4.2.8; and see Syrianos, *Strategy*, 18.18–36; and 18§§38–73 with commentary below. Here are two variants on a tactic. An excellent example comes from a campaign in the year 970. In the spring of that year a large Rus' force under Svyatoslav (*PmbZ* 27440) invaded Thrace, sacking the fortress of Philippoupolis (mod. Plovdiv) and moving on down the road to Constantinople.

The emperor John I Tzimiskes appointed Bardas Skleros, together with the *patrikios* Peter, both experienced commanders, to take command of a small force and reconnoiter the enemy dispositions (*PmbZ* 22778, 20785, and 26496). Informed of the imperial army's presence, Svyatoslav dispatched a substantial force consisting of both Rus' and Bulgar troops as well as a powerful detachment of allied Pechenegs to drive the Romans off. On receipt of the information that the enemy force was quite close, in the region of Arkadioupolis (mod. Lüleburgaz), Bardas divided his force into three: two divisions were concealed in the rough scrub and woodlands on either side of the track leading toward the enemy position; the remaining division he led himself, launching a furious surprise attack against the Pecheneg contingent. Although heavily outnumbered he was able to draw the enemy out of their encampment and feign a gradual withdrawal. After some fierce fighting Skleros ordered the prearranged signal to be given for the whole force to fall back. The Pecheneg force rushed in pursuit, but was then caught between the two concealed forces lying in wait, who attacked the unsuspecting enemy from both flanks and the rear. Within a few minutes the Pechenegs had received such a savage mauling that they turned and fled, while their allies, the Rus' and Bulgars, who had been hastening to catch them up in their pursuit of the supposedly defeated Romans, were caught in the panic and suffered similarly heavy casualties as the rout became general. According to a contemporary, the Romans lost some 550 men and many wounded, as well as a large number of horses, which fell to the archery of the Pechenegs. The combined enemy force, however, lost very many thousands. See Leo the Deacon, *Historia*, 109–11 (Talbot and Sullivan, 159–60); Skylitzes, *Synopsis historiarum*, 288.23–291.4 (trans. Wortley, 276–77); and the discussion in McGeer, *Sowing the Dragon's Teeth*, 298–300.

251–52 τοῦτο δὲ ... αὐτοῖς   Here Leo adds "the northern tribes" and "like the Turks" to the original. *Turks* was the standard term for the Magyars (cf. 18§40.210–17, for example, and Moravcsik, *Byzantinoturcica*, 2:321–22), although whereas in this constitution Leo regards the Turks as an undifferentiated body, in Const. 18 he distinguishes between the Magyars (along with the Bulgars) and other nomads in respect of tactics and fighting organization. "Northern tribes" may refer to the Russians, since they are bracketed together with the Scythians, which by the late ninth and tenth centuries was a term also applied to the Rus' as well as to genuine steppe nomad peoples: Moravcsik, *Byzantinoturcica*, 2:279–83.

253–69   §39   Maurice, *Strategikon*, 4.3.2–19, although Leo omits Maurice's remark on the Hephthalite ploy against the Persian king Firuz (Perozes). See Christensen, *L'Iran sous les Sassanides*, 289–94; and on the Hephthalites, also called the White Huns, see Moravcsik, *Byzantinoturcica*, 2:127–28. The best-known example of digging a trench before battle, although not quite the same as the tactic described here, is that employed by Belisarius in the battle near Dara in 530, cf. Procopius, *Wars*, 1.13.13–14, and discussion in Greatrex, *Rome and Persia*, 171–73 with Lillington-Martin, "Dara Gap."

256–57   ἀλλὰ μηδὲ τὸ ἐπαρθὲν χῶμα ἐάσῃ   Read: "He ought then *not* to allow...."

270–76   §40   Maurice, *Strategikon*, 4.3.21–29.

277–92   §41   Maurice, *Strategikon*, 4.3.35–51. Leo emends the original, omitting (a) the width of the band of caltrops to be scattered ("as wide as the battle line" in Maurice); (b) the width of the four or five passageways to be left clear (300–400 ft. in Maurice); and (c) the fact that the ambushed enemy force can neither retreat nor advance due to the caltrops. Apart from the *Strategikon*, followed by Leo, the other treatises make no mention of the warning markers (large branches, spear heads with odd shapes, etc.) to be set up as an indication of the presence of caltrops, although they all advocate their use. There is very little evidence of the use of caltrops in battlefield situations, apart from their mention in the *Vita S. Philareti iunioris* at the battle of Troina in the eleventh century (see on Const. 5.29), and an account in Tabarī of their use by Sasanian forces as a defense against Muslim armies in 638 and 641: Tabarī, *Ta'rikh al-rasul wa'l-muluk*, 12:140 (2356); 13:180–81 (2597), 187–88 (2603–2604). See also on Const. 5§4.33 above and 20§147 below.

293–305   §42   Maurice, *Strategikon*, 4.3.52–65; and cf. Const. 11§13.77–79. These small pits were known as *lilia*, "lilies" in Latin, and were regularly employed to support defensive positions. See Webster, *Roman*

Imperial Army, 176, 239–40; also Wheeler, "Rome's Dacian Wars," 1218, n. 79, and, for an excavated example from Hadrian's Wall, pl. XXVa. See also discussion in Bidwell, "Systems of Obstacles" and Woolliscroft, "Excavations at Garnhall," 142–45, 162–63. Leo notes that "the older authorities," i.e., Maurice, "referred to (them) as horse-breakers," whereas the *Sylloge tacticorum*, 22.5 simply refers to them as the "so-called" horse-breakers. The word is not found in treatises or other sources before the *Strategikon*; see Stephanus, *Thesaurus*, 6:642; Trapp, *Lexikon*, 4:714: ἱπποκλάστης. Similar small pits with spikes or stakes in them are described in the anonymous treatise on *Campaign Organization*, 2.21–23, this time referred to as "foot-breakers," ποδοκλάσται; and in the possibly ninth- or tenth-century *Apparatus bellicus* (ed. Zuckerman), §§75–76.90: φόσσας μικρὰς ἐχούσας ἔνδοθεν σκόλοπας, ποδάγρας ἢ ἱπποκλάστας λεγομένας, "small pits with stakes inside, called foot-traps or horse-breakers." For an example of the use of pits in battlefield contexts, esp. against cavalry, see Bachrach, *Merovingian Military Organization*, 135–36; and for the late Roman context, with Roman and ancient examples and further discussion, see Rance, *Strategikon*, 4.3, note 8.

306–22   §§43–45   Maurice, *Strategikon*, 4.3.65–83.

318   σφιγκτοὶ   Here Leo changes Maurice's term from δρουγγιστί, which in the early tenth century would have to refer to the tactical unit, or *droungos*, rather than a specific formation or arrangement of the soldiers. Where he does use the adverb, he glosses it as "in a mass without regular files," as here (and see ll. 369, 381), or an equivalent phrase: cf. 7§34.231–32; 18§142.752; and cf. *Sylloge tacticorum*, 45.10: δρουγγιστὶ καὶ ὡς μάζα: "in a crowd like a clump/mass." Cf. Trapp, *Lexikon*, 3:412.

323–41   §§46–48   Maurice, *Strategikon*, 4.3.84–102.

340   προσκουλκεύειν   Leo copies the verb from Maurice (see also *Strategikon*, 9.5.90), but it is otherwise unattested apart from one reference to προσκουλκάτορες to describe soldiers sent ahead as an advance guard, employed by Malalas in a passage taken from a (fragmentary) fourth-century historian, Magnus; see Malalas, *Chronographia*, 253.68–69 (trans. Jeffreys, Jeffreys, and Scott, 179); *FHG* 4.4–6.

342–64   §§49–52   Maurice, *Strategikon*, 4.4.

365–407   §§53–57   Maurice, *Strategikon*, 4.5.

370   κατ' ὄρδινον ἀκίας ἢ δεκαρχίας ἢ πενταρχίας. Leo follows Maurice exactly throughout, referring here to the standard linear formation of a cavalry *bandon*, either 10 men deep, headed by the dekarch, or 5 deep, with each dekarchy in 2 files of 5 side-by-side, headed by the dekarch and the pentarch. As noted above on 4.77–81, the position of pentarch was probably purely functional rather than an actual grade or rank. The degree to which regular provincial forces were in fact able to manage the sort of discipline and order prescribed in the *Strategikon* and copied or assumed by Leo remains in doubt; although it is clear that Byzantine forces saw themselves as better ordered and disciplined in battle than most of their enemies. See on 12§§2.9–28.212 above.

381   τότε δρουγγιστὶ καὶ ὁμοῦ ἄνευ ὀρδίνων σφιγκτοὺς τάσσεσθαι See on l. 318 above.

391–402   Although again taken directly from the *Strategikon,* Leo's concerns here surely mirrored contemporary anxieties about military discipline and the ability of the soldiers to carry out more complex battlefield maneuvers.

395–98   οὗτοι οὖν . . . ἐργασίαν The idea of the commander as a diligent wrestler, taken directly from Maurice, in fact goes back to an earlier tradition; see Onasander, *Strategikos*, 41.1152 (42§6).

408–532   §§58–76 Taken from Maurice, *Strategikon*, 12.B.11–13, 17–18.

408–11   §58 Leo's connecting paragraph, referring to the material in Const. 7 on infantry formations which he has already presented.

414–25   Maurice, *Strategikon*, 12.B.11.3–18.

423–24   On the *kampidouktor*, see above, on Const. 7§38.266.

427–28   παρατάξεις . . . τόπους The MS tradition of the *Taktika* agrees on τόπους here, but in the *Strategikon*, 12.B.12.3 the MSS have τρόπους. This is presumably Leo's own emendation, perhaps to make more sense, for him, of the following instruction on the disposition of the archers in a line of heavy infantry.

427–39   Maurice, *Strategikon*, 12.B.12.3–15.

437   ῥικτάρια ἢ τζικούρια ἢ τι τοιοῦτον Leo changes Maurice's βηρύττας ἤτοι μαρτζοβάρβουλα (*Strategikon*, 12.B.12.12) to a more contemporary and meaningful phrase. See on Const. 6§22.131–32 above. The archers were usually placed behind the heavier infantry for

protection, and would hence shoot over the heads of the troops in front of them. Although Leo also copies Maurice's instruction for a mixed formation, it appears somewhat impractical in an actual combat situation. Placing slingers on the wings and archers to the rear seems to have been standard practice from the later Roman period and earlier; see on ll. 616–22, below; Aelian, *Tactica theoria*, 7.4–5; Arrian, *Techne taktike*, 9.1–2; Onasander, *Strategikos*, 17.625–35 (17); and Vegetius, *Epitoma*, 2.17.1–2; 3.17.7–9, where their role is to harass the approaching enemy formations at long range before retiring through the main battle line to the rear and continuing to shoot from there. See Wheeler, "Firepower" with the discussions in idem, "Legion as Phalanx"; cf. Syrianos, *Strategy*, 35.24–34.

440-63   §§61–63   Maurice, *Strategikon*, 12.B.13; and on maintaining the line and not breaking ranks to attack, see also *Rhetorica militaris*, 41.3–4.

464-70   §64   Another connecting paragraph from Leo, referring again to Consts. 7 and 12 in which the various formations and maneuvers are detailed.

469-70   A maxim repeated from ll. 157–58 above (14§22) (q.v.), taken from Maurice, *Strategikon*, 7.B.12.

471-510   §§65–72   Maurice, *Strategikon*, 12.B.17.17–55.

476   This ancient metaphor (see Aelian, *Tactica theoria*, 7.3 = Arrian, *Techne taktike*, 8.4; Asklepiodotus, 2.5) was well-known to the military treatises: cf. Syrianos, *Strategy*, 15.21–24. It is repeated here from Maurice by Leo, and a little later in the *Sylloge tacticorum* 32.5, 35.11, 45.20, although the editor retains the reading in M and W, ὀφθαλμὸν, which is clearly a copyist's error, noting the correct reading in A (ὀμφαλὸν), which occurs also in Maurice and later in this constitution (14.591).

493   μέση δὲ τάξις   Read: "The average formation..." The translation here makes little sense in the context—given the details of the maximum and minimum depths of the line, and the absence of any reference to "sections" of the line, the word μέση is best translated as "middling," i.e., average or usual: the standard formation of the battle line should be 8 deep, i.e., the same as the standard tent group or mess of 8 soldiers. See Maurice, *Strategikon*, 20.B.9.20–29 = Const. 4§74.313–19. Cf. also *Sylloge tacticorum*, 43.1–2.

493-94   τῶν ... σκουταράτων λεγομένων ἤγουν ὁπλιτῶν   This is the older reading (M and W), whereas the paraphrase in A and the early eleventh-century MSS VBE retain the original *skoutatoi*, as at *Strategikon*, 12.B.17.39. *Skoutaratos* appears to be a slightly later term—it is absent from the *Sylloge tacticorum*, the *Praecepta* of Nikephoros, and the anonymous *Campaign Organization*, but appears as the standard word for heavy infantry soldier in the *Taktika* of Nikephoros Ouranos (see the index, s.v. σκουταράτος). The presence of this term here (and also of *lorikatos* for the older *zabatos* at l. 549, see below) may suggest that a different compiler or scribe, with a slightly different and maybe more "demotic" technical vocabulary, was at work on this Constitution.

497   μετὰ τῶν ἀστιλιῶ   ἀστίλιον = Lat. *hastile*, shaft (of a spear): Mihăescu, "Éléments latins," 491; Trapp, *Lexikon*, 2:219. The passage reflects the duties of the "noncommissioned" officers of the Roman legion: Goldsworthy, *Roman Army at War*, 182 (with earlier literature), although the extent to which the practice described by Maurice and repeated by Leo was in force is not known. See 7§39, 12§§51–52.

511-32   §§73–76   Maurice, *Strategikon*, 12.B.18.

518-20   ἤτοι τὰς ἔχουσας τὰς λεγομένας τοξοβολίστρας, καὶ τὰ μαγγανικὰ ἀλακάτια   A gloss inserted into Maurice's original text here to clarify the type of artillery meant. See on Const. 5§6.38–39 above.

533-43   §§77–78   Maurice, *Strategikon*, 12.B.23.3–13 (and cf. 12.B.20.17–20)

547-69   §§80–83   Maurice, *Strategikon*, 12.B.23.14–36 and cf. Const. 20§211. Note that at l. 549 Leo substitutes the term *lorikatoi* (cf. Lat. *loricatus*) for Maurice's original *zabatoi*. See on ll. 493–94, and 6§2.9, above. There is little evidence of drinking before battle in order to stiffen resolve (in contrast to that for the effects of alcohol on troops after a victory); see, e.g., Wheeler, "Land Battles," 203 n. 69.

570-76   §84   See also 13§4.17–19 (and to the regulations and punishments outlined in Const. 8).

577-79   §85   Here Leo introduces the material he has taken from the older treatises of Aelian, Onasander, and Arrian. For the most part he has curtailed and paraphrased his originals by bringing elements from different subsections together thematically. Up to l. 615 he remains more or less historical, referring to "the ancients"; thereafter he returns to the style of the previous sections of this constitution, giving advice as though recommending particular tactics to his generals.

580–84   Cf. Aelian, *Tactica theoria*, 8.3.

584–92   Cf. Aelian, *Tactica theoria*, 7.2; Arrian, *Techne taktike*, 8.1–5.

597–600   Cf. Aelian, *Tactica theoria*, 7.4; Arrian, *Techne taktike*, 9.1–2, 14.1–2; Onasander, *Strategikos*, 17.625–29 (17).

600–604   Cf. Aelian, *Tactica theoria*, 8.3, 15.2.

605–8   §88   Cf. Aelian, *Tactica theoria*, 10.1–4.

609–15   §89   Cf. Aelian, *Tactica theoria*, 7.5; 8.13; 15.1; 18.3–4; Arrian, *Techne taktike*, 9.2–4; Onasander, *Strategikos*, 16.616–21 (16).

616–22   §90   Cf. Aelian, *Tactica theoria*, 7.4–5; Arrian, *Techne taktike*, 9.1–2; Onasander, *Strategikos*, 17.625–35 (17).

623–30   §91   See Const. 7§§48 and 54; and cf. Arrian, *Techne taktike*, 13.1–2; Onasander, *Strategikos*, 20.666–75 (20§§1–2).

631–39   §92   Cf. Arrian, *Techne taktike*, 13.1–2; 15.1–5; Onasander, *Strategikos*, 18.638, 643 (18); 19.655–56 (19§2).

640–48   §93   Cf. Onasander, *Strategikos*, 19.645–62 (19).

649–60   §§94–95   Cf. Onasander, *Strategikos*, 21.679–94 (21§§1–2).

661–66   §96   Cf. Onasander, *Strategikos*, 21.701–4 (21§4).

667–74   §97   Cf. Onasander, *Strategikos*, 23.765–80 (23§§1–3); and, e.g., Polyaenus, *Strategika*, 1.35.1; 2.1.3 (*Hypotheseis*, 14.3); 5.7 (= *Hypotheseis*, 14.17) and esp. *Rhetorica militaris*, 19–23 (with commentary: Eramo, *Siriano*, 142–43) on a range of false rumors or reports to be deployed according to circumstances before or during a battle. Cf. Frontinus, *Stratagems*, 2.4.9–11.

675–89   §98   Cf. Onasander, *Strategikos*, 28.827–31 (28); and cf. on Const. 14§§33.211–34.228, above.

690–702   §§99–100   See on 7§40 and 14§3.12–13 above; and cf. Onasander, *Strategikos*, 33.

703–8   See above on 12§57, 13§§1.3–4.17, and below on 20§78 and 110; and cf. Onasander, *Strategikos*, 23 and *Rhetorica militaris*, 19, 23–24. The encouragement to use deception, portents, and so forth has a long history. See, e.g., the mid-tenth-century compilation known as the *Parekbolai*, chap. 44.35, 114, which recommends the commander announce that the victories to come were already prophesied in Scripture; see Dagron, "Byzance et le modèle islamique," 226 and

Dain and de Foucault, "Stratégistes byzantins," 368 (and see ll. 106–7 above on the fatalism suffered by defeated troops). For spreading false rumors to encourage the soldiers or dishearten the enemy; see above, ll. 123–27, 667–74; and 20§§8, 13–14. Leo's reference here to the final Constitution, 20, consisting of the collection of gnomic sayings, illustrates the unitary composition of the main body of his treatise; see in particular 20§141.

CONSTITUTION 15

This constitution may have had, or may have been intended to have, an independent existence as a separate treatise, inserted (together with Constitutions 17 and 19) in the oldest manuscript M, composed in the middle decade of the tenth century without a chapter number and as an appendix after the epilogue and the rest of the *Taktika*. In W these constitutions are inserted into the position they currently occupy, probably also the position they had in the original structure of the treatise and which the redactor of M changed under instruction from the person who commissioned the manuscript (see chap. 2 for more detailed discussion). Siege warfare, both offensive and defensive, was recognized by Byzantine historians, generals, and writers of military handbooks as an essential element of the empire's military effort. Without adequate defenses many major fortresses could not have survived hostile attack, and the territory they controlled and hence the empire as a whole would the more easily have succumbed to the invasions and attacks which it had to confront throughout its history. Detailed accounts of the preparations necessary to withstand a siege are given in the military handbooks, especially in the tenth-century *De obsidione toleranda*, "On Resisting a Siege," a text which in turn forms part of a much longer tradition of poliorcetic literature going back at least to Aeneas Tacticus in the fourth century BCE, although Leo does not appear to draw directly on the text of Aeneas. By the same token, the treatises also include information on how to carry out aggressive siege warfare and reduce enemy strongholds. On medieval East Roman siege techniques during this period: Haldon, *Warfare, State and Society*, 183–89 with sources and literature. From before Leo's time one of the most graphic accounts are the two sieges of Thessalonike described in the first and second collections of the *Miracula S. Demetrii*; and contemporary with Leo (although problematic in its date and attribution; see *ODB* 2:1098–99), the account of the siege and sack of Thessalonike by John Kaminiates (see *PmbZ* 22904 but see Frendo, "Thessaloniki"). For the middle of the ninth century there are also accounts of the siege, fall, and sack of Amorion in Arabic and later Syriac accounts; see Tabarī, *Ta'rikh al-rusul wa'l-muluk* (also in Vasiliev, *Byzance et les Arabes*, 1:302–9); Michael Syr., 97–101; *De XLII martyribus Amoriensibus*, 2.11; 4.42–44; 7.65 and 71; Symeon *magistros*, 227.240–41. A clear account is found in Treadgold, *Revival*, 302–3.

3–4   Leo's constitution on siege warfare is very selective and omits reference in detail to several important contemporary practices. The section is drawn from several sources, including Maurice, *Strategikon*, 10, Onasander, and a body of material which may reflect a lost treatise, dubbed by Dain the *Antipoliorceticum* (Dain and de Foucault, "Stratégistes byzantins," 349–50, 366–67), composed probably after the end of the sixth century (see Sullivan, "Byzantine Instructional Manual," 140), elements of which survive only in late tenth-century manuals such as the anonymous *De obsidione toleranda*, and the treatise ascribed to Syrianos *magistros*, compiled possibly in the mid-ninth century rather than earlier, and which Leo also knew. The major Byzantine sources on the subject are Maurice, *Strategikon*, 10; Syrianos, *Strategy*, 13; *Sylloge tacticorum*, 53; and the anonymous *De obsidione toleranda*, ed. van den Berg. See also *Campaign Organization*, 21; Nikephoros Ouranos, *Taktika*, 65. A number of minor treatises or fragments of treatises have also survived, largely derivative of the Roman and Hellenistic manuals. For discussion, see McGeer, "Byzantine Siege Warfare" and Sullivan, "Byzantine Instructional Manual," 140–45. For a general survey of some aspects of late Roman and Byzantine siege warfare, see Sullivan, "Byzantium Besieged"; Kučma, "Osady"; and Purton, *Early Medieval Siege*, 7–32, 110–23, and 123–33 on Islamic siege techniques. For a tenth-century account of siege techniques and the construction of a range of devices used to protect and attack walls and towers, see the *Parangelmata poliorcetica*, although it would seem that many of the machines described were either impractical or no longer general employed. See also the commentary to Maurice, *Strategikon*, 10 (Rance) for historical examples and precedents (up to the seventh century) as well as explanatory discussion of technical terms. It is noteworthy that Leo barely mentions (cf. 15§28.172–74) mining the walls of a fortress or town, in contrast to some of the ancient and some more recent treatises (cf. Syrianos,

Strategy, 12.30, 40–42; 13.3–42 [in great detail]; *Sylloge tacticorum*, 53.9; 54.3; *De obsidione toleranda*, §§180–97; *Parangelmata poliorcetica*, 13–15 and commentary), and in contrast to the comments of the late tenth-century writer Nikephoros Ouranos, who states in his *Taktika* (65.22, with details of the method in 20–24) that this was the main means of capturing a fortress or town that could not be induced to surrender through fear or starvation (but see below on ll. 175–82); nor does he mention the *laisa*, a Slavic term describing a light, portable, house-shaped structure of woven branches with a steep roof and several entrances covered by matting or wicker screens, again varying in size or dimensions, and designed to protect the men carrying them or working behind or beneath them, and which occurs in several of the treatises written within half a century of the *Taktika*; see McGeer, "Tradition and Reality," 135–38; Sullivan, commentary to *Parangelmata poliorcetica*, §9; and on ll. 161–74 below. Leo makes no reference to the outer defenses of a fortress or town, such as the deepening of ditches or the digging of additional defensive earthworks, including the pits to stop cavalry, and so forth (all mentioned, for example, by Kekaumenos, *Strategikon* [Spadaro], 2:79 [116.21–29; Litavrin, 178.12–17]), although he does mention these in his discussion of marching camps (see on Const. 11 above). The reason for these omissions is certainly his reliance on Onasander and Maurice, neither of whom goes into greater detail. Although written later and representing a summary of much of the earlier material, the late tenth-century *Campaign Organization*, 21 is a useful comparative account of siege warfare, relying, like Leo, on a fairly general statement of tactics and methods, but largely avoiding technical detail.

9–21   §§2–3   Cf. Onasander, *Strategikos*, 38–39 and esp. 40; Maurice, *Strategikon*, 10.1.4–8; and Vegetius, *Epitoma*, 4.28. The dangers of being attacked by surprise while besieging an enemy force were widely recognized—see Haldon, *Warfare, State and Society*, 185–86 and McGeer, *Sowing the Dragon's Teeth*, 358. On the technical terms for ditch or fosse, see on Const. 11§8.40–42 above; and for the digging and fortifying of entrenchments around the fortress in question, see, for example, Theophanes continuatus, 68–69; Skylitzes, *Synopsis historiarum*, 39.44–45; trans. Wortley, 42 (Michael II's measures at Thessalonike in 823); Symeon *magistros*, §135.10 (Symeon of Bulgaria's attempt to lay siege to Constantinople in 913); Leo the Deacon, *Historia*, 71.17–20 (Nikephoros Phokas's siege of Arka in Syria; see Talbot and Sullivan, 122 and n. 98).

18   Leo omits Maurice's reference to events of the Persian wars in the sixth century: *Strategikon*, 10.1.7–8.

22–26   §4   Cf. Onasander, *Strategikos*, 41.1112–14 (41§1).

27–35   §5   Cf. Onasander, *Strategikos*, 41.1115–27 (41§2–42§1).

36–40   §6   Cf. Onasander, *Strategikos*, 41.1129–34 (42§2).

41–44   §7   Cf. below, ll. 135–38 and ll. 256–58; Maurice, *Strategikon*, 10.1.8–12; cf. also Syrianos, *Strategy*, 9.37–39; 10; *Campaign Organization*, 21.3–17; Nikephoros Ouranos, *Taktika*, 65.1, 3, 7–10; and *De obsidione toleranda*, §3. Ensuring that supplies were present in sufficient quantities, or that the besieged enemy's supplies could be cut off, was essential and was generally the first means of attack. In 823 Michael II had a trench dug around the city of Adrianople in order to force the surrender of the remaining supporters of Thomas the Slav; see on 9–21 above. By the same token, Basil I's sieges of both Tephrike and Melitene failed because the cities in question were well provisioned, whereas the besieging force ran out of supplies: *Vita Basilii imperatoris*, 37.24–26, 40.38–42 (138, 146). See the recommendations of the *De obsidione toleranda*, §§1–2. The Arab attack on Amorion in 838 was successful because the defenses were breached within ten days of the bombardment beginning, but the attackers had first of all dug a ditch around the city to cut it off; see Symeon *magistros*, 227, 235; and the accounts in the Arabic and Syriac traditions (see above, introductory paragraph).

45–57   §§8–10   Maurice, *Strategikon*, 10.1.12–22.

58–72   §§11–14   Cf. Maurice, *Strategikon*, 10.1.23–31.

73–90   §§15–16   Cf. Onasander, *Strategikos*, 41.1156–86 (42§§7–14); Maurice, *Strategikon*, 10.1.32–42. Cf. *Sylloge tacticorum*, 54.2. For the division of the besieging forces into relays to maintain a continuous attack, found in ancient sources, see Sullivan, commentary to *Parangelmata poliorcetica*, §4.10–11, with literature and sources and see, e.g., Aeneas Tacticus, 38.1–2.

87   ἐκλυόμενοι   Read "fatigued" or "exhausted" rather than "unstrung" (cf. also on 10§13.62 and 13§8.46).

88   λογισμῶν   "Calculations" or "calculated plans" might be better than simply "stratagems."

93-95   §18   Onasander, *Strategikos*, 41.1188-89 (42§14).

96-108   §19   Cf. Onasander, *Strategikos*, 41.1141-54 (42§§4-6); cf. Maurice, *Strategikon*, 10.3.28-35 for the defenders' perspective.

109-21   §20   Cf. Onasander, *Strategikos*, 41.1192-1207 (42§§15-17); Polyaenus, *Strategika*, 7.6.10 (= *Hypotheseis*, 54.21; cf. Frontinus, *Stratagems*, 3.8.3); 3.9.1-3 (*Hypotheseis*, 5.2, 45.3, 32.5). Note that where Leo retains the verb σαλπίζω, following Onasander's σαλπιγκταί, the Ambrosian paraphrase has σαλπιγκτὰς ἤγουν βουκινάτωρας, the latter being the more usual term. See Sullivan, commentary to *Parangelmata poliorcetica*, §4.12 with literature (and additionally on trumpets see on Const. 11§19.105-7, §23.145; 13§11.54-57); and cf. *Hypotheseis*, 54.21, 55.1.

122-31   §21   Cf. Onasander, *Strategikos*, 41.1211-28 (42§§18-22).

132-38   §22   Cf. Onasander, *Strategikos*, 41.1237-46 (42§23); cf. Maurice, *Strategikon*, 10.3.4-8 for the same point, but from the defenders' perspective.

139-60   §§23-26   Maurice, *Strategikon*, 10.1.38-54.

146-48   See the editor's comment, 361, n. 2. This story recalls also Judges 9:52-54, where Abimelech is injured by a millstone thrown down by a woman on the battlements of the fortress he was attacking. Cf. the death of Pyrrhus at Argos (Plutarch, *Pyrrhus*, 34.2-3), likewise stunned by a rooftile, falling from his horse and thus enabling his enemies to slay him.

154-55   διὰ πυρφόρων σαγιττῶν   For firearrows, see James, "Archaeological Evidence" and esp. Kiechle, "Brandwaffen im Altertum." Firearrows were employed by the defenders of Thessalonike to damage the stone-throwing machines of the besieging Slav forces in the late sixth century: *Miracula S. Demetrii*, 154.18. For other incendiary devices, see Sullivan, *Parangelmata poliorcetica*, 161-62 (comm. to §2.9) and §49.20 (and see on Const. 19§6.32-33 and §§63.354-64.361, below).

157-58   τῶν πέτρων πυρὸς πεπληρωμένων δι' ὕλης   For the various types of siege artillery listed here see above on Const. 5§6.38-39. Leo changes Maurice's *chouzia* (literally a gourd; see Dagron and Feissel, *Inscriptions de Cilicie*, 171, no. 108.6 and in detail Rance, *Strategikon*, 10.1, note 7), which appears otherwise only in Leo, *Problemata*, 10.6, to *petrai* (stones), the older term presumably having dropped out of use or being less readily understood, although the "stones" in question must refer to ceramic containers of some sort.

161–74   §§27–28   Onasander, *Strategikos*, 41.1136–40 (42§3); cf. Vegetius, *Epitoma*, 4.13–17; *Parangelmata poliorcetica*, §2 and §§13–14 with commentary (159–63, 182–87). For the wooden towers, see on Const. 15§45.273 below. At 15§28.172–74 Leo refers to mining the walls of a fortress, his only reference to such activity, even though it was widely practiced (for late Roman practice see Rance, *Strategikon*, 10.1, notes 10–11). Perhaps the best-known ancient mine and countermine are those uncovered during the excavations of the Roman fortress at Dura Europos (mod. Salihiya) on the Euphrates, taken by the Sassanians in 256; see Leriche, "Techniques de guerre," 84–85; also James, "Stratagems, Combat and 'Chemical Warfare'." Countermining was intended to cut off the enemy sapping activity before it became a danger. A good example is provided by Procopius's account of the Roman counter-sapping activities at the siege of Dara: Procopius, *Wars*, 2.13.20–28. Kekaumenos, *Strategikon* (Spadaro), 2:79 (116.35–118.10; Litavrin, 178.24–32) describes the process of countermining in detail. For an earlier account, see Syrianos, *Strategy*, 13; for later accounts see, for example, Leo the Deacon, *Historia*, 25.11–27.5 (Talbot and Sullivan, 77–79; the Byzantine siege of Chandax on Crete in 961, although his account is based largely on Agathias; see Sullivan, "Byzantine Offensive Siege Warfare," 181–82); Basil II's siege and capture of Moglena (Skylitzes, *Synopsis historiarum*, 352.27–31; trans. Wortley, 334); Anna Komnene, *Alexiad*, 11.1.6–7 (trans. Sewter, 335 / Frankopan, 299) for the siege of Nicaea; and 13.3.4–6 (trans. Sewter, 400–402 / Frankopan, 362–64), for the Normans' attempt to mine the walls of Dyrrhachion.

175–82   §29   This is Leo's own comment; but see also Vegetius, *Epitoma*, 4.16–17. Although Leo glosses over many of the technical details of siege warfare with respect to types of artillery and their placement, the positioning of artillery was especially important. On occasion, and as dictated by circumstances, commanders built earth-and-timber embankments upon which to place stone-throwers or other projectors. The embankments could be extended forward to the walls, so that both siege engines and soldiers could be positioned to shoot down onto the battlements and the defenders' positions. This was a labor-intensive and time-consuming but well-established practice, recorded by Vegetius, by Procopius (as used by both Persians and Romans), and by Kekaumenos, for example. In his account of Basil II's siege of the Bulgarian fortress of Moreia, between Philippoupolis and Triaditza, Kekaumenos describes how the Bulgars were able to sally out and penetrate the timber framework of the mound, and set it on fire from within (so that the Romans failed to see what had happened), and destroy it: Kekaumenos, *Strategikon* (Spadaro), 2:81 (120) (Litavrin, 180.32–182.12). See Vegetius, *Epitoma*, 4.15.7 (trans. Milner, 121 and n. 10). Contrary to what Nikephoros Ouranos suggests (that mining the walls of a fortress or city was the usual practice in his day), Leo the Deacon's account frequently mentions also the use of artillery to bombard the walls; see, e.g., Leo the Deacon, *Historia*, 15.22–23; 25.16–17; 71.18–20 (Talbot and Sullivan, 68, 77, 122); and note that the Cretan expedition of 949 took both artillery and the materials for constructing a siege tower with it; see on 267–69 and 273 below. Earlier, Michael II is reported to have been reluctant to construct siege machines while besieging Thomas the Slav in Adrianople in case his Bulgarian allies learned the same skills (although this may well be a later Byzantine cultural prejudice written into the account): Theophanes continuatus, 68.13–17; Skylitzes, *Synopsis historiarum*, 42.39–42 (trans. Wortley, 42), since a siege tower and artillery of various types were constructed by the Slavs who attacked Thessalonike in the early seventh century and later by the Avars (*Miracula S. Demetrii*, 2.1 and 2). Contemporary with Leo, the account of the siege and capture of Thessalonike in 904 also refers to siege artillery: *De expugnatione Thessalonicae*, 29.3, 5 (seven machines, constructed by the attackers on Thasos and brought across to the attack). The late tenth-century anonymous treatise *Campaign Organization*, 21.87, 105 assumes that in a siege engines of some sort will be set up (the text recommends that they be set up quickly to avoid delays which would entail the consumption of supplies): 21.106–17.

183–90 See Const. 15§§35 and 39; cf. Onasander, *Strategikos*, 38.1055–60 (38§§7–8).

194–200 §§31, 34 Onasander, *Strategikos*, 38.1038–54 (38§§1–6); cf. Maurice,
and 210–16 *Strategikon*, 10.1.23–29.

201–6 §32 This reference is to the campaign led by Nikephoros Phokas in
southern Italy—Calabria—in 885. See *Vita Basilii imperatoris*, 71.12–
29 (244–46); Skylitzes, 160.69–78 (trans. Wortley, 154); *PmbZ* 25545;
and see Cheynet, "Phocas," 292.

207–9 §33 Another of Leo's own comments, but following a standard east
Roman line on the justification of warfare. See above, on pr.§4.32–36.

222–28 §36 Onasander, *Strategikos*, 39.1064–79 (39§§1–3; cf. Polybius, *Hist.*,
9.14.5–18.7).

229–37 §§37–38 Onasander, *Strategikos*, 39.1081–95 (39§§4–7).

238–49 §39 Onasander, *Strategikos*, 35.984–90 (35§§4–5), 41.1209–35 (42§§18–
22), 1246–55 (42§§24–26). Treatment of prisoners was a major concern
of Byzantine writers, in particular in view of the need both to conserve
manpower and to ensure the possibilities for prisoner exchanges.
Although Leo does not specifically mention these activities, and
although his text here follows Onasander, it was clearly a concern. For
discussion, see Patoura, Αἰχμάλωτοι and Kolia-Dermitzaki, "Some
Remarks"; see also Const. 15§§31–35, 16§9, 19.226–27; and Campagnolo-
Pothitou, "Échanges de prisonniers." For the longer-term background of Christian attitudes to prisoners and ransoming, see Osiek,
"Ransom of Captives," Rotman, "Byzance face à l'Islam arabe" and
idem, "Captif" (who argues for important changes in attitude to
Muslim prisoners by the later eighth century); and for the Islamic perspective Guemara, "La libération et le rachat des captifs." Note also,
for more senior prisoners: Simeonova, "Arab Prisoners-of-War."

250–55 §40 These appear to be Leo's own words, and may well reflect his
own reading from chronicles or his generals' experience: the clearest example of failure to defeat an enemy in the field prior to a siege
is, of course, the case of Amorion in 838, when Theophilos attempted
to keep the two invading enemy columns separated and defeat them
piecemeal, but failed when his own, numerically superior force, was
broken at the battle of Anzen in July 838 (see Treadgold, *Revival*, 299–
300; Haldon, *Byzantine Wars*, 82–85).

256–85 §§41–45 Maurice, *Strategikon*, 10.3.3–27; cf. *Sylloge tacticorum*, 53.1.
At ll. 258–61 the advice that those unable to contribute effectively to

the defense should be evacuated was a standard procedure where practicable, as in Italy in 537 when Belisarius evacuated noncombatants from Rome (Procopius, *Wars*, 5.25.2–4; cf. Vegetius, *Epitoma*, 4.7.10), or in Anatolia in 838, when the population of Amorion was evacuated to refuges in the Emirdağ before the Arab forces arrived to besiege the city: *Theophanis Chronographia*, 388 (Mango and Scott, 539). During the siege of Adrianople by Michael II in 823, Thomas the Slav evacuated all the nonmilitary population and animals. General preparations for resisting a siege are also outlined by Kekaumenos, *Strategikon* (Spadaro), 2:79 (116.21–118.25) (Litavrin, 178.12–180.13). For descriptions of the various weapons and tools employed in siege warfare, see, e.g., the mid-tenth-century treatise *Parangelmata poliorcetica* in Sullivan, *Siegecraft*.

263 κιλίκια   See on 5§5.36 and Rance, *Strategikon*, 10.3, note 243. For the following lines see also Syrianos, *Strategy*, 13.72–135, and cf. *Parangelmata poliorcetica*, §39.30–35 and commentary, 217–18; Vegetius, *Epitoma*, 4.6.

264 σάρκινα   Lat. *sarcinae* ("luggage," "packs"). Cf. *Parangelmata poliorcetica*, §12.21 and comm. (182) and Trapp, *Lexikon*, 7:1530. It is not always clear what the term meant at this time. It can refer either to fresh hides (see *Sylloge tacticorum*, 53.5, which paraphrases σάρκινα with βύρσας νεωδόρων βοῶν, "hides of newly flayed oxen"; cf. the much earlier Vegetius, *Epitoma*, 4.15.4; 17.1: *crudis ac recentibus coriis*) or to a type of climbing-netting; see *Parangelmata poliorcetica*, §12.21: αἵ κατασκευάζονται διὰ πλοκῆς καὶ ῥαφῆς δεσμούμεναι, δικτυωταὶ οὖσαι ὡς τὰ λεγόμενα σάρκινα ("which are constructed by being bound together with plaiting and stitching, net-like, similar to the so-called *sarkina*"). Leo follows Maurice in listing these items as separate articles; see Rance, *Strategikon*, 10.3, note 24.

264 σχοινία εἰλημμένα   "Woven cable netting" rather than "coiled ropes"; see Syrianos, *Strategy*, 13.115–20: σχοινία εἰς σχήματα δικτύων πλέξαντες, which were to be hung from the battlements to absorb the momentum of stone shot; Philo of Byzantium, *Poliorcetica*, 3.3: ἐκ τῶν σχοινίων πλέξαντας δίκτυα ("nets woven from ropes/cables"); and *Parangelmata poliorcetica*, §12.21.

264 πόντιλα   Lat. *pontilia*, referring in general to timbers or bridging timbers, here referring apparently (following Maurice, *Strategikon*,

10.3.11; with Rance, *Strategikon*, 10.3, note 26; 12.B.21, note 221; see Trapp, *Lexikon*, 6:1351) to boards or planks suspended from or fixed to the battlements as a means of protecting against enemy missiles. According to the *Sylloge tacticorum*, 53.5 this consisted of "timbers joined together to form a trellis (the Romans call these *pontila*)": ξύλα συνηρμοσμένα ὡς ὕφασμα (πόντιλα ταῦτα Ῥωμαῖοι καλοῦσι), suggesting a method for raising the height of the wall with an additional wooden breastwork, similar perhaps to the measures described by George of Pisidia (*Bellum Avaricum*, ll. 271–73) during the siege of Constantinople in 626: φραγμούς τε ποιεῖν καὶ πεπηγμένους πάλους/προσαντιβάλλειν καὶ πλέκειν τεῖχος νέον ("[to make] . . . fencing and erect posts, and to build a new wall . . ." Cf. Vegetius, *Epitoma*, 4.19.

267–69 **§42** Leo rephrases the original phrase in Maurice, *Strategikon*, 10.3.13: τοὺς παλλίωνας ἤτοι ἐμβόλους; cf. *Parangelmata poliorcetica*, §10.2–3: τῶν ἐμβόλων . . . ἤτοι τῶν χελωνῶν. . . . *Embolos* originally referred to either a type of structure like the "tortoise" or shed concealing a ram, cf. Du Cange, *Glossarium Graecitatis* 378; Sophocles, *Greek Lexicon*, 453; Lampe, *Patristic Greek Lexicon,* 453; cf. Mihăescu, "Littérature byzantine," 270–71 (and Trapp, *Lexikon*, 3:487); and Sullivan, commentary to *Parangelmata poliorcetica*, §2.5 and §7.15–35, or the ram or "beak" itself. By the tenth century the term was apparently no longer clearly understood in this latter sense. The term *palliones* in Maurice, which is an equivalent of the shed or "tortoise" containing the ram, applied to the whole device—shed with ram—remains obscure (see Trapp, *Lexikon*, 6:1181, and Rance's commentary at Maurice, *Strategikon* 10.3.13, note 290), but is probably a version of *spaliones*, a wicker cover or shed used to protect soldiers or workers, described by Agathias during his account of the siege of Onoguris and by Menander the Protector in similar terms; see Suidas, 1:4.901 (415); Menander Prot., *Frg.*, 40 (trans. Blockley 247–49); Agathias, 3.5.9–11 (trans. Frendo 73); Trapp, *Lexikon*, 7:1589. Leo also elaborates on Maurice's account (which entails deploying grappling hooks and throwing down pitch, fire, and stones on the enemy), to state that grappling hooks and sharpened beams covered in pitch should be employed. See Syrianos, *Strategy*, 13.61–71 and cf. Vegetius, *Epitoma* 4.20. For *chelonai* with rams and shelters for men undermining the walls, see *Miracula S. Demetrii*, 148.27; 152.10–11, 18–20; 186.23; they are listed among siege equipment for the Cretan expedition of 949, see Haldon, "Chapters II, 44 and 45," 225.116, 118 (= *De cerimoniis*, 670.10–12). The grappling hooks were known as "wolves," and were hitched to oxen behind the wall so that once the enemy *chelona* or tortoise was secured the oxen would be driven back from the wall, lifting and overturning the shed as they pulled. If the terrain was unsuitable for the oxen, then the "wolf" was to be firmly attached to the battlement or tower so that it could not be pulled back; see the description in Kekaumenos, *Strategikon* (Spadaro), 2:79 (116.28–34; Litavrin, 178.18–23); and for an earlier account, in which the prepared "wolf" forced the attacking troops to withdraw their tortoise and ram, see *Miracula S. Demetrii*, 152.11–17 (where the device is described as having had an effect on the attackers "like that of a hobgoblin on small children"). Such siege sheds or "tortoises" are frequently described in accounts of sieges, used both to shelter the men operating the ram or undermining the walls. See, for example, Anna Komnene's account of the sieges of Nicaea and Dyrrhachion: *Alexiad*, 13.1.6 (trans. Sewter, 335 / Frankopan, 299) and 13.3.1 (trans. Sewter, 400–401 / Frankopan, 362–63), where large wheeled *chelonai* were employed. For various types of tortoise, and their practicality or not, see *Parangelmata poliorcetica*, §2.2, §§7–11, and §13, and commentary (160, 173–80, 182–86); with Lendle, *Schildkröten*.

273　τοὺς ἐπαγομένους πύργους　Siege towers mounted on wheels, or rollers, appear in several earlier accounts of siege machinery. They are attested at the Roman siege of Amida in 503 (Joshua the Stylite, §56 [45]; trans. Trombley and Watt, 67), at the Persian siege of Martyropolis in 530 (Malalas, 470; trans. Jeffreys, Jeffreys, and Scott, 273–74), at the Goths' siege of Rome (Procopius, *Wars*, 5.21.3–4, 14); and during the siege of 626, where twelve wooden towers were constructed by the Avars (*Chronicon Paschale*, 720.1–3; trans. Whitby and Whitby, 174). They appear also at the siege of Thessalonike in 616–18: *Miracula S. Demetrii*, 180–82; see Rance, *Strategikon*, 10.3, note 33. The standard term by the tenth century seems to have been ξυλόκαστρον or ξυλόπυργος. The list of equipment for the Cretan expedition of 949 mentions the various iron components for a *xylopyrgos*, probably a siege tower with wheels; see Haldon, "Chapters II, 44 and 45," 225.116 (= *De cerimoniis*, 670.10–11) and commentary: "For equipment for siege warfare: a wooden tower, tortoises, large bow-ballistae with pulleys and silken strings, traction-powered stone-throwers, lambda-framed stone-throwers, artillery pieces and their fittings: rams for the tortoises, and for the various items of artillery ring-clamps, shackles and bolts, leather-covered iron slings, plates for covering the sheaves of the various pulleys, crowbars, mallets, pick-axes, weights, hides, felts, spades, cauldrons, levers/shafts, shovels, various ropes, nails. . . ."; and compare with *Parangelmata poliorcetica*, §30 (and commentary, 208–10). The hides and felts were to protect the wooden structures— siege towers, artillery, and *chelonai* (see, for example, *Miracula S. Demetrii*, 148.28; 152.23, 31; 154.19–21; 186.23; 188.34). Cf. *Parangelmata poliorcetica*, §2.6 for a portable ξυλοπύργιον (with commentary, 160); the mid-tenth-century Ambrosian paraphrase of Onasander refers to ἐλεπόλεις, ἤτοι ξυλόκαστρα: Korzensky and Vári, 75.628. Alexios I used siege towers against the forces under Bryennios during the siege of Kastoria, as did Robert Guiscard in the siege of Dyrrhachion: Anna Komnene, *Alexiad*, 4.1.2; 4.4–8; 6.1.1–2 (trans. Sewter, 135, 141–43, 181/ Frankopan 109, 116–17, 153); and Bohemund later built a huge wooden siege tower, 5 or 6 cubits (2–3 m) higher than the towers of the city defenses, during his attack on the same fortress: *Alexiad*, 13.3.7–12 (trans. Sewter, 403–4/ Frankopan, 365–66). Such towers were protected by hides to ward off the danger of fire, which was the main means of their destruction (see *Miracula S. Demetrii*, 148.28–31; *Alexiad*, 3.12.2, 5; 4.1.2; 11.1.6; 13.3.1 (trans. Sewter, 131, 132, 135, 335, 400 / Frankopan, 106, 107, 109, 299, 362) for the fire that destroyed Bohemund's giant tower); *chelonai* or tortoises were similarly protected: *Parangelmata poliorcetica*, §15.18–19 and comm. at 189. There is no reason to think that they were not employed where the ground was level and the conditions favorable and where a full-scale assault was necessary. To what extent an eleventh-century MS illumination showing a siege tower reflects actual construction remains uncertain; see Sullivan, *Siegecraft*, Ill. 15 (Vat. gr. 1605, fol. 26). The towers were generally both time-consuming and costly to build, and in many accounts only one is mentioned in a siege. A siege tower is mentioned in both the Slav and Avar attacks on

Thessalonike in the early seventh century: *Miracula S. Demetrii*, 186.24, 219.1. Note the 949 Cretan expedition was also equipped with the parts and materials for just one such device: Haldon, "Chapters II, 44 and 45," *loc.cit.* and cf. *Alexiad*, 13.3.7 (trans. Sewter, 402/Frankopan, 364), where Anna reports that the giant siege tower brought by Bohemund to the attack on Dyrrhachion had been constructed a year before the attack (and was presumably carried in sections and rebuilt once the siege had begun).

277–85 ἀναγκαῖον δὲ ... ἠνεῳγμένα. See Maurice, *Strategikon*, 10.3.17–27. The construction of Byzantine towers and fortifications more broadly still requires a general treatment, in turn based on a fuller analysis of the known standing monuments in the Balkans and Turkey. Leo adds nothing to Maurice, however; see Rance, *Strategikon*, 10.3, nn. 34–36; and in particular Dunn, "Heraclius's 'Reconstruction of Cities'," esp. 800. See also Lawrence, "Skeletal History"; Saunders, "Qal'at Seman"; and Sinclair, "Byzantine and Islamic Fortification." For general discussion, see Foss and Winfield, *Byzantine Fortifications*; Pringle, *Defence of Byzantine Africa*; and the brief survey in Purton, *Early Medieval Siege*, 119–23. For *paraportia* and *parapyle* see on Const. 11§14.81 above.

286–92 §46 Cf. *De obsidione toleranda*, §§68; 149–50; *Sylloge tacticorum*, 53.7. For collecting stones on the walls see also Kekaumenos, *Strategikon* (Spadaro), 2:79 (116.24–25; Litavrin, 178.15). The practice was the usual means of defending walls from ladders or other attack; see Leo the Deacon, 16.1 (Talbot and Sullivan, 68), for example, and *Parangelmata poliorcetica*, §2.8–9, with commentary, 161.

293–97 §47 Maurice, *Strategikon*, 10.3.28–35.

298–305 §48 Maurice, *Strategikon*, 10.3.33–35. Leo changes and expands this passage and places the emphasis on possible dissension among the population.

299 ταῖς τοῦ τείχους πεδατούραις *Pedatoura* referred originally, and probably into the seventh century, to a section of the walls, and this is the meaning Leo retains here. For discussion see Dietz, "Cohortes, ripae, pedaturae," and in detail Rance, *Strategikon*, commentary to 10.3.33. By extension it seems by Leo's time to have come to mean also the soldiers of a patrol along the walls, or indeed any patrol in a garrison; see Haldon, *Byzantine Praetorians*, 541; "Strategies of Defence," 145. See also on Const. 11§16.91–92 above.

306–31 §§49–53 Maurice, *Strategikon*, 10.3.36–56.

325 πλέθρου See Trapp, *Lexikon*, 6:1312; Schilbach, *Metrologie*, 81–83: literally a pond or pool, but the term *plethron* also referred to a measured distance of under 100 Roman feet. In order to emphasize the way in which a limited supply should be managed, Leo alters Maurice's text, which states simply, "If the drinking water comes from a cistern or from a pond/pool, it must be. . . ."

332–42 §§54–55 Leo here paraphrases some of the general advice in the older treatises. Cf. Onasander, *Strategikos*, 40, for example.

343–402 §§56–64 Maurice, *Strategikon*, 10.4.3–62, with Rance, *Strategikon*, 10.4, and notes. See also *Sylloge tacticorum*, 55; and see especially Syrianos, *Strategy*, 9, on forts (*phrouria*), and 12, on building a city (*polis*). The first thing that strikes the reader of the latter text is that the "city" in question is in fact a (small) fortress: "Suitable sites for building a city, especially if it is going to be fairly close to the border, are those on high ground with steep slopes all about to make approach difficult. Also suitable are sites with large rivers flowing around them or which can be made to do so, and which, because of the nature of the land, cannot easily be diverted. Finally, there are sites on a promontory in the sea or in very large rivers connected to the mainland only by a very narrow isthmus" (11.3–9; trans. Dennis). The section on *phrouria* is especially interesting. The text tells us that their function is to observe the approach of the enemy, to receive deserters, and to prevent people from fleeing to the enemy. Such installations should be near the routes they are meant to observe, but not obvious enough to attract attention. They should exploit natural features for their defense, they should have a small garrison of men without their families, and the troops posted there should be relieved at regular intervals. All these characteristics are repeated in the later, mid-tenth-century treatise on shadowing or guerrilla warfare, where outposts should be situated on "high and rugged mountains," some three or four miles apart and relieved every two weeks. What is meant by the term *phrourion* thus becomes clear—a small, defended post with minimal facilities, relatively unconspicuous, easily abandoned and recovered after the enemy have moved past, and essential for maintaining a watching brief on hostile incursions (Syrianos, *Strategy*, 9; *Skirmishing*, 1). The policy outlined here is reflected in a number of actual developments along the empire's frontiers. During the reign of Constantine V, for example, the emperor built a series of forts and strongholds in the regions bordering the Bulgarians, populating and garrisoning them with migrants transferred from Asia Minor; see *Theophanis Chronographia*, 429.19–22, 25–30 (Mango and Scott, 593–94); Nikephoros, *History*, chap. 73 (145). See the discussion in Ditten, *Ethnische Verschiebungen*, esp. 184–86. In the early 780s the empress Eirene undertook the construction or refurbishment of a line of fortified posts at Philippoupolis, Beroea, Markellai, and Anchialos; see *Theophanis Chronographia*, 457.6–11 (Mango and Scott, 631). These activities, esp. those under Constantine V, seem to have been carried out in the sort of circumstances envisaged in this passage of Leo, taken from Maurice but still reflecting political-military practice on the frontier. Note that the descriptions of a frontier context, of the forts and "cities" which are situated there and their function in watching for enemy movements, and a number of related topics, as described in Syrianos, *Strategy*, all fit the context of the eighth–tenth centuries far better than that of the late Roman period, thus reinforcing the argument that this text may be a product of the ninth-century Byzantine world. See chapters 1 and 2 above, pp. 18–20, 39, 70.

378–80   §61   Onasander, *Strategikos*, 41.1236–45 (42§23).

381–84   §62   See Rance, *Strategikon*, 10.4, note 53: Roman cavalry were allowed to receive issues of fodder after 1 August, a reflection of the decreasing availability of adequate pasturage from that point in the year (see *CTh* 7.4.8, promulgated in the year 362)

385–93   §63   Maurice, *Strategikon*, 10.4.41–51.

386   βουττία τέλεια   "Complete" or "full-size" rather than "well-built." The exact meaning of the term is not always clear. In the Cretan expeditionary documents for 949 thirty bronze *boutia* are listed as being provided for the *droungarios* of the fleet, but bronze barrels are an unlikely interpretation and buckets or tubs seems more probable here; see Haldon, "Chapters II, 44 and 45," 233.221. *Bout(t)ion* derives from Lat. *buttis* or *butta*; see Trapp, *Lexikon*, 2:293; Mihăescu,

"Éléments latins," 490; and on water-containers in general (and particularly in ships) at this period, see Pryor and Jeffreys, *Age of the Δρόμων*, 361–78. Note that Leo omits the taps (*epitonia*) that should be inserted into the casks (= Maurice *Strategikon*, 10.4.47), suggesting he did not really grasp the arrangement described in his exemplar.

395   στυππίου   For *Styppeion* as "tow," see Mihăescu, "Éléments latins," 488; Trapp, *Lexikon*, 7:1626; and Pryor and Jeffreys, *Age of the Δρόμων*, 147–52, 484–87. Leo omits Maurice's ingredient *kampsarika* (*Strategikon*, 10.4.55) presumably because he did not recognize the word. See Trapp, *Lexikon*, 4:757.

394–402   §64   Leo copies directly from Maurice, and the dimensions given measure roughly 6 m × 3 m × 3 m, holding up to 54 m³ of rainwater (54,000 litres = 11,878 imp. gallons). See also Rance, *Strategikon*, 10.4, note 59, with discussion and late ancient examples with sources.

CONSTITUTION 16

This very short constitution summarizes what the general is supposed to take into account after the fighting is over, and is drawn almost entirely from Onasander, with a few elements from Aelian and Maurice, *Strategikon*, 7.B.11–12. In the first version of the *Taktika* it would have been the penultimate constitution, preceding the section on the tactics and mores of foreign peoples and the epilogue (since Const. 17 on surprise attacks was not, apparently, integrated into this original scheme), and following on directly from Const. 14, dealing with the day of battle. It is very general and contains very little technical or specialist information, and several of the points made are repeated, albeit with different wording, from earlier sections. In its content, which deals with the aftermath of a war rather than an actual battle, it serves in many respects as a conclusion to the main part of the treatise, with the section on foreign peoples which comes after it, although taken directly from Maurice, fitting in some respects somewhat uncomfortably into the overall structure, except for the sections on the Saracens, which appear also to have had an independent existence (see below on Const. 18). Const. 16 ends rather suddenly with a paragraph on the importance of the general being accessible and open to approach, and with the somewhat vague prescription to "observe these matters also after war as well as whatever else may occur to you along these lines." In general, the constitution leaves the reader with the impression of both vagueness and incompleteness.

6-32 §§2–5 Onasander, *Strategikos*, 34, and see esp. *Rhetorica militaris*, 55.1: Δεῖ δὲ ἐπινίκιον λέγοντας πρῶτον μὲν εὐχαριστεῖν τῷ θεῷ ὡς προξένῳ τῆς παρούσης νίκης; and 55.2 for the words of the prayer of thanks (and see Eramo, *Siriano*, 189 n. 149). For booty and plunder, fame, glory, rewards, and promotions as an encouragement in war, see esp. ibid., 45.5–6: καὶ τί οὕτω τῶν πάντων ἐπιδοξότερον, ὡς νίκη κατὰ βαρβάρων, ἣν πόνος ποιεῖ καὶ ὁμόνοια καὶ ἡ πρὸς τοὺς ὁμοφύλους ἀγάπη; διὰ ταύτην δωρεαί, διὰ ταύτην στέφανοι, διὰ ταύτην βραβεῖα, τιμαί, θρίαμβοι, ᾄσματα ἐπινίκια, καὶ τἄλλα, οἷς χαίρει βίος καὶ δι᾽ ὧν τὸ τῆς ἱστορίας πλατύνεται στόμα. See on Const. 12§57.420 above and on 20§192 below.

33–38 §6   Aelian, *Tactica theoria*, prologue 6 (Köchly and Rüstow, 234).

39–47 §7   Onasander, *Strategikos*, 35.974–82 (35§§1–3).

48–55 §9   Onasander, *Strategikos*, 35.984–90 (35§§4).

56–63 §10   Onasander, *Strategikos*, 35.992–94 (35§5), although Leo considerably expands the sentiments expressed in his exemplar; cf. the Ambrosian paraphrase, 64.470–73. See also Maurice, *Strategikon*, 8.2.27.

64–73 §§11   Onasander, *Strategikos*, 36.998–1004 (36§§1–2), again expanded by Leo: cf. the Ambrosian paraphrase, 64–65.475–81; Maurice, *Strategikon*, 7.B.6; and Const. 14§31.204–8; 20§72.

66–67 Cf. 2 Macc. 12:43–46.

74–86 §§12–14   Onasander, *Strategikos*, 36.1006–17 (36§§3–6). Leo's reference to "one of the ancients" in ll. 84–85 is to this passage in Onasander.

78–80 Cf. Maurice, *Strategikon*, 8.1.32.

87–105 §§15–16   Onasander, *Strategikos*, 37.1020–35 (37§§1–5); cf. Maurice, *Strategikon*, 8.1.23, 37. Here the advice is to honor agreements with the enemy, but somewhat different advice is given at 17.25–29, for example, likewise taken uncritically from Maurice, see *Strategikon*, 9.1.16–20.

106–13 §17   Onasander, *Strategikos*, 11.540–46 (11§6); and cf. Consts. 2§19.133; 13§9.47–51.

CONSTITUTION 17

This substantial constitution is based largely on Maurice, *Strategikon*, 9, with some extracts from Onasander. Maurice's sources remain uncertain. Nevertheless it is readily seen that what Leo borrows from Maurice still has direct relevance to his own times, especially in light of the sort of raiding warfare typical of the eastern frontier well into the tenth century. This constitution, like Const. 15, was also taken out of the body of the *Taktika* by the redactor of M (see chapter 2 above), but must nevertheless have been conceived by Leo as part of a single work, since he refers explicitly on two occasions within the text to the constitution on marches (see below on l. 314). The contents of Const. 17 are also reflected, in a different style, language, and arrangement of the material, in the later treatise *Skirmishing*. The fact that the constitution also appears as a separate constitution in M suggests the importance of this type of frontier warfare in both Leo's eyes and in the opinion of whoever commissioned this manuscript. Indeed, the content of some of the passages dealing with this type of warfare in this constitution is close enough to that in *Skirmishing* to suggest that the writer of the latter treatise used Leo as a starting point for the organization of his own material and advice. Similar considerations regarding the importance of Consts. 15 and 19 apply, as well as the (also independently circulating) section from Const. 18 on "How to Fight the Saracens" (see the Introduction, above). The importance of this section is very clear from the way in which Leo takes sections from different books of the *Strategikon* to create his own variant on the scenarios and advice he wishes to give, an important indication both of the fact that he is intending to produce a handbook of contemporary relevance and of the actual historical—tactical and strategic—situation with which his officers had to contend. While Leo does not differentiate between different types of raid, since he is basically repeating the text of Maurice, Const. 17 does reflect the military situation along the frontiers of the empire in both the Balkans and Anatolia, and deals, albeit very briefly, with almost the same topics as the author of the treatise on guerrilla or skirmishing warfare, writing in the 970s. This gives a much more detailed account (while pointing out that the heyday of this strategy is past: *Skirmishing*, pr.3–12), and highlights three types of hostile raid into Roman territory, differentiated either by size or by timing. The first involves small, fast-moving parties of cavalry, which might invade Roman territory at any time, and whose entry should be communicated to the local commanders as quickly as possible by the border scouts and watchposts, so that they might be met, ambushed or hemmed in, and turned back—where possible without any substantial gains in booty (*Skirmishing*, 4, 6, and compare Const. 17§76.448–55, §§77.456–81.479). The second category is the major raid, usually mounted in August and September, involving quite large forces made up of volunteers for the *jihād* along with regular troops from the Arab borderlands—Aleppo, Tarsos, Antioch—which fulfilled both an economic and an ideological function: economic in terms of seizing livestock and other booty; ideological in respect of the desire of many Muslims to participate in the *jihād*. The local commander was enjoined to use every means at his disposal to find out when such raids would begin, by which route, and how numerous the enemy host would be (*Skirmishing*, 7; cf. Const. 17§69.395–400). The invading force should then be shadowed; smaller raiding parties that were sent out once it had reached Roman territory should likewise be shadowed and attacked or harassed, as appropriate. The landscape should be deprived of provisions and livestock, where possible, to maximize the invaders' logistical difficulties (Const. 17§§66.377–67.389); and the enemy force should be subject to constant harassment as it moved, foraged for supplies, set up camp, or attempted to collect booty (Const. 17§59.341–47). The passes through which it would return should be occupied and ambushes laid; the water supplies should be held by Byzantine forces; and the enemy should be attacked as they return, laden with captives and captured livestock or other booty (Const. 17§60.348–51; *Skirmishing*, 3–5, 8–11). The Roman response was not always effective, and there are many examples where enemy leaders were able to outwit and out-general the Roman commanders (see, for example, Howard-Johnston, "Byzantine Anzitene," esp. 241–45, a brief analysis of a successful raid into Anzitene in eastern Asia Minor mounted by the emir Sayf ad-Daula in 956). Finally, Roman commanders had to be alert to the possibility of surprise raids, launched before the local population had been evacuated or any sort of ambush or shadowing-party organized (see below on 17.341–56). In this case, a series of emergency measures are set out, particularly involving the general in preparing a feint attack to distract the enemy from pillaging the villages while they are being hastily evacuated. Thereafter, the strategy of harassment, ambush, feint attacks by day and by night, and a whole range of other guerilla tactics, come into play: *Skirmishing*, 12, for surprise attacks. For analysis of the "guerilla" strategy described in detail in the treatise, see Dagron and Mihăescu, *Traité sur la guérilla*, 195–237.

3–7   §1   Leo's own introductory words.

8–14   §2   Cf. Maurice, *Strategikon*, 9.1.3–9. Dennis's note 2, p. 393 (1st ed.), is misplaced (see on ll. 64–69 below).

15–22   Maurice, *Strategikon*, 9.1.9–15.

25–63   §§5–9   Maurice, *Strategikon*, 9.1.16–53. The passages about being prepared to launch surprise attacks reflect a long-standing Roman tradition that continues well beyond Leo's time, and are reflected in the suspicions voiced by Kekaumenos about enemy embassies and the dangers they may conceal; see, e.g., Kekaumenos, *Strategikon* (Spadaro), 2:35 (74.14–20); 2:36 (74.21–33; Litavrin, 142.24–144.8). At Gaza in 634 the Roman commander is reported in one tradition similarly to have attempted to seize ʿAmr b. al-ʿĀs and other Muslim commanders during a parley; see Kaegi, *Byzantium and the Early Islamic Conquests*, 94–95. This is, of course, one of the aspects of east Roman methods of waging war or dealing with enemies that westerners professed to find most distasteful, although similar tactics were generally employed when occasion offered. By the same token, advice which might be seen as somewhat contradictory was also given; see, e.g., Const. 20§37, §39 (Maurice, *Strategikon*, 8.1.32, 37 = Onasander, *Strategikos*, 38) and §97 (Maurice, *Strategikon*, 8.2.36) where the general is specifically advised to beware the potential treachery of an enemy after oaths have been given, but on no account to betray his own promises.

39–41   §6   Maurice, *Strategikon*, 9.1.29–31. Joshua the Stylite, §§56 (trans. Trombley and Watt, 68) and 75 (trans. Trombley and Watt, 92–93) describes a similar trick employed twice by Roman commanders during the war with the Persians in 502–4.

44–63   §§7–9   Maurice's account is followed by Leo, but at ll. 56–59 (*Strategikon* 9.1.32–53) has only (*pace* Dennis, *Taktika*, 395 n.) the crossing of a river in common with those of Polyaenus, *Strategika*,

1.29.1 (*Hypotheseis*, 29.1) and 3.9.61 (no mention of bridges or encampments); cf. Vegetius, *Epitoma*, 3.7.2–3.

46–47   πύργους τε ξυλίνους ἢ ἀπὸ οἰκοδομῆς ἐκ λίθων ξηρῶν ἢ χώματος ἑκατέρωθεν τῶν ἄκρων τῆς γεφύρας ἐγείρουσιν   Read, "At both ends of the bridge they raise either towers of timber or ramparts of dry stone or earth." See Rance, *Strategikon*, 12.B.22, note 238.

45–46   τῶν λεγομένων μονοξύλων   A much-debated term. In this context it is introduced by Leo as part of his paraphrase of Maurice and means pontoons (see also Rance, *Strategikon*, 12.B.21, note 220) although whether this means a boat constructed from a hollowed-out tree trunk, or alternatively a boat whose hull was constructed from a series of single strakes or planks remains unknown. See *DAI* 2:23–25 (comm. to chap. 9.2–3); Strässle, "*To Monoxylon*"; Crumlin-Pedersen, "Schiffe und Schiffahrtswege."

64–69   §10   Maurice, *Strategikon*, 9.2.2–7, although Leo omits the example of the general Lusius Quietus from Roman history (see Maurice, *Strategikon*, 307 n. 29; Rance, *Strategikon*, 176 n. 15). The references to Dio given at Dennis, *Taktika* 393, n. 2 (1st ed.), are confused: Dio 68.8.3 concerns Quietus actions dutring Trajan's first Dacian war; reference to Edessa and Nisibis (in connection with the Parthian war) is at 68.30.2.

64–139   §§10–25   Maurice, *Strategikon*, 9.2. Cf. Syrianos, *Strategy*, 39 for several similar passages, which Leo certainly knew, although the account of "dark lanterns" used to guide the column marching by night at Syrianos, *Strategy*, 39.26–30 is not referred to (see 155–58), and instead Leo refers merely to the setting-up of waymarks to guide the troops, following Maurice (186–90 = Maurice, *Strategikon*, 9.3.45–49). There are few accounts of marches by night, but in 1159 the imperial army made a night-march, lighting its way through the winter snow by affixing cressets or portable torches onto the spears and lances of the cavalry: Kinnamos, *Epitome*, 195–96. See Elbern, "Leuchterträger."

Note the continued use of the term "encampment"—*aplikton*—to refer also to the distance between marching camps, from Maurice and Theophylact Simocatta (see, e.g., *Historia*, 6.6.5: the distance from Heraclea in Thrace to Drizipera was 4 camps; that from Drizipera to

Dorostolon on the Danube was 20 camps). The tenth-century treatise *Skirmishing* likewise speaks of marches in terms of a day or the distance between two camps: e.g., 22.15, 23.17; much later Akropolites, 119, 120, 126 refers to an imperial *stathmos* both as a station en route and as a standard distance between two such stations; see Schilbach, *Metrologie*, 36. One day's march naturally varied according to the conditions (the troops, the terrain, the weather, etc.) and distances of up to 16 Roman miles/23–24 km per day (1 Roman mile = 1480 m; there may have been a slightly shorter Byzantine mile of 1437 m: see Schilbach, *Metrologie*, 33–35) are suggested by medieval sources. For the rates of march, see on Const. 9§7.33 above and McGeer, *Sowing the Dragon's Teeth*, 340–41.

70–78   §§11–12   Maurice, *Strategikon*, 9.2.11–14, with n. 30 (307); and cf. Th. Sim., *Historia*, 6.5.6 (cf. Whitby, *Emperor Maurice*, 155; Rance, *Strategikon*, 9.5, note 17). This attack in fact took place in 592, during the reign of Maurice. Leo appears not to have had a text such as Theophylact Simocatta at hand, for if he had read it, it would have been quite apparent that this attack did not happen under Heraclius. Theophanes, *Chronographia*, 270 (Mango and Scott, 393) merely copies Theophylact. It is possible that Leo confused this event with the better-known "Avar surprise" (misdated in Theophanes to 617/18: *Theophanis Chronographia*, 301.26–302.4; Mango and Scott, 433–34) near Herakleia in 623, when the emperor Heraclius and his retinue were ambushed in Thrace, outside the Long Wall; see Mango and Scott, 434 and n. 1 and Whitby, *Emperor Maurice*, 130–32 for discussion; see also *Chronicon Paschale*, 712.12–713.14; trans. Whitby and Whitby, 165–66 and n. 451; Nikephoros Patr., chap. 10.15–30. Since Leo was clearly working from Maurice's text, he or his redactor/scribe must have read in some other source (the *Chronicon Paschale* or Theophanes), or been told at some point, of this later event.

79–83   §13   Cf. 17§32; 20§§15, 44. For further references to false deserter stratagems, see Brizzi, *I sistemi informativi*, 252 and n. 381.

84–87   §14   Cf. *Skirmishing*, 4 for similar advice.

96   ἔξπληκτον   See on 9§46.217 above.

96–139   §§17–25   Recalls some of the advice in *Skirmishing*, 8–11, although the operations in the latter are conceived as defensive maneuvers whereas in Leo (following Maurice) the assumption is that the Roman forces are on the offensive and marching into enemy territory. See below.

99   ὡς ... σημείου   See on 9§59.292–93 above: a milestone and by extension a (Roman/Byzantine) mile (approx. 1480 m; Schilbach, *Metrologie*, 32–34).

110–13   μὴ ἐπὶ μέτωπον ... φυλλατομένου   Leo glosses some of Maurice's technical phrases here, suggesting that the formal language of the Roman and pre-Roman military establishment was no longer as transparent as it had once been. The phrase ἐπὶ μέτωπον is the standard ancient term for "in line" or "by the front," just as ἐπὶ κέρας refers to an order drawn up in columns, "by the flank," and both expressions were usual in the older treatises; see, for example, Onasander, *Strategikos*, 16; Aelian, *Tactica theoria*, 30.1; Arrian, *Techne taktike*, 26.1; Rance, *Strategikon*, 9.2, note 27 and 12.B.20, note 203; and on Const. 9§66.331–34 and §70.355–56 above.

117–19   ἵνα μὴ ... φεύγειν   The tactics outlined in these sections—leaving the enemy force an obvious exit through which to escape and thus encourage flight, sounding trumpets to panic them—were well established, recommended by both earlier and later Byzantine treatises; see especially the detailed account in *Skirmishing*, 24; and cf. Syrianos, *Strategy*, 39.36–46; *Sylloge tacticorum*, 48.2. An older example: Polyaenus, *Strategika*, 2.1.4 (*Hypotheseis*, 45.2). On "battle from desperation," see Wheeler, "Polyaenus," 39–41. Leo's sentiment is echoed in many other contexts: the early ninth-century chronicle of Regino of Prüm notes, of a defeat of a Breton king at the hands of the Vikings after an initial victory, "when he reashly pursued them further than he should have, he was killed by them, because he did not know that while it is good to win, it is not good to push your victory too far; for the enemy's desperation is dangerous" (trans. MacLean, *Reginald of Prüm*, 209, s.a. 890); and see on 14§21.147–48 above.

136–39   §25   Cf. the late tenth-century treatise *Campaign Organization*, 25 for similar advice, drawn from the common stock of prescriptions in the military treatises.

140–43   §26   Leo's own comment on the *tourmarchai*, reflecting perhaps contemporary attitudes toward the provincial military; see also above, on Const. 3§1.8. One of the hallmarks of the later treatise on skirmishing is the importance placed upon the judgment and independence of these local commanders in the frontier regions. They were to organize regular, small-scale raids into hostile territory, and they should be prepared to attack an invading force whenever an appropriate opportunity arose, and not necessarily wait for the arrival of reinforcements or the local senior commander (for detailed analysis and discussion, see Dagron and Mihăescu, *Traité sur la guérilla*, 161–71, 177–93). It was these senior officers, next in rank below the thematic *strategos*, but based in their own provincial headquarters, together with the lower-ranking *droungarioi*, who managed much of the military activity in each military province. The majority of such officers for whom the sources provide information seem to have come from relatively well-off social backgrounds, not unexpectedly, given the advantages attached to the possession of a certain degree of literacy, the ability of such men to pay their way, support their greater expenses, and, in the *themata* certainly, provide for the poorer soldiers in their units. See Haldon, *Warfare, State and Society*, 270–74.

144–202, §§27–37, 44–57   Maurice, *Strategikon*, 9.3, apart from ll. 62–74. See
254–331 also *Strategikon*, 8.35–36.

144–49 §27   Leo alters Maurice's text to omit references to the "Slavs and Antes and other peoples of this sort" who fight without regular order and discipline (*Strategikon*, 9.3.7–8). In Book 11 of the *Strategikon* there is a full account of the customs, habits, and tactics of the Slavs that Leo omits from his own version of that book (Const. 18). Instead he merely describes some of their customs, and refers back to his own Const. 11 as though the fighting methods of the Slavs were described there (18§102.490), although he does repeat information from the *Strategikon* further on in this constitution; see on ll. 205–53 below. In fact they are not mentioned at all, suggesting that Leo, or the redactor of the text, had forgotten, or did not know, that this reference to "Slavs and Antes and other peoples of this sort" had been excised. The implications of this for the way in which the *Taktika* was composed are unclear, but would suggest that the editing process was not thorough. Leo also omits reference to the term *kleisoura* here, which appears in Maurice with the generic meaning of a pass or narrow passage; see on Const. 9§27.122 above.

151   διὰ βασταγῆς δημοσίας   The term is taken directly from Maurice, and although the public transport system operated through the middle Byzantine *dromos* seems to have been a much less extensive institution than its late Roman and pre-Justinianic precursor, it still supplied transport for the army in certain circumstances. It is the only occasion on which Leo refers to a state transport of military supplies, although it is clear that within Roman territory the local provincial officials, under the thematic *protonotarios*, was responsible for collecting provisions and other supplies and moving them to points at which the army could collect them. Originally under the authority of the Praetorian Prefects, by the 760s, and probably by the middle of the seventh century, the *dromos* was an independent department under its logothete, a high-ranking officer for whom numerous seals survive; see Laurent, *Corpus*, 2:195–243; Oikonomidès, *Listes*, 311–12; and Hendy, *Studies*, 608 and n. 238. The operations of the *dromos* were closely associated with those of the *logothetes ton agelon* (logothete of the herds), the officer in charge of the imperial stud ranches, in particular the *metata* of Asia and Phrygia (Oikonomidès, *Listes*, 338; and Laurent, *Corpus*, 2:289–91); *CPTT* 161 and 184. On transport and logistical arrangements more generally see Haldon, *Warfare, State and Society*, 139–45; and for the supply train, see on Const. 10, above.

155–58   §29   See above on ll. 64–139.

161–74   §§31–32   For some comment on and examples of this for the late Roman period, see Rance, *Strategikon*, 9.3, note 38.

166–74   §32   On the untrustworthiness of defectors and deserters, see below on ll. 559–69.

171–74   Following Maurice, *Strategikon*, 9.3.31–34, in turn based on Onasander, *Strategikos*, 10.427–35 (10§15); and see on Const. 18.661–68 below with further literature. Maurice's advice resembles that of Vegetius, *Epitoma*, 3.6.5–7, in particular the suggestion that unexpectedly useful or important information can be obtained from local inhabitants and captives, a result of common sources, in particular Onasander (via a Latin intermediary, probably the lost work on military theory by Frontinus). For a survey of one aspect of intelligence gathering operations and methods before the launch of a campaign, see Koutrakou, "Diplomacy and Espionage"; and see on Const. 18§132.661–68 below.

183-85   εἰ δὲ ... ταύτας   An obvious point but which occasionally was ignored or nearly resulted in disaster. Lack of water and scarcity of provisions cut short John I Tzimiskes' campaign into Ecbatana in 974: Leo the Deacon, *Historia*, 163.12–24 (Talbot and Sullivan, 204). In 1068 Romanos IV discovered in time that the district of Melitene, to which he was marching, could not support his troops because of the ravages and devastation of the previous year: Attaleiates, *Historia*, 136.5–8 (trans. Kaldellis and Krallis, 247–49), and examples discussed by McGeer, *Sowing the Dragon's Teeth*, 357–58.

186-90   §35   See above on ll. 64–139.

189   τοὺς ὑστερίζοντας ἐκ τοῦ στρατοῦ   "stragglers" rather than simply "troops marching along later"; see 1.324 for exactly the same phrase, where it can only mean stragglers.

191-97   §36   Although Leo does not refer to it, this arrangement of supporting foragers or raiders with a second group of units in battle order ready to drive off enemy attacks (standard practice since ancient times, of course; see Onasander, *Strategikos*, 10.387–91 [10§8]) was referred to by the tenth century as a φοῦλκον; see *Skirmishing*, 8.29; 9.87–94; 10.122, etc.; *Campaign Organization*, 22. Following his exemplar Maurice, Leo also uses the word in its older sense, of a densely ordered body of soldiers; see on Const. 7§54.370 above (and cf. also on Const. 4§3.20).

198-202   §37   Cf. Const. 14.76–96. Failure to guard against a surprise attack while foraging or dealing with the pack train, on the one hand, and the failure of the foragers to locate and secure adequate provisions, could prove disastrous: it was by surprising the Paulician troops while they were attending to such matters that the *domestikos* Christopher (*PmbZ* 21258) defeated the forces of the Paulician leader Chrysocheir at Bathys Ryax in 872/73; see on Const. 14§36.234–39. By the same token, the ill-disciplined Roman troops campaigning against the Bulgars in 707/8 were caught while foraging, and the army put to flight (*Theophanis Chronographia*, 376, trans. Mango and Scott, 525). Basil II's ill-fated first campaign against the Bulgars in 986 failed in part because the foraging parties were inadequately protected, and were consequently ambushed and cut to pieces by the enemy. Within a few days the imperial army was forced to retire in disorder, losing much of its baggage train in the process (Leo the Deacon, *Historia*, 171–73; Talbot and Sullivan, 213–15; *PmbZ* 20838). Anna Komnene describes how careless foragers were caught and killed or captured by Pecheneg scouts during a campaign in 1087: *Alexiad*, 7.3.1 (trans. Sewter, 222/ Frankopan, 191); Choniates notes that Manuel I defeated Turkish attempts to disrupt the work of his foraging parties in the 1175 campaign to Dorylaion by establishing a regular system of pickets and counterraids on the Turkish forces, something which the military treatises themselves advise (Choniates, *Historia*, 176–77). Equally, of course, it was clearly understood that attacking the enemy while they were out foraging (or collecting booty) offered a good opportunity to break up an invading force; see below on ll. 386–89.

203–53 §§38–43 Maurice, *Strategikon*, 11.4.175–224. Leo takes this section from a different book of Maurice, dealing with the Slavs and Antes, but restructures it slightly, prefaces it with a comment on attacking "Scythians or similar nations," and brings the technical terms for the officers up to date (thus the *hypostrategos* becomes the *merarches*, for example; see on 4.69 above). The presence of this section here suggests that some effort at rationalizing the contents of Const. 17 were made, to bring together similar topics from disparate sections of the *Strategikon*. The instructions for a raiding force divided into two columns are reminiscent of those set out in the treatise on *Skirmishing*, esp. 20. While based on Maurice's text, Leo's account certainly had direct relevance to the situation in his day, and reflects, in a far more generalized form, the more detailed information included in the later treatise on frontier raiding. According to the author of the treatise on skirmishing or guerilla strategy, local commanders maintained small bands of raiders, specially selected for their physical prowess and bravery, whose task it was to raid deep into enemy territory in order to foment insecurity and uncertainty. These soldiers, referred to in the text as *trapezitai* or *tasinarioi*, were also to take prisoners who could inform the Byzantine command of the movements of troops, the intentions of the Arab commanders, and so forth (cf. Const. 17§87). Similar soldiers were employed on the northern front in the Balkans, called *chonsarioi*. Both groups appear as irregular but paid soldiers, and required payment, rewards, largesse, and so forth to keep them loyal; they were also not to be trusted, but should be regularly checked by other agents of the commander, who should on no account reveal his own plans to them. In many cases they probably represented some of the more autonomous populations that were within or along the borders of the empire and whose marginal situation between two cultures suited them ideally for this task; but they were also potential enemies, and needed careful handling. In the tenth century, and with the large-scale migration of Armenians into the Taurus and North Syrian regions, later into Cilicia as well, these irregulars were drawn from among these newly settled migrant populations. But there is every likelihood that they were drawn at an earlier period from similarly marginal groups—Isaurians, for example, a mountain people who had always been difficult to control or administer from Constantinople; or the Mardaites, during the seventh century. See *Skirmishing*, 2.3; 7.1. *Trapezites* derives from a Persamenian word, *darpaspan*; *tasinarios* is a transliterated Armenian term meaning one of a group of 10; *chosarios* derives from a Bulgar term for robber, and evolves into the later *hussar*. See Dagron and Mihăescu, *Traité sur la guérilla*, 252–57 for discussion and literature; and Litavrin, *Soveti i rasskazi*, 353–54; Spadaro, *Cecaumeno*, 65 n. 2. It was from such sources that the regular light cavalry, especially those who formed the advanced scouting parties of larger forces, were ideally to be recruited, for they had detailed knowledge not just of the regular routes, but also of the side paths, hidden tracks, watering and camping places in the mountains, as well as the habits and customs of the enemy; see McGeer, *Sowing the Dragon's Teeth*, 212, 300; and *Campaign Organization*, 18.

254–331 §§44–57 These sections, especially §50, repeat or summarize some of the material already presented in 11§§13–20 and §§39–40. Note that at ll. 283–84 and at l. 314–15 (on which see also 9§69.344–51) and ll. 333–34, Leo refers explicitly to his sections on camps and on marches. For the protection of camps while setting up the defenses see, for example, Vegetius, *Epitoma*, 3.8.5.

284 δι᾽ ἑνὸς δρούγγου In the original (Maurice, *Strategikon*, 9.3.103–5) the term refers simply to "a body" of troops; by Leo's time *droungos* had the more technical meaning of a specific subdivision of a thematic army, which is how he understands it here. See on Const. 4§3.20 above.

308–10 ἀλλὰ μηδὲ ... γίνεται   See the editor's comment for ll. 309–13 for a sixth-century example (at Maurice, *Strategikon*, 9.3.125–27). Dennis (*Taktika*, 415 n. 6) somewhat misleadingly cites a similar anecdote about poisoned barley found in the seventh-century John of Nikiu's *Chronicle* (ed. Zotenberg, 96, p. 408; trans. Charles, 156–57) and in the context of Maurice's Persian wars (in fact, the war to restore the Persian king Khusru II to his throne), as though it were a source for Leo's account. But Leo merely copies Maurice here, and there is no evidence that he had access to, or even knew of, the *Chronicle*. Manuel I's expeditionary army in 1176 suffered from the fact that the Turks polluted the water supply along the route and destroyed any supplies that the Roman forces might have consumed: Choniates, *Historia*, 178–79. See Grmek, "Ruses de guerre."

314–18 §55   See Const. 9§69.344–51, which would appear both to be more rational and consistent with maintaining order than the instructions given here, with which it appears to conflict.

322 νωτοφύλακας   Leo substitutes the somewhat more technical term, used for the rear division in the battle line (see 12§22.158; §23.170; §28.211 etc.), for Maurice's less formal ὀπισθοφύλακες (*Strategikon*, 9.3.135).

323 The distance specified, taken from Maurice, probably represents a day's march, dependent upon conditions.

324 ὑστερίζοντας ἐκ τοῦ στρατοῦ   See above on l. 189.

329–31 Although Leo takes this passage straight from the *Strategikon*, the military codes included sanctions against soldiers who left their unit and disobeyed orders when in Roman lands. See Const. 8§§2 and 4; and see above on Const. 8§8.25–27 and §11.33–36. The extent to which discipline was actually enforced depended on the character and quality of the commanding and subordinate officers as much as on any abstract moral absolutes.

332–34 See above on 9§27.121–29, §§36.161–38.175, §§42.188–51.254.

341–56 §§59–61   Maurice, *Strategikon*, 10.2.3–14 and (for 352–56) 9.44.139–44; and cf. *Skirmishing*, 3–6, 8–11, where the sort of strategy alluded to by Leo (following Maurice) is expounded in great detail and in the context of the warfare along the eastern front in Asia Minor in the tenth century. The strategy did not always succeed: in 770/71, for example, the emperor was informed of an Arab raiding force and ordered the cavalry armies of Anatolikon, Thrakesion, and Boukellarion to occupy the pass through which the raiders would have to return, and at the same time he issued instructions to the Kibyrrhaiotai troops and their ships to occupy the harbor of the fortress town of Syke, which the Arabs had attacked. Although demoralized by their failure to take the town, the raiders nevertheless were able to force their way back through the pass, with most of their booty intact: *Theophanis Chronographia*, 445 (trans. Mango and Scott, 615). The treatise on skirmishing warfare warns the general to be on the lookout for ambushes set by retreating Arab forces, and notes that apparently victorious pursuits of enemy forces have resulted in ignominy and disaster through failure to take such matters into account (*Skirmishing*, 9.41–56; 16). In 787/88 Roman forces marched separately to trap and destroy an Arab force at Podandos, but were themselves ambushed and defeated: *Theophanis Chronographia*, 463 (trans. Mango and Scott, 637).

352–56 §61 Maurice, *Strategikon*, 10.2.10–14, and Rance, *Strategikon*, 10.2, note 13.

356–61 §62 Maurice, *Strategikon*, 10.2.15–18. Leo changes Maurice's text to say that the general should keep himself safe (see, e.g., Onasander, *Strategikos*, 33)—in Maurice it is the army that the general should keep safe. But then the passage makes little sense, since the point is that with the army safe the enemy will not risk dispersing in case of counterattack.

362–65 §63 Maurice, *Strategikon*, 10.2.19–22.

366–69 §64 Maurice, *Strategikon*, 10.2.23–25, but Leo changes the wording again, making it Roman territory attacked by the enemy whose location is suitable, rather than enemy territory suitable for a Roman counterattack. Cf. Syrianos, *Strategy*, 6.14–24.

370–73 Maurice, *Strategikon*, 10.2.25–29; see also *Strategikon*, 9.B.75.

373–76 See above, on Const. 11§21.121–27 with further literature. The editor mistakenly suggests (Dennis, *Taktika*, 419 n. 9) that the campaign in question was in Calabria, although it was in Cilicia; see *Skirmishing*, 20.14–44; Cheynet, "Phocas," 293–95.

377–89   §65–67   Maurice, *Strategikon*, 10.2.30–38 and cf. Vegetius, *Epitoma*, 3.3.3–6. Leo paraphrases the *Strategikon*, but in his terms the reference to sending away the local horses (or animals) would undoubtedly be understood as underlining the nature of the mobile raiding warfare in the frontier regions (and given his example, above, he probably had the Anatolian frontier in mind). Although described in much greater detail in the late tenth-century treatise on *Skirmishing* (2.1), Leo's account here reflects the established practice of evacuating men and materials, including the civil population, from the path of invaders. See *CPTT* B.37–38 and comm., 158; Dagron and Mihăescu, *Traité sur la guérilla*, 228–29. For harassing the enemy while they forage for supplies, see Syrianos, *Strategy*, 35.33–35; *Skirmishing*, 6.4–5; 9.9, 11, 13; 16.5–7. By depriving the enemy of fodder and other provisions, they might be encouraged to send out foraging parties farther and farther away from the main force, thus providing even better opportunities to destroy the invading forces piecemeal.

390–94   §68   Leo's summation of the foregoing paragraphs.

395–96   Leo's own words, prefacing the following section taken from Maurice.

397–427   Maurice, *Strategikon*, 9.5.4–25; and Rance, *Strategikon*, 9.5, note 86 with Roman and Hellenistic sources. This exercise in calculating the area taken up by a particular formation, which Leo takes here from Maurice, is a standard feature in older Roman and Hellenistic treatises, although Maurice develops it to calculate the space taken up by a particular formation of cavalry. See, e.g., Aelian, *Tactica theoria*, 11.1–6; Asklepiodotos, *Techne Taktike*, 4.1, 4; and Vegetius, *Epitoma*, 3.15. The figures are intended to illustrate how the general might extend or contract his formation to conceal or emphasize the numbers of his force and how one might calculate the size of an enemy army. Obviously the size of the force—a cavalry army of three hundred thousand, occupying an area of over 0.25 square miles (0.68 sq. km)—is meant purely as an illustration; see also Jähns, *Handbuch*, 1:155. The space allocated to each horseman in closed order is three (Roman) feet by eight feet, although this was not sufficient space to allow movement; but this does facilitate a calculation of the size of late Roman and Byzantine cavalry mounts. A Roman foot was 31.23 cm, marginally longer than a modern foot (Schilbach,

Metrologie, 20). In the *Sylloge tacticorum*, 43.6, mounted soldiers are allocated just over 1 *orguia* (or 2.16 m length: there were three basic *orguiai*, of respectively 1.87, 2.10, and 2.16 m—the most likely measures here are the second and third; see Schilbach, *Metrologie*, 22–24) of space laterally, when on the march, and half an *orguia*, or just over 1 m, when drawn up in close order for battle. This allows just over three feet per man and horse and, as noted above, is the standard spacing for cavalry in the ancient treatises.

While this spacing does not allow precise calculations, it is possible to suggest on the basis of the figures given that the average height of the late Roman or Byzantine cavalry mount cannot have been greater than about 15.2 hands (1 hand = 4 in. or 10.15 cm), and would support conclusions drawn by scholars of the Roman and medieval military mount that a well-built animal with a strong back was greatly preferred over a taller horse, and that this standard of animal, regardless of particular breed, remained the norm well into the twelfth century across western, central, and southeastern Europe and Byzantium, although variations in weight (as opposed to height) were certainly present. For further literature and discussion see on Const. 10§13.56 above. The sturdy Roman military horse appears to have disappeared from the central and eastern European regions after the fifth century, but the evidence from equine skeletal remains suggests that it was replaced relatively quickly by an equally sturdy animal, averaging around 14 hands, by the late ninth century, partly a result of the introduction of such animals to the region by the Avars and later the Magyars. The very limited Byzantine evidence suggests that specialist breeding of the larger military horse continued, although whether the numbers were sufficient to supply the cavalry armies of the late seventh and eighth centuries may be doubted. While the Roman and Byzantine warhorse will thus have weighed on average some 1000 lbs (455 kg), there is reasonable evidence to suggest that from the late ninth century in eastern and central Europe and from the late tenth century in Italy a slightly heavier animal was beginning to be bred, of similar height but possibly up to 1200 lbs (545 kg). This may be reflected in the widespread appearance of the horseshoe from the ninth century (see on 5§3.29 above); although probably not at this time in response to new cavalry tactics (which seem only from the very late eleventh or twelfth century at the earliest to have become more generally adopted): Bökönyi, *History*, 270–72; Benecke, "Zur Kenntnis." It is clear from both Latin and Arabic sources of the Crusading period that the Turkish cavalry were largely mounted on slightly smaller and certainly lighter animals of between 13.5 and 14.5 hands, with a weight average of some 900 lbs (400 kg). The evidence, based on skeletal remains, bit sizes, and literary references, is relatively uncontentious (although skeletal evidence from Byzantine locations is still scarce and statistically contentious); see esp. Kroll, *Tiere*, 168–71 with further literature and eadem, "Groß und stark?"; see also Gladitz, *Horse Breeding*, 130–35; Dixon and Southern, *Roman Cavalry*, 165–73; Hyland, *Medieval Warhorse*, 32, 85–86; Gillmor, "Brevium Exempla"; Bartosiewicz, "Lovak a Kárpát-medencében"; and the important contributions to Lazaris, *Cheval*. Byzantine cavalry mounts of the tenth century were probably thus of the standard height—approx. 15 hands—and weighed some 1000 lbs (455 kg), thus heavier than the mounts of many of their eastern opponents but probably a little lighter than the horses of the Lombards, Franks, and, later, Normans. For comparison, the generally recommended height for heavy cavalry mounts in eighteenth- and nineteenth-century European armies was 15 hands and for light horse between 13 and 14 hands. Smaller horses were lighter and faster and thus more maneuverable in the sorts of skirmishing context in which they would be deployed, and this rule seems to have been the norm for all except the larger, heavier warhorses bred for battle in the medieval west after the twelfth century. See in general also Hyland, *Horse*; Vigneron, *Cheval dans l'Antiquité*; anonymous, *Horse and Stable Management* (HMSO, London, 1904); Gilbey, *Small Horses*; Rogers, *Mounted troops*; and Priskin, "A Kárpát-medence."

427–32 Aelian, *Tactica theoria*, 11.2, with Leo's own comment: ὥστε καὶ ... τὸ πλῆθος.

430–31 σκουλκάτορα See above, on 4.110.

433–79 §§73–81 Maurice, *Strategikon*, 9.5.30–74; for the difficulty associated with eyewitness calculations of large mounted forces, see Golden, "War and Warfare," esp. from 145; and of large numbers of people or animals in general, France, "Size." For modern emphasis and discussion of the difficulties entailed in estimating crowd sizes or large bodies of people or animals, see Watson and Yip, "How Many" and Aik and Zainuddin, "Curve Analysis."

456-512 §§77–87 Once again prefigures the later treatise on *Skirmishing*, esp. 6–8, 15, 17; the sections on scouts and watch posts are close to *Skirmishing*, 1–2 in detail.

449-51 εἰς δὲ . . . διαστήματος Read "In unobstructed and open country set up many more patrols in different places in relays (κατὰ συνέχειαν) and at intervals (ἀπὸ διαστήματος)." See l. 453, where ἀπὸ διαστήματος is correctly understood as "at intervals."

459 ἐξπλοράτωρας The only time Leo uses this archaic term, borrowed directly from Maurice, from the standard Latin term for a spy who actually gets among the enemy to collect intelligence; see Vegetius, *Epitoma*, 3.6.10–12 and cf. 3.6.26; and Trapp, *Lexikon*, 3:544; but see Arrian, *Ektaxis*, 1.1, where a unit of *exploratores* is referred to as κατάσκοποι ἵππεις, suggesting a less clearly defined distinction The gloss Leo adds, *kataskopous*, shows that he understands the difference between spies and scouts or members of patrols—*biglatoras*, even if he retains Maurice's *kataskopoi* in l. 456 to refer to both categories. See *Skirmishing*, 2.23–31 for the activities of such scouts.

475 ἀλλεπαλλήλους Better translated as "one after another" rather than "constantly changing." The "different and constantly changing" refers to the patrols, not the terrain, as the rest of the passage makes clear.

480-91 §§82–83 Onasander, *Strategikos*, 10.402–8 (10§§10–11). Concern with guards and maintaining a vigilant watch is a frequent theme in all the treatises. See the prescriptions in Const. 11, esp. §§2, 8, 13–14, 17, 20; and, for example, *Campaign Organization*, 4; *Skirmishing*, 1.30–34; Syrianos, *Strategy*, 7; and *CPTT* C.420–39, 573–79.

492-539 Maurice, *Strategikon*, 9.5.74–124.

500 Leo omits the obsolete term καλκατούρα (Lat. *calcatura*) used by Maurice for the tracks or traces of the horses' hooves, which he probably did not understand. See Mihăescu, "Éléments latins," 275–76 and Trapp, *Lexikon*, 4:745. See *Skirmishing*, 6.20–21; 14.59 for scouts estimating the size of a raiding force from the tracks of the animals. It is interesting that Leo did not make a guess at *kalkatoura*, which could readily have been judged from the context—an indication of his complete lack of any practical experience or knowledge in such matters.

517–33   §89   The detection of spies by such means is a common motif in the gnomic literature; see Polyaenus, *Strategika*, 3.13.1 (*Hypotheseis*, 7.1); 5.28.2 (*Hypotheseis* 7.2), with Const. 20§216; and Vegetius, *Epitoma*, 3.26.27–29. This appears to have been a well-established means of discovering infiltrators into a camp; see also Nikephoros, *Praecepta*, 6§1 and Syrianos, *Strategy*, 2.4–13, where it is noted as one of the beneficial results of properly observed camp discipline, "as was the custom among the ancients" (l. 8).

519   περὶ δευτέραν ἢ τρίτην ὥραν τῆς ἡμέρας   Roughly between seven a.m. and eight a.m. (during the summer; two hours later in midwinter). See Grumel, *Chronologie*, 163–65. Periods for particular duties or activities appear also to have been signaled by trumpet blast, but the way in which the military day was thus divided is not clear. See on Const. 13§11.54–57 above. The Byzantine day was divided into 12 equal hours, as was the night, but since the days and nights vary in length according to the seasons, the hours vary accordingly: daytime hours being longer in summer and shorter in winter, nighttime hours longer in winter and shorter in summer. These varying hours were referred to as καιρικαί (Lat. *temporales/inaequales*). The Roman day was divided into 4 equal periods: *mane, ad meridiem, de meridie, suprema*; and the night likewise into 4 *vigiliae* or watches. From the fifth century the chief hours of the day (at which Christian communities prayed) were as follows:

hora prima (ὥρα πρώτη)	sunrise
hora tertia (ὥρα τρίτη)	mid-morning
hora sexta (ὥρα ἕκτη)	midday
hora nona (ὥρα ἐνάτη)	mid-afternoon
vespera (ἑσπέρα/ἑσπερινόν)	one hour before sunset
completorium (ἀπόδειπνον)	after sunset.

534–36   Γίνεται... ποιεῖν   Translate: "The spies we have been discussing may be recognized through other signals such as these and similar, which it is necessary to practice, so as to employ different signals or commands."

539-42   Leo's own comment, but on passwords (*synthemata*); see also Polyaenus, *Strategika*, 3.13.1 (*Hypotheseis*, 7.1).

542-44   Maurice, *Strategikon*, 9.5.125–27.

545-58   §91 Cf. Const. 13§5; and Onasander, *Strategikos*, 10.393–400 (10§9);
Polyaenus, *Strategika*, 8.16.8 (*Hypotheseis*, 7.5); and Maurice, *Strategikon*, 7.A.5.

559-70   §92 Onasander, *Strategikos*, 10.428–36 (10§7) and Const. 17§32; note
also *Rhetorica militaris*, 45.8 and commentary in Eramo, *Siriano*,
183–84. Byzantine experience of betrayal by false deserters who then
leaked information to the enemy was keen: cf. the advice on this given
by Kekaumenos, *Strategikon* (Spadaro), 2:29 (68.24–70.2) (Litavrin,
138.1–12). The disastrous loss of Amorion in 838 was ascribed to just
such a deserter, who fled to the Byzantines, accepted Christianity,
but then informed Mu'tasim of a weak point in the city defenses; see
the Arabic account in Tabarī, *Ta'rikh al-rusul wa'l-muluk* (also in
Vasiliev, *Byzance et les* arabes, 1:302–9), together with that of Michael
Syr., 97–101 and the *De XLII martyribus Amoriensibus*, 2.11, 4.42–
44; 7.65 and 71; Symeon *magistros*, 227.240–41; and the discussion in
Treadgold, *Revival*, 302–3.

CONSTITUTION 18

This is the second-longest constitution in the *Taktika* (after Const. 20, which has over 200 subsections) with 150 sections (154 in the PG version) devoted to the tactics and customs of the empire's neighbors and enemies, but with introductory paragraphs outlining Roman methods as a background and in order to set the following sections into context. As has been widely recognized, this constitution was the first attempt at a comparative ethnographic account of the empire's neighbors since the *Strategikon*, and while basing his work firmly in the latter text Leo nevertheless introduces a substantial amount of new information (on the Saracens) and updates somewhat the material relevant to the Balkans and the empire's northern neighbors; see Shepard, "Byzantine Writers on the Hungarians," 101–2. Const. 18 was briefly commented on and edited with a Hungarian translation and valuable critical annotations by Vári in 1900 ("Bölcs Leo hadi") The first part of the constitution is devoted to established Roman battlefield practice with additional variations and deployments and military exercises, together with some points of guidance to the general (§§1–39), including a section on steadfastness and self-sacrifice (§§16–21) and three paragraphs on archery (§§22–23, 33). Much of this is taken from Maurice, and for a detailed technical and historical commentary on the original text Rance's notes are an essential aid. Particularly important in this opening section are Leo's comments on the Saracens at §24 (on which see below). After these introductory remarks Leo goes on to describe various aspects of the tactics and customs of the Turks (by whom he means the Magyars), Scythians (Turkic peoples, including both the Magyars and the Bulgarians) (§§40–73), the Franks and Lombards (§§74–92), the Slavs (§§93–102), all again drawn largely from Maurice.

§§103–49 represent new material reflecting Leo's own efforts to offer advice for dealing with the armies of the Caliphate and the various emirs along the Syrian border, and contain important information on the Byzantine and Leo's own attitude to this enemy, as well as (one version of) Byzantine understanding of their faith and cultural traditions in warfare. Of this section, §§103–25 had an independent existence and circulated separately, or were excerpted and incorporated, as noted in chapter 2 above, after the main body of the text in M, and under the title: Λέοντος ἐν Χριστῷ βασιλεῖ αἰωνίῳ βασιλέως Ῥωμαίων πῶς δεῖ σαρακηνοῖς μάχεσθαι. The last sections, §§136–49 (§150 is based on Maurice again), offer further tactical advice on how to deal with the Saracens in battle.

This section was not originally a unitary composition, however. It is noteworthy, for example, that the content and tenor of the information from §103 up to and including §125 is slightly different from that in the sections that immediately follow (§§126–35): the former passage deals with the general mores and habits of the Saracens, including their recruitment methods and related matters; the latter, while including some similar material (but only in §130, on why the Saracens volunteer to fight, which merely amplifies on the point already made in §126), is far more historically and geographically specific and conveys seemingly contemporary and recent information, referring as it does to Syria, Palestine, Cilicia, the Taurus mountains, the campaigns of Basil I, the cities of Tarsos and Adana, and the Kibyrrhaiotai fleet. It is thus possible that §§126–35 represent an addition to an original text, which would have ended at §125—which concludes, appropriately, with the words: Εἰ γὰρ . . . καὶ τὴν θείαν ἐπὶ πᾶσιν ἕξωμεν συμμαχίαν, καὶ εὐκόλως τὴν κατ' ἐκείνων κατορθώσωμεν νίκην.

More significant by far, however, is the wording of §135, which looks even more like a concluding paragraph, especially since what follows in §§136 and thereafter is quite different in character from the content of the previous sections and has every appearance of an additional section tacked on to an originally shorter version of Const. 18.

The text of §135 runs as follows:

Συνελόντα δὲ εἰπεῖν, ἅπαντα τὰ προειρημένα περὶ τῆς τακτικῆς θεωρίας ἀπ' ἀρχῆς ἄχρι τέλους, ὅσα τε διά τε τὰ ὅπλα καὶ τὰς ὁπλίσεις καὶ τὰς γυμνασίας καὶ τὰς παρατάξεις τὰς πολεμικὰς καὶ τὰς ἄλλας στρατηγικὰς μεθόδους εἴρηται, ἕνεκεν τοῦ Σαρακηνῶν ἔθνους ἡμῖν καὶ παρηγγέλθη καὶ διατέτακται. τοῦτο γὰρ γειτονεῦον τῇ ἡμετέρᾳ πολιτείᾳ, οὐδὲν ἧττον τοῦ πάλαι Περσικοῦ ἔθνους τοῖς ἀρχαίοις βασιλεῦσι, τὰ νῦν ἡμῖν ἐνοχλεῖ καὶ παραλυπεῖ τοὺς ἡμετέρους ὑπηκόους τὸ καθεκάστην, οὗ χάριν καὶ τὸν παρόντα τῆς πολεμικῆς διατάξεως ἀνεδεξάμεθα πόνον.

εὕρηνται δὲ ἡμῖν πρὸς τοῖς εἰρημένοις καὶ ἕτερα παρατάξεων σχήματα, οἷς χρησάμενος, ὦ στρατηγέ, κατὰ τοῦ τοιούτου βαρβαρικοῦ ἔθνους εὐδοκιμήσεις. εἰσὶ δὲ ταῦτα.

"To sum it up, all that we have written about tactical theory from the beginning to the end, all that was said about weapons, armament, drills, battle formations, and other military methods in connection with the Saracen people has been transmitted and set forth by us. This people that borders on our commonwealth causes us no less trouble now than the Persian people of old did to former emperors. They cause harm to our subjects every day. It is for this reason that we have undertaken the present task of formulating instructions for war. In addition to what we have already said, we have found other models of battle formations that you may well consider employing, O general, against this barbaric people. They are the following."
(trans. Dennis, *Taktika*, 489)

Here we have a summary and concluding sentiment, and I would suggest that the paragraph originally ended with the clause: οὗ χάριν καὶ τὸν παρόντα τῆς πολεμικῆς διατάξεως ἀνεδεξάμεθα πόνον ("It is for this reason that we have undertaken the present task of formulating instructions for war"), which both rounds off the section on the Saracens and acts as a specific reference to the reasons for the composition of the constitution. Looked at from this perspective, the following sentence ("In addition to what we have already said. . . .") reads as an awkward addition, tacked onto §135 in order to smooth the transition to the sections which follow, sections which deal with somewhat different material, namely the various battle formations that the Byzantines themselves should deploy (but note that while the context described in §135 implies the Arabs, no specific enemy is named in any of the remaining paragraphs of Const. 18). And while §§136–49 reflect material taken largely from Const. 12, Leo incorporates information and advice which certainly appears to refer to contemporary practice in the provincial armies.

There are thus several possibilities for the original composition of Const. 18. A first draft may originally have consisted simply of §§103–25 as a separate treatise on the Saracens (hence the existence of these paragraphs as a separate text in M), incorporated at some stage after its composition into Const. 18, with §§126–35 representing a slight expansion of an original. Equally, and perhaps more likely, one

version of Const. 18 almost certainly ended with §135, which not only states explicitly that it concludes the text (which includes tactical theory "from the beginning
to the end"), but which also explains why the constitution was written in the first
place. It is, in fact, a perfect concluding paragraph. The inclusion at some point
after the completion of Const. 18 up to §135 (or of §§103–35) of §§136–45, dealing
with the Byzantine tactical response to Saracen tactics, is thus a final addition
to a text which appears to have gone through two, even three, drafts before the
form in which it was finally transmitted in the *Taktika* crystallized. This would
also explain why the justification for the sections on the Saracens appears where
it does, where one would hardly expect it, or a "concluding" paragraph, if it were
the original reason for the composition of the text of either the *Taktika* as a whole
or of Const. 18 in the final form in which they are extant.

While the constitution is derived in its basic format and inspiration from
Maurice, Leo thus brings it up to date in several ways, and the sections on the
Saracens, or rather on dealing with Islamic armies, is especially significant. The
extent to which this marks either something new, or a break with traditional
approaches to the Islamic enemy to the east, has been discussed at length by
Dagron ("Byzance et le modèle islamique"; with the summary in Dagron and
Mihǎescu, *Traité sur la guérilla*, 145–49), and will be taken up again at the appro-
priate points in the commentary, below.

This constitution contains the most references in the *Taktika* to the divinity and
to divine support, to attitudes of barbarians and Christians to God, and to the need
to fight for one's faith (Const. 20 and the Epilogue have more references), a point
discussed in chapter 1 above. For discussion of Byzantine approaches to "outsiders"
more broadly, see Zástěrová, "Ethnika"; "Ethnology," in *ODB*, 2:734; the neglected
but valuable Lechner, *Hellenen und Barbaren*; Wiita, *The Ethnika*; Malamut,
"L'image byzantine des Petchénègues"; and especially Dagron, "'Ceux d'en face'."

3–9   §1   Leo's own introductory words, perhaps reflecting his reading of
Onasander, *Strategikos*, pr.9–10; see Const. 14§96 and cf. Xenophon,
Cyropaedia, 1.6.38. For "shrewdness" in the ancient military treatises see Wheeler, *Stratagem and the Vocabulary of Military Trickery*,
46–48 and idem, "Polyaenus: *Scriptor Militaris*," 29.

10–29   Ἡ μὲν οὖν συνεχὴς ... χρειώδης   Maurice, *Strategikon*, 6.pr.3–17.

29–84   Leo repeats Maurice, *Strategikon*, 6.1–5, but omits the late sixth-
century names of the four formations (the "Scythian," the "Alan," the

"African," and the "Italian"), and for the "Italian" represents it, with its double battle line, as the standard and preferred Roman formation described in Const. 12. Much of the descriptive information connected with the four battle formations is summarized from material present in Const. 7. The exercises described in ll. 30–36 recall the exercises described in the *Mirac. S. Anastasii Persae* 1.3.mir8.3–6 (135 with notes) as taking place in March. See Kaegi, "Some Seventh-Century Sources,"177–81 and Wheeler, "Occasion," esp. 357–60; for Maurice's "African drill," see Kaegi, *Muslim Expansion*, 135–40. Similar exercises were reported to have been carried out under Nikephoros II Phokas and John I Tzimiskes, who drilled their troops rigorously and regularly, and the *Praecepta* of Nikephoros refer to the exercises and drills practiced by the troops (e.g., Nikephoros, *Praecepta*, 1§10.108–9; 4§10.103–6); see McGeer, *Sowing the Dragon's Teeth*, 217–19. The better Byzantine commanders appear generally to have been quite aware of the limitations of the different sorts of troops under their command, and this is reflected in the *Praecepta* of Nikephoros, which sets out quite simple, easily managed tactical maneuvers for the great bulk of the thematic infantry, who were on the whole neither well equipped nor very unreliable.

37–56   §§7–9   Cf. Leo the Deacon, *Historia*, 63.14–18 (Talbot and Sullivan, 112) for comparable exercises carried out by troops under Nikephoros Phokas at Constantinople. See above and *Mirac. S. Anastasii Persae* 1.3.mir8.3–6.

57–61   §10   The standard and recommended "Roman" formation, as emphasized at length in 12§§2.9–16.118. See above.

78   ἐπὶ δόρυ κλίναντας   Along with ἐπὶ σκουτάριν κλίναντας (l. 82), these are contemporary and ancient technical expressions. The cavalry equivalent for ἐπὶ σκουτάριν was ἐφ᾿ ἡνία (to the reins); see Aelian, *Tactica theoria*, 24.2; 25.1–4; 19.12; and compare Syrianos, *Strategy*, 22.

83   ἀποκαθίστασθαι   Another technical term from the ancient treatises, which Leo takes from Maurice; see Aelian, *Tactica theoria*, 29.3.

85–93   §§14–15   Leo's own introduction to the next section, based very loosely on *Strategikon*, 11.pr.7–12 and 11.4.225–39. In the sections that follow, dealing with the moral values to be observed by the Romans in addition to their basic battle formation and the best responses to various enemy formations, Leo takes the paragraphs in the *Strategikon* which deal with the Persians—essentially all of *Strategikon*, 9.1—and quite simply restructures or rephrases the material to make it apply to the contemporary Byzantine armies; see detailed commentary on Maurice's section on the Persians in Rance, *Strategikon*, 11.1, notes 2–30. At a later stage of this section, in contrast, where he might have employed the material from the *Strategikon* in his characterization of the Saracens, he merely refers to the latter as a "race" as pestilential as the Persians of old; see below, and Dagron, "'Ceux d'en face'," 222–23.

94–103   §§14–15   The themes here—loyalty to the fatherland, obedience, steadfastness, and self-sacrifice—are Leo's version of Maurice, *Strategikon*, 11.1.2–5. Leo inverts Maurice's evaluation of the Persians' fear of their leaders, which is what lends their armies order, cohesion, and obedience (a topos of Roman and Byzantine commentators on their enemies), into Roman discipline and willingness to die for their faith (a reflection of his views about Islamic beliefs later in this constitution, see below), with an emphasis on the correct faith of the Christians and on the responsibility of the general to inculcate similar attitudes in those who are not yet trained to think in this way, probably a reference to his wish to suggest a new emphasis on Christian faith as the core of Roman identity and resistance to their enemies (but particularly Islam).

104–6   §18   Cf. Maurice, *Strategikon*, 11.1.6–8. Again Leo takes the account of Persian approaches to warfare and strategy, which privilege planning and generalship over boldness or recklessness, and applies it to the Roman general whom he addresses.

107–13   §19   The emphasis on bearing up under adverse conditions, whatever may befall, is now applied by Leo to the Romans, and recontextualized in this very clearly Christian context. Note also the emphasis on the emperor sharing the sufferings of the general and his soldiers, a motif that recurs in the military harangues generally (if problematically) ascribed to Constantine VII and is also a topos of the historiography of the period, associated also with the notion of the emperor as father of his soldiers; see *CPTT* C.452–54, 466–73 and commentary, 242–44; and above on Consts. 4§1.12, 12§57.409–20. For the Persians, see Rance, *Strategikon*, 11.1, note 3.

114–17 §20 Cf. Maurice, *Strategikon*, 11.1.4–10, taking Persian virtues and applying them to the Romans.

118–21 §21 Maurice, *Strategikon*, 11.1.11–14. Leo draws directly from his exemplar in the *Strategikon*, but referring now to the "Persian tribes" (rather than, as in Maurice, the Persian *ethnos*); see also Rance, *Strategikon*, 11.1, note 6.

122–32 §§22–23 Maurice, *Strategikon*, 11.1.16–17, and cf. 59–63 (with Rance, *Strategikon*, 11.1, note 9 on Persian archery). Arab pride in their horses is repeated on several occasions later in the constitution; see on ll. 640–46 below.

133–36 §24 Leo's first comment on the ways in which Islamic armies are recruited, introduced here to vary his otherwise evident reliance on Maurice. See below on the text from §103 onward.

137–209 §§25–39 These lines all follow Maurice's account of Persian tactics and preferences in battle, although Leo misses the lines at *Strategikon*, 11.1.38–42: καὶ τὰς συμβολὰς δὲ ἐκ τοῦ πρᾴως καὶ προσεχόντως περιπατεῖν ἴσας καὶ πεπυκνωμένας κατ' ὀλίγον ποιεῖται. Λυπεῖ δὲ αὐτῷ ψύχος καὶ βροχὴ καὶ νότου πνοή, διαλύουσα τὴν τῶν τόξων δύναμιν.... ("and they then gradually engage by advancing at a steady and deliberate pace in an even and dense formation. Cold and rain and a southerly wind cause them trouble, by weakening the strength of their bows...."); but see on ll. 573–75 below; and Rance, *Strategikon*, 11.1, note 21, although rain is reported to have weakened the archery of the Turks at the battle of Anzen/Dazimon in 838; see Theophanes continuatus, 128.8–11 and Genesios, 3.14 (48; trans. Kaldellis, 64). The omission reflects both Leo's concern to generalize in this passage and to exclude the older references to the Persians as the subject of the section from which he was taking his text; and more importantly because he applies this stereotype to the Saracens later in the same constitution (see on ll. 573–76 below). By the same token, at ll. 144–45, Leo retains the comment in the *Strategikon* about how the Persians station a body of 400–500 additional elite cavalry in the center division, but applies it to the Byzantines (this is not a formation recommended in the *Strategikon* for Roman armies). Is he perhaps thinking of a formation similar to the wedge of *kataphraktoi* stationed in the center division according to the treatises of the following half century? See, for example, *Sylloge tacticorum*, 48 and Nikephoros, *Praecepta*, 3–4 (I owe this observation to Philip Rance). At l. 170 Leo omits *Strategikon*, 11.1.51–53, where Maurice comments on the disorganization of the interior of Persian camps. Leo alters the text so that the points made in the *Strategikon* now apply either to the Romans or to an unnamed enemy.

156–59   ἀλλὰ καὶ . . . παύσηται   To what extent this tactic, associated in Maurice with the Persians, was used by eastern Roman commanders in the ninth and tenth centuries is unknown. See Rance, *Strategikon*, 11.1, note 20.

210–349   §§40–72   This whole section on the Turks follows Maurice, *Strategikon*, 11.2 on the armament, tactics, and customs of the Avars, more or less exactly, with the difference that the introductory section opens with Leo's own comments noting the fact that the Bulgars are now Christian and that the Turks had brought them to peace when they broke the treaty arrangements in 894; see chap. 2 above, p. 60. For detailed commentary on the Avar section in Maurice, see Rance, *Strategikon*, 11.2, notes 31–63, and for an argument that there existed archived Byzantine sources of information about the Hungarians, perhaps established during the reign of Leo VI, see Howard-Johnston, "*De administrando imperio*," 316 and 324–25. By "Turks," of course, Leo means the Magyars, and the term *Tourkia* in Byzantine writers usually refers, from this time until the appearance of the Seljuks, to Hungary. See in particular Shepard, "Byzantine Writers on the Hungarians," 101–3; Moravcsik, *Byzantinoturcica*, 2:320–21 (no. 3), and Stephenson, *Byzantium's Balkan Frontier*, 38–45; Zuckerman, "Les Hongrois"; and Macartney, *Magyars*, 125–34. The Bulgars had by the late ninth century been established in the Balkans for almost three hundred years and were thoroughly intermixed with the indigenous Slavic and pre-Slavic population of the lands they conquered. The Bulgarian elite retained some Turkic traditions in their tactics, particularly when fighting along their northern borders and the Danube plain, more suited to mounted nomadic styles of warfare, whereas facing the Byzantines they had largely adapted to the mountainous and wooded landscape of the southern Balkan region; see especially Fiedler, "Bulgars" and Iotov, "Note." In general see Shepard,

"Slavs and Bulgars" and Halperin, "Bulgars and Slavs." According to Byzantine traditions, the Bulgars were usually at an advantage in wooded or mountainous terrain, whereas the Byzantines preferred open country, where they held the advantage: *Scriptor incertus*, 42.87–90 (ed. Iadevaia). Byzantine defeats at the hands of the Bulgars in the late ninth and early tenth centuries, however, which were all in open country, suggest that Bulgarian forces could more than hold their own, and that the differences in tactics and equipment between the two were minimal.

212–13   On the Bulgarian attack and the subsequent campaign, culminating in their defeat at the hands of the Magyars, the Bulgarian counter-attack and defeat of Byzantine forces at Bulgarophygon in 896, and their involvement of the Pechenegs against the Magyars, see Shepard, "Bulgaria," 570–71; esp. Macartney, "Bulgaro-Greek War"; Čankova-Petkova, "Der erste Krieg"; Obolensky, *Byzantine Commonwealth*, 105–6; and Stephenson, *Byzantium's Balkan Frontier*, 18–123 for background. See also Shepard, "Symeon of Bulgaria." Leo diplomatically omits the Byzantine defeat. For the Magyar attacks on Bulgaria see Dimitrov, "Bulgaria and the Magyars," 67–68; Györffy, "Landnahme, Ansiedlung und Streifzüge"; and Curta, *Southeastern Europe*, 188–91 For general context see the essays in Csernus and Korompay, *Hongrois et l'Europe*; Shepard, "Byzantine Writers on the Hungarians."

222–26   §41 Maurice, *Strategikon*, 9.2.4–11, except that Leo replaces "the Turks and the Avars" with "Bulgarians . . . and . . . the Turks" as representing "Scythian" peoples who give thought to organized order of battle and tactical formations which permit them to engage in close combat. Leo adds the phrase "καὶ μοναρχούμενα," "under a single leader," reflecting his own understanding and knowledge of the tactics of these foes.

227–32   §42 Again Leo's own words on the current situation. Clearly, however, Leo's account of the Magyars ("Turks") that follows is intended also to elucidate Bulgarian tactics and fighting methods, in spite of his pious statement to the contrary at ll. 230–32: δι' ὅπερ οὔτε . . . προθυμούμεθα.

233–34   Leo adds to Maurice in order to include the Bulgars in what follows (having just remarked that he will not describe Bulgarian tactics

and battle order because of their shared faith with the Byzantines).
See Moravcsik, *Byzantinoturcica*, 1:406–8 and Stouraitis, "Byzantine
War against Christians," esp. 93–95 and 99 on the Byzantine view of
(Christian) Bulgarian aggression against the empire.

242 κρύπτοντα τὴν βουλὴν αὐτῶν Leo's translation of the odd term
κρυψίβουλα in Maurice (*Strategikon*, 11.2.16), which seems to mean
either dissembling or deceitful rather than merely concealing their
intentions. Cf. Trapp, *Lexikon*, 4:890, s.v.; for these attributes of the
Turkic peoples (Huns, Avars) see also Th. Sim., *Historia*, 1.3.2; and for
the much older antecedents of Roman notions of the deceitfulness of
"barbarians" in general see, for example, Polyaenus, *Strategika*, 7.pr.
On their avarice and lust for treasure, see Sinor, "Greed."

250 Leo updates Maurice's *zabai* with the contemporary *lorikia* (although
Maurice does use the term once, of Persian equipment; see Rance,
Strategikon, 11.1, note 7); see on 5§3.19 and 6§2.9 above. On the armor
and weaponry of the western steppe zone in the eighth–eleventh century see Golden, "War and Warfare," 128–52 with sources and further literature; and for the Magyars in particular, Bálint, *Archäologie
der Steppe*; Kovács, "A kalandozások"; and the contributions to
Mende, *Research on the Prehistory*, esp. the brief survey by Langó,
"Archaeological Research."

250–56 §§47–48 In general on the Magyars see Macartney, *Magyars*.
Whether Leo's account, taken straight from Maurice, applies in
every detail to the Magyars is unclear (see the comment in Shepard,
"Byzantine Writers on the Hungarians," 101–3), although the archaeological evidence for Magyar weaponry at this period would suggest that, apart from local and regional variations in style of sword,
saber, bow, and so forth, there was no real difference in either weaponry or defensive equipment across the western Eurasian steppe,
and that there was substantial continuity of tactics across the western steppe from the fifth century onward. See Golden, "War and
Warfare," 148–51; Kazanski and Sodini, "Byzance et l'art 'nomade'";
Iotov, "Note"; Kiss, "Frühmittelalterliche byzantinische Schwerter";
Stadler, "Chronologie"; Mavrodinov, *Trésor protobulgare*, 120–21 (and
see Bálint, "Der Schatz"); László, "Contribution," 172–74; and the
illustrated overview (albeit without annotations and bibliography) in

Gorelik, "Arms and Armour," 127–47. For archaeological-historical background: Kazanski, "Armes." But the archaeology also suggests that Leo was unaware of some specific traits in Hungarian weaponry and tactics, since he ignores the use of the axe, which is found in fairly substantial numbers in Magyar graves of the period. Of course mortuary practices may not necessarily reflect daily practice, and the archaeological record can be deceptive regarding combat tactics; and the absence of the axe (absent also in Maurice's account of the Avars) does suggest that Leo relied wholly on Maurice here, and that no up-to-date information was incorporated into his account. Leo's account of the "Turks" is, of course, full of the standard topoi about steppe nomads found in ancient and late ancient literature. These characterizations or caricatures were not confined to the Byzantine world, and it is worth citing *in extenso* some passages from the account of the Hungarians given in the *Chronicon* of Regino of Prüm, completed in about 908 or very shortly thereafter:

> "In the year of the Lord's incarnation 889, the Hungarian people, who were extremely warlike and more savage than any beast . . . emerged from the Scythian kindoms. . . . After they had been forced to flee by the violence of the Pechenegs . . . they roamed the wilderness of the Pannonians and then the Avars, and sought their daily food by hunting and fishing. Then they attacked the lands of the Carinthians, Moravians and Bulgars with the infestation of constant raids, killing a very few with the sword and many thousands with arrows, which they fire from their bows made of horn with such skill that it is almost impossible to avoid being hit by them. But they know nothing about fighting hand-to-hand in formation or taking besieged cities. They fight by charging forward and turning back on their horses, often indeed simulating flight. . . . [T]hey leave the battle at the height of the fighting and soon afterward come back from their retreat to fight again, so that just when you think you have won, the critical moment has to be faced. . . . They do not live like humans, but like beasts. For, so it is rumored, they eat their meat raw, drink blood, chop

up the hearts of captives and swallow them bit by bit just as if
they were medicine; and they are not swayed by compassion
nor moved by any stirrings of pity. . . . They ride their horses
all the time; they are accustomed to travel, halt, think and talk
on them. They put a lot of effort into teaching their children
and slaves horse-riding and archery. By nature they are puffed
up, quarrelsome, deceitful and insolent. . . . They are always
restless . . . taciturn, and more given to action than to words."
(trans. MacLean, *Regino of Prüm*, 202–6, s.a. 889)

Regino's account is heavily based on a much earlier account taken
from the second-century historian Justin's epitome of the *Philippic
History* of Pompeius Trogus, where the nomads and mounted archers
are in fact the Parthians. Like Leo, Regino modifies his source and
adds some contemporary detail, but does not change the basic picture
of the steppe nomads as bestial, faithless, pitiless savages. If anything,
Leo's account is a good deal less bloodthirsty than Regino's.

255–56 οἱ ἵπποι . . . σκέπονται Archaeological evidence from central Asia
strongly suggests that the elite members of nomadic clans could
afford to breed and maintain larger and more robust animals than
the majority of ordinary warrior herdsmen, horses which were thus
more capable of bearing the additional weight of protective bard-
ing; see the material and palaeozoological statistics in Čalkin,
Drevneyi životnovodstvo plenyon, whose work largely relates to the
earlier Scythian cultures of the S. Russian steppe; and, for the evidence from horse skeletons in Avar and Magyar burials, see Bökönyi,
History, 267–77; Khudyakov, *Vooruženie srednevokovykh kočevnikov*,
136–38; most recently Priskin, *A Kárpát-medence*; and Szentpéteri,
"Archäologische Studien." I am greatly indebted to Henrietta Kroll
(Römisch-Germanisches Zentralmuseum and Univ. Mainz) for information on archaeozoological matters.

257–58 §49 Regino of Prüm notes the effectiveness of the Hungarian
archery: "killing a very few with the sword and many thousands with
arrows, which they fire from their bows made of horn with such skill
that it is almost impossible to avoid being hit by them" (see above, on
Const. 18§47.251–56). See also Gombos, *Catalogus*, 2:843, a prayer to

St. Geminianus from the inhabitants of the town of Modena: "Defend us from the arrows of the Hungarians.") See in general on the archery of the steppes Kaegi, "Contribution of Archery," with extensive references to Byzantine accounts of the effectiveness of steppe archery; the older but still useful Darkó, "Influences touraniennes"; and Golden, "War and Warfare," 151.

259 ἱππαρίων καὶ φοραδίων Leo's equivalent for the ἀρρένων τε καὶ θηλειῶν of the *Strategikon*, 11.2.31, adding to the text slightly and changing it (from "males and females" or "stallions and mares," to qualify the foregoing "horses") to a list—"horses, ponies, and mares." The majority of horse skeletons (or partial skeletons) buried in Avar and Magyar graves were stallions or geldings, suggestive of the practical importance of maintaining a good stock of mares; see Bökönyi, *History*, 268–69.

261 See on Const. 11. Leo repeats the information that the "Turks" do not set up entrenched camps from Maurice. In fact steppe warriors did on occasion set up fortified encampments, sometimes using the spear-and-shield type of palisade and digging trenches (see on 11.40–42, above, and Ibn Rusta, *Kitāb al-A'lāq an-Nafīsa*, 143), more usually drawing their wagons and carts into a defensive circle which could serve both as a battlefield redoubt and a refuge. See the twelfth-century Byzantine accounts of the Pecheneg wagon laagers in the late eleventh and early twelfth century: Anna Komnene, *Alexiad*, 7.3.7 (trans. Sewter, 224/Frankopan, 193); Kinnamos, *Epitome*, 8; Choniates, *Historia*, 1, p. 15 (trans. Magoulias, 10). But these refer to specific situations when warfare was endemic or expected, and the norm for steppe peoples does indeed appear to have been as Maurice—followed by Leo—described; see also Rance, *Strategikon*, 11.2, note 42.

268–69 Leo retains Maurice's original wording (11.2.41) here rather than replacing *moira* with *droungos*, presumably because the original is clearer.

273–77 τὸν δὲ τοῦλδον . . . εἰς φυλακὴν αὐτῆς Leo omits the original ἀδέστρατα (Strategikon, 11.2.45; and see Rance, *Strategikon*, 11.2, notes 46, 47), referring to the spare mounts ready for action, and states merely that the baggage train is kept behind the battle line. In the next paragraph (l. 276) he retains the reference to the extra or spare horses, because he seems not to have understood the technical value of *adestrata* (see on Const. 10§6.25–26 above) and thus the difference between the two groups of animals, the latter representing the large body of animals accompanying every nomadic army under arms.

278–80 The sentence might better be rendered as "they leave the depth of their files . . . unspecified" rather than translating ἀορίστως as "irregular."

283 Leo again omits a late Roman term (κοῦνα, from Lat. *cuneus*, a wedge formation: Trapp, *Lexikon*, 4:873) which was no longer in use: *Strategikon*, 9.2.54; and on feigned flight and the "wedge" formations attributed to nomadic armies, see Rance, *Strategikon*, 11.2, note 49.

284–88 §57 The relentless pursuit practiced by Turkic warriors once the enemy battle formation was broken is well illustrated by that to which the emperor Theophilos was subject after the battle of Anzen/ Dazimon in 838; see Theophanes continuatus, 127–29 and the Arabic accounts in Vasiliev, *Byzance et les Arabes*, 1:275, 299–301, 309; 333–34.

295–98 §59 Ταῦτα μὲν . . . συναποβαλόντες Leo is keen to stress the rapprochement between the Bulgarians and the Romans—perhaps also a gesture to his former companion in the imperial palace Symeon of Bulgaria. See Shepard, "Ruler," esp. 350–58.

299–316 §§60–65 Leo repeats Maurice's list of conditions unfavorable to nomadic forces; see 12§106.710–12 (= Maurice, *Strategikon*, 7.A.pr. 27–28). He importantly omits the advice to attack in February or March, when the enemy's horses are weakened from the rigors of winter and lack of pasturage, presumably because the Byzantine state had at this time no shared frontier with nomadic peoples and attacks were both irrelevant and impractical. Mounted nomad armies were especially vulnerable to lack of pasture, since they did not normally collect fodder for their animals but put them to graze, and winter, especially when the steppe might be covered in snow, was an especially dangerous period, when thousands of animals could be lost; see Gladitz, *Horse Breeding*, 19–20, 65–66. The vast herds of animals they took with them required considerable amounts of open grazing, and although there were, and are, several varieties and subtypes of animal, bred for different purposes and in different regions of the steppe from Mongolia through Turkestan and across through the Ukrainian steppe, their basic requirements for feed and water are much the same. For example, a typical Akhal Teke or Pazyryk-type steppe horse requires some 120 acres (this is a very crude and highly variable estimate, of course) of grazing per annum, roughly ⅓ acre per day, far more than the barley- (or similar) fed horses of a sedentary culture; and much less water than the non-nomadic horse: for general characteristics see Sinor, "Horse and Pasture" and for a general overview of types and characteristics see Čalkin, *Drevniyi životnovodstvo plenyon*; Roberts, *Horse*, from 360; Bökönyi, *History*, 267–71 and 275–79 (on Magyar horses of the conquest period); Benecke, "Zur Kenntnis"; and Priskin, "A Kárpát-medence." This placed natural limits on their ability to campaign, except where, as in the European theater, they could rely on subject peoples to fight on their behalf. Nevertheless the enormous advantages held by nomad warriors over sedentary foes were to a degree offset if such limitations could be exploited, and although Leo is copying Maurice, Byzantine familiarity with Eurasian steppe warfare over many centuries means that Leo's repetition and occasional updating of his original retains its immediacy and value for his own times.

Leo's (and Maurice's) point about level, unobstructed ground as being appropriate for dealing with mounted archers depends, of course, on the Roman forces retaining their cohesion, as stated, but also on rapidly closing with the enemy in order to minimize the effects of their superior archery. See 18§§33, 35. The inability of the Byzantine forces to close with the Turkish archers at the battle of Anzen in 838 for example, was a major reason for the dissolution of the Roman line and the disintegration of the emperor Theophilos's army (on the confusion in the sources for the battle of Anzen, see on 14§8.40–47 above). But note that at 18§26.149–52 Leo recommends drawing up the battle line of a force of mounted archers in difficult and rugged terrain precisely because this will break up any charge mounted by lancers—the difference depends on who takes the initiative in the attack.

303-4 οὐδὲ γὰρ ... ἵπποις   For the stereotype, see the comments in Rance, *Strategikon*, 11.2, note 52. Note also the editor's comment, citing Ammianus Marcellinus, *Historia*, 31.2.6, regarding the Huns.

310–13 §64 Nomad fickleness and duplicity is frequently singled out by Byzantine writers, and indeed there are several examples of sudden shifts in allegiance—most notably among the Pecheneg allies of Romanos IV at Mantzikert in 1071. See Attaleiates, *Historia*, 117.14–120 (trans. Kaldellis and Krallis, 287) and the more general comments in Golden, "War and Warfare," 126–27. On the topos of nomadic fickleness and untrustworthiness, see also Rance, *Strategikon*, 11.2, note 56.

323 Refers back to 10§§3, 6, 9, and 15–16 (but taken directly from *Strategikon*, 11.2.85, where it refers to bk. 5).

326–27 κατὰ τὸν ... τρόπον See Maurice, *Strategikon*, 11.2.87–88. Leo emends the text to omit the reference in Maurice to the diagram at *Strategikon*, 12.A.7.

330–31 ἐν τῷ περὶ τάξεως λόγῳ Again taken directly from Maurice, *Strategikon*, 11.2.90–91, where it refers to bks. 2 and 3; in the case of the *Taktika* this material is incorporated into Consts. 7§§41–45 and 12. If Leo intended this phrase to refer to sections of his own *Taktika*, he presumably meant Const. 7, although it is just as likely that his redactor/copyist simply failed to edit Maurice's phrase out. He cannot be referring here to the *Sylloge tacticorum*, as suggested by the editor (Dennis, *Taktika*, 461 n. 16), and if Leo meant it to refer to anything other than Maurice or his own treatise, the reference is much more probably to the work of Syrianos *magistros*. See also on Const. 19§72.397–99 below.

343–46 See on 17§19.117–19 for a similar view from a contemporary western writer (albeit, like Leo, using a much older account as the basis of his remarks).

348–49 Maurice, *Strategikon*, 11.2.107–8, repeating the standard warning about the dangers of foraging when the enemy is close by; see on Const. 17§37.198–202 above.

350–59 §73 Leo's own comments, noting that the Bulgars and Turks—Magyars—differ only in minor respects, and noting that there is no immediate threat from the latter. Lines 354–59 appear to be Leo's words (or are based on an as yet unidentified source).

360–69 §74 Leo's introductory remark to the following section, reflecting the political situation in Italy and the Balkans: for general surveys, see Auzépy, "State of Emergency," esp. 257–58 on the Byzantine "recovery" of the Peloponnese in the late eighth and early ninth century; McCormick, "Western Approaches"; Shepard, "Western Approaches (900–1025)"; and Loud, "Byzantium and Southern Italy."

370–76   §75   See on 17§27.144–49, and below on 18§§93.441–102.494.

377–440   §76–92   The section is based on Maurice, *Strategikon*, 11.3 (on which see Rance, *Strategikon*, 11.3 and notes 64–75), although Leo frequently rephrases or reorders the original. For the most part it seems that the Byzantine court was relatively ignorant of Frankish affairs and those of lands to the west and north of Italy, whereas Italian matters, closer to hand, were more closely observed. See Shepard, "Aspects"; Lounghis, "Adaptability," esp. 347–58; Harris, "Wars and Rumours"; and Malamut, "Constantin VII," 273–79.

378   ἀλλ' οἱ μὲν Λογγίβαρδοι . . . ἀπώλεσαν   Leo's own observation on the current state of the Lombards in Italy. For general context and background see Brown, "Byzantine Italy (680–876)"; McCormick, "Western Approaches," 408–23; and Delogu, "Lombard and Carolingian Italy." For Byzantine Italy in the late ninth century and beyond, see the essays in Martin, *Guerres, accords et frontières*; von Falkenhausen, *La dominazione bizantina*; and Martin, "Thèmes italiens."

388–90   §78   For Frankish armament in Leo's time, which the archaeological and pictorial evidence suggests had changed only slightly from the late sixth century, see Coupland, "Carolingian Arms and Armour"; Nicolle, *Medieval Warfare Sourcebook*, 1:78–91; Friesinger, "Waffenkunde des 9. und 10. Jahrhunderts"; and Oakshott, *Archaeology of Weapons*, 164–80. More generally for Frankish/German military organization and warfare, see Werner, "Heeresorganization und Kriegsführung"; critical of Werner's approach to numbers (and with earlier literature), Reuter, "Carolingian and Ottonian Warfare"; Bowlus, *Franks, Moravians and Magyars*; idem, "Militarische Organization"; and in general, Bachrach, *Warfare*. Leo retains Maurice's remark on the "shorter" Frankish sword (but see Rance, *Strategikon*, 11.3, note 66, suggesting that this may in fact be a misunderstanding of the original text of the *Strategikon* by a later copyist), and archaeologically this seems to be born out in some cases—the average length of western swords in the late ninth–mid-tenth centuries seems to have been between 0.85 m and 0.95 m in length, although some were slightly longer, similar to their Byzantine equivalent (for Byzantine swords and sabers see above on Const. 5§2.9, 17–18); see also Coupland, "Carolingian Arms and Armour," 42–46. He also updates his information by noting that some Frankish warriors also carry their swords slung from a waist belt, as well as from a baldric across the shoulder; see in general Menghin, *Schwert* and Bruhn-Hoffmeyer, *Middelaelderens tvaeggede svaerd*.

393 καθάπερ Ῥωμαῖοι Leo's addition to Maurice's original text, with the implication that the Franks of Leo's day do have an order of battle but that it is not the same as the Roman. While Leo's statement, from Maurice, that they "take more pleasure in fighting on foot" was certainly no longer true of Frankish royal armies at this period, and might suggest that neither he nor those who advised him in the compiling of the *Taktika* had much idea of actual Frankish practice, at ll. 419–20 he does note that the Franks have been trained to charge swiftly and on level ground, a comment absent from his source. On the other hand, the field armies of the German emperors of the tenth century often included very substantial numbers of trained infantry; see Bachrach, *Warfare*, 70–101. On the highly tendentious nature of late Roman accounts of Germanic armies, especially with regard to their lack of cavalry and their tactical simplicity, see Rance, *Strategikon*, 11.3, notes 67–69. Leo attempts to bring his comments up to date, since it was by his time, and given Byzantine campaigning in Italy, impossible to reproduce the stereotype quite so simply.

401–5 καὶ μάλιστα ... αὐτῶν Leo adds this to his original, suggesting some accurate reports on the nature of Frankish and Italian levies for campaigns, although by his time a more efficient system of trained infantry levies and mounted vassals commanded by local lords was replacing the traditional levy on all freemen; see Fleckenstein, "Adel und Kriegertum"; Goetz, "Social and Military Institutions," esp. 479–80; the literature to ll. 388–90 above; and Bachrach, *Warfare*, 70–101 and 193–15.

407–8 Another indication of the fact that Leo's information is uncertain— this statement is taken from Maurice, but could no longer be said to apply to the mounted forces of the past-Carolingian world, although it may describe the forces of the Lombard duchies more accurately; see Sergi, "Kingdom of Italy," 349–50 and Rasi, *Exercitus italicus*.

409–12 §84   Here Leo adds his own comments on the venality of the Italians, based on much better information and experience of dealings with envoys and others from the peninsula. He may have in mind recent diplomatic and associated dealings with both the papacy and with the southern and central Italian dukes, on the one hand, and with "Frankish" rulers such as Arnulf of Carinthia (*PmbZ* 20578), attempting to assert their rights in the north, on the other. Leo reportedly betrothed his daughter Anna in about 899–900 to Louis III of Provence, although it is not clear whether the marriage actually took place; see Kresten, "Zur angeblichen Heirat Annas" and the survey in McCormick, "Western Approaches," 418–23; *PmbZ* 20430, 24756.

419–20   διὰ τὸ ... ἐγγυμνασθῆναι   See Maurice, *Strategikon*, 11.3.29–30. Leo adds again his own comment to bring his original up to date. Here we have a clear reference to the sort of cavalry tactics with which the Franks in general came during the course of the tenth and eleventh centuries to be associated. For "Frankish", i.e., German weapons and military equipment in the tenth century, see Bachrach, *Warfare*, 135–51. The archaeological and pictorial evidence suggests that still at this time Frankish/Italian cavalry were not very heavily armored, the soldiers protected by short-sleeved mail shirts or leather hauberks reaching to the knee, a circular shield, and armed chiefly with sword and lance, although the bow is occasionally mentioned (see Nicolle, *Medieval Warfare Sourcebook*, 1:77–78) and that the effects of a Frankish cavalry charge were the result of a lack of cohesion among unformed or disorganized infantry rather than to the disciplined attack of well-organized "wedges" of cavalry as typified by the Normans from the early and middle eleventh century. Even here the use of the "couched" lance, held under the arm against the body so that the impetus of the horse provides the thrust, rather than the arm of the soldier, seems to have been relatively rare until the twelfth century; see Bachrach, "Verbruggen's 'Cavalry'"; Cirolot, "Techniques guerrières"; and esp. Flori, "Encore l'usage"; with the useful comments on various aspects of this issue in Bennett, "Medieval Warhorse," 34–36. It has been suggested that this style of attack was in fact adopted by the Normans of S. Italy from their Byzantine allies or foes in the early eleventh century: Nicolle, *Medieval Warfare Sourcebook*, 1:78, although the evidence is circumstantial.

421–40 §§88–92   Taken with minor variations in wording from Maurice, *Strategikon*, 11.3.30–46; although again the stereotype of Frankish cavalry being easily led into disaster by a feigned flight should be treated with caution, for any period; see Rance, *Strategikon*, 11.3, note 71.

441–94 §§93–102   Maurice, *Strategikon*, 11.4. For discussion and further literature and sources, see in detail Rance, *Strategikon*, 11.4, notes 76–132. Leo reproduces here a very much shortened version of Maurice's text, interspersing it with his own remarks on occasion. While Maurice spends a great deal of time on a very detailed account of Slav customs, weaponry, and tactics, a reflection of personal experience and the campaigns conducted by other Roman officers in the Balkans in the second half of the sixth century, much of the material was by Leo's time entirely outdated, although no doubt Leo and his readers accepted much of it as being still applicable to some Slav peoples south of the Danube. See also on Const. 17§27.144–49 above. Thus at l. 443 here, where Maurice has "when in their own land," which Leo supplements with "when they dwelled across the Danube"; and in the subsequent lines (443–49) brings the account up to date by noting that the Slavs had crossed over onto Roman territory ("from there to here"). It is not quite clear to whom "the ruler of their own race" is intended to refer—possibly to their subjection to the Bulgar khan after the 680s; see Curta, *Southeastern Europe*, 59–110. At ll. 453–60 he then describes the successful missionary and conversion activity carried out during the reign of Basil I, again alluding to their being freed from "slavery to their own rulers," reflecting perhaps a shift in the form and means of paying local dues, and possibly also to the establishment of the newly organized theme of Dalmatia ca. 870; see Oikonomidès, *Listes*, 353; *ODB* 1:578–79; Curta, *Southeastern Europe*, 144–47; with the older literature cited by the editor at 471 n. 21. The exact implications of the expression γραικώσας (l. 454; see Trapp, *Lexikon*, 2:328) "made them Greek" has the sense of "Romanized," probably by bringing them further into the Byzantine cultural and ecclesiastical orbit (see

·

McCormick, "Western Approaches," 418–22), rather than literally suggesting that they had to speak Greek or dress in a "Greek" manner, although placenames may have been changed and Greek speakers from elsewhere in the empire may have been settled among them to assist in the process of acculturation; see Koder, "Anmerkungen" and Vasmer, *Slaven*, 325.

475   Leo moves the reference in Maurice to the millet cultivated by the Slavs to the next paragraph, but omits their livestock (the *Strategikon* lists two types of millet: 11.4.18), presumably since this no longer has logistical value for Roman armies in the Balkans.

475–77   §98   Although taken from Maurice, this appears to reflect actual mourning practice and ritual suicide in some Slav communities as reported by later commentators, including St. Boniface in the eighth century, Thietmar of Merseburg, and later Arab commentators such as Masʿūdī and Ibn Fadlan; see Jopson, "Early Slavonic Funeral Ceremonies," 59–67, and in general Barford, *Early Slavs* and Cross, "Primitive Slavic culture." The practice may be in part a topos, which appears in several earlier writers (Procopius, Theophylact Simocatta), as suggested by Zástěrová, "Zur Problematik," 16–17, but clearly seems to have some basis in the observations recorded by visitors to some Slav groups.

478   See on l. 475 above. Here Leo refers to their food (millet), but otherwise elaborates his own brief summary of the much more detailed passages in Maurice about the lifestyle of the Slavs, omitting the information about their preference for dwelling in woods and marshes, their seminomadic lifestyle, the way they hide in streams by using reeds through which to breathe, and other details of their fighting methods (see Maurice, *Strategikon*, 11.4.23–43).

482–87   §100   Maurice, *Strategikon*, 11.4.44–50, with Rance, *Strategikon*, 11.4, notes 87–90, for a detailed commentary on sixth- and seventh-century Slav weaponry and military equipment.

488   Leo inserts this information from *Strategikon*, 11.4.23, previously omitted.

489–94   §102   A reference to Const. 17; see chapter 2 above on the structure of the *Taktika*; and on 17§27.144–49 and §§38.205–43.253. Leo does not in fact include much information on the Slavs at all except here and in Const. 17, where he uses some of Maurice's material from bk. 11

for his discussion of ambushes, but not with specific reference to the Slavs alone.

495–500 §103   Leo's opening paragraph on the Saracens, to whom the rest of Const. 18 is now devoted. For his motive in compiling this section, see §135.690–93; below, on §122 and thereafter (from l. 592); and the introductory discussion to this Constitution, above pp. 331–34.

495–615 §§103–25   This passage (PG §§109–31) was, as noted, excerpted from the main body of Const. 18 and appended as a separate text at ff. 401–2, 404, after the end of the *Taktika*, and provided by the copyist (as noted in chap. 2) with its own title: Λέοντος ἐν Χριστῷ βασιλεῖ αἰωνίῳ βασιλέως Ῥωμαίων πῶς δεῖ σαρακηνοῖς μάχεσθαι. The differences between this version of these paragraphs (denoted by the editor as M¹) and the version within Const. 18 are minimal, as the critical apparatus clearly shows. See further on this question the commentary to ll. 686–95 below. For the general political history background, see Kaegi, "Confronting Islam" and Shepard, "Equilibrium to Expansion."

While much of the material on the Saracens in this constitution is new, some of it is drawn from the section in Maurice on the Persians, *Strategikon*, 11.1 which, as we have seen, Leo also converts to use in the opening paragraphs of this constitution, describing elements of the Byzantine practice of war. For the derivation of the term Saracen see *Der kleine Pauly*, 4:1548; MacDonald, "Quelques réflexions"; Graf and O'Connor, "Origin"; and Bowersock, "Arabs and Saracens." For the appearance and evolution in the use of the term from the third–fourth centuries onward in Greek and Latin sources, see Lenski, "Captivity and Slavery," 239–40; and on the general evolution of Byzantine views of the Arabs and Islam from the seventh century see Jeffreys, "Image of the Arabs"; Rotman, "Byzance face à l'Islam arabe" and idem, "Captif"; Koutrakou, "Image of the Arabs"; and Christides, "Names"; with the background comments of Hoyland, *Seeing Islam*, 12–26. For matters more particularly relating to Leo, the Saracens, and the late tenth-century texts, see Dagron, "'Ceux d'en face'," 221–24 and Kolias, "*Taktika*." In what follows, Leo does not distinguish between "Saracens" and the various other groups who actually fought in the armies of the Hamdanids or of Tarsos (in contrast, e.g., to the treatise on *Skirmishing*; see Dagron, "'Ceux d'en face'," 224), except to note the presence of "Ethiopians" (see on ll. 528–30, below), whereas other Byzantine authors note the differences in tactics and equipment employed by Daylamites, Kurds, Bedouin, and Turks, all of whom might appear in "Saracen" armies of the period; see McGeer, *Sowing the Dragon's Teeth*, 230–42, whose account relates to the second half of the tenth century but remains valid for the preceding half century or longer. For surveys of ninth- and tenth-century Byzantine warfare with Islamic enemies in both the western Mediterranean and in the east, see for the former Blysidou et al., *Βυζαντινά στρατεύματα στην Δύση*, 263–92, and for the latter the relevant narrative sections of the chapters on the eastern *themata* in Blysidou et al., *Μικρά Ασία των Θεμάτων*.

501–7 §104   The details given by Leo here differ in several details from those in the *Chronographia* of Theophanes, providing information about Arab fighting techniques and beliefs not found in the latter, and thus probably deriving from a different tradition or from oral sources. But Leo may well also have had access to the *Chronographia*, since it was certainly available to Constantine VII, or his copyists, some time later when he worked on the relevant sections of the *De administrando imperio*. See *DAI* 1: chaps. 14ff.; 2:70–71 (and see Bury, "Treatise," 525ff.; Moravcsik, *Byzantinoturcica*, 1:366; and Mango and Scott, *Theophanes*, 97–98). For the complex historiographical tradition underlying ninth- and tenth-century Byzantine knowledge of the Islamic conquests, see Brandes, "Frühe Islam"; Hoyland, *Seeing Islam*, 32ff., and 387–453; and Conrad, "Theophanes and the Arabic Historical Tradition."

508–13 Εἰσὶ δὲ . . . ἀτιμάζοντες   Leo's brief summary of what he had read, or been told, about Muslim beliefs; see on ll. 620–25 below and the discussion on just and holy war in chapter 1 above. The four propositions Leo lists—that Muslims cannot bear to call Christ God; that God is the cause of both evil and good, and that he rejoices in war; that they observe their own laws as inviolable; and that they are hedonistic and licentious—reflect a mixture of elements of an established Christian polemic directed against Islam, found in various forms; see Hoyland, *Seeing Islam*, 454–519 for a survey of Christian apologetic and disputation literature up to the early ninth century (and 480–501 for the

Greek material); and *ODB* 2:1017–18. The Byzantine Greek tradition tended to be the least well-informed and the most distorting in its representation, but certain features were always singled out, as here. See John of Damascus, *De haeresibus* 100–101 (PG 94:677–780, at 764A–773A), or (less contemptuous) Theodore Abu Qurra, *Opuscula*, in PG 97:1461–1601 (trans. Khoury, *Théologiens byzantins*, 93–105). For examples of the less well-informed versions, see Theophanes, *Chronographia*, 334.20–22 (trans. Mango and Scott, 465) and later derivative chroniclers such as George the monk (697.11–13). In a similar vein, the *Rhetorica militaris* (10.1, 26.1) specifically singles out the facts that "the enemy" fight for their faith and attack the Romans on account of their faith; that they are godless or fight against God. See also Ducellier, "Byzance, juge cruel."

513–16 τῇ οὖν … ἀντιστρατευόμεθα   Note again the similarity between Leo's sentiments here and *Rhetorica militaris*, 26.1: αλλ' ἡμᾶς … εὐσεβεῖς τε ὄντας καὶ νόμου διὰ παντὸς φύλακας ("but we … are pious and entirely respectful of the law"). Leo's emphasis on Christian piety and orthodoxy, and the observance of the law, both sacred and secular, is the foundation of his critique, as it is in that of the *Rhetorica militaris* (see also ibid., 37.7 [6]), and is found throughout the *Taktika*, echoing in turn the prooemium to the *Procheiros Nomos*. The emphasis on the law recalls in addition the precepts enunciated in the prooemium; see the commentary to the prooemium, above. The importance of divine support, and of the soldierly piety and the righteousness of the Romans' cause by which this was to be secured, is a recurring motif; see Const. 12§55.396–98, §§56.409–57.419; 13§1.3–5; 14§1.3–7; §101.704–5; 16§15.95–105; 18§19.107–13 (and associated commentary); and although it is found likewise in Maurice's *Strategikon* (see *Strategikon*, pr.1–9, 36–49; 2§18, for example), it is far more sharply stressed in the *Rhetorica militaris* (e.g., 9.3, 10.1, 28.1–5). The ninth-century context of Christian-Muslim warfare in the east has sharpened the real differences between the two, and lends Leo's injunctions a greater cultural and moral force; see the discussion in Cosentino, "Syrianos's 'Strategikon'," 267–75.

517–22 §106   Practical information gleaned from Leo's generals and others. See also 18§134.678–81 below. The impact of camels on horses was recognized in ancient times, since horses are notoriously shy of camels and need to be acclimatized to their presence: cf. Polyaenus, *Strategika*, 7.6.6 and Frontinus, *Stratagems*, 2.4.12, an account of the failure of Croesus of Lydia's elite cavalry when confronted by Persian camels. Cf., for example, the battle of Mammes in N. Africa in 535, when the Roman commander Solomon was forced to dismount his cavalry since the horses would not face the Moorish camels surrounding their camp: Procopius, *Wars*, 4.11.42–55. Vegetius, *Epitoma*, 3.23.1–2 notes that camels are regarded as ineffective in war; but the Romans used them in the eastern provinces as beasts of burden or supply animals (see Jones, *Later Roman Empire*, 768, 841–42; and see Maurice, *Strategikon*, 12.B.22.115–16), and there were also scouting units of *dromedarii* posted at strongpoints along the eastern frontier. See Wheeler, "Army and the *Limes*," 258–59 and commentary on Const. 11§38.215–16, above. For the Byzantines the camel also had a symbolic importance, partly a result of the lowly status of camel drivers or herdsmen (and the *DAI* 1: chap. 14, 10–14, describes Muhammad as a camel driver), and was associated with humiliation or shame; see *ODB* 1:368.

Islamic armies made effective use of cymbals, drums, and other percussion instruments (see, for one example, *Vita Basilii imperatoris*, 51.20–24 [186], where forces from Tarsos launch a night attack on the Byzantine encampment and, using drums and other effects, cause a panic in the imperial army). Byzantine commanders were advised to warn their troops about this, if they were not accustomed to them already; see Const. 20§76, for example. But Byzantine armies also employed similar techniques, both on the march, and during the battle, at least in the middle and late tenth century; see Leo the Deacon, *Historia*, 24.15–17, 110.11–13, 133.6–8 (Talbot and Sullivan, 76, 160, 179). According to Leo (following Maurice) the Roman and Byzantine practice was the reverse—to maintain silence as long as possible, thus unnerving the enemy, and shouting the war cry at the last minute as the battle line closed with their foe; see, e.g., Const. 12§§54–55, a tactic prescribed also in the *Praecepta* of Nikephoros, 4.11.110–12. See commentary on Const. 12§55.395–400 above. The use of drums and cymbals was—at least by the second half of the century—clearly a tactical option and would depend upon the immediate situation, although experience on the eastern front may well have contributed to or stimulated their use; see Leo the Deacon, 36.5–7 (Talbot and Sullivan, 87), where the soldiers are trained both to make such a din with trumpets, drums and cymbals, but also to withstand its effect. For the historical context in Asia and the Near East see Nikonorov, "Musical Percussion Instruments."

523–26 §107 A tactic universally practiced, with either soldiers and noncombatants, or horses (and other animals) and banners; see, e.g., 12§38.421–25 and comm. (see Maurice, *Strategikon*, 2.20.1–13).

527 §108 (Maurice, *Strategikon*, 11.1.8–9, with Rance, *Strategikon*, 11.1, note 4). The idea that climate and geographical location, associated with the humors of the body and the four elements, directly affected temperament and constitution is ancient and formed the basis also of western medieval understanding of ethnic and other differences between the peoples of the earth; see Mango, *Byzantium*, 182–87 for comments on the Byzantine understanding, and more broadly Nutton, *Ancient Medicine*, 75ff.; Wallace-Hadrill, *Greek Patristic View*. Anastasius of Sinai in the late seventh century explicitly relates temperament and constitution to climate; see, for example, Anastasius of Sinai, Qu. 26.1–2, 28.16 (52, 68). Leo's immediate inspiration here is the *Strategikon*, 11.1.8–9, 41–42, which he had already employed in the opening section of this constitution. See also below, on ll. 573–76. Muslim stereotypes of the Byzantines were no less powerful: Byzantines were treacherous, but also fierce and courageous; monogamy and celibacy were found particularly puzzling or deviant; castration was seen as a peculiarly Byzantine practice and regarded as abhorrent; miserliness was associated with the Byzantines also. See Shboul, "Byzantium and the Arabs," esp. 53–58; el Cheikh, *Byzantium*, esp. 120–29. By the ninth and tenth centuries, Christianity, with its trinitarian theology, was seen as both irrational and also responsible for the demise of ancient Hellenic learning: ibid., 100–111. See also Guemara, "La libération et le rachat des captifs."

528–30 §109 Leo's "they say" is indicative of his reliance on reports from his commanders or others. These "Ethiopians" were African volunteers and mercenaries in the Tūlūnid armies, first recruited in substantial numbers in the 860s, from upper Egypt and Sudan (Arabic *sūdān* = black); see Shaban, *Islamic History*, 109–11; Kennedy, *Armies of the Caliphs*, 157–58; and Bacharach, "African Military Slaves," although as Shaban has argued, it is most unlikely that these *sūdān* were slaves. See also Christides, "Image of the Sudanese"; *ODB* 1:732–33. Together with volunteers and soldiers from other parts of the Islamic world they constituted one element of the razzias that assembled every summer in Cilicia or northern Syria to raid Byzantine territory; see— albeit describing the situation in the middle of the tenth century— *Skirmishing*, 7.1. "Ethiopians" were hardly an unusual feature at the Byzantine court: Harūn b. Yahya refers to "black Christians" forming a unit of the palace guard at Constantinople in the late ninth century; cf. Vasiliev, *Byzance et les Arabes*, 2.2:385 and note at 435 and idem, "Harun ibn Yahya's Description." They may be the same as the unit of "Ethiopians" referred to for the 830s in a late ninth-century addi- tion to the *Acta* of the forty-two martyrs of Amorion, who appear by name in this text alone: *De XLII martyribus Amoriensibus*, 27.9–11. Their origins are obscure—freed slaves or even converts captured in battle, or volunteers who had made their own way to Byzantine terri- tory. Although Leo bemoans the inadequacy of Byzantine archery, it was in fact only from the time of Ma'mūn's and especially Mu'tasim's recruitment of Turkish mounted archers from the mid-830s onward (see Kennedy, *Armies of the Caliphs*, 118–24) that archery appears to have become a serious problem for the Byzantines. The Sudanese infantry referred to by Leo used a different type of bow, and although we have very little other evidence for their armament, Leo's comment suggests that they were also a source of concern for Byzantine troops. See al-Tahir, "Nubian Archers" (useful for the sources).

531–37 §110 See below, Const. 18§114.553–56. Leo's comment reinforces the point, apparent from both textual and archaeological evidence, that while there were significant differences in points of detail, certainly of style, and in some cases of fabrication, differences in the fundamental types of defensive equipment and weaponry between the Byzantines and their Islamic neighbors were minimal, to a large extent a result of both familiarity and constant warfare over the preceding centuries. See Zaki, "Medieval Arab Arms"; Nicolle, *Islamic Arms and Armour*;

idem, "Byzantine and Islamic"; Schwarzer, "Arms"; and esp. the excellent survey of Hamdanid forces in McGeer, *Sowing the Dragon's Teeth*, 229–48, which, while dealing with the second half of the tenth century, applies in essentials equally to the preceding century. The rapidity with which the Tarsiot forces struck into Byzantine territory and withdrew again is certainly to be ascribed to the fact that their infantry forces were largely mounted while moving. Leo later recommends that Byzantine commanders adopt the same tactic to increase their troops' mobility: Const. 20§206.1060–62.

538–40 Another generalization applied to the Saracens, apparently drawn from Maurice on the Persians (*Strategikon*, 11.1.46–47), which takes no account of the regional or tactical variations in the make-up of Islamic armies. The army of Tarsos consisted of several different bodies, and indeed the history of Byzantine-Tarsiot conflict in the ninth century hardly bears Leo's simplistic comment out. Many of Leo's comments about the "Saracens" are, however, valid for the lightly armed and fast-moving Bedouin horsemen, against whose hit-and-run tactics Nikephoros Phokas later warns: *Praecepta*, 2§10.104–11, 126–29 (μὴ καταδιώκειν τοὺς Ἀραβίτας); 4.180–89 and the detailed account in McGeer, *Sowing the Dragon's Teeth*, 238–42 with sources.

541–49 §112   Comments that refer back to ll. 510–12 and expand a little on what Leo saw as Islamic fatalism, which contrasted strongly with the Byzantine view—evident in Maurice's *Strategikon*, in the *Taktika*, and in the later treatises, that God's support was essential to victory, but that men made their own fate and exercised their own freedom of action to make good or bad decisions. Indeed, fatalism in war was seen as strongly to be discouraged, to the extent that—following ancient precedent again—false rumors of success or reinforcements or enemy failures were to be circulated, if necessary; negative reports were to be withheld; and troops who had suffered defeat were not to be engaged in battle again until they had had a chance to rebuild their morale and confidence. See, for example, Const. 14§§15–18, 97, 101 and commentary to 14§97.667–74, §101.703–7; Maurice, *Strategikon*, 8.13–14; and cf. *Rhetorica militaris*, 21.1–4 and 22.1–8. Yet Leo also recognized that Roman troops were just as liable to a certain defeatist fatalism if worsted in battle—note the advice offered at Const. 14§15.

Leo's remarks on the Saracens' fear of nighttime fighting, as well as how best to deal with their camps in night attacks, is later taken up in the treatise on *Skirmishing*, 20.80–86 and 24 (and cf. Syrianos, *Strategy*, 39 on night combat). It is, however, clear that any attack at night is dangerous and that the Byzantines were to be well prepared for this eventuality; see, for example, Syrianos, *Strategy*, 29.13–50 on the measures to be taken to defend a camp, particularly at night; and introductory remarks to the commentary to Const. 11, above, at pp. 237–39.

550–56 §§113–14 This rectangular formation was very different from that employed by the Byzantine armies of Leo's time and before, and indeed similar defensive formations had been employed since ancient times, see, e.g., Polyaenus, *Strategika*, 1.49.2 (*Hypotheseis*, 46.2) and the extensive comment in Rance, *Strategikon*, 7.B.11, note 78 with Hellenistic and Roman sources. At Const. 9§§45.207–47.233 and 14§20.141–45 (and see comm., above), and following Maurice, Leo describes a formation that amounts to a hollow square as the standard defensive marching formation; but it has no tactical battlefield role. As has been shown, a fully formed square formation was adopted at some point before the middle of the tenth century, possibly under the leadership of the commander John Kourkouas in the course of his eastern campaigns in the 920s–940s. It is described in detail in two treatises, the *Sylloge tacticorum*, 47§§3–16 written at some point after Leo and by the 950s, and the so-called *Syntaxis armatorum quadrata*, probably, but not certainly, slightly later in date, before appearing in the *Praecepta* of Nikephoros; see McGeer, "*Syntaxis armatorum quadrata*." An account of one of Alexios I's campaigns, in 1116, particularly notes the strength of this marching arrangement, which Anna describes as like a moving city (Anna Komnene, *Alexiad*, 15.6, 7; trans. Sewter, 480–87, 491 / Frankopan, 440–44, 450–51); the armies of the First Crusade employed a comparable arrangement when in hostile territory—described similarly by a later Muslim commentator as being "akin to a walking city"—and this may suggest some Byzantine influence (Smail, *Crusading Warfare*, 156–62; the Muslim comment is from al-Mankali, see McGeer, *Sowing the Dragon's Teeth*, 278–79). McGeer suggests that it was based on the basic Byzantine marching camp (*Sowing the Dragon's Teeth*, 257–64), but it seems equally if not more likely that it was also inspired by direct familiarity with the square Arab formations, to which Leo attributes in this passage both strength and flexibility in battle and on the march— exactly the qualities later ascribed to the Byzantine quadrangular formation. Apart from this battle order, the Saracens also used standard linear formations (noted by Nikephoros, *Praecepta*, 1§13, for example), which Leo incorrectly attributes to their experience of Roman tactical order, although linear battle lines were the universal norm. Leo does not here differentiate between infantry and cavalry, so it is not entirely clear whether the formations he is describing apply to both arms, or only one.

557-66 §§115-16 An expansion on the Saracens' battle order, emphasizing the stability and solidity of their (square) formation—again very similar, if very much briefer, to the accounts in the later Byzantine texts of the "new" Byzantine infantry formation. See *Sylloge tacticorum*, loc. cit., and Nikephoros, *Praecepta*, 1§§5–13. Leo's comment on naval warfare reflects his interest in this aspect of the confrontation with the empire's Islamic enemies and prefigures Const. 19 as well as later comments in this Constitution (e.g., §131); see esp. Const. 19§§15–16. His final sentence is perhaps an indication of the reports of military men with whom he had conversed, and certainly suggests the respect and wariness with which the Byzantines had come to treat their foes on the eastern frontier.

567-72 §117 Cf. Maurice, *Strategikon*, 11.1.6–7 on the Persians. It is not entirely clear what Leo means here: presumably a reflection of reports that Saracen commanders planned their battles, as did Byzantine commanders (and as Leo himself exhorts them to do: Const. 3§1; 5 etc.; 12§75; 13§9–10), but something which (according to Roman and Byzantine prejudices) other "barbarian" commanders did not. The evidence is very limited, but suggests that this was the norm for Byzantine commanders. At the battle of Versinikia in 813 an originally agreed plan was later ignored by some officers because of the emperor's delay in launching his own attack; see *Theophanis Chronographia*, 500 (trans. Mango and Scott, 684); *Scriptor incertus* (ed. Iadevaia), 42; Theophanes continuatus, 14–15; at the battle of Anzen in 838 a war council on the eve of battle was certainly held to plan tactics: Genesios, 3.14 (48; trans. Kaldellis, 63) and Theophanes continuatus, 127; and the same is true of other ninth-century battles, such as Bathys Ryax in 872/73; see on Const. 14.234–39 above. At a somewhat later date, Romanos IV is reported to have assembled his generals and gone over the battle plans before the confrontation at Mantzikert in 1071 (Attaleiates, *Historia*, 158.20–22; trans. Kaldellis and Krallis, 289); tactical plans were discussed before a battle with Hungarian forces in a campaign of 1167 (Choniates, *Historia*, 152); and the emperor Michael IV's expedition against the Bulgarian rebels in 1040 is described as following the proper rules of strategy, advancing in the correct order, pitching camp according to the regulation method, with the emperor calling a council of war before battle to plan tactics: Psellos, *Chronographia*, 4.44 (trans. Sewter, 112). The point is emphasized in Attaleiates' account of the commander of Byzantine forces campaigning against the Pechenegs in 1049; he both failed to prepare his troops or consult his officers before battle and suffered a disastrous defeat: Attaleiates, *Historia*, 32.13–15 (trans. Kaldellis and Krallis, 57); Kekaumenos, *Strategikon* (Spadaro), 2:64 (96.26–97.15) (Litavrin, 162.12–28). Skylitzes, in contrast (*Synopsis historiarum*, 468.18ff.; trans. Wortley, 437), describes how the commander (Constantine Arianites, *magistros* and *rhaiktor*: on whom see in detail Litavrin, *Soveti i rasskazi*, 395–400) set up a well-defended camp, but refused to accept the advice of his two subordinate commanders, Katakalon Kekaumenos and the Frank Hervé, which was given in a planning meeting to discuss the best approach to the Pecheneg forces. Again, in the following year, the *hetaireiarches* Constantine was sent against the Pechenegs at Adrianople, and again there took place a discussion about tactics in the commander's tent before battle was joined (Skylitzes, *Synopsis historiarum*, 470.73–76; trans. Wortley, 438). Leo remarks that he has heard about this aspect of the Islamic commanders' leadership; but nothing of the emperor Basil's own accounts of his wars with the Saracens or Paulicians is known, and the *Vita Basilii*, while offering some vivid if partial accounts of Basil's own battles, is no more informative.

573–76   Λυπεῖ δε ... εὑρεθήσεται   Leo adapts this directly from Maurice, *Strategikon*, 11.1.41–42, although as noted already the damp and rain is indeed reported to have adversely affected the Turkish archery at Anzen in 838; see on ll. 137–209 above. While this is still a stereotype, therefore (see Dagron, "'Ceux d'en face'," 221ff.), it is not entirely inaccurate (the advice regarding wet or damp weather and slackening of bowstrings is ancient; see, e.g., Frontinus, *Stratagems*, 4.7.30 and Dio, *Roman History*, 40.15.4), the more so as the Byzantines clearly relied far less heavily on archery—in spite of Leo's and, earlier, Maurice's insistence upon expanding Roman competence and use of this arm.

576–78   πολλάκις γὰρ . . . διεφθάρησαν   Leo pursues the stereotype, even if it is quite apparent that Arab raids could occur at any season (see below) and that wintering raids in the seventh century in particular had caused serious problems for the Romans—although the Arabs had never been able to establish a permanent foothold beyond the Taurus and Anti-Taurus. See Kaegi, "Confronting Islam," 375 with sources and, for the potential effects of such raids, Haldon, "Cappadocia" and England et al., "Historical Landscape Change." The *Miracula S. Theodori tironis* records a wintering raid, probably in the early 660s, for example: ed. Delehaye, mir. 5; such raids were clearly unusual. In February of the year 811 a winter raid caught the Byzantine commander (the future emperor Leo V) and his troops by surprise. Winter campaigning was universally regarded as difficult and dangerous, and both Byzantine and Arab sources comment on this. Michael Psellos makes particular mention of the fact that Basil II was often successful in war because he launched attacks in the winter, when the enemy was least expecting it. In 625/26 the emperor Heraclius opened a campaign in September, as winter set in, in the mountainous region of northwest Iran, which the historians themselves note was unusual and took the Persian forces by surprise: Psellos, *Chronographia*, 1.32 (trans. Sewter, 25); for Heraclius: *Theophanis Chronographia*, 317–19 (trans. Mango and Scott, 448–49). It is significant that when Amorion was taken in 669, the successful Byzantine counterattack was launched during the winter of the same year, when the Arabs were least expecting it and when resupplying their troops and relieving them was most difficult: Lilie, *Byzantinische Reaktion*, 72–74. During the period ca. 663 to 678 wintering raids were an annual event; see Lilie, *Byzantinische Reaktion*, 69–88; thereafter they continued, but far less frequently (ibid., 95–96), and in the period 720 and afterward very few are registered in the sources; see Haldon and Kennedy, "Arab-Byzantine Frontier," 113 and n. 154. But while we possess little specific information, it is not unlikely, either, that Byzantine troops had on occasion been able to defeat raiding parties in the winter season.

579–83 §119 On the Arab side of the frontier, its defenses, and the main military centers from which attacks into Byzantine territory were launched, see Haldon and Kennedy, "Arab-Byzantine Frontier," 106–11; Bosworth, "City of Tarsus"; and below on l. 618. Raids into imperial territory came from several possible directions in N. Syria (Aleppo, Antioch), E. Anatolia (Melitene/Malatya, Germanikeia/Maraş) and Cilicia (Tarsos, Adana), and it seems that it was chiefly for the large summer raids in September that the volunteers from other parts of the Islamic world—Egypt, Palestine, and further afield—would travel to these launching-points. The Tarsiots were a particular thorn in the Byzantine side because, as Leo notes further on, they combined land with maritime raids and presented a double threat. As the editor points out (Dennis, *Taktika*, 481 n. 26) the strategy described by Leo here was still operating in the middle of the tenth century, as described in *Skirmishing*, 7.1, and it is clear that the author of the latter treatise took his inspiration for the strategy for dealing with enemy raids from the *Taktika*. According to the Arab geographers, winter raids were the most difficult to undertake, and the troops in question could stay in enemy territory for only a very limited period. The geographer Kudāma ibn Ja'far notes that a winter raid (*sātiyah*: in February/March) into Byzantine territory should not take up more than twenty days there and back, since that is the maximum time for which they can carry supplies with them (and the pack animals whose loads had been consumed could then be used to carry the booty). This was in strong contrast to the spring raid, which lasted about thirty days, and the summer raid (*sā'ifah*), lasting up to sixty days. In the second and third cases, fodder and grain for the animals and provisions for the soldiers and camp attendants were extracted from the areas through which the army marched: Kudāma ibn Ja'far, 199–200.

There were several major routes of access from the Cilician and north Syrian regions into Asia Minor. North of Tarsos, in the gorge of the Yeşiloluk, the defile of the Cilician Gates led through the Taurus to Podandos and either westward to Loulon and Herakleia, and eventually turning off to the north, Ikonion; or northward, either directly or via Tyana, to Caesarea. A second route led northward from Germanikeia (Maraş) to Koukousos and then westward via the Kuru Çay pass to Kaisareia; other routes led from Adata, to the northeast of Germanikeia, across the Anti-Taurus past Zapetra to Melitene; from Melitene via a series of defiles and passes either to Kaisareia via the pass of Gödilli Dağ (the Byzantine *kleisoura*, or frontier pass, of Lykandos), or to Sebasteia via the valley of the Kuru Çay. In addition, there were a number of minor routes through mountain passes which were covered by Arab and Byzantine forts, and were the scene of frequent clashes. That from Mopsouestia up to Anazarba and through the defile to Sision, thence north to Kaisareia or, further to the east, the routes which led from Melitene eastward to Arsamosata (Šimšat) and on to Chliat/Ahlat on Lake Van, and northward; see Ramsay, *Historical Geography*, 270–89 (passes over the Anti-Taurus) and 349–46 (over the Taurus); Anderson, "Road System." Arab sources describe in some detail the two major routes that an invader could follow across the Taurus and Anti-Taurus mountains. From Kaisareia, several alternative roads were taken—northwest up to Ankara, north to Basilika Therma, and on up to Tabion (at the crossroads of this north-south route and the major west-east military road), and thence to the road running between Gaggra and Amaseia; or northeast up to Sebasteia, then west and north to Dazimon and Amaseia. Alternatively, a series of easterly routes leads from Sebasteia across to either Kamacha, or to Koloneia and Satala; see, e.g., Ibn Khurradādhbīh, 73–75, 82–83, 85–86; with Kudāma ibn Jaʿfar, 193–96 for the details of the Islamic frontier regions and their defenses.

584–87 Χρὴ οὖν ... διαφθεροῦσιν αὐτούς Leo describes the classic shadowing and hit-and-run strategy presented in detail in the later treatise on *Skirmishing*. Although it is not clear when this strategy first evolved (and it is not at all certain that it had been in place since the late seventh century), it was certainly practiced by the late eighth century, at least on occasion. In the year 779, Theophanes records that the emperor Leo IV ordered each *strategos* to select 3,000 troops to harass an invading Arab army which had reached Dorylaion (rather than attempt a direct confrontation), and that this was specifically to prevent the enemy sending out pillaging raids. He also reports the emperor's order to destroy pasture and other supplies, with the result that the Arab force had to retire after fifteen days: *Theophanis Chronographia*, 452.6–12 (trans. Mango and Scott, 624); and Lilie, *Byzantinische Reaktion*, 171–72. Theophanes' account suggests that this strategy was not usual, however, and it may be that it was developed only from this time, perhaps in conjunction with the creation of new frontier districts, the *kleisourai*. See on 9.122 above. The general Andrew the Scythian (*PmbZ* 20351) was apparently successful in applying such a strategy: *Vita Basilii imperatoris*, 50.4–12 (178–80) and Hild and Restle, *TIB* 2:261–62 with Arabic sources, but the strategy was not always successful. In 770/71 the cavalry forces of Anatolikon, Thrakesion, and Boukellarion occupied the pass through which a substantial raiding party would have to return, but the raiders were able to force their way back through the pass, with most of their booty intact: *Theophanis Chronographia*, 445 (trans. Mango and Scott, 615).

Counterattacking into enemy territory at the same time as an enemy attack into Roman territory had become a standard strategy by the early tenth century, although in *Skirmishing*, 1.2–7 and 20 (Dennis, 146.12–45 and 218–22) the author implies that it was practiced before this period. See Const. 17§§64–65 and cf. Syrianos, *Strategy*, 6.14–24; see also below on ll. 669–74 and Dagron, "Combattant byzantin."

587–88   ἢ ὅτ' ἄν ... ἐξηρτισμένοι   Leo presumably speaks here from some knowledge of the major victories thus won over Arab armies in the late ninth century. A substantial operation had been conducted by the commander Petronas (*PmbZ* 5929) in 863, when the forces of the emir of Malatya (Melitene) were trapped by a combined Byzantine army of some thirteen different corps, which had marched by separate routes to meet near the point at which the action took place (Theophanes continuatus, 179–83; cf. Genesios, 4.15 [68–69; trans. Kaldellis, 85–86]; Skylitzes, *Synopsis historiarum*, 99–101 [trans. Wortley, 100–101]). The defeat of the Paulicians at Bathys Ryax in 872/73, for example, involved the convergence of the forces of Charsianon and Armeniakon, some of which had shadowed the enemy for several days: *Vita Basilii imperatoris*, 41.24–31 (150); Genesios, 4.36 (86–88; trans. Kaldellis, 108–9); and Skylitzes, *Synopsis historiarum*, 138–40 (trans. Wortley, 136–37). In both cases the enemy forces were almost annihilated. In 781/82, when Hārūn al-Rashīd led a major incursion into imperial territory, several Byzantine divisions combined to meet his columns. One Arab force was left to besiege Nakoleia on the borders of the Opsikion and Anatolikon *themata*, a second diverged into the Thrakesion region, and the main force under Hārūn proceeded to Chrysoupolis on the Bosphorus. The first detachment was surprised and routed by a Byzantine division, while the attack on the Thrakesion *thema* was turned back after an apparently drawn battle. The *tagmata* and other unnamed corps (but including the Boukellarion division) were able to encircle the main raiding force in the Sangarios valley on their return march; see *Theophanis Chronographia*, 456 (trans. Mango and Scott, 629); Brooks, "Byzantines and Arabs," 737–39; and Tritle, "Tatzates' Flight." The strategy did not always work. In 787/88 Roman forces marched separately to trap and destroy an Arab force at Podandos, but were themselves ambushed and defeated: *Theophanis Chronographia*, 463 (trans. Mango and Scott, 637).

589–91   §121   The extreme caution urged by the treatises was not always followed in practice, but it was certainly justified, as the example of the failed ambush in 770/71 above (see on ll. 584–87), and many others, illustrate. This attitude is especially clear in this section of the *Taktika*, and it may be indicative of the uncertain morale or confidence of Byzantine commanders and soldiers—something which Leo attempts to address directly, in the case of the "Saracens," in this constitution; see below.

592   στρατευόμενοι   While Leo uses the word quite correctly in a standard sense, the terms στρατευόμενος and στρατιώτης also came to bear a technical meaning, the former the individual enrolled into active military service, the latter the owner of property registered as "military land" subject to an obligation to furnish a soldier or a contribution toward furnishing a soldier. See in detail Haldon, *Recruitment and Conscription*, 41–62.

593   πανοικεί   "En masse" or "altogether" rather than "with his whole household," which would make little sense in the context (cf. Dagron,

"Byzance et le modèle islamique," 221 n. 13 for a parallel usage in the
Rhetorica militaris, 12 [51.1]).

592–98 §122   See the commentary to 18.621–22, below. This passage has been
the subject of a detailed analysis by Dagron ("Byzance et le modèle
islamique," 221–22; and Dagron and Mihǎescu, *Traité sur la guérilla*, 145–49; see also Stouraitis, "Methodologische Überlegungen,"
282–85), who has noted Leo's awareness in §122 of the very different
basis for recruiting soldiers and for stimulating the zeal of the Muslim
soldiers from that in the Byzantine world, perhaps reflecting also an
understanding of the ways in which religious foundations—*waqf*s—
functioned to support the activities of fighting volunteers; see Cahen,
"Réflexions"; Stouraitis, "Jihād and Crusade"; and Pitsakis, "Guerre et
paix." The term μισθός is taken by Dagron, certainly correctly, to mean
a heavenly reward rather than a monetary or material one (cf. Const.
12§57.410 and commentary), again an equivalent of contemporary
Arabic terms for spiritual recompense, *ajr* (see Bonner, *Aristocratic
Violence*, 41–42, 122–25), and Leo contrasts these characteristics of
Islamic military recruitment favorably with the Byzantine situation,
wishing in the next paragraph (§123) to have a similar practice in his
own realm. Leo is impressed by the voluntary nature and enthusiasm
for warfare demonstrated by the Muslim population, noting that there
is no register of soldiers, and underlining the "patriotism" of those
who feel that dying in the cause of their nation, or providing equipment and weapons for the soldiers, is reward in itself. Dagron noted
that this passage also alludes obliquely to the notion of "holy war"—
jihād—which Leo clearly saw as a major difference between Islam and
Christianity, but an idea which, in a suitable Christian guise, he wanted
Roman soldiers to emulate—not because they would go to paradise if
they died in battle, but rather, as he expresses it at ll. 621–22, for the salvation of their souls: ὑπὲρ τῆς ψυχικῆς ἡμῶν σωτηρίας. But this does
not mean that Leo, or the Byzantines more widely, evolved a notion
of "holy war." Because the Byzantines fought under the symbol of the
Cross, and because they saw themselves as soldiers of Christ fighting to
preserve God's kingdom on earth, no theory or doctrine of "holy war"
was necessary, since warfare was almost by definition of a religious
character: the eastern Roman empire was the sole Orthodox polity

fighting to preserve and extend the Christian faith. For Leo's views on this as expressed in the *Taktika*, see Dagron, "Légitimer la guerre"; Stouraitis, "Jihād and Crusade," 19–23; idem, "Methodologische Überlegungen," 275–90; Haldon, *Warfare, State and Society*, 27–33; and Michailidis-Nouaros, "Δίκαιος πόλεμος," 411–34.

Jihād was itself an idea which crystallized only slowly and emerges in its established form in texts only of the late eighth century, promoted in particular as one aspect of the ideology of the frontier and as an element of good Islamic rulership—as exemplified by leaders such as the caliph Hārūn al-Rashīd; see esp. Bonner, "Some Observations," with sources and literature; idem, *Jihad in Islamic History*; and Mottahedeh and al-Sayyid, "Idea of the Jihād."

In treating this aspect of Islamic warfare, Leo idealizes the elements he wants to stress—the reality was not quite as he describes (see Kennedy, "Financing" and idem, *Prophet*, 173–209, 265–76). He also tacitly ignores the "regular" element in the Islamic armies' composition and resourcing. Kudāma ibn Jaʿfar, 193–95, writing in the first half of the tenth century, notes that the revenues of the provinces which face Cappadocia and Anatolikon amount to 100,000 dinars and are spent on public defensive works, border guard units, spies, couriers, guards for the passes, and defiles, as well as on fortresses. The cost of the summer and winter expeditions is usually some 200,000 dinars although it can be as much as 300,000. For the provinces facing Charsianon and (parts of) Chaldia the revenues amount to 70,000 dinars, of which 40,000 is spent on the maintenance of the fortresses; in order to pay the soldiers and irregulars for the frontier garrisons, the government at Baghdad has to spend an additional 120,000 or 170,000 dinars annually, not including the additional expenses required for the additional troops who participate in the annual raids. As for the provinces facing the Armeniakon and remaining parts of Chaldia, the annual revenues amount to 1,300,000 dirhams (between 65,000 and 108,000 dinars), whereas the maintenance of the fortifications, the garrisons and related costs amounts to 3,000,000 dirhams (150,000–250,000 dinars), so that the state has to provide 1,700,000 dirhams (85,000–142,000 dinars) annually (the dinar was a gold coin, worth between 12 and 20 silver dirhams, depending on the weight and purity of the latter); for discussion of the figures given by Kudāma, see Vasiliev, *Byzance et les Arabes*, 1:96–97; also see Kennedy, *Prophet*, 268–70. Whatever their reliability as absolute figures, they give an indication of the proportional costs as understood by a contemporary.

The provinces and the government at Baghdad were thus responsible for the expensive fortifications and garrison forces, and this was clearly a very considerable burden. Leo is quite correct that substantial numbers of volunteers, supported by their kin and through religious foundations—*waqf*s—were a major component of the summer raids; but it should be recalled that the main burden of military expenditures fell on the provinces, drawn from both their tax revenues and from religious foundations set up for this purpose, and on the central government income, which subsidized the core units of *ghulam* or military servitors/slaves in fortress cities such as Tarsos. The volunteers—*muttawi'ah*—were housed in dwellings maintained by their various home provinces and supported by *waqf*s, and although equipped through voluntary contributions, while on campaign they were in fact supported by central funds. See Haldon and Kennedy, "Arab-Byzantine Frontier," 110–11.

599–600 Ῥωμαίους δὲ . . . συνεκστρατεύειν   A slightly better translation might be: "The Romans must not only put these principles into practice but, resolute in purpose, both the soldiers and those not enrolled for military service must campaign together. . . ." The verb συνεκστρατεύειν has a metaphorical value regarding the noncombatants. See *Rhetorica militaris*, 10.1 and 26.1, where the need for the Romans to mobilize to defend their faith and to fight together, united, against a common foe is given expression. Leo takes up the same ideas in this passage, expressing the wish that ordinary Romans who may have no direct association with the military should contribute to the war against Islam in the same way as Muslims do to their war effort, with contributions in money and materials, a point he makes earlier (see above on Const. 4§1.3ff.).

As far as the evidence can take us (see chapter 3 and comm. to Const. 4, above), the established arrangement by this time was one whereby territorially recruited units (*banda*) were recruited from specified districts (*topoteresiai*). The military service of the less well-off (thus by no means all) of those obligated to military service and entered on the registers had been, at least since the time of Nikephoros I, supported ἀλληλεγγύως, that is to say, by other tax-paying members of the fiscal community from which the soldiers were drawn, in respect of their equipment and other elements of their support; see Brubaker and Haldon, *Byzantium in the Iconoclast Era*, 744–52. In Const. 4 Leo urges the commander of a provincial army or *thema* to select only those fit for service and well able to support that service, whose households can be managed for them by their families or others during their absence, and who are liable only for the major public taxes (thus no extraordinary levies and impositions). Until later in the tenth century the obligation to serve in the army of a province remained personal and, apparently, hereditary for those whose familes were inscribed in the military registers, but not directly associated with their property—this was a development that appears during the reign of Constantine VII, and transforms the system of recruitment and financing of the provincial armies; see Haldon, "Military Service" and the commentary to Const. 4§1.3–14 above. There were also obligations on private bodies, such as monasteries and imperial official, to supply pack animals for imperial (but not provincial) military undertakings, and these appear to have been long standing; see *CPTT* C.67–116 and comm., 187–95. But it appears not to have been the practice to impose on nonmilitary households, nor for those not involved in the military effort to volunteer their support. Leo's admonishments in this respect appear to have had little impact, for the later evidence shows that no such system of voluntary support ever materialized. Leo also ignores the fact that the resident provincial populations in fact already bore a substantial fiscal burden in supporting the armies, since an elaborate system for raising and moving military supplies of all kinds—food, weapons, and so forth—already operated, and all those households which were not exempt by virtue of their military obligations of their service for the *dromos* had to contribute in many different ways. Naturally this burden tended to fall most heavily on the provinces where the military was most active, but there were few parts of the empire entirely free of the military; see Haldon, *Warfare, State and Society*, 139–48 for a survey of the evidence with further literature. The final sentence of this paragraph also suggests that the system of communal solidarity established by Nikephoros I was not working particularly effectively, since in theory at least, as noted already, such a system already operated, even if the ninth-century evidence strongly suggests that it was flawed; see Haldon, "Military Service," 26–28. Basil I, Leo's father, is himself reported to have refused to address the problems which had evolved over the years in the fiscal assessment of the empire and to carry out a reassessment of tax burdens, on the grounds that suitable honest and trustworthy officials were not available to fulfill this task, and that it might bring hardship. Whether this may have impacted also on the support levied for soldiers on active service, no text makes this explicit; see *Vita Basilii imperatoris*, 99 (320–24).

There was no system of voluntary support for soldiers, although on occasion, and in an emergency, the church and monasteries sometimes "volunteered" to subsidize the government's need for cash. But the threat of direct imperial impositions or appropriation always lay close to the surface. Shortly after Leo's reign, during the crisis of war with the Bulgarians in 920, the patriarch Nicholas I wrote to a number of senior churchmen, encouraging them to provide funds for the war effort (because if they did not, imperial officials would come and appropriate the necessary cash regardless; see Nicholas I, *Letters*, 92.10–26, 94.31–40); yet a few years earlier Nicholas had protested energetically against the levies demanded of the church, and also against the conscription of monks into military service (*Letters*, 150, 164, 183). In this general context, therefore, it is not difficult to see why Leo hoped to induce a change of heart in the Byzantine population at large. On Byzantine propaganda aimed at soldiers, see also Koutrakou, *Propagande*, 354–86.

600–602 κατὰ . . . τῶν ἐθνῶν   Leo reinforces the message (cf. ll. 620–25) that this fighting is to defend Christianity against Islamic blasphemy (recalling in the expression κατὰ τῶν ἐθνῶν the struggle of the children of Israel, the biblical forebears of the Byzantines as God's Chosen People). Compare with *Rhetorica militaris*, 10.1, 26.1.

603-6   Leo's list of the ways in which he hopes Romans will support their soldiers, both materially—equipment, weapons, horses, etc.—and also spiritually. The latter was not lacking, as the letters of the patriarch Nicholas I referred to above testify, since the church regularly prayed for success in battle against those who threatened the empire, and rejoiced when that success was forthcoming; see Nicholas I, *Letters*, 44, for example; and see ep.§62.266–72, where Leo notes that one task of the clergy accompanying armies is constantly to pray for them and their success. The comment on looking after the households of those who go to war reflects the relative ineffectiveness of the established system for communal solidarity, at least beyond the need to redistribute or support the fiscal burdens of the soldiers. Indeed, the evidence from hagiography and letters suggests that there was little or no support for the families of those liable to military obligations who had died; see a letter from the patriarch Nicholas I, for example, asking for release from military obligations on behalf of a widow who cannot afford to equip her son: Darrouzès, *Épistoliers*, 2:50.130–31; and a similar case in the *Vita Euthymii Iunioris*, 172. The author of the treatise on guerilla warfare along the eastern frontier remarks that soldiers, in spite of their fiscal privileges and favored status, might still be oppressed and reduced to penury by state officials (referring to all soldiers, not just those in possession of "military holdings"). See *Skirmishing*, 20.6 (ed. Dennis, 216–17).

   As for equipment, those registered for military obligations were supposed to provide for themselves, and if unable to do so, were to be supported by their fiscal community. Under Nikephoros I this support amounted to a total of 18 ½ nomismata; no source refers thereafter to the cost of equipping soldier, although as is well known the late tenth-century legislation and other evidence mentions property valued at up to three or four pounds of gold as the basic requirement to support military service; see Haldon, "Military Service," 25–26 and idem, *Recruitment and Conscription*, 41–65. The difficulties of maintaining a well-equipped army through such means became increasingly apparent across the ninth and tenth centuries, however, as emperors came to rely more and more on the recruitment of professional "mercenary"—indigenous alongside foreign—soldiers and units, a shift which placed ever greater pressure on the administration to fiscalize and commute military service so that the resources thus released could be used to purchase or maintain and equip such troops. Leo's emphasis on the *strategos* selecting only those who could afford their equipment and mounts etc. from the military register of the *thema* is, as noted already, an indication of the situation; see on Const. 4§1.5–14 above.

607–11   §124 Leo's somewhat pious hope. What he was perhaps unable to perceive was that the real issue faced by the Byzantine military was leadership, on the one hand, and discipline on the other—a survey of conflicts across the period from the eighth to eleventh centuries would demonstrate that the east Roman army, when properly led by competent and experienced commanders in whom the soldiers had confidence, was well able to defeat its various enemies. See the discussion in Haldon, *Warfare, State and Society*, 228–33; and idem, "Approaches." But discipline was a major issue, and since the late seventh century—by the latest—the traditional system of training soldiers and the unit discipline associated with that training, along with the selection and education of noncommissioned officers, had been completely transformed. In spite of Byzantine consciousness and insistence upon their Roman past and their Roman military traditions—order, cohesion, etc.—this was a medieval army, not a Roman one. Leadership, discipline, and morale were absolutely crucial ingredients in success on the battlefield. Occasional efforts to establish such discipline were often successful, but there seems to have been very little institutional continuity built into the system, so that each generation of commanders had to start over, so to speak.

612   καὶ μάλιστα τόξοις καὶ βέλεσι   As noted in the introductory comments to Const. 18 (pp. 332–33 above), this paragraph may have concluded an earlier version of the section on the Saracens. The emphasis on archery returns to an earlier theme, and in this case appears to have borne fruit, although probably a "natural" tactical response of Byzantine commanders from Leo's time and a reflection of the sources of recruitment of the armies, rather than a direct result of his pleas: by the middle and later years of the tenth century archery had clearly become a more significant element in Byzantine armies,

to the extent that the imperial government intervened to ensure that adequate supplies were available. Already for the seaborne expedition of Himerios to the Syrian coast in 910–11 the order went out to several provincial commanders to commission and deliver several tens of thousands of arrows; see Haldon, "Chapters II, 44 and 45," 209.96–101 (= *De cerimoniis*, 657.12–13, 17–18). In the *Praecepta* of Nikephoros, 1§14.138, the field army is accompanied by pack animals carrying "imperial arrows," no doubt those commissioned by direct imperial order in the same way as for the expedition of Himerios; see also McGeer, *Sowing the Dragon's Teeth*, 65.

616–19   §126 Leo returns to his main theme, that of motivation. Here, it is the desire of the large number of volunteers for the summer raids to win booty (see l. 594 above; and §130.647–48) that Leo highlights, thus contrasting the numbers of volunteers for the fighting on the Islamic side with the reluctance of Romans to put themselves forward. This may well have been a realistic appraisal, of course, although in Islamic theory a warrior who fought merely for personal gain or sought maryrdom for prestige would be eternally damned; see Bonner, *Aristocratic Violence*, 24–42, 130ff., and more generally idem, *Jihad*. Booty is, of course, an important factor for Roman soldiers too—the topic appears on several occasions in various constitutions of the *Taktika*, mostly where Leo follows Maurice in advising commanders to beware of letting troops run after booty when they should remain in order and are still facing or pursuing the enemy: 12§101; 13§15; 16§4, 23–25; 17§§25, 37. It does appear also as an inducement to the troops, but it is never mentioned as a motive for recruitment.

618   "Inner" Syria and Palestine are the sources of volunteers for the summer raids—Leo has in mind the Syrian hinterland from which Cilician cities such as Tarsos and Adana drew their volunteers. Arabic sources designated the frontier zone until the eighth century as *al-dawāhī*, "the outer lands," but after the resettlement and refortification of the major centers at Mopsouestia (al-Massīsah), Adana, and Tarsos the whole front-line zone is described as *al-thughūr* ("the passages"), with the towns lying behind it referred to as the *'awāsim* ("the defenders"). These bands of territory were then further divided laterally into two broad regions, that of Syria and, further east, the more mountainous region of al-Jazīrah; see Wellhausen, "Kämpfe der Araber" and Haldon and Kennedy, "Arab-Byzantine Frontier," 106–9 for sources and a more detailed description of these regions and their major cities and fortresses. The division between these different elements, especially between *thughūr* and *'awāsim*, was far more fluid in reality than the Arab geographers' accounts would suggest; see Bonner, "Naming of the Frontier" and idem, *Aristocratic Violence.*

620–25 §127 See 14§31.206–8 and compare *Rhetorica militaris*, 9–13, 37.7 (6), for example, which emphasize that the Christian Romans fight their enemies on behalf of their faith, their fellow Christians, and their country. Here again Leo's debt to this text seems clear. This is an important paragraph, summing up Leo's hopes as outlined in the preceding passages, and discussed in detail by Dagron, "Byzance et le modèle islamique," 224–32; see also on ll. 508–14 above and discussion in the introduction. Closely related ideas were current in the western church too. Pope Leo IV (*PmbZ* 4240), for example, wrote in 853, "any believer who dies fighting in this war . . . will not be denied his place in heaven. The Almighty knows that, whoever of you died, died for the sake of faith, salvation of the soul, and the defense of the fatherland of the Christians" (MGH Ep 3: *ep.* 28, p. 601). Similarly, at some time ca. 877–79, Pope John VIII (872–82; *PmbZ* 23470) wrote to the bishops of west Francia in response to their question as to whether those "who have recently died in war, fighting in defense of the church of God and for the preservation of the Christian religion and of the state, or those who may in the future fall in the same cause, may obtain indulgence for their sins." The pope was quite clear: those who, out of love for the Christian religion, "die in battle fighting bravely against pagans or unbelievers, will receive eternal life" (MGH Ep 5: *ep.* 150, pp. 126–27). The notion of both absolution for sins and of a heavenly reward is thus given clear expression. To what extent such ideas (in the west) can be traced further back, to ideas generated during the wars against the Saxons in the late eighth century, and the possible knowledge in the west of Islamic notions of holy war brought from Spain by Christian exiles or travellers, remains debated; see Hen, "Charlemagne's Jihad" and for background, idem, *Roman Barbarians*, esp. 172–76; and more generally on "just" and "holy" war, see Stouraitis, "Just War."

It is unlikely that Leo knew of the papal position, and the fact that the Frankish bishops needed the pontiff's guidance may itself indicate the ambivalence in Christian thinking on this issue (and note *Rhetorica militaris*, 36.9, which encourages Christians to take up arms against enemies of the faith in spite of Gospel discouragement). Leo takes up the message in the *Rhetorica militaris* by contrasting Christian views and attitudes with those of Islam. And even if both here and at ll. 612–15 the impression is one of a pious hope on Leo's part, rather than an assumption that the idea, and the practice which should accompany it, would be taken up, his repeated emphasis on and promotion of the religious motivation of the Roman war effort, fighting in particular for the salvation of souls, kin, and Chistian brothers in general (see esp. on Const. 12.409–20 and commentary above), is undeniable. On the religious motif running through the *Taktika* more broadly, see chapter 1; and text and commentary above, at e.g., Consts. 11§19.107; 12§55.396–98; §92.637, 641; 13§1.3–5; 14§1.3–7, dealing with the chanting of the *trisagion*, the blessing of standards, prayers for the army by the accompanying clergy, shouting out the *niketerion*, or victory cry of the cross, etc.

Following the *Rhetorica militaris*, therefore, Leo gives a renewed and heightened emphasis to the established idea that soldiers who fell in the fight for the true faith would receive the appropriate spiritual reward, and that the Christian Romans are the Chosen People, to be distinguished from all others and, in this case, from Islam as the particular enemy of the Christian faith. This is a set of ideas that is also closely linked to the cults of soldier saints and their evolutuion from the seventh century onward in particular. See esp. White, *Military Saints*, 40–93. For the evolution of the notion and general background see Munitiz, "War and Peace"; Dagron, "Légitimer la guerre" with idem, "Byzance entre le djihad et la croisade"; Laiou, "Just War"; eadem, "On Just War"; further discussion in Haldon, "Fighting for Peace," with sources and older literature; Dennis, "Defenders"; Cheynet, "La guerre sainte"; Pitsakis, "Guerre et paix"; and esp. Stouraitis, *Krieg und Frieden*, 344–62; idem, "Methodologische Überlegungen", and for the continuing debate: Kolia-Dermitzaki, "Holy War."

626–85   §§128–34   Concerned specifically with the two major sections of the frontier, Cilicia/Syria and Syria/Mesopotamia. The strategy for dealing with enemy raiders in the passes of the Taurus is described very briefly, but prefigures the much more detailed account in the later treatise *Skirmishing*, esp. 3–6, 11, 17; and 20–23, and Leo no doubt has in mind recent examples of the successful ambushing or cutting off of enemy forces in the mountains of the border region; see on ll. 584–87 above.

626–34   §128   Here Leo turns to pragmatic issues of combat along the frontier.

635–46   §129   This appears to be genuinely new information on "Saracen" tactics, rather than drawn from the *Strategikon* and revised or emended to suit Leo's purpose, and is taken up again in §§136 and afterward. The information in the first section—up to §134—is at a fairly general level; thereafter it becomes detailed enough to suggest that Leo may well have been working from a written memorandum on enemy tactics, perhaps compiled by one of the commanders who had experience of fighting the Arabs. Essentially Leo's advice (or that of his informant) is that an initial Roman attack against the enemy formation, which is not to be broken easily, should be employed for long enough to draw out the enemy against a solid Roman defensive position, at which point Roman spearmen will present a solid wall from behind which archers will so sufficiently damage and demoralize the attacking troops that they will be easily broken by a second Roman attack. Leo goes into no further detail of the formation he describes here. How deep the line should be is specified only later, when files of ten men are recommended (see §143.770).

639–46   Arab warriors' pride in their horses and horsemanship was proverbial, and Leo—and the Byzantine soldiers who fought with them—were fully aware of this. At some point after Leo compiled the *Taktika* the term *pharia* (or variants such as φάρης, φάρας, φαρίον, the confusion in accent reflecting the fact that it was a loanword), to refer to a high-bred warhorse of Arab type, becomes common in medieval Greek texts; cf. Theophanes continuatus, p. 480, 4.7.9 and Du Cange, *Glossarium Graecitatis*, s.v., 1665–66. It survives in modern dialects as a synonym for ἄλογο; cf. Cypriot τὸ φαρίν (Giagkoulis, *Θησαυρός Κυπριακής Διαλέκτου*, s.v., 386). Warhorses were officially recognized as part of the booty from a successful battle, although there were several breeds in use. The term *faras*, which Leo renders as φάρια, was used only of pure-blood Arabians (with terms such as *hejjīn*—half-Arab—for other types), which were the most valued and expensive animals, and were short-backed, light, and fast (see also on Const. 17§§69.397–72.427 above). See Hyland, *Medieval Warhorse*, 40–44, 106–8. The Arab horse seems to have evolved by the seventh century, although the evidence is ambiguous and the issue remains open. Their qualities were appreciated widely, and their introduction after 711 into the Iberian peninsula meant that they were soon in great demand. In 876 the pope, John VIII, wrote to king Alfonso of Galicia requesting that he be sent some of the fine horses known as *Alpharaces* (Joannis Papae VIII, *Epistolae et decreta*, in PL 126:664A–B, *ep.* 19: "aliquantos utiles et optimos Mauriscos cum armis, quos Hispani cavallos Alpharaces vocant" [not in MGH Ep 5]). The importance of the horse in Arab Islamic culture is demonstrated by the large number of treatises devoted to horses and to equitation. By the middle of the ninth century Hisham ibn al-Kalbi and Muhammad ibn al-Arabi, among several, were writing treatises on horses and horsemanship in which the *faras* was treated in detail; see Ibn al-Kalbi, *Livres*. Al-Kalbi's work, *Ansab al-khayl fi al-Jahiliyah wa-al-Islam wa-akhbaruha* (The genealogies and histories of horses in the era before Islam and after the rise of Islam), is usually referred to as *Kitab al-Khayl* (Book of horses). See also F. Viré, "Faras," in *EI* and Viguera and Sobredo, "Hippology." The tactic of using poisoned arrows against horses is mentioned only here (although Leo repeats Maurice's comment about the Slavs' use of poisoned arrows; see on ll. 482–87 above).

647–50   Leo returns to the theme of booty and plunder as a motivating force in Islamic warfare (see on ll. 616–19 above; and compare an earlier stereotype from the Latin tradition on the Saracens: Ammianus Marcellinus, *History*, 14.4.3–6). His apparent lapse into somewhat crude generalizations about the Saracens—they do not know how to farm but live by the sword alone—is in fact not entirely incorrect. The military forces based in the fortress towns of the *thughūr* were maintained primarily through government subsidies, thus not engaged in farming or in any other economically productive activity. At Tarsos in the middle of the tenth century, for example, over 60% of the population were single military men whose livelihood depended entirely on subsidies from *waqf*s, from voluntary contributions, and from plunder—emphasizing Leo's point in respect of the latter, of course—and the core units of the Tarsiot army was made up of slave soldiers maintained by subsidies from Baghdad. See Canard, "Quelques observations," 46–52. The situation at military centers such as Adana or, further east, al-Hadath, was similar: ibid., 46–47. Leo's main purpose here is to belittle the "Saracens" by emphasizing their greed for material as opposed to spiritual gain, thus contrasting their approach to warfare with that of the morally upright Romans, whose wars are justified as a defense of their faith and of Roman territory and people, rather than by a mere base desire for plunder. See the comments of Stouraitis, *Krieg und Frieden*, 327–44 and Dennis, "Defenders of the Christian People," 32.

650-53   ὅθεν ... ὑπὸ τῶν ἰδίων   An optimistic view—perhaps a reflection of Leo's genuine naïveté in respect of Islamic warmaking.

654-60   §131   The Cilicians, and especially the army of Tarsos, were a major thorn in the side of the empire throughout the second half of the ninth century and well into the tenth century; see Bosworth, "City of Tarsus." The Byzantines were able to defeat them and confine them beyond the Taurus for while after a victory in 879, but were heavily defeated in 883. Thereafter into the early 890s regular raids from Tarsos and Adana crossed into imperial territory and carried off vast amounts of movable wealth in people and livestock. In 895 Leo was able to conclude a truce with Tarsos, but war soon broke out again and in 897 Tarsiot forces raided as far north as Euchaita, and in 898 were able to defeat both the Kibyrrhaiotai fleet and attack the province by land. Probably in the 890s (the exact date remains problematic; see Cheynet, "Phocas," 293–95 and above, on Const. 11§21.121–27), Nikephoros Phokas the elder defeated Arab forces near Adana, which stemmed Cilician raids for a short while. But they continued to present a problem throughout the rest of Leo's reign, and the problem was only finally resolved only when Nikephoros II Phokas took the city in 965, having devastated its hinterland over several campaigning seasons. See Ostrogorsky, *History*, 237f., 257; and the detailed discussion and narrative in Vasiliev, *Byzance et les Arabes*, 2.1:79–94, 99–103, 120–26, 137–42 and Pryor and Jeffreys, *Age of the Δρόμων*, 62–63. The threat was especially serious because of the maritime aspect, since, as Leo notes here and as is clear from contemporary or near-contemporary sources, the war fleet of Tarsos could reach deep into Byzantine territorial waters and was a constant threat to the southern and central Aegean region—a signal defeat of the Kibyrrhaiotai fleet in 898 opened up the whole Aegean to a disastrous series of raids, for example; see Vasiliev, *Byzance et les arabes*, 2.1:132–33 with 157–81 on the maritime raids of 902–4, culminating in the capture and sack of Thessalonike. The only advantage the Byzantines had was, as Leo states in the next paragraph, the fact that the Cilicians usually could not campaign by land and sea at the same time. For the situation in 904, see also Skopelite, "Ναυτικές δυνάμεις."

656–57   τῶν λεγομένων κουμβαρίων   Derived either from Arabic *qunbār* (see Pryor and Jeffreys, *Age of the Δρόμων*, 513 n. 61 with literature) or alternatively from the word *qabīr*, pl. *qubār*, "big," usually associated with the word for a ship, *marqab*; see Christides, "Two Parallel Naval Guides," 61–62 with app. A, 95–98; and see Theophanes continuatus, 196.17; Du Cange, *Glossarium Graecitatis*, 688–89; Trapp, *Lexikon*, 4:872; Antoniadis-Bibicou, *Études*, 167–69; Trapp, *Lexikon*, 4:872. Generally understood as a larger and slower vessel than the Byzantine warships. See commentary to Const. 19.428 below. On the Arabic terms for warships and various other vessel, see Christides, "Naval Warfare," 140–41.

661–68   §132   See on 17§32.171–74 above. The *Kibyrraiotai* command was established probably before 732, although the issue is debated; see Grigoriou-Ioannidou, "Κιβυρραιωτῶν" and Oikonomidès, *Listes*, 351 with literature. See also Hild and Hellenkamper, *Kilikien und Isaurien*, 45–59. The region was divided into smaller units during the ninth century; see Ahrweiler, *Mer*, 108–9 and *DOSeals* 2:109–11, 150–51. Leo outlines the basic strategy to be employed, determined by the information received from spies in Tarsos and Adana—the Byzantines took great pains to maximize the intelligence received from informants in Islamic territory.

Muslim sources note the importance the Romans attach to spies and intelligence-gathering. Writing in the middle of the tenth century the geographer Ibn Hawqal noted that the Byzantines acquire information through "agents and spies," including merchants, which allows them to find out the weaknesses in the Caliphate's defenses: Ibn Hawqal Abu'l-Qasīm, *Kitāb Sūrat al-Ard*, 98, 193. See *De cerimoniis*, 657.3–12 (Haldon, "Chapters II, 44 and 45," 209.87–99); for a comparison with preparations for an imperial expedition, see *CPTT* B.3–33. Byzantine monks and holy men as well as merchants played a key role in gathering intelligence; see Dagron and Mihăescu, *Traité sur la guérilla*, 248–50; Koutrakou, "Spies of Towns"; eadem, "Diplomacy and Espionage"; in general Dvornik, *Origins*, 121–87; and J. Preiser-Kapeller, "Geheimdienst." For a maritime aspect, see Christides, "Military Intelligence." Leo's final remark concerning the limited numbers of troops serves as a useful reminder that the flow of volunteers and soldiers to the Cilician front was not as great as either Byzantine or Arab sources often claimed. See above on ll. 616–19 and 647–50.

669–74 §133   A further elaboration on the strategy described in the previous paragraph. For the maritime context, see Pryor and Jeffreys, *Age of the Δρόμων*, 50–71. Leo cites here the campaign led by his father over twenty years earlier (see Honigmann, *Ostgrenze*, 62–63; above, on Const. 9§21.92–96; and Hild, *Straßensystem*, 134–35), although he might also have mentioned in the same context the more recent campaign of the 890s into Cilicia under Nikephoros Phokas the elder (which he does refer to at Const 11§21; Honigmann, *Ostgrenze*, 82–83). He makes no mention of the successful Byzantine naval expedition against Tarsos and Cilicia in 901, at a point when the Tarsiot fleet had been partially destroyed in an internecine conflict. See Vasiliev, *Byzance et les Arabes*, 2.1:141–42. It is difficult to know to what extent this strategy, along with that of counterattacking into Cilicia or N. Syria when the Arab forces invaded the regions of Cappadocia or Anatolikon, was already regularly practiced on the eastern frontier, to which Leo is now offering formal imperial encouragement, and to what extent Leo is suggesting something new—the evidence from both Byzantine and Arabic sources for the period from the reign of Basil I into that of Constantine VII would suggest the former, and the remarks in *Skirmishing*, 20.6 (Dennis, 220.45–49) reinforce this impression: Πρὸς τούτῳ δὲ καὶ οἱ στρατηγήσαντες τῶν Ἀνατολικῶν καὶ Καππαδοκίας ἐν τοῖς ἄνω χρόνοις, τῶν Ταρσιτῶν κατὰ τῆς Ῥωμαϊκῆς χώρας τὴν ἐξέλευσιν ποιουμένων. . . . ("Long ago, furthermore, when the Tarsiots were attacking Roman territory, the commanders of Anatolikon and Kappadokia. . . .") See also on ll. 584–87 above.

670-71 ἅμα καὶ ἑτέροις συστρατήγοις τοῖς ἀρκοῦσι Here Leo refers to the combined forces of several regions, by which several smaller armies could be united into a much more substantial force, either to encircle and defeat an invader; see preceding paragraph; and on 9§21.92–96; 12§§17.119–28.212; 14§36.234–39; 18§120.584–87, above.

675-78 Τοὺς δὲ . . . βασιλείᾳ A reference to the campaign led against Theodosioupolis in 902 by Leo Katakalon Abidelas, the commander who had replaced Nikephoros Phokas the elder as *domestikos* of the *Scholai* in 896 and who had been defeated in the same year by the Bulgars at Bulgarophygon in Thrace; see Vasiliev, *Byzance et les arabes*, 2.1:117. On Leo Katakalon (probably to be identified with the *magistros* Leo Katakylas/Katakoilas), see *CPTT* commentary at 180–81 with sources and literature. It is not clear whether the commander of the 902 expedition is the same, of course. His Arabic epithet, Abidelas—Abd al-ʾAs—suggests that he may come from a collateral branch of the family. On the Katakylas/Katakoilas/Katakalon family (although there is no certainty that they represent a single family or clan), which first appears in the reign of Michael II, see *PmbZ* 3639 and 24329; *PBE*, Katakylas 1; and Cheynet and Theodoridis, *Sceaux byzantins*, 112–14. Although Leo states that Theodosioupolis was taken in 902, it was in Arab hands again soon after and was finally taken only in 949; see *DAI* 1: chap. 45 and 2:173; Honigmann, *Ostgrenze*, 79.

678-84 ἐπειδὴ δὲ . . . τῇ θέᾳ αὐτῶν On horses, camels and cymbals or drums, see on ll. 517–22 above. The advice here probably derives from reports from officers or from his father and elder brother Constantine, who had died in early September 879; see Halkin, "Trois dates historiques."

686-95 §135 This paragraph represents the conclusion of Leo's detailed account of the nature of Saracen warmaking and how to confront it, and may indeed have been the concluding paragraph of an original version of the Saracen section, whether as an independent text or as part of Const. 18; see the discussion at pp. 332–33 above. It is frequently assumed that Leo's comment at ll. 692–93, that it is "for this reason that we have undertaken the present task of formulating instructions for war," applies to the whole of the *Taktika* (Sullivan, "Byzantine Military Manuals," 153; Dagron, "Byzance et le modèle islamique," 220; and Kolias, "*Taktika*," 130). But Leo is quite explicit here—the words τὸν παρόντα τῆς πολεμικῆς διατάξεως . . . πόνον cannot refer to the whole treatise, but to this particular constitution (διάταξις), or even this section of Const. 18. As suggested above, therefore, §§103–35 might thus be understood to represent an extended and updated version of a composition on the Saracens which originally consisted of §§103–25 only, which may explain why it was precisely these paragraphs that appear, reproduced under a separate heading, as a separate text in M (fols. 401–2, 404). Indeed, from the language of the text, it is possible to argue that this extended version of the section on the Saracens, before its incorporation into Const. 16 (18), originally ended here, at l. 693, and that the following words (ll. 693–95: εὕρηνται δὲ ἡμῖν πρὸς τοῖς εἰρημένοις . . . ταῦτα ["In addition to what we have already said. . . . the following"]) were an addition whose function was to connect the paragraphs on the Saracens that were part of an earlier version of this section of the constitution with the following paragraphs on battlefield dispositions, paragraphs which make no mention whatseoever of the Saracens, or indeed of any named enemy. It seems likely that, between the extant, full version of the section on the Saracens, which survives in Const. 16 in M (18 in W) on the one hand, and the separate *diataxis* at fols. 401–2, 404 in M on the other, there existed another, intermediate version, composed by Leo himself or on his order, represented by the addition of §§126–35 (as far as l. 693) to the original §§103–25. When this was eventually incorporated into Const. 16, Leo then added the final sections on battlefield formations (§135.693–§150.855). Note the allusion at ll. 767–69 to this constitution being intended for an army of brave volunteer soldiers rather than one of slaves, a reference to his earlier suggestions for improving the quality and comments on the nature of Islamic warfare.

In contrast, the similar wording in ep.§71.317–21 uses the phrase τὸ παρὸν ... βιβλίον, "the present book," and refers to "as we have said," which can refer back only to his comments in Const. 18 (M 16); there is nowhere else in the *Taktika* where he expresses the idea that the work or section was composed on account of the threat from the Saracens. It also suggests that this paragraph of the epilogue might itself have been updated at the same time as Leo incorporated the "Saracen section" into Const. 18, and reflects his revised view of the reason for composing the treatise. That Leo's words certainly represent vividly the concern he felt at the threat from the Saracens in the east, and that his inclusion of the paragraphs on the Saracens represent both original and contemporary material is not to be doubted, as has been pointed out by several commentators. See Dagron, "Byzance et le modèle islamique," 220–32. On the other hand, if this analysis of the process by which the Saracen section in Const. 16/18 came into being is correct, then the argument that the *Taktika* was *originally* composed *specifically* with the Saracen threat in mind (even if it seems that Leo did then make this adjustment in a later version) must clearly be abandoned, the more so in light of the fact that, as noted already, Leo makes no mention at all of this issue, either in passing or as a motive for composing the *Taktika*, in his prooemium (see chapter 1 above).

696–855   §§136–50   The final paragraphs of this Constitution represent once again new material, most probably derived, as he himself notes at various points (and cf. Const. 19.5), from information passed on by those with actual experience of warfare—perhaps the commander Nikephoros Phokas the elder himself, whom Leo refers to as "our" general. The tactical arrangements outlined in the following paragraphs, while reflecting the more general prescriptions set out in Const. 12, which are themselves copied from Maurice, go into much greater detail about the exact composition of the different elements of the army, and the numbers reflect both Leo's own up-to-date perspective as well as the more realistic possibilities for a thematic cavalry force open to front-line commanders. The figure of 4,000 mounted soldiers to be selected from each *thema*, the recommendation that joint operations with at least two other thematic commanders, to produce an army of 12,000 (see §§147 and 149), may be necessary when facing a strong enemy force, and the remark that if one assembles all the effective troops of the *themata* one can put a force of as many as 30,000 into the field, give some idea of the numbers of effectives that the Byzantine provincial armies could actually field at this period. Of course, this excludes the less reliable and poorly equipped infantry, and as Leo states, it leaves these and any others registered for active service in each *thema* at the disposal of the commander for other tasks. For more detailed discussion of numbers and the size of armies, see Haldon, *Warfare, State and Society*, 99–106 and idem, "Chapters II, 44 and 45," 305–34.

696–723 §136 This section summarizes Const. 12§§19–28, and while it is difficult to prove that it reflects more nearly the actual tactics employed by Byzantine commanders in the field, this and the following paragraphs do appear to reflect a closer familiarity with practice than Const. 12, derived almost entirely from the *Strategikon*.

724–30 §137 Cf. 12§23.

731–34 §138 Cf. 12§52 and below, ll. 739–44.

735–38 §139 Cf. 12§28.

739–44 §140 Cf. 12§52.

745–47 §141 Cf. 12§6 and §51.

748–59 Τὸ δὲ . . . ὑποδεχόμενοι   Cf. 12§20 and §80.

759–63 οὐ γὰρ μοι δοκεῖ . . . παρατάξεις   Cf. 12§§1–3, 6–11 for Leo's detailed account of the value of a double or triple battle line (following the *Strategikon*).

765–67 καὶ τῶν πλαγιοφυλάκων . . . ὀφείλει   Cf. 12§21.

767–69 Here Leo seems to be alluding to his earlier suggestions for improving the morale and quality of the Byzantine armies, where he contrasted the current state of affairs with the way in which Islamic armies were recruited.

770–73 Τὰς δὲ . . . πεντάρχης   Cf. 12§30, §34, and see on 4§35.139–44 above. Leo seems confused on the role and identity of the pentarch. Following the *Strategikon*, at Const. 4§15, the *ouragos* is the tetrarch, correctly identified as the last man in the file. In contrast to the account in the *Strategikon*, however, where the pentarch is placed second in the file behind the dekarch, Leo places him "in the middle" of the file (Const. 4§14). But at Const. 18§146, no tetrarchs are listed at all, whereas there are 400 dekarchs for 800 pentarchs. Leo seems thus to conflate pentarchs with tetrarchs, and assumes a pentarch also acts as *ouragos*. Whether this is the result of Leo's real unfamiliarity with the formations and functions he is describing, or a result of his playing with numbers, following the precedent of the Hellenistic treatises he has before him (cf. the figures reproduced at Const. 4§§58–59 or 14§§85–87, for example), or both, remains unclear.

773–76   οὓς καὶ . . . πλευράν   Cf. 12§36.

777–91   §144   Cf. above, §§136–39.

792–97   §145   This piece of advice does not appear elsewhere and reinforces the impression that Leo is working here from reports from experienced officers.

798–802   Συναχθήσονται . . . ͵ατμς′   Again not found elsewhere in the *Taktika*. The enumeration is reminiscent of the lists of soldiers and officers of varying grades in the account of campaign payments issued to the soldiers mustered for the expeditions of 910–11 and 949; see Haldon, "Chapters II, 44 and 45," 209.79–84, 221.36–62, 223.92–97. Unlike the latter, however, which shows very considerable variations in the proportion between the various grades of officer and soldiers, both among similar units and between different types of unit, Leo's figures are very regular, illustrative again of the somewhat theoretical nature of his advice, but reflecting probably also the desired arrangements where enough men were at the commander's disposal. The figures from the expeditions of 910–11 and 949 represent actually mustered numbers who assembled and were sent on campaign, illustrative in this case of the gap between the troops a commanding officer might hope to raise and those whom he could actually muster.

802–7   καὶ οὕτως . . . καιρόν   Leo stresses again the importance of the general selecting the best-equipped soldiers from the military register, and emphasizes the difference between these (by describing them as the "military *thema*") and "the other troops from the *thema*."

803–4   τὸ λεγόμενον στρατιωτικὸν θέμα   The expression does not to my knowledge appear in any other source of this or an earlier period, but is clearly intended to differentiate the militarily effective provincial army from the remaining registered persons, and this passage in particular makes it clear that, whatever the actual number of registered *stratiotai* in a *thema*, the actual number of combat-ready and adequately equipped soldiers was, at least by this time, and probably for some time in the preceding century, quite limited. See above, comm. on 4§1.3–14 and below on ll. 841–43.

808–18   §147   See 12§§20–24, although Leo here offers a slightly different tactical formation (in Const. 12, following the *Strategikon*, he had advised against dividing up the second line of the battle formation into more than a single division, for example), again on the basis of there being two or three thematic commanders with their "military *thema*" at hand to form up as a single army.

819–29   §148   A strategy whereby a smaller enemy force is "pinned" by an initial attack, while another division attacks from the flank or rear. The final line suggests Leo's continued awareness of Byzantine anxiety about the fighting power of their enemy.

830–38   Εἰ δὲ . . . διωρισάμεθα   In effect a repetition of the details given in §§136–45.

838–46   εἰ δὲ καὶ . . . τῶν ἐχθρῶν   On the basis of Leo's figures, an army of up to 30,000 would have comprised some 6 or 7 *themata*, and as we have seen there are several examples of such forces assembling to deal with enemy invasions and raids in the eighth and ninth centuries, apart from expeditions led by an emperor, such as Theophilos's campaigns of 837, for example. In 778 the general Michael Lachanodrakon is reported to have marched against Germanikeia with a force consisting of troops from Thrakesion, Boukellarion, Armeniakon, Anatolikon, and Opsikion: *Theophanis Chronographia*, 451.11–14 (trans. Mango and Scott, 623). Leo's figures here enable us to make better sense of such forces, which did not reach the huge numbers some commentators have imagined, but which may have attained as many as 30,000 on exceptional occasions. The numbers and logistical constraints were the key determining elements here. Even at the height of the victorious warfare of the second half of the tenth century a field army of some 30,000 was regarded as unusually large: Leo the Deacon, *Historia*, 132.19–20 (Talbot and Sullivan, 179). See further discussion on numbers in chapter 3 above.

841–43   καὶ διακρίναντες . . . ὀλιγότητα   It remains unclear to what extent this situation was long standing, especially since Leo's wording makes it seem like a recent problem, as do his remarks in the prologue (pr.§5). But this is unlikely, especially as there are occasional references in sources of the ninth century and before to emperors replenishing the ranks of the army, while hagiographies, although certainly relying to some extent on received topoi, contain a number of stories about impoverished soldiers. Given the references in this treatise to the process of selecting battle-ready troops as being old and well established, to the evidence for the relative inefficiency of the system of maintaining provincial troops at their own expense, and to the evidence from the legislation of the reign of Constantine VII that there was until his reign no formal connection between registered soldiers and the property upon the income from which their service was supposed to be supported, it seems more likely that while Leo complains about the situation with which he was familiar, it was not new. See commentary on 4§1.3–14 above.

847–55   §150   See Maurice, *Strategikon*, 11.4.225–39, on which Leo's concluding paragraph is loosely based. In one original version of the *Taktika* (extant in M) this paragraph would have been immediately before the *gnomika* or "concise sayings" which now form Const. 20, and the epilogue, and which seems to summarize not only Const. 18 but the treatise as a whole. Since Leo takes bk. 12 of the *Strategikon* as the source for material incorporated into earlier constitutions of the *Taktika*, one wonders whether in one early version of the *Taktika* the treatise in fact ended here, with the *gnomika* and the epilogue added at a later stage, especially in light of the penultimate paragraph of the epilogue (see above, on ll. 686–95).

CONSTITUTION 19

Leo's motives for including a constitution on naval warfare and related matters can reasonably be assumed to reflect the current situation in his own lifetime and reign, during which seaborne attacks on Byzantine territory in S. Italy, the Adriatic, the Aegean, and the eastern Mediterranean basin were both frequent and damaging (see, for example, the discussion in Lev, "Mediterranean Encounter," esp. 132–33; and the literature cited below).

Originally an integral part of the *Taktika*, this constitution was incorporated into the treatise as 19, which is how it appears in W and derivative manuscripts. As we have seen (chapter 2, pp. 56–66 above), it had, probably at an early stage, an independent existence, as its internal and original section numbering would indicate. For the version in M (which antedates both A and W) 19 was not included in the main body of the treatise but retained as a separate short text, beginning at fol. 394. In A, the text has been transmitted in a slightly differently worded version, under the title Ναυμαχικὰ Λέοντος βασιλέως (hereafter Leo, *Naumachika*). It was similarly taken out of the main body of the *Taktika* and placed at the end of the treatise, at fols. 323r–331r, where it is followed by (a) an extract of six paragraphs from the *Taktika* under the title Ἐκ τοῦ κύρου Λέοντος τοῦ βασιλέως (= Const. 20§§196, 201, 220; ep.§§44, 45, 47), at fol. 331r–v, in turn followed by (b) a section taken from Maurice's *Strategikon*, 12.B.21 on crossing rivers (fols. 331v–332r: Ἐκ τοῦ Μαυρικίου πῶς δεῖ διαπλέειν τοὺς ποταμοὺς . . .; see Rance, *Strategikon*, 12.B.21, note 218), (c) the treatise known by the title Ναυμαχίαι Συριανοῦ μαγίστρου (hereafter Syrianos, *Naumachiai*) at fols. 333r–338v (a folio which would have contained the first page is missing; see Dain, *Naumachica*, 16, 40, 43–44); and (d) that composed for the *patrikios* Basil in the later 950s (at fols. 339r–342v: hereafter Basil, *Naumachika*). The treatise of Syrianos survives only in A, and while the others were copied into much later medieval or postmedieval manuscripts, A remains the only collection of these naval treatises. See Dain and de Foucault, "Stratégistes byzantins," 384–85.

Although Leo claims in his opening paragraph (ll. 3–4) that he had not been able to find any treatise specifically dealing with naval warfare in the older tactical treatises (that is to say, in writers such as Aelian or Onasander), he does refer to "what we have read here and there," probably meaning the work of Syrianos, among others—who anyway counts as more recent, given the generally accepted ninth-century date for his writing; and many of the stratagems Leo describes can be found in Polyaenus, *Strategika*, which he exploited extensively. Whether Leo also had access to a text such as the late third-century BCE treatise known as the *Mechanike syntaxis* ascribed to Philo of Byzantium, for example, is not at all clear and would require a separate analysis. Philo's treatise contains at the end of book 5 a brief account of the use of warships in besieging coastal cities and of how to deal with an enemy fleet, some of which is very close to passages in the *Taktika*. Philo was certainly known in the late ninth and tenth centuries, and some of his writing can also be seen in the sections on siegecraft and fortifications in Syrianos; they were likewise a source for ps.-Heron, *Parangelmata Poliorcetica*. See Dain and de Foucault, "Stratégistes byzantins," 323–24; Dennis, *Three Byzantine Military Treatises*, 37 n. 3; Sullivan, *Siegecraft*, 2 and 7. The text survives in three manuscripts of the late tenth and early eleventh centuries (V = Vaticanus gr. 1164; E = Scorialensis gr. Y-III-11; and P = Parisinus gr. 2442 with its complement B = Barberinianus II 97 [276]; see chapter 2 above and the editor's introduction, in Garlan, *Recherches*, 281–86), so that it was transmitted, together with Leo's *Taktika*, with the same collection of tactical treatises assembled in the middle of the tenth century. Pryor and Jeffreys, *Age of the Δρόμων*, 176–80 list possible sources for Leo, before dismissing them all apart from the extract on crossing rivers, taken from the *Strategikon*, and the treatise on naval warfare of Syrianos.

Yet for some of his naval examples Leo certainly drew on the *Strategika* of Polyaenus, of which, as noted in chapter 2 above, the emperor appears to have had access to an unabridged or original version. Thus a number of his examples for naval tactics and strategy echo accounts from the latter, as at *Strategika*, 3.10.1–17, for example. It is noteworthy that the late tenth-century *Hypotheseis* (or *Excerpta Polyaeni* as the text is alternatively known), gathers the extracts on naval warfare together in its chapters 57 and 58, and it may be that the compiler of this source took his inspiration for the organization of the material from Leo. Indeed, Leo notes at 19§72 that if someone were to search through the "volume of ancient tactics and stratagems" more information on naval warfare could be found, and this is a description which fits the Polyaenus text well.

As we have said, Leo almost certainly had at his disposal a version of the treatise on naval warfare ascribed to Syrianos *magistros*, which constituted one part of a much longer treatise, and included the separate texts known as the *Rhetorica militaris* and the Περὶ στρατηγικῆς (or Syrianos, *Strategy*) (see Zuckerman, "Military compendium" and Dain and de Foucault, "Stratégistes byzantins," 342–44), thought by some to have been compiled in the middle of the ninth century: S. Cosentino, "Syrianos's 'Strategikon'" (although Zuckerman, "Military Compendium," 209–24, prefers a sixth-century date; as does Eramo, "Ῥωμαῖοι e Ἄραβες"). Whether or not the revised date is correct, some of the information in Constitution 19 would support the contention that Leo was familiar with this text, or part of it (although Eramo, "Compendio," 215–17 is less persuaded). But in light of Zuckerman's arguments, following which Constantine VII certainly had a version of the full treatise of Syrianos, this conclusion seems reasonable, even if not certain, the more so since neither Maurice nor Onasander, nor Aelian's *Tactical Theory*, all of which were used by Leo, contain information on naval matters. Aelian, *Tactica theoria*, 2.1, stated that he would deal with naval warfare at a later point in his treatise, but this appears to be part of a conventional statement of intent in such treatises, rarely fulfilled, based on the fact that Aeneas Tacticus had intended to include such a section, now lost, at the end of his own treatise (see below; *Aeneas Tacticus*, 40.8; and Wheeler, review of Charles, *Vegetius*). No such section was included in Aelian's treatise, of course, and Leo must mean another source. Given the similarities in certain paragraphs between Leo and Syrianos, it seems most likely that it is to this text that he is referring. See Pryor and Jeffreys, *Age of the Δρόμων*, 175–81.

Interestingly, as we will see below, some of Leo's prescriptions are remarkably similar to those in Vegetius, *Epitoma*, 4.31–46, which in turn bear a similarity to some of those of Syrianos, *Naumachiai*. This is most probably the result of the fact that Vegetius (whose vocabulary in many places reveals a familiarity with Greek technical terms and writings), indirectly and through earlier Latin authors (see *Epitoma*, 1.8 and 3, praef.), and Syrianos, possibly via a lost intermediary (see Dain and de Foucault, "Stratégistes byzantins," 320), may have drawn on a common source, no longer extant. According to a traditional view, this was the supposedly lost section on naval warfare composed by the fourth-century BCE writer Aeneas Tacticus. But this seems unlikely. Aeneas was the first of the classical writers on tactics and warfare (see Dain and de Foucault, "Stratégistes byzantins," 319–21 for the manuscript tradition and history of the text), but in the absence of all references to such a work in later writers the very existence of this naval section may be doubted. For a supposed association between these three texts, see Lammert, "Seetaktik," esp. 285ff.; for a more sceptical view, see Rance, "Aineias' *Poliorketika*."

The only other comparable text on naval affairs—a (probably) early tenth-century guide to maritime matters in Arabic, incorporated into the *Kitāb al-Harādj* of Kudāma ibn Jaʿfar—was not known to Leo, although it offers, in summary form, a good deal of comparable advice on crew organization, weaponry, and armament; see Christides, "Two Parallel Naval Guides." Leo's naval section was later translated into Arabic, both in full and in summary form; see Serikoff, "Leo VI Arabus"; Christides, "Naval Warfare"; and Pryor and Jeffreys, *Age of the Δρόμων*, 645–66.

The treatise Ναυμαχικὰ Λέοντος βασιλέως has been most recently edited and translated by Pryor and Jeffreys, *Age of the Δρόμων*, 484–516, and Dimitroukas, *Naumachika*, 38–77 with brief notes at 252–70, and although as noted already the text of this version and of Const. 19 vary in places, they are close enough to be sure that one derives from the other, although which is the earlier version remains unclear. As is apparent from the critical apparatus in the Dennis edition, the text of M, the earliest of the three manuscripts, is closer to that of W than to that of A, which makes it likely that the text in A is later and represents a slightly revised version compiled specifically for that manuscript collection and for the *patrikios* Basil the *parakoimomenos* ca. 959–63 (see chap. 2, pp. 55–66). Given the similarities between the two traditions, however, the commentaries by Pryor and Jeffreys and Dimitroukas serve as a valuable base for the commentary presented here, and in order to avoid repetition, I will comment on items only where an alternative interpretation or additional information is required, giving otherwise simply the references to the Pryor and Jeffreys and/or Dimtroukas discussion.

For literature on Byzantine naval warfare and organization: Pryor and Jeffreys, *Age of the Δρόμων*; Pryor, "Shipping and Seafaring," with literature; Ahrweiler, *Mer*; Eickhoff, *Seekrieg*; Kolias, "Byzantinische Kriegsmarine"; Makrypoulias, "Navy"; Pryor, *Geography, Technology and War*; Lewis, *Naval Power*; Toynbee, *Constantine Porphyrogenitus*, 323–45; Christides, "Raids"; and idem, *Conquest of Crete*. The older article of Bury, "Naval Policy," offers valuable insights, as does the more recent Stanton, *Norman Naval Operations*. For general comments on Byzantine ships and sea-communications: Dimitroukas, *Reisen und Verkehr*, 2:413–544; Makris, "Ships"; Avramea, "Land and Sea Communications," esp. 77–88; Koder, "Aspekte der Thalassokratia"; and for the broader historical-anthropological and geographical context, see Horden and Purcell, *Corrupting Sea*, 133–43.

5   παρὰ τῶν πλωΐμων στρατηγῶν   In Leo's time there were three maritime *themata*, namely Kibyrrhaiotai, Aegean Sea, and Samos, in addition to land themes with a maritime force attached to them, such as those of Kephallenia, the Peloponnese, and Hellas; see Brubaker and Haldon, *Byzantium in the Iconoclast Era*, 729–30, 733–34, 739–40, 758–59. All provided ships and men for the campaigns of 910–11 or 949, for example. The Kibyrrhaiotai were established probably before 732, although the issue is debated; see Grigoriou-Ioannidou, "Κιβυρραιωτῶν"; Oikonomidès, *Listes*, 351; and Hild and Hellenkamper, *Kilikien und Isaurien*, 1:45–46. For the location of the Kibyra in question (southern Caria or on the Pamphylian coast, to the east of Side): Yannopoulos, "Cibyrra et Cibyrrhéotes." For the theme of the Peloponnese: Živković, "Date"; older literature and discussion: Oikonomidès, *Listes*, 350; Winkelmann, *Rang- und Ämterstruktur,* 104–5; Belke, "Einige Überlegungen". And for Kephallenia: Soustal, *Nikopolis und Kephallenia*, 52–53; and Oikonomidès, *Listes*, 352. For a brief historical survey of both Nikopolis and Kephallenia, see Soustal, *Nikopolis und Kephallenia*, 54ff.

5–6   ἀνεμάθομεν ... πεπονθότων   Again Leo confirms his reliance on contemporary commanders for much of his information. He is quite straightforward about Byzantine defeats at sea, the more recent of which had occurred in 888 and 901; some successes had also been achieved relatively recently, in 879 and 880 and again in 901: Vasiliev, *Byzance et les Arabes*, 2.1:141–42; Pryor and Jeffreys, *Age of the Δρόμων*, 62ff., 385; and Dimitroukas, *Naumachika*, 252 n. 10 on Himerios, the most successful Byzantine naval commander of Leo's reign. Equally successful and famous naval commanders had been Niketas Ooryphas and his successor, Basil Nasar, in the 870s; see on ll. 181–86 below.

8   δρομώνων   See also on 61–68 (§§9–10) below. By the ninth century the standard nontechnical term for an oared warship with two banks of oars, of varying size, as here at ll. 61–68. For the evolution of the ship type and the term from its earlier single-banked meaning, see Pryor and Jeffreys, *Age of the Δρόμων*, 123–304 with detailed discussion of the sources and earlier literature; Dimitroukas, *Reisen und Verkehr*, 2:413–26; and Rance, *Strategikon*, 12.B.21, note 219. Although Leo does not use it, an alternative term for the standard bireme (two

oar-bank) warship of the imperial fleets was *chelandion*, which was borrowed into Arabic as *shalandi* (see Agius, *Classic Ships*, 337–38) and into Latin and Italian as *scelandrium, salandria, zalandria* or similar, and which Pryor and Jeffreys note derives ultimately from classical Greek κέλης, a "courser" or fast horse, applied then to fast-sailing galleys but also to horse transports. See Pryor and Jeffreys, *Age of the Δρόμων*, 166–73, 188–92; Dimitroukas, *Naumachika*, 252, n. 15; and Karapli, "Βυζαντινόν χελάνδιον." For Byzantine ship terminology, see also Eickhoff, *Seekrieg*, 136–37; and Ahrweiler, *Mer*, 408–18. There was a range of different types of vessel, described by their oarage system, their size, or their hull-shape, giving rise to terms such as *pentekonter, akation,* and *galaia* (or *galea*) for lighter, single-bank oared warships, or *saktoura* or *myoparon* for larger, round-hulled sailing vessels, for example, and many others; along with generic terms such as anc. Greek *holkas* and early Byz. Greek *karabos/karabion* (the latter was certainly Greek; see Du Cange, *Glossarium Graecitatis*, col. 589; Trapp, *Lexikon*, 4:763; *LSJ, s.v., pace* Pryor and Jeffreys, *Age of the Δρόμων*, 164–65). See Ahrweiler, *Mer*, 414–15.

9–15 §2 See also §31 and Syrianos, *Naumachiai*, 5.1 3. As Pryor and Jeffreys, *Age of the Δρόμων*, 392–93 note, Leo's injunction that the overall commander of a naval force should have an expert knowledge of the sea, winds, navigation, and so forth is unrealistic, since such commanding officers were frequently not professional or practiced seamen. They must have relied on the actual seamen among their ship commanders—the sailing masters, so to speak—for such matters, as Syrianos, *Naumachiai*, 5.3 makes clear; see below on 19§§29.170–31.186. There existed handbooks dealing with weather and winds, reference to some of these appearing in a short treatise on imperial military expeditions by land composed during the reign of Constantine VII (*CPTT* C.200–202; see Pryor and Jeffreys, *Age of the Δρόμων*, 191. By the same token there were also guides to routes, of which the so-called *stadiodromikon* attached to the various documents associated with the Cretan expedition of 949 may be an abridged or mutilated example, although some of the places mentioned can hardly be understood as either ports or navigation points, still less as stopping points; see Pryor, "Σταδιοδρομικόν"; also Avramea, "Land and

Sea Communications," 80–82; Dagron, "Firmament"; and Lambros, "Τρία κείμενα." For Byzantine navigation and the portulan tradition, see Ahrweiler, *Mer*, 164–69, 451; Huxley, "Porphyrogenitan Portulan"; Koder, *Aigaion Pelagos*, 102–3, with a more accurate reckoning of the distances involved; and Pryor and Jeffreys, *Age of the Δρόμων*, 264–66. For some more general considerations, see Kislinger, "Verkehrsrouten zur See"; Udovitch, "Time, the Sea and Society."

16–19 §3 The general vagueness of Leo's instructions here suggests his distance from the realities of fleet construction and ship maintenance, which was a major undertaking requiring considerable expense and time; see Pryor and Jeffreys, *Age of the Δρόμων*, 394. Byzantine governments maintained, as need demanded, a regular program of shipbuilding: in the documents for the expedition of 949 reference is made to the crew of a warship of the *thema* of the Aegean Sea, left to "cut the wood for the eighth indiction." This wording implies that an annual levy of timber for shipbuilding was imposed on coastal properties, whether private or belonging to imperial estates, and that supplies for shipbuilding were likewise raised by levies on the appropriate craftsmen or others. Nails and a variety of metal fixtures, for example, were thus raised from the naval *themata* of Samos and Kibyrrhaiotai, and from the Thrakesion region, in 949; see Haldon, "Chapters II, 44 and 45," 211.114–33 (the text makes no mention of the metal, but Vegetius notes in the late fourth or early fifth century that bronze is far superior to iron for ships' fastenings and nails, since it does not corrode like iron: *Epitoma*, 4.34.2–3). How regular or frequent this sort of levy actually was is impossible to say, although it seems, from the amount of imperial naval activity in the tenth century, that a sustained program of ship construction did exist. See the accounts in Ahrweiler, *Mer*, 419–39; Eickhoff, *Seekrieg*, 87–88; and Pryor and Jeffreys, *Age of the Δρόμων*, 71ff.; and cf. Christides, "Second Arab Siege," 520–22 for Egyptian levies in timber etc. and preparations for the construction of a fleet in the years preceding the siege of Constantinopple in 717. The government could also raise ships (presumably for transport, not fighting vessels) by imposing a levy of vessels on ports and maritime communities; see Stavridou-Zaphraka, "Ἀγγαρεία," esp. 38–40. Vegetius notes that cutting and seasoning the timber at the appropriate time of the year (from July through to January) was crucial to the quality of the wood: *Epitoma*, 4.36.1–2.

20–25   §4   See Pryor and Jeffreys, *Age of the Δρόμων*, 192–238 on the hull and basic structure of the *dromon*. By the fifth century a shift in hull construction, from shell-first to frame-first techniques, is evident in the archaeology. While lighter Byzantine warships in the ninth–eleventh centuries appear—on the basis of the limited archaeological evidence from the Yeni Kapı excavations—to have still been constructed on the shell-first principle (whereby the frames were fitted after the planking had been completed and affixed to the keel), heavier vessels were probably constructed using a frame-first technique, with the planking added to the skeleton of the vessel. The evidence for this is based partly on archaeological evidence, partly on written sources, and partly on deduction from the ways in which the ships were fought (i.e., not with rams—see the next entry) after the sixth century. See the useful short summary and discussion in Mor, "Socio-economic Implications," with older literature; and Pulak et al., "Shipwrecks," 33; Pomey et al., "Transition."

25   συγκρουόμενος   Read "struck," rather than rammed, as translated by Dennis, since medieval warships did not have rams but were equipped with spurs attached to the stem of the vessel; see Leo, *Naumachika*, ed. Pryor and Jeffreys, §4 (p. 484) and pp. 134–52 (with sources and older literature) on the ram and the spur; with Mor, "Socio-economic Implications," 43–47.

26–29   Ἐχέτωσαν … ἀπαιτεῖ   See Pryor and Jeffreys, *Age of the Δρόμων*, 276–304 on the oarage system of the *dromon*, and 238–54 on masts and sails.

27   αὐχένας … κάρυα   See Trapp, *Lexikon*, 2:244 and 4:769 and Dimitroukas, *Naumachika*, 253–54 nn. 18 and 19 for etymology and discussion.

29   ξύλα τινὰ ἐγκοίλια   Pryor and Jeffreys, in Leo, *Naumachika*, §5 (486) translate this as "floor timbers"; given the uncertain technical knowledge of the authors of these texts it is difficult to know which is the more accurate; see Pryor and Jeffreys, *Age of the Δρόμων*, 202 with earlier specialist literature; Trapp, *Lexikon*, 3:439.

30–31   ναυπηγόν … ὁμοίων   The editor omits the words for "one of the oarsmen"; see Leo, *Naumachika*, 5 (p. 486). The shipwright was therefore not a specialist position, but one of the regular crew members; see Dimitroukas, *Reisen und Verkehr*, 2:428. For crews in general: Pryor and Jeffreys, *Age of the Δρόμων*, 266–76; and Dimitroukas, *Reisen und Verkehr*, 2:426–31.

32–33 Ἐχέτω δὲ ... ἀκοντῖσαι   The question of how exactly "liquid fire" was projected and what its basic fuel was has now been resolved, at least in its broad outlines, although since there is as yet no archaeological or material evidence for the device its exact form and size cannot be determined; see detailed discussion with sources and earlier literature in Haldon, "Greek Fire Revisited" with the account in Pryor and Jeffreys, *Age of the Δρόμων*, 607–31. Each *dromon* was equipped with one such projector, according to Leo, although in the records from the Cretan expedition of 949 it appears that *dromones* and *chelandia* could have two or three such devices; see *De cerimoniis*, 671–73 (Haldon, "Chapters II, 44 and 45," 227.141, 229.157–58 with commentary at pp. 278–81).

34–37 καὶ ἄνωθεν ... ὅπλων   The liquid fire projector and its crew were thus stationed below a fighting platfrom and were protected from enemy missiles by planking around their position. See Pryor and Jeffreys, *Age of the Δρόμων*, 203, 620–21.

38–39 Ἀλλὰ καὶ ... σανίσιν   These structures appear to have been two "castles" or raised and fortified platforms, constructed on either side of the main mast (the foremast in such a lateen-rigged vessel) similar to that at the bows. The phrase "around the middle of the mast" is, according to Pryor and Jeffreys, in all versions of the text, from Leo VI to Nikephoros Ouranos, a corruption or misunderstanding, and should refer not to "around the middle of the mast" but rather "around [or at] the middle of the masts," i.e., between the taller foremast and the second mast amidships. This makes much better sense of the text and is also practical: Pryor and Jeffreys, *Age of the Δρόμων*, 229–36; Dimitroukas, *Naumachika*, 256 n. 28; and Karapli, "Καστέλλωμα."

45–51 Ἕκαστος δὲ ... ἄνδρας ρ′   On the oarage system of the warships see the detailed technical discussion in Pryor and Jeffreys, *Age of the Δρόμων*, 276–304. Their estimate of the approximate maximum length of a bireme *dromon* of the type described by Leo, of some 31–32 meters, should probably be revised in light of the findings from the excavations at Yeni Kapı. Here, the length of the smaller, lighter, single-oarbank *galeai* has been calculated at ca. 30 m, so that the larger *chelandia* and *dromones* will probably have been several meters longer: see Pulak et al. "The Shipwrecks," 26, and on 61-68 below. The rowing crew of these warships were both oarsmen and soldiers. See §§14 and 75; Haldon, "Chapters II, 44 and 45," 225.107-8 and n. 68; 254ff., 334ff.; Pryor and Jeffreys, *Age of the Δρόμων*, 255; and Dimitroukas, *Naumachika*, 256 n. 29. Evidence for the arrangement of oarsmen's benches from the Yeni Kapı excavations demonstrates that Pryor's calcuations for the dimensions and oarage of such vessels are based on plausible arrangements, and probably accurate to within a relatively small margin of error. See Kocabaş, "Technological and Constructional Features," esp. 176–83 for the dimensions and details of wreck YK 16, a (purported) *galea*.

51-57   ἔξω δὲ . . . κεντάρχου   Here Leo lists the main "officers" of each warship: the *kentarchos* (also called the *nauarchos;* see l. 57 and Pryor and Jeffreys, *Age of the Δρόμων*, 268) was the commander, and under him were the two *protokaraboi* or helmsmen, each handling a steering oar at the stern; two oarsmen at the bows of the vessel served as *proreis*, one as *siphonator* (or *siphonarios:* cf. *De cerimoniis*, 669), in charge of the liquid fire projector; the other in charge of the anchor. In addition the *kentarchos* had at his disposal a standard bearer, who presumably had other duties (such as signaling or conducting orders to the crew) in addition to handling the ship's flag or pennant. In fact the evidence suggests that there were more than this basic minimum of subordinates, to ensure adequate manning both of the liquid fire projector and of the steering-oars; see Haldon, "Chapters II, 44 and 45," 225.101. For discussion see Pryor and Jeffreys, *Age of the Δρόμων*, 269-76. Leo does not mention *kometes* at this point in his constitution on naval matters, but they do appear later, in §25, as commanders of a group of three or five vessels; they appear likewise among the officers of the naval *themata* in the records for the 949 Cretan expedition (Haldon, "Chapters II, 44 and 45," 215.189).

55   τὰς ἀγκύρας   Iron anchors of varying sorts were carried, and were attached by various types of cable. See Pryor and Jeffreys, *Age of the Δρόμων*, 210–14; Dimitroukas, *Naumachika*, 257 n. 32. Examples of

Y-shaped and T-shaped iron anchors have been excavated from the Theodosian harbor at Yeni Kapı in Istanbul: for a tenth-eleventh-century Y-shaped example from shipwreck YK1, see Kızıltan, *Stories*, 32 and Fig. 16; 218, Pl. 281.

57 ὁ . . . κράββατος   Although translated by the editor as "pallet" (and see Trapp, *Lexikon*, 4:878–79), the term seems rather to refer to a covered structure or half-cabin of some sort, the berth of the *kentarchos*, from which he commanded the vessel and which also offered some protection against enemy missiles; see Dimitroukas, *Naumachika*, 257 n. 34 and Pryor and Jeffreys, *Age of the Δρόμων*, 215–16. See also *CPTT* C.322 and trans., 115, where *krabbatos* is translated as "couch," although a better term might be "cabin" or an equivalent.

61–68   §§9–10   Consensus on the different sizes and types of Byzantine warship has not yet been reached. Leo alludes here to larger *dromones* and also to smaller faster vessels with a single bank of oars, called *galaiai*. It is possible that some of the Yeni Kapı wrecks are examples of such vessels; see Pulak et al., "Shipwrecks," 25–28; Mor, "Socio-economic Implications," 56; Kocabaş, "Technological and Constructional Features," 176–83. From the details recorded in the various expeditionary documents for 911 and 949, it appears that there were at least three categories of regular warship: *ousiaka chelandia* (i.e., "crewed" *chelandia*), which carried either 108 or 110 men, *pamphyla chelandia* (i.e., "fully crewed" *chelandia*) which carried 120 and up to 160, and *dromones*, which seem to have been larger and heavier vessels, carrying 220 and up to 300. The *ousiakon chelandion* does not appear in the 911 documents, but its complement of 110 was probably not a maximum; see *De cerimoniis*, 652.10–653.16, 664.19–665.13, 670.3–6 and Haldon, "Chapters II, 44 and 45." It is difficult to know exactly what type of vessel is implied by these terms. *Dromon* and *chelandion* were clearly synonymous in certain contexts, and the difference between ships which were *ousiaka* and ships which were *pamphyloi* seems to have lain primarily in the size of the crews, and only secondarily in the size of the vessels themselves. On the other hand, the eight *pamphyloi* of the imperial fleet mentioned in the 949 documents appear to have carried three liquid fire *siphones*, as did the *dromones* of the fleet, so the distinction is even more blurred. See the discussion in Ahrweiler, *Mer*, 411–13 and see Const. 19§42: the text makes it clear that a *pamphylon* was simply a *chelandion* which was filled with the best and most able soldiers and oarsmen in the fleet, and which had a larger-than-usual complement. Such vessels are described as being "made the *pamphylon*," and the verb παμφυ-λεύω is used of the process of filling them up. The term *ousiakon chelandion*, in contrast, must therefore be understood to represent the standard oared fighting ship with the basic crew of 108–10, described here as a *dromon* with a crew of 100 oarsmen/soldiers, a *kentarchos*, a standard bearer, 2 *protokaraboi* (steersmen), a servant or assistant to the *kentarchos*, two first oarsmen (one to act as *siphonator*, one in charge of the anchor tackle) and a *proreus* (Leo's *Naumachica* and the relevant chapter of the *Taktika* of Nikephoros Ouranos repeat the description). This makes a crew of 108, exactly as described in the *De cerimoniis*. This could then be increased to "make into a *pamphylon*" of up to 160 crew when required; although it is also clear that some vessels were built on slightly larger lines and designated as *pamphyloi* from the start (*De cerimoniis*, 664.7–8). But the practicality of these crewing arrangements remains to be established. In ancient warships they were very finely tuned in order to achieve optimal maneuverability and speed under oars, and simply increasing the size of the crew (or the cargo—for example, supplies of water) would have dislocated the system, given the relatively low freeboard; see Morrison et al., *Athenian Trireme*, 191–230. The lighter "galleys" were smaller and used for scouting operations, as mentioned here by Leo and as described in the records of the expeditions in 911 and 949 (and cf. Syrianos, *Naumachiai*, 6.2). For detailed discussion of all these, with sources and older literature see Haldon, "Chapters II, 44 and 45," 334–37; Pryor and Jeffreys, *Age of the Δρόμων*, 188–92, 254–62, 283–86; Spanoudes, "Δρόμωνα"; and Dimitroukas, *Naumachika*, 257–59 n. 35.

66–68 §10 Cf. Maurice, *Strategikon*, 12.B.21.3–10 (= Ἐκ τοῦ Μαυρικίου, §1) and Syrianos, *Naumachiai*, 6.4–5.

69–72 §11 For the transportation of horses and the problems of supplying warships at sea, see Pryor and Jeffreys, *Age of the Δρόμων*, 304–33.

73–77 §12 This paragraph appears to react to Syrianos, *Naumachiai*, 9.8–9, which states that the commander contemplating war should have a good knowledge of both his own and the enemy's strength in ships and men, based on reports from several sources including scouts and deserters. At §21, however, Leo proposes the even more unlikely scenario of the commander knowing the quality and ability of every soldier and oarsman under his command.

78–84   §13   A paragraph modeled on 5§§6–7 and Const. 10. Cf. also 5§9; and Maurice, *Strategikon*, 12.B.21.15–16 (= Ἐκ τοῦ Μαυρικίου, §3).

85–107   §§14–16   The only point in the treatise where Leo discusses hand-thrown stones as weapons.

85–95   §14   Cf. 6§§21–22 for the various types of armor and weaponry carried by infantry.

96–107   §§15–16   Cf. 18§116, esp. ll. 564–66, where the wording is very similar. For the same general principle, although a different context, cf. Polyaenus, *Strategika*, 3.10.6 (*Hypotheseis*, 57.5), 12.

108–12   §17   Cf. 18§129, esp. ll. 637f.

113–18   §18   See Polyaenus, *Strategika*, 3.10.10–11 for a possible inspiration for this section. This is an important paragraph, since Leo alludes here to a major problem for naval forces, namely the maintenance of adequate supplies of food and water. The text suggests that the soldiers and sailors of an inadequately provisioned fleet could easily turn to looting and plundering within their own territory; see on 19§40.232–37 below and Pryor and Jeffreys, *Age of the Δρόμων*, 385–86. The passage also points to the differences between land-based and maritime campaigns. On land, and within imperial territory, Byzantine armies were supported by a relatively sophisticated logistical support system, a system which does not appear to have been available to support naval forces, even when in home coastal waters, except where the individual commander set up a system of provisioning, as Leo's text here implies. Instead, the commander is urged to get his forces as quickly as possible into enemy waters, from where he can pillage the coastlands for supplies; see 19§§32–33—although again the practicality of such a strategy, given the difficulties inherent in carrying it out, may be doubted. The supplying of naval forces was thus infintely more problematic than it was for land armies. On some aspects of this, see Pryor and Jeffreys, *Age of the Δρόμων*, 333–78; Gertwagen, "Harbours and Facilities"; Erdkamp, *Hunger and the Sword*, 52–62;

and Christides, "Second Arab Siege"; and cf. Syrianos, *Naumachiai*, 5.1–2 for the importance of the fleet having men on board who know the coasts and watering places.

114–16   ἵνα μὴ ... ἀδικοῦσι   Better rendered as "so that, lacking these things, they neither mutiny nor, if they are in their own country, oppress and mistreat their fellow-citizens, our subjects. ..." See the preceding entry; commentary to ll. 232–37 below; Kekaumenos, *Logos nouthetetikos pros Basilea* (Litavrin), 276.21–278.7 for soldiers who are not paid or provisioned in a timely manner; and ibid., 294.1–8, an account of a provincial fleet that, on the pretext of offering protection, effectively plundered Roman territory, especially in the Cyclades and around Cyprus and Crete, presumably because no regular resupply arrangements had been set up, and because, once at sea, the fleet was far less accountable to the imperial government. Supplying an army from enemy territory—where practicable—was far to be preferred; see discussion in Haldon, *Warfare, State and Society*, 171–73.

119–23   **§19**   It is not clear whether officials and officers in the maritime forces were more susceptible than those of the land forces to receiving gifts or sums of money in return for excusing sailors from their duties or for properly equipping themselves. Similar sentiments exactly are expressed nearly two centuries later in Kekaumenos, *Logos nouthetetikos pros Basilea* (Litavrin), 292.18–296.8. The issue clearly affected the land armies as well, as the legislation of the Macedonian emperors makes clear; see Const. 8§26, and commentary to 8§26.82–87 above. In general on the process of social subordination and dependency which evolves at an ever-increasing pace across the tenth century, see Morris, "Powerful and the Poor." The rendering of "customary dues," συνηθείαι, was a standard element in the tax-collection process, but was usually extended also to a range of other "services" rendered by imperial officials in the provinces and at the expense of the tax-payer; see Oikonomidès, *Fiscalité et exemption*, 76–78, 86–105. By extension the term *synetheia* was also applied to unofficial gifts of cash (or other goods) in return for an exemption from some duty or other. In the case alluded to by Leo this probably meant either exemption from service or from appearing for duty without the appropriate equipment (which is what Kekaumenos states in the late eleventh century).

The final warning suggests that such "customary" gifts were a major problem among the senior officers.

124–29   §20   According to Pryor and Jeffreys, *Age of the Δρόμων*, 232, this passage is proof that the warships were fully decked and that the lower bank of oars was below the main deck of the vessel.

130–41   §§21–23   A repetition and expansion on the instructions given in §§11–13; for ll. 136–38 cf. also Maurice, Ἐκ τοῦ Μαυρικίου, §5.

142–47   §24   Cf. 13§§1 and 4; Dimitroukas, *Reisen und Verkehr*, 2:512–13, and, more generally on religiosity and superstition in Byzantine maritime affairs, ibid., 510–19; Dimitroukas, *Naumachika*, 261 n. 46; and Dennis, "Perils of the Deep." We should nevertheless beware of ascribing such anxieties as are expressed in literary and rhetorical contexts to those who actually worked on the sea, at least in the form in which they are presented in the literary evidence.

148–51   §25   Cf. 13§2, and cf. Maurice, *Strategikon*, 12.B.21.18–21 (= Ἐκ τοῦ Μαυρικίου, §4). The implication of the clause ὅστις ναύαρχός τε . . . ἕκαστα is that each such squadron commander was responsible for both the logistical and the tactical aspect of his command, an impression reinforced much later, if somewhat obliquely, by Kekaumenos, *Logos nouthetetikos pros Basilea* (Litavrin), 294.1–10.

152–54   Οἱ δὲ . . . πλωΐμου   Cf. 12§§75 and 80; 13§18.

154–61   ἐπὶ δὲ . . . τάξεσιν   For the command structure of the fleets in the late ninth century and later see Pryor and Jeffreys, *Age of the Δρόμων*, 266–68; Oikonomidès, *Listes*, 352–53; Haldon, "Chapters II, 44 and 45," 258 and n. 67; 334–39; together with the older but still relevant sections of Ahrweiler, *Mer* and Eickhoff, *Seekrieg*. In the expediton of 949 the imperial fleet was divided into four squadrons, referred to as *themata*: *De cerimoniis*, 667.14f.; Haldon, "Chapters II, 44 and 45," 223.64; 258.

162–86   §§28–31   The naval equivalent of the instructions given by Leo in Const. 12.

175   Cf. Maurice, Strategikon, 12.B.21.30–33 (= Ἐκ τοῦ Μαυρικίου, §7)

176–77   οἰονεί τις παράταξις γεγυμνασμένη   Translated by Pryor and Jeffreys, *Age of the Δρόμων*, 497 (Leo, *Naumachika*, §30; and see *Age of the Δρόμων*, 395) as "according to the formation which has been exercised," although Dennis's rendering "like a well-trained battle line" (an alternative might be "like a practiced battle line") is more accurate and fits the context better.

177–78 καὶ ἐν ταῖς ὁρμησίαις … ποιείτωσαν Read "and in the moorings of the *aplekta* let them make their landfall in good order," where ὁρμήσια is taken as "mooring," or "beaching" (cf. Trapp, *Lexikon*, 5:1150–51) and κατάπλουν as "landfall" or "landing," thus rendering the editor's conjectured καί in l. 178 unnecessary: cf. Maurice, *Strategikon*, 12.B.21.16–21 (= Ἐκ τοῦ Μαυρικίου, §4); Dimitroukas, *Naumachika*, 262 and n. 50; and Pryor and Jeffreys, *Age of the Δρόμων*, 197 (Leo, *Naumachika*, §30). The implication of the term *aplekton* is that there were established ports or bases along the coasts within imperial territory, which as with the permanent or semipermanent marching camps for the terrestrial forces were locations at which supplies and provisions would be delivered by the local populations: Dimitroukas, *loc. cit.*, with sources and literature, who notes the difference between an *aplekton* or base/harbor on the one hand, and ὁρμησίαι or σκάλαι, which were beaching places or harbors with a quay and the possibility of mooring rather than beaching the ships (and for the comparable arrangements for supplying fleets and armies by sea/river in the Roman republic, see Erdkamp, *Hunger and the Sword*, 46–52). Very little is known of these Byzantine fleet-bases (such as Phygela, Kepoi, and Attaleia), in contrast to the information available on the Aghlabid *Ribats* of the N. African coast, for example; see Eickhoff, *Seekrieg*, 122–24. For general discussion of Byzantine sea-routes and ports: Kislinger, "Verkehrsrouten zur See" and Gertwagen, "Harbours and Facilities."

181–86 §31 For earlier and more detailed advice on navigation, winds, weather, and tides, see Vegetius, *Epitoma*, 4.38–43, drawing on authorities such as Timosthenes and Varro and a longer Latin tradition of writing about naval warfare and related maritime matters; see Charles, *Vegetius*. No comparable ancient Greek tradition has been identified and neither fragments nor allusions to such texts exist, apart from the topos in some of the ancient treatises to accompanying or planned works on naval matters (see p. 391 above and Rance, "Aineias' *Poliorketika*"). A Byzantine tradition of guides to navigation, winds, and weather did, however, evolve, along with itineraries and geographically ordered lists of ports, watering places, harbors, and so forth, and this probably included some form of charts; see Hunger, *Profane Literatur*, 1:525–27. It is likely that Leo, or his informants, were aware of its existence; see Dagron, "Firmament," and the literature cited below. It may be compared with that which evolved in the Islamic world under Fatimid auspices in the late tenth and eleventh centuries, although the stimulus for the compiling of such guides and maps has been associated (as with the Italian portolan tradition) in particular with commercial expansion in the middle and late tenth century and beyond; see Rapoport, "View from the South." Arab geographical texts of the ninth and tenth centuries sometimes included maps, however, and it is possible that similar maps were made by Byzantines at the same period. Here Leo's advice seems largely unnecessary if addressing practical sea officers, for whom his somewhat simplistic remarks can hardly have been of much value. The last sentences (ἀλλ' ὑφορᾶσθαι . . . πορείαν) read as though the commander of the fleet might be a complete novice. But on the other hand Leo's advice may be intended to remind senior naval commanders that they should rely on the sea officers who had real, long-term experience at sea, rather than on their own intuition or judgment of such matters. This is clearly the advice (and presumably the standard practice) given in the somewhat earlier treatise of Syrianos, *Naumachiai*, 5.1–3.

Although it is the case that on occasion command of Byzantine seaborne expeditions was given to landsmen, some of whom might even be nonmilitary (e.g., the expedition of John, a deacon of Hagia Sophia, in 714/15, during the reign of Artemios: *Theophanis Chronographia*, 385 [trans. Mango and Scott, 535]; and that of 949 against Crete, commanded by the incompetent and inexperienced Constantine Goggyles; see *PmbZ* 23823; Pryor and Jeffreys, *Age of the Δρόμων*, 71 and n. 128; Vasiliev, *Byzance et les Arabes*, 2.1:331–41; and Eickhoff, *Seekrieg*, 325; although it should be noted that Goggyles was defeated after he had landed, rather than in a naval engagement), this seems only very rarely to have been the case. Yet able commanders such as Adrianos (see *PmbZ* 20122; *PLBH* 1:58), Niketas Ooryphas (*PmbZ* 5503 and 25696; *PBE*, Niketas 66; cf. Ooryphas 2, 3), Nasar (*PmbZ* 25490; see Vasiliev, *Byzance et les Arabes*, 2.1:96–99;

2.2:100, 137, 215–16; *ODB* 2:1439), Himerios (see *PmbZ* 22624; *ODB* 2:933) or Romanos Lekapenos (*PmbZ* 26833; see Runciman, *Emperor Romanus Lecapenus*, 63), many of whom came from inland regions of the empire and had made their careers on land long before appointment to a maritime post, clearly relied on experienced seamen, even when they had themselves occupied their posts for more than a few years, as several texts suggest; see Ahrweiler, *Mer*, 398, note 8; Dagron, "Firmament," 145–46; Avramea, "Land and Sea Communications," 81; Christides, "New Light"; and for major sea routes in the Byzantine world see the survey in Dimitroukas, *Reisen und Verkehr*, 2:432–505

The large number of seaborne military undertakings mounted by the empire across the period from the seventh to the tenth century—many of which were in fact very successful, even if there were a number of signal defeats (see Pryor and Jeffreys, *Age of the Δρόμων*, 385; Eickhoff, *Seekrieg*, 65ff., 173ff., 235ff.; and Ahrweiler, *Mer*, 35–44, 93–97, 111ff.)—argues for both a strong tradition of practical seamanship in imperial and provincial fleets and a well-understood need for the sort of technical knowledge required by fleet commanders. This entailed careful navigation through Aegean or Adriatic island waters and through the difficult and often dangerous coastal waters of the central and eastern Mediterranean, knowledge of watering places, safe havens in bad weather, and so forth—see the account of the naval campaigns of Niketas Ooryphas and Nasar in the 880s, for example: *Vita Basilii imperatoris*, 60–63 (214–26) with the comments of Pryor and Jeffreys, *Age of the Δρόμων*, 388–90; Pryor, "Σταδιοδρομικόν"; and Dimitroukas, *Reisen und Verkehr*, 2:520–37.

For the general context of the warfare between the empire and its various Islamic enemies in the central and western Mediterranean, see in particular the survey by Blysidou in Blysidou et al., *Βυζαντινά στρατεύματα στην Δύση*, 263–392 and Lev, "Mediterranean Encounter." For further general remarks on the Arabic evidence for seamanship, navigation and seaborne commerce in the 9th–10th centuries in particular, see also Christides, "Raid and trade."

187–90 §32 Again Leo stresses the need not to plunder the indigenous tax-paying population; see ll. 113–18.

191–98   §33   Cf. Ἐκ τοῦ Μαυρικίου, §6; Syrianos, *Naumachiai*, 5.2–3; 6.1–4; 7.2; 9.43. The warnings Leo issues here may derive from his reading of extracts from ancient naval encounters in Polyaenus or a version thereof. See for possible parallels, for example, *Strategika*, 1.45.2 (*Hypotheseis*, 31.1, Athenian defeat at Aegospotami); also 1.48.4 (*Hypotheseis*, 47.2 and 57.3); 5.32.1 and 2 (*Hypotheseis*, 31.10 and 11) for other examples of inadequate scouts or sentries in naval conflicts. The fact that most naval activity took place within a reasonable distance of the shore is emphasized here and elsewhere by Leo's insistence on the use of scouts from the fleet on land and by the assumption that the fleet will often land troops to fight ashore; see §§23, 40, 57, for example, and Pryor and Jeffreys, *Age of the Δρόμων*, 389–90. An example of the success of the strategy of preparedness described at ll. 197–98 occurred in the early 880s, when Basil I, alerted by imperial spies and other informants in Syrian coastal cities and ports, fitted out a fleet at Constantinople in readiness for an attack on the Aegean provinces of the empire. Arab spies reported back that the empire was prepared for the attack, which was abandoned as a result; see *Vita Basilii imperatoris*, 68 (pp. 234–36). It is possible that Leo has this incident in mind here (although on other occasions he made explicit mention of his father Basil's involvement; see Const. 9§14; 18§95 and §133). For discussion of Byzantine and Arab naval strategy and the absence of major sea fights on the open sea, see Christides, *Conquest of Crete*, 60–61, and on scouting and spying, idem, "Two Parallel Naval Guides," 90–91.

202   ἄνω που διεταξάμεθα   See Consts. 12, esp. §§17–28; 14§§59–69.

203–9   §35   Cf. 14§1 and §84 and Syrianos, *Naumachiai*, 9.15 (and see also 9.16ff.). For the regulations themselves see Const. 8.

210–21   §§36–38   Cf. Syrianos, *Naumachiai*, 9.10–12 (which offers more detailed advice than given by Leo here); Const. 14§35; and cf. Thucydides 7.61.3. This clear admonition to avoid pitched battles at sea wherever possible and except where the Byzantine fleet commander has an absolute superiority in vessels and manpower is indicative both of Byzantine distrust of the sea as a medium for successful warfare and of remembered defeats. Indeed, this was the standard approach for both sides to maritime conflicts; see Christides, *Conquest of*

Crete, 60. For Byzantine attitudes toward, and fears concerning, the sea, see Dennis, "Perils of the Deep"; Vryonis, "Θάλασσα καὶ ὕδωρ"; and for the caution typical of all the naval treatises, Pryor and Jeffreys, *Age of the Δρόμων*, 387–88. For examples of effective Byzantine ambushes or surprise attacks, see *Vita Basilii imperatoris*, 61.9–26 (216–18): in 874 the admiral Niketas Ooryphas was able to surprise a Cretan fleet in the Gulf of Patras by portaging his vessels across the isthmus of Corinth and catching the enemy force unawares, some of their vessels being burned while beached, others sunk or taken at sea; *Vita Basilii imperatoris*, 63 (224–26): the fleet commander Nasar is able to surprise a Saracen fleet while beached, and while the crews were scattered to pillage the surrounding districts in 879. The ships were burned, the raiders for the most part killed or captured. Further examples and discussion: Christides, *Conquest of Crete*, 61.

215–16 Possibly derived from Syrianos, *Naumachiai*, 9.10–12 (but not Demosthenes, *pace* Dennis, *Taktika*, 517 n. 8; see above, chapter 2, p. 50)

222–28 §39 Leo returns to the theme of divine assistance and favor, earned through the purity of the motives of those fighting for their Christian faith. See on 18§123.599–602, §127.620–25 above and the general discussion in the introduction. For treatment of prisoners, see on 15§39.238–49 above.

229–39 §40 See 20§45 and Maurice, *Strategikon*, 8.1.44. Pragmatic advice and an illustration of Leo's awareness of the serious problems of discipline and motivation facing east Roman forces at sea. Note in particular his final sentence, ὀλίγοι γὰρ . . . Ῥωμαίοις, a refreshingly practical admission.

233–34 καταξυλώσαντες See *LSJ*, s.v. and Trapp, *Lexikon*, 4:788. The exact meaning remains unclear, although the implication is, as the editor translates, that pieces of wood or rafts served as a means of getting ashore. Pryor and Jeffreys, *Age of the Δρόμων*, 501, translate it as "abandon ship," which gives the meaning if not an exact translation.

232–37 μὴ ἐν τῇ ἰδίᾳ . . . προτιμήσεται See Syrianos, *Naumachiai*, 9.16–18, where the issue of desertion and the punishments it merited are presented in detail; and ibid., 9.23, where ships placed in a formation near to a friendly coast are described as being the more likely to desert, a passage which may underlie Leo's advice here. But note Syrianos, *Naumachiai*, 9.42 and 44, where the opposite advice is given, on the grounds that if the Roman fleet is defeated it can take refuge on its own shores. There are several instances of desertion or mutiny known from the historical record; see *Vita Basilii imperatoris*, 62 (220–24) (Skylitzes, *Synopsis historiarum*, 154.10–155.34; trans. Wortley, 149), where some of the soldiers/sailors from an imperial fleet sailing toward the Adriatic desert in order to avoid fighting (and where the outcome of the punishment meted out to those whom the fleet believed to have deserted—but who were in fact Muslim prisoners—was to encourage the sailors and soldiers to resume their duties, as described in Syrianos, *Naumachiai*, 9.18); or the *Vita S. Nili*, 9.60 (PG 120:105–8), where in 965 the registered sailors/soldiers of the *thema* at Rossano (Rousianon) destroyed their ships in order to avoid serving at sea: von Falkenhausen, *Untersuchungen*, 127. Reluctance for battle, along with desertion, are clearly major concerns; see Syrianos, *Naumachiai*, 9.21, 23, 29.

240–46   §41   On planning before battle, see on 18§117.567–72 above. See Polyaenus, *Strategika*, 1.48.2, 5 (*Hypotheseis*, 57.2, 4); 3.10.4 (*Hypotheseis*, 14.14); 3.10.12, 16 (*Hypotheseis*, 57.7) for the expectation that the signal to attack or retreat will be given by the commander's vessel.

247–52   §42   The term *pamphylos* may originally have been associated with Pamphylia, although this cannot be proven; see above on 19§§9.61–10.68. Ahrweiler, *Mer*, 415, suggested that *pamphylos* referred to a round-ship rather than a warship, employed for transporting material and livestock. But *pamphyloi* are explicitly described as a subtype or variant of *chelandia* or *dromon*, while round-ships were by definition wind-driven, not oar-driven; see Pryor, *Geography, Technology and War*, 27ff. and Pryor and Jeffreys, *Age of the Δρόμων*, 192. The term could thus apply either to fully manned warships or to specially selected crews for such vessels, as described here by Leo; see Haldon, "Chapters II, 44 and 45," 336–38.

253–58   §43   Cf. Syrianos, *Naumachiai*, 9.7.

259–80   §§44–47   Cf. 12§87 (and Maurice, *Strategikon*, 7.B.16.27–31); on signals see also Syrianos, *Naumachiai*, 7.1–2; Syrianos, *Strategy*, 30.17–20; and above on ll. 240–46.

278 καμελαύκιον For signals at sea and in naval contexts see also Pryor and Jeffreys, *Age of the Δρόμων*, 397–99 for discussion with further sources and literature. The word usually referred to a felt or leather cap or headgear (see Trapp, *Lexikon*, 4:754). In the documents relating to the Cretan expedition of 949 each *dromon* is furnished with a total of 50 *kamelaukia*, with the standard meaning of a cap or headcover (listed along with 50 surcoats, probably for the lower level oarsmen): Haldon, "Chapters II, 44 and 45," 225.106 (= *De cerimoniis*, 670.2) and Dimitroukas, *Naumachika*, 264 and n. 60. For standards and flags, see on Const. 6§16.79–86 above.

281-86   §48 Cf. Syrianos, *Naumachiai*, 8.1.

287-91   §49 See §40 and on the importance of forming a fleet up in an appropriate tactical order Dimitroukas, *Naumachika*, 264 n. 62 and Christides, "Naval Engagement."

292-331 §§50-57 Describes a range of formations that may be based loosely on book 5 of the late third-century BCE *Mechanike syntaxis* of Philo of Byzantium, esp. §§103–9 (ed. Garlan, 326–27; see above, introduction to commentary to Const. 19, p. 390), but more probably on sections of Polyaenus. See, for example, *Strategika*, 3.10.1–2, 3 (*Hypotheseis*, 13.5), 4 (*Hypotheseis*, 14.14), 5 (*Hypotheseis*, 4.3), 6 (*Hypotheseis*, 57.5), 7–12, 13 (*Hypotheseis*, 57.6), 14–16, 17 (*Hypotheseis*, 57.8); 3.11.3 (*Hypotheseis*, 16.1; 57.9), 10; 4.7.6–7 (*Hypotheseis*, 57.14–15); 5.22.1–4 (*Hypotheseis*, 33.1, 16.1 [with 57.9 and 18]; 57.19, 54.18 [with 57.20]) (and compare with Const. 19§§53–56). See also Syrianos, *Naumachiai*, 9.20–41, where similarities are apparent in respect of items such as the crescent-shaped formation (Const. 19§50; Syrianos, *Naumachiai*, 9.30–41), the line abeam (Const. 19§51; Syrianos, *Naumachiai*, 9.24–29), ambushes and outflanking maneuvers (Const. 19§§36, 53; Syrianos, *Naumachiai*, 9.12–13, 25–26, 33). Syrianos makes no mention of feigned retreats (*Taktika*, Const. 19§§54–56), nevertheless the similarities in content remain striking; see Dimitroukas, *Naumachika*, 265–66 and note 65. For discussion of Byzantine naval warfare tactics in general: Pryor and Jeffreys, *Age of the Δρόμων*, 382–406, esp. 399ff.; Eickhoff, *Seekrieg*, 158–70; Christides, "Arab-Byzantine Struggle," esp. 94ff.; and Lammert, "Seetaktik." See also Stanton, *Norman Naval Operations*, 269–72 for similar naval tactics in eleventh- and twelfth-century Norman fleets in the Mediterranean.

292–98   §50   Cf. Syrianos, *Naumachiai*, 9.30.

294–95   ἐν δὲ … ἐνδοξότητα   Cf. Syrianos, *Naumachiai*, 9.6.

299–301   §51   Cf. Syrianos, *Naumachiai*, 9.35–41; Dimitroukas, *Naumachika*, 265, note 65.

302–3   Ποτὲ δε … δρομώνων   Cf. Maurice, Strategikon, 12.B.21.11–12 (= Ἐκ τοῦ Μαυρικίου, §3.1)

310–15   §54   See, for example, Polyaenus, *Strategika*, 1.40.9 (*Hypotheseis*, 19.1); 1.48.2 (*Hypotheseis*, 57.2); 3.10.6 (*Hypotheseis*, 57.5)

316–27   §§55–56   Cf. Syrianos, *Naumachiai*, 9.27 for the same tactic, and see Polyaenus, *Strategika*, 3.10.6 (*Hypotheseis*, 57.5), 12, 16 (*Hypotheseis*, 57.7) for similar examples.

328–31   §57   Attacking an enemy fleet while under construction or undergoing repair, or when beached, was an effective and less costly means of neutralizing their naval power; see on ll. 210–21 above and comments in Dimitroukas, *Naumachika*, 266–67 nn. 69–70. Such advice derives from ancient sources, of course; cf., e.g., Vegetius, *Epitoma*, 4.45.1–2.

332–37   §58   Cf. §§36 and 39.

339–41   οἶον τὸ … καπνίζον αὐτά   Several of the characteristic features of the liquid fire device; see Haldon et al., "Greek Fire Revisited," 293, 310–11. The verb καπνίζω should in this context be understood as "to burn" or "set on fire/ignite," rather than "to blacken with smoke" as the translation reads.

342–43   τοξοβολίστραι … μυίας   See on 5§6.38–39 above and cf. Maurice, *Strategikon*, 12.B.21.12–15 (= Ἐκ τοῦ Μαυρικίου, §3). For ship-mounted antipersonnel artillery in the ancient and Roman world, see Marsden, *Greek and Roman Artillery*, 169–73.

343   μυίας   See Kolias, *Byzantinische Waffen*, 242–45 with sources and literature; Nishimura, "Crossbows"; and Chevedden, "Artillery in Late Antiquity," 146–48 and n. 77. Dennis, "Flies," suggested that the correct term for the bolts used by these weapons was "mice" (μύαι), but Kolias (loc. cit.) has shown that "flies" (μυίαι) was in fact the original form, confused in the manuscript tradition. Leo makes no mention of the sort of artillery mounted on ships used in besieging coastal cities, described by a number of poliorcetic texts; see, e.g., *Parangelmata poliorcetica*, §53, and commentary at 235–40.

344-47   **καὶ θηρία . . . τῶν πλοίων**   None of the surviving Hellenistic or Roman naval treatises or sections make mention of this technique; neither does Vegetius in his section on naval warfare and ship-to-ship fighting (*Epitoma*, 4.44–46), although the idea may have some basis in reality. Malalas records that Cleopatra carried "asps and other snakes in containers" in her warship for this purpose: Malalas, *Chronographia*, 9.10 (ed. Thurn 167.32–34; trans. Jeffreys et al., 116), a tale probably derived from Vergil (*Aeneid*, 8.675–713), although the latter does not give as much detail as Malalas, so another source may be concealed behind this tale. Similarly, Herodian records that the besieging army of Septimius Severus was met by the Parthian defenders of Hatra in 198/99 by, among other things, clay jars filled with highly poisonous scorpions, in which the desert hinterland of the fortress-city was rich: *History*, 3.9.3–8 and commentary (although the editor regards the story as unlikely). There are other purportedly historical examples, although again it is difficult to know in each case to what extent popular legend and the historian's imagination have played a role; see Mayor, *Greek Fire, Poison Arrows*, esp. 176–86. It is most likely that Leo simply borrowed the idea from Malalas without any real understanding of its impracticality at sea. While such a story is not found in Polyaenus, *Strategika*, which includes numerous anecdotes about naval warfare and battles, Frontinus (*Stratagems*, 4.7.10–11) reports a tale in which Hannibal advised Antiochus to hurl jars filled with vipers into enemy ships (although according to another version it was advice given to Prusias: Nepos, *Hannibal*, 11.5–6).

351   **τρίβολοι δὲ σιδηραῖ**   Cf. esp. Philo of Byzantium, *Mechanike syntaxis*, §§104–5; and note *Parangelmata poliorcetica*, §39.8 with commentary, pp. 216–17. The documents for the Cretan expeditionary force in 949 include caltrops for each warship; see Haldon, "Chapters II, 44 and 45," 227.130–31; 229.166–67.

354-61   **§§63–64**   These two passages refer to an age-old tradition of incendiary missiles, on the one hand, and to what is claimed here to be a newer technique, "small 'siphons' thrown by hand from behind iron shields . . ." (and see also on 15.155, above). The term *siphon* had a number of meanings, including a hose or water-lead, a tube or siphon (used to draw wine from a cask), a small water reservoir, or a small earthenware pot (the last two derived from the second); see Mastoropoulos, "Σίφων—σ(ι)φούνι." It could also refer to a tube through which liquids could be projected under pressure, as a very clear description in the *Poliorketika* of Apollodorus (possibly early second century CE, but with subsequent additions by later writers; see Whitehead, *Apollodoros*, 17–34 and critical discussion in Wheeler, review of Whitehead, *Apollodoros*), ed. Whitehead, 56 and 60 (ed. Wescher, 174.4–5, 183.5–6), makes apparent, and hence its use to describe part, or all of, the liquid-fire projecting device. In the context of liquid fire, it appears to have the specific meaning of a hose or tube through which or from which the liquid fire was projected (See *LSJ*, s.v. and Kahane, "Abendland," 408). One possible interpretation of this passage is that this type of pot could be modified so that the spout acted as a type of lead for a fuse by which the incendiary material was ignited. In this case these hand *siphones* were thus a type of hand-hurled pot or "grenade," distinct from the fixed-position liquid fire projectors mounted on the warships, and their particular characteristic—whatever it was—would account for their being referred to as "recently devised" or "prepared."

The issue is complicated by the other evidence, however. One source, although paraphrasing a much earlier third-century treatise (Julius Africanus, *Kesti*, F12.11.22–27), notes that it can be used to project both liquid or prepared fire or noxious juices against horses (*Sylloge tacticorum*, 65: Πρὸς φυγὴν δὲ ῥᾳδίως οἱ ἵπποι τρέπονται ἄν τινες τῶν ψιλῶν καλουμένων πεζῶν τῶν ἀσπιδηφόρων ἱππέων ἑστῶτες ὄπισθεν χειροσίφωνά τε κατέχοντες χυλὸν ἐφορβίου ἔχοντα τοῖς μυκτῆρσι τῶν ἵππων ἐμβάλωσον. . . . "Horses are turned easily to flight if some of the so-called light infantry stand behind the shield-bearing horsemen holding a hand-siphon containing euphorbium juice [Engl. milkweed/spurge: see Wallraff, *Julius Africanus, Cesti*, at 67 n. 55] and hurl it into the nostrils of the horses. . . ."), which suggests strongly that it was in effect a simple single-piston syringe (as in the original third-century text). Fuel or other liquids were drawn up from a reservoir and squirted out; alternatively, a reservoir attached to the body of the syringe provided fuel, which was drawn up by a simple reciprocating valve mechanism. Hand-siphons are likewise listed among the equipment to be made ready for defending a city or fortress against enemy attack, deployed to burn the enemy's siege machines: *De obsidione toleranda*, §113 (ed. van den Berg, 64.8; trans. Sullivan, 189), while "hand-held swivel fire-throwers" are described in the mid-tenth-century *Parangelmata poliorcetica*, §49.20 (with commentary at 231). Given that simple syringes were certainly available before this time, it may be that the novelty consisted in the attachment to the syringe itself of a small tank or tube-shaped reservoir to give a slightly longer-lasting or more effective supply. It is difficult otherwise to explain the use of the phrase "newly devised" in Leo's writings; and an illustration of such a device in an eleventh-century Vatican manuscript certainly supports this interpretation (if it can be relied upon, of course): Cod. Vat. graec. 1605, fol. 36 (repr. in Sullivan, *Siegecraft*, pl. 22). See Dimitroukas, *Naumachika*, 268–69 n. 77; Haldon, "Greek Fire Revisited," 278–80; and Pryor and Jeffreys, *Age of the Δρόμων*, 618ff.

362–65   §65   Cf. Vegetius, *Epitoma*, 4.44 for similar weapons and tactics; Philo of Byzantium, *Mechanike syntaxis*, §§104–5.

369   γερανίων   A type of crane or derrick, cf. γερακαραία, Haldon, "Chapters II, 44 and 45," 211.126 and commentary at 270; Trapp, *Lexikon*, 2:313–14; *Parangelmata poliorcetica*, §2.8–9, §54.6, and commentaries, pp. 161, 240.

373–83   §68   On the practicality or not of the tactics described here, Pryor and Jeffreys, *Age of the Δρόμων*, 203–8.

384–89   §§69–70   All commentators are baffled by these two tactics, which seem at best quite impractical and at worst extremely dangerous; see the editor's comments, 531 n. 14 and Pryor and Jeffreys, *Age of the Δρόμων*, 405–6.

390–96   §71   There was a longstanding Roman and Byzantine tradition of not revealing specialist military technologies or weaponry to outsiders or potential enemies. This is perhaps most obvious where liquid fire was concerned; see *DAI* 1: chap. 13.73–103, with commentary, 2:66–67, and the somewhat romantic but useful discussion in Roland, "Secrecy, Technology, and War." But it applied likewise to other materials, and a general prohibition was repeated in Leo's novel 63 (ed. Troianos, 206–8 and Noailles and Dain, 231–33). How practical, and how strictly observed this control actually was, and to what extent it remained merely a topos of foreign relations, is impossible to say. The export of weapons was strictly prohibited, although once more it is impossible to know the extent to which such prohibitions—repeated in late ninth- or tenth-century codifications derived from sixth-century legal collections—were actually applied or, indeed, practicable. The government at Constantinople clearly was able to enforce the prohibition on the export of certain commodities, such as particular types of silk garment. How far this applied to weapons is not known. Marcian issued a decree forbidding the export of weapons: *CJ* 4.41.2 (a. 455–57) (and see Thompson, *Attila and the Huns*, 180 for the Huns' desire to purchase Roman weapons and metal for weapons production). And in the sixth century Justin II ordered the confiscation of weapons purchased by Avar emissaries in Constantinople while on a diplomatic mission in 562: Menander Protector, *Fragmenta*, 5.4 (trans. Blockley, 53). The military codes retain this prohibition (see Ashburner, "Mutiny Act," §451; Korzensky, "Leges," §45; not repeated in Const. 8, however). In the years 814–820 the emperor Leo V and the doge of Venice attempted to prohibit Venetian merchants from trading with the Saracens, a measure also restricting the commerce in weapons and arms. The doge Ursus I repeated this ban in 876: Tafel and Thomas, *Urkunden*, vol. 1, no. 3 (3); no. 7 (5). Later, in 971, a Byzantine embassy threatened to burn Venetian ships transporting wood to the Arabs, upon which the doge forbade Venetian merchants to sell military supplies, including timber for ship-building, mail, shields, swords, lances and other weapons: Tafel and Thomas, *Urkunden*, vol. 1, no. 14 (27); while Basil II was able to impose an embargo on Byzantines (and by implication Byzantine goods of all kinds) traveling to Muslim lands in 1016 (Felix, *Byzanz und die islamische Welt*, 68, 80-81), although this was almost certainly a temporary measure. But such orders can have been applied only with great difficulty, and the archaeological record from the Balkans and Italy certainly suggests that items of military equipment, both weaponry and armor, traveled widely beyond the nominal borders of the eastern Roman world; see, e.g., Haldon,

"Some Aspects." Weapons were certainly sent as gifts to foreign potentates (see Kolias, *Byzantinische Waffen*, 134); and when necessary the empire was willing to send weapons on a larger scale to its allies—as in 1161–62, when Manuel I sent weapons to help the pro-imperial party at the Hungarian court: Kinnamos, *Epitome*, 223. For the extent of imperial control over "prohibited" goods, see the comments of Antoniadis-Bibicou, *Recherches*, esp. 50–51, 78–79. This was also a concern reflected in other sedentary cultures where "barbarians" were concerned. The export of iron and weapons to the Hsiung-Nu nomads was prohibited by Han Chinese authorities, although the prohibition seems frequently to have been ignored; similar prohibitions were maintained by the T'ang and Ming, with equally limited effectiveness; see references and discussion in Sinor, "Inner Asian Warriors," 142–43; more generally Long, *Openness, Secrecy, Authorship*.

397–99   §72   The editor suggest that this "encyclopaedic volume" was the *Sylloge tacticorum*, but since this was almost certainly a compilation of the years after Leo's reign (see chapter 2 above), it seems much more likely that Leo is in fact referring here to the collection of stratagems of Polyaenus (rather than the tripartite treatise of Syrianos *magistros*, which is probably what Leo is referring to in Const. 18§68.331) of which Leo made extensive use in this constitution and in Const. 20. The reference to an "unlimited number" of stratagems for naval warfare itself might suggest as much. In contrast, the *Hypotheseis*, compiled at some point between the third and ninth centuries and derived from Polyaenus, groups many of the naval stratagems together in section 57 (Melber ed., 498–503; Krentz and Wheeler, *Polyaenus*, 2:990–1000), so that they are easily read and could hardly be described as "unlimited" (although they are not all in this section and several more are to be found in earlier sections).

The stratagems themselves included the employment of a range of pole weapons for cutting rigging or for reaching and attacking the enemy crew from one's own vessel—some are described, for example, by Vegetius, *Epitoma*, 4.44 and 46; many (or their parts) are listed in the records of the expedition of 949; see Haldon, "Chapters II, 44 and 45," 227.125, 131 and 229.168 and Pryor and Jeffreys, *Age of the Δρόμων*, 403–4.

400–424 §73–76   Summarizes §§9–24 above.

425–31 §77   For the *koumbaria,* see on 18§131.656–57 above. These "Scythians" are, of course, the Rus' (Russians), later so called (in A, which paraphrases the *Taktika,* as also in the treatise of Nikephoros Ouranos; see the editor's comment, 533 n. 16). On Rus' vessels, see *DAI* 2:23–25. Leo refers only twice in the *Taktika* to the Rus', and obliquely, as "Scythians" or as "more northerly" peoples, here, and at 14.251–52. On their appearance in Byzantine sources from the ninth century and after, see *DAI* 2:20–23; and on the Rus' journeys down into the Black Sea via the various river barrages, ibid., 2:31–61 and 1: chap. 9, pp. 56–62. For general background, see Franklin and Shepard, *Emergence of Rus,* 91–11.

432–35 §78   See §50 and commentary on Const. 20§59.292–98 below. Note that Leo once more refers to "the testimony of the ancients who made use of this method," in this case probably Polyaenus (see *Strategika,* 3.10.13 and 17 = *Hypotheseis,* 57.6 and 8), possibly Syrianos (and maybe another, unidentified, author). Pryor and Jeffreys, *Age of the Δρόμων,* 515 n. 63 point to Thucydides, *Peloponnesian War,* 3.78.3 and 7.70.4 as examples of the crescent formation used in a withdrawal from a naval engagement; two further examples in Polyaenus, *Strategika,* loc. cit.; and see Stanton, *Norman Naval Operations,* 52.

436–40 §79   Leo refers to his earlier prescriptions in Const. 16, largely drawn from Onasander and Maurice (and see the commentary, above; and Const. 20§§191–92).

441–52 §§80–81   Summarizing earlier instructions in §§9–13 and §§20–22.

453–60 §§82–83   See 2§§18, 21, 34.

CONSTITUTION 20

This constitution is the longest in the *Taktika*, and a considerable proportion of its 221 paragraphs are taken from a similar list of sayings in the *Strategikon*. Many others are drawn in particular from Polyaenus and Onasander, while a good number refer back to advice and counsel given in earlier constitutions. Maurice's collection of maxims is in its turn almost entirely derived from similar older collections such as that of Polyaenus, among others (see detailed notes in Rance, *Strategikon*, 8). The result is that a substantial number of the basic items in the first part of the constitution, taken from Maurice, are repeated in the second half, albeit with different wording, and borrowed from Polyaenus and Onasander. There is thus very little original here, except that Leo often expands on his exemplar, clarifying with examples or parallels the basic concept and, again, referring back implicitly to earlier constitutions in his *Taktika*. There is no doubt that he had at hand a manuscript or manuscripts that included the *Strategikos* of Onasander and the *Strategika* of Polyaenus. From §136 to the end most of the sayings Leo uses are to be found in stories from Polyaenus, although some are also found in Maurice and Onasander. In the material from Polyaenus he takes out, for the most part, the historically specific names and places, so that he can present what were originally anecdotes about particular ancient commanders and their campaigns, battles, tactics, and so forth in an anonymous and neutral guise (on this, see chapter 2, above; and Wheeler, "Notes").

Leo clearly considered his collection of 221 "concise sayings" to represent a concluding section in which the key precepts outlined in the preceding nineteen constitutions were to be given added force and summed up, as the opening words make quite apparent. As has been shown, the opening letters of each paragraph except the first form an acrostic: ἐν ὀνόματι τοῦ πατρὸς καὶ τοῦ υἱοῦ καὶ τοῦ ἁγίου πνεύματος τῆς ἁγίας καὶ ὁμοουσίου καὶ προσκυνητῆς τριάδος τοῦ ἑνὸς καὶ μόνου ἀληθινοῦ θεοῦ ἡμῶν λέων ὁ εἰρηνικὸς ἐν χριστῷ αὐτοκράτωρ πιστὸς εὐσεβὴς εὐμενὴς ἀεισέβεαστος αὔγουστος καὶ *τοοθπννιοα* βασιλεὺς ῥωμαίων. The letters τοοθπννιοα have been shown to replace the expunged name of the emperor Alexander, Leo's brother, a change most likely carried out at the order of the emperor Constantine VII; see Grosdidier de Matons, "Trois études," 3; with Karlin-Hayter, "Emperor Alexander"; but see Schminck, *Studien*, 97 n. 271. It should be noted also that the acrostic referred to Alexander as *basileus*, as opposed to Leo's preferred title of *autokrator*, illustrative of his slightly subordinate position in Leo's eyes. Whether or not the random arrangement of the maxims in Const. 20 was, as suggested tentatively by Grosdidier de Matons, "Trois études," 3, a mnemonic device, remains to be determined.

The acrostic incorporated into Const. 20 emphasizes the strongly Christian, orthodox, and trinitarian aspects of Byzantine identity and belief, underlining the repeated invocations of God and reliance on divine support (for which the general and the army are, however, equally responsible in respect of the piety and respect for the divinity that they demonstrate in their actions and behavior more generally), which in this constitution are referred to on twenty-eight occasions, more than in any other constitution in the *Taktika*. The acrostic was a regular feature of later Roman and Byzantine parainetic and gnomological literature, into which category Leo's Const. 20 fits very well, as well as in Fürstenspiegel, in liturgical poetry, and in various forms of verse (Hunger, *Profane Literatur*, 1:157–65; 2:94, 105–7, 147, 165; *ODB* 1:15). Leo was certainly familiar from his youth with such devices, as the dedication to him of a 66–chapter "book of advice"—Κεφάλαια παραινετικά—with the individual sections bound together by an acrostic from his father Basil I (while there are clear indications of Basil's involvement the body of the text was almost certainly composed by Photios) demonstrates: Basil I, Κεφάλαια παραινετικά (ed. Emminger); Markopoulos, "*Chapitres parénétiques*"; and esp. Čičurov, "Gesetz und Gerechtigkeit," esp. 40–44. In the case of Const. 20, the acrostic serves in particular to reinforce the parainetic value of the text, underlining the legislative authority of the treatise as a whole as presented in the prooemium; by the same token, it reflects Leo's final paragraph in the epilogue (§73; and see also the opening §§2–7 and commentary, below) which follows, in which once again it is the divinity which is the ultimate source of wisdom and advice, and to whom the general—as well as the God-appointed emperor—should look for guidance (see discussion above in chapter 1).

3-8 §1 See pr.§6.

7-8 κατὰ τὸν σόφον. . . . ἀπεργάζεται Cf. Proverbs 1:15, 20:5. Cf. the opening words of text C of the *Three Treatises on Imperial Military Expeditions*, compiled at the order of Constantine VII, modeled on Proverbs 1:8: Ἄκουε, υἱέ, λόγους πατρὸς σου, Σολομῶν σοι παρακελεύεται. . . . The image of Solomon and the attribute most associated with him, that of wisdom, became almost a hallmark of Macedonian imperial propaganda, although it is present in the work of such as Agapetos and George of Pisidia, among others. It was employed in Basil's Κεφάλαια παραινετικά addressed to Leo, for example: μὴ κατόκνει τὰ τῶν παλαιῶν διεξέρχεσθαι γνώμας... καὶ πλέον πάντων τά τε Σολομώντεα (ed. Emminger, 73, §66.18–22), and also by Photios in his praise of Basil I; see Photios, *Epistulae*, 241.91–91; Markopoulos, "Anonymous Laudatory Poem." It is used here and elsewhere by Leo, and in particular by Constantine VII; see *CPTT* 178–79; Magdalino, "Observations," 58; more generally, Magdalino and Nelson, "Introduction," esp. 22ff.; and Tougher, "Wisdom of Leo VI." On Davidic and Solomonic imagery in the imperial image of both Basil I and Leo VI, see discussion in chapter 1, pp. 13 and 27 above.

The following paragraphs, §§2–135, are for the most part repetitions of, or elaborations and paraphrases of, Maurice, *Strategikon*, 8.1.1–44; 2.1–101, with some additional and more contemporary material interpolated by Leo. Leo reproduces the 124 sayings in *Strategikon* 8.1.1–44 and 2.1–80, omitting nine maxims (8.1.17, 19, 31; 8.2.10, 17, 18, 50, 59, 70), and he omitted all the maxims in *Strategikon* 8.2.81–101. Maurice's section 8.2.1–101 appears to have been drawn from Greek translations of older and well-established Latin collections of sayings and advice, partially reproduced in Vegetius, *Epitoma*, 3.26 (although not certainly an integral part of the original treatise. See the introduction in Lang, ed., *Vegetii Epitoma rei militaris*), which Leo takes from Maurice and rewords, expands, or paraphrases in his own §§46 ff.; see Rance, *Strategikon*, notes to 8.1 and 2 and ibid., "Sources." The final sections of this constitution, §§136–221 refer back to earlier parts of the *Taktika*, esp. Consts. 1 and 2, 8, 9, and 12–14, but include additional material from the ancient authorities upon whom Leo draws, notably Onasander and Polyaenus, and bear many similarities to the comparable collection of historical and legendary examples in the *Sylloge tacticorum*, 76–102, also known as the *Leonis imperatoris strategemata* (in Polyaenus, *Strategika*, ed. Melber, 21–22 and 505–40; and *Stratagems of the Emperor Leo VI*, in Krentz and Wheeler, *Polyaenus*, 2:1005–73; see chapter 2, p. 46, n. 16 above). The material is organized only very loosely around three overlapping themes, focusing chiefly on the nature of generalship and command, on the personal qualities of the commanding officer, and on a range of stratagems and tactical ploys to be used against an enemy.

9–17 See Onasander, *Strategikos*, 41.1129–34 (42§2) and 32.899–911 (32§§9–10); Maurice, *Strategikon*, 8.1.1. At l. 25, τὸ γὰρ μέλλον ἀόρατον echoes Isocrates, *Ad Demonicum*, 29.3, although Leo's own source for the maxim was almost certainly Basil I, Κεφάλαια παραινετικά, p. 3, §38.24–25: διότι τὸ μέλλον ἀόρατον. See also Const. 2§6; 12§105; 14§3; §§99–100; 15.6; and cf. *Rhetorica militaris*, 36.2.

26–29 §3 Loosely derived from Maurice, *Strategikon*, 8.1.2–3, although Leo represents the general idea and adds the emphasis on reverence for God.

30–34 §4 Maurice, *Strategikon*, 8.1.2; and Const. 2§28.

35–48 §5 Cf. Maurice, *Strategikon*, 8.1.3.

49–54 §6 Cf. Maurice, *Strategikon*, 8.1.2; Onasander, *Strategikos*, 2.154–57 (2§2). Leo elaborates on the original. The editor's note (541 n. 6) relates in fact to the next paragraph.

55–57 §7 Maurice, *Strategikon*, 8.1.4; Onasander, *Strategikos*, 1.64–65 (1§4); and Const. 2§4.

58–61 §8 Maurice, *Strategikon*, 8.1.8; Consts. 14§97; §15.57; and Rance, *Strategikon*, 8.1.8, note 10. Wrongly identified by Dennis, *Taktika*, as *Strategikon*, 8.1.9.

62–65 §9 Maurice, *Strategikon*, 8.1.5; see Consts. 2§26; §3.8.

66–68 §10 Maurice, *Strategikon*, 8.1.6 and cf. Onasander, *Strategikos*, 13.

69–73 §11 Maurice, *Strategikon*, 8.1.7; and see *Taktika*, pr.§8; 12§108.

74–77 §12 Maurice, *Strategikon*, 8.1.9.

78–82 §13 Maurice, *Strategikon*, 8.1.10.

83–86 §14 Maurice, *Strategikon*, 8.1.12; cf. Const. 14§97. See Onasander, *Strategikos*, 23; and cf. Polyaenus, *Strategika*, 1.35.1.

87–90 §15 Maurice, *Strategikon*, 8.1.11; cf. Const. 17§13.

91–94   §16   Maurice, *Strategikon*, 8.1.13; cf. Const. 14§17; Cf. Polyaenus, *Strategika*, 2.1.3 (*Hypotheseis*, 14.3).

95–98   §17   Maurice, Strategikon, 8.1.14. But see also *Rhetorica militaris*, 56 and esp. 57 for a detailed account of a variety of rhetorical strategies for encouraging defeated soldiers.

99–102   §18   Maurice, *Strategikon*, 8.1.15. See Const. 13§6.

101   Where the first edition reads ὅσον, the editor upon which Dennis based his decision has since preferred the reading of MW, ὡς ἄν.

103–5   §19   Taken from Maurice, *Strategikon*, 8.1.18, but contradicting several of the regulations, likewise taken from Maurice, at Const. 8.20–21, 23, 25.

106–8   §20   Maurice, *Strategikon*, 8.1.16; see also Rance, *Strategikon*, 8.1.16, note 15; Const. 20§72.

109–112   §21   Maurice, *Strategikon*, 8.1.27. Cf. Onasander, *Strategikos*, 10.413–18 (10§13). See on Const. 11§21.116–21 above; and Rance, *Strategikon*, 8.1.27, note 23.

113–20   §22   Based on Maurice, *Strategikon*, 8.1.20 but greatly expanded: cf. Polyaenus, *Strategika*, 1.36.2, 2.10.3 (*Hypotheseis*, 8.4), 4.3.15; and see Const. 20§161. See Wheeler, "Polyaenus: *Scriptor Militaris*," 38 with n. 119.

121–24   §23   Maurice, *Strategikon*, 8.1.21.

125–29   §24   Maurice, *Strategikon*, 8.1.23; cf. Const. 16§15.

130–33   §25   Maurice, *Strategikon*, 8.1.22; cf. Const. 12§104.

134–37   §26   Maurice, *Strategikon*, 8.1.24.

138–42   §27   Maurice, *Strategikon*, 8.1.26, 2.36; and Consts. 18§25, 16§§15–16, 20§97. See Polyaenus, *Strategika*, 3.9.17 (and cf. *Hypotheseis*, 1.19–22) and cf. Syrianos, *Strategy*, 20.6–7; the maxim "I did not expect that" is found in Polybius (10.32.11–12), for example, whom Leo copies from Maurice, loc. cit. The author of the *Vita Basilii imperatoris*, 49.5–7 (174) was familiar with it, and comments on Basil I's campaign in the summer of 878: "Since, however, he expected some kind of attack in the narrow passes (for he knew that it was a paltry excuse for a commander to say, 'I should never have expected it' [οὐκ ἂν προσεδόκησα],) he laid ambuscades at suitable places. . . ." (trans. Ševčenko). See Wheeler, "Πολλὰ κενὰ τοῦ πολέμου."

143–47 §28 Maurice, *Strategikon*, 8.1.25; and see on Const. 17.117–19 above. This is a standard piece of advice which can be found in most of the ancient and later authors: cf., e.g., Onasander, *Strategikos*, 38.1049–54 (38§§5–6) and Syrianos, *Strategy*, 24.16–18, 42.43–46. It is repeated in *Skirmishing*, 24.2 (234 Dennis). See Rance, *Strategikon*, 8.1.25, note 21.

148–54 §29 Again Leo offers a more elaborate and detailed version of Maurice, *Strategikon*, 8.1.28.

155–66 §30 Maurice, *Strategikon*, 8.1.29 and cf. ibid., 9.3.62–74. Again, Leo considerably expands on the original.

167–69 §31 Maurice, *Strategikon*, 8.1.30.

170–73 §32 Maurice, *Strategikon*, 8.1.32; cf. Const. 16§§13–14; and Onasander, *Strategikos*, 11.514–37 (11§§1–5)

174–80 §33 Maurice, *Strategikon*, 8.1.33.

181–85 §34 Maurice, *Strategikon*, 8.1.34; cf. Const. 15§§3–4.

186–89 §35 Maurice, *Strategikon*, 8.1.35; Const. 14§99; cf. Onasander, *Strategikos*, 10.483–85 (10§24)

190–93 §36 Maurice, *Strategikon*, 8.1.40; cf. Onasander, *Strategikos*, 1.75–79 (1§8); and Polyaenus, *Strategika*, 3.7.1 (*Hypotheseis*, 52.2)

194–98 §37 Maurice, *Strategikon*, 8.1.32; Const., 16§§15–16; Onasander, *Strategikos*, 37.

199–203 §38 Maurice, *Strategikon*, 8.1.36; see Const., 17§32.

204–8 §39 Maurice, *Strategikon*, 8.1.37. See on Const. 15§§15–16 and on Const. 17.8–63, above. The contradiction between this and earlier statements is simply repeated from Maurice, *Strategikon*, 9.1.16–20; 11.3.39–40.

209–18 §40 Maurice, *Strategikon*, 8.1.38; cf. Const. 15§§2–4. For ll. 216–18: Plutarch, *Moralia*, *Apophthegmata Lakonika*, 210E (Agesilaos), with further references (Loeb ed.).

219–23 §41 Maurice, *Strategikon*, 8.1.39; cf. Onasander, *Strategikos*, 9.324–33 (9§§2–3).

224–28 §42 Maurice, *Strategikon*, 8.1.43; cf. Const. 14§§21–22.

229–31 §43 Maurice, *Strategikon*, 8.1.41.

232–36 §44 Maurice, *Strategikon*, 8.1.42; cf. Syrianos, *Strategy*, 41.

237–47 §45 Cf. Const. 19.40. An expanded version of the original: Maurice, *Strategikon*, 8.1.44; cf. also Const. 17§61.

248–50   §46   Maurice, *Strat.*, 8.2.15. See also Onasander, *Strategikos*, 9.324–33 (9§§2–3); Polyaenus, *Strategika*, 3.9.35. See Const. 7§2 with 9§4, which also implies that idleness leads to harmful thinking; also 11§5.

251–55   §47   Maurice, *Strategikon*, 8.2.1; Onasander, *Strategikos*, 5.

256–57   §48   Maurice, *Strategikon*, 8.2.2; Const. 7§§1–2; and cf. Vegetius, *Epitoma*, 3.26.2.

258–61   §49   Maurice, Strategikon, 8.2.3; see Const. 19§21; cf. Vegetius, *Epitoma*, 3.26.6.

262–64   §50   Maurice, *Strategikon*, 8.2.5 (and cf. ibid., 3.26.5); and Const. 19§71.

265–68   §51   Maurice, *Strategikon*, 8.2.4; cf. Vegetius, *Epitoma*, 3.26.4. See Const. 17§4; 19§36 and cf. 18§36, §121.

269–71   §52   Maurice, *Strategikon*, 8.2.9; cf. Vegetius, *Epitoma*, 3.26.13 and Polyaenus, *Strategika*, 3.9.35. See Const. 7§2, 11§5.

272–74   §53   Maurice, *Strategikon*, 8.2.6; cf. Vegetius, *Epitoma*, 3.26.7.

275–78   §54   Maurice, *Strategikon*, 8.2.7; cf. Vegetius, *Epitoma*, 3.26.9. See also Consts. 12§18, 13§3, 17§69, and 19§21.

279–84   §§55–56   Maurice, *Strategikon*, 8.2.8; cf. Vegetius, *Epitoma*, 3.26.10–11. For §55 see also *Taktika*, pr.§8 and 12§3.

285–87   §57   Differently worded but also apparently based on Maurice, *Strategikon*, 8.2.9; cf. Vegetius, *Epitoma*, 3.26.12. Note that Maurice 8.2.9 is two sayings, as in the same material in Vegetius (*Epitome*, 3.26.12–13) and Leo (Const. 20§52 and §57), rather than one as in the Dennis-Gamillscheg edition. For discussion of this passage in the context of Leo's views on the justification for war more generally see esp. Stouraitis, "Jihād and Crusade," 19–23; and see on Const. 18§105.508–16, §122.592–98, and §127.620–25 above. Cf. 20§169.

288–91   §58   Maurice, *Strategikon*, 8.2.12; cf. Onasander, *Strategikos*, 4 and above, on pr.32–36; 15§33.207–9; 18§105; and 2§§29–31.

292–96   §59   Maurice, *Strategikon*, 8.2.11; cf. Vegetius, *Epitoma*, 3.26.16. See Consts. 12§§10 and 104, 14§§7 and 23.

297–300   §60   Maurice, *Strategikon*, 8.2.13; cf. Vegetius, *Epitoma*, 3.26.7; Frontinus, *Stratagems*, 1.3.1–10. For related paragraphs: Consts. 12§4, §103, §106; 13§14; 14§24; 17§§3–4; 18§1, §14; 19§2, §31.

301–4   §61   Maurice, *Strategikon*, 8.2.14; Const. 9§§3–4.

305–10   §62   Maurice, *Strategikon*, 8.2.16. Leo will have been aware of cases in the history of his own empire where allies introduced to support

Roman armies were readily turned against them by bribes or other inducements.

311–13   §63   Maurice, *Strategikon*, 8.2.19; cf. Const. 13§8; and Vegetius, *Epitoma*, 3.26.17; with Rance, *Strategikon*, 8.2.19, note 51.

314–18   §64   Maurice, *Strategikon*, 8.2.20 and 21 (317–18); cf. Vegetius, *Epitoma*, 3.26.25. See also Consts. 9§§42–43, 14§77, 18§34, 20§189.

319–22   §65   Maurice, *Strategikon*, 8.2.22; cf. Vegetius, *Epitoma*, 3.26.28.

323–25   §66   Maurice, *Strategikon*, 8.2.23; partly based on a translation from Vegetius, *Epitoma*, 3.26.29. See also Const. 3§9; and cf. *Rhetorica militaris*, 50.1 with commentary, Eramo, *Siriano*, 186 n. 141.

326–34   §§67–68   Maurice, Strategikon, 8.2.24–25; refers back to details given in Const. 17 in particular.

335–37   §69   Maurice, *Strategikon*, 8.2.30; cf. Onasander, *Strategikos*, 36.1006–17 (36§§3–6). See also on the topic of the general's words raising morale Const. 2§12, §18.124ff.; 16§12 and comments.

338–43   §70   Cf. ep.§8. This paragraph seems to be Leo's own composition, but echoes strongly one of the opening sections of the first *parainesis* addressed to him by his father, the emperor Basil I; see Basil I, Κεφάλαια παραινετικά (ed. Emminger), p. 51, §3, enjoining Leo to protect and honor the church and the clergy.

340–43   διὰ τοῦτο . . . καταφρονοῦνται   It is not clear here whether Leo has in mind a particular event or set of events, but the archbishop of Caesarea, Arethas (*PmbZ* 20544), wrote two letters on the subject, the first to Leo VI, the second to the *magistros* Kosmas (*PmbZ* 24110), on the issue of the right of asylum for murderers. The second letter was certainly written after Leo's death, but the date of the first is problematic; see Karlin-Hayter, "Aréthas et le droit d'asile" and Arethas, *Scripta minora* 1: epp. 29 and 30, pp. 257–59, 260–64. Karlin-Hayter thinks (p. 617) that the letter should be dated ca. 899, about the time of the death of Stylianos Zaoutzes (*PmbZ* 27406), who was largely responsible for Leo's collection of *novellae*, since Arethas seems to be responding to a request for an opinion on asylum, but that it might be later. In view of this hitherto unnoticed reference to the issue in Const. 20 (which does not appear anywhere else in the *Taktika*), the letter may well be later, toward the end of Leo's reign, and thus might suggest that Leo was spurred to include in his final constitution—which, as we have seen, was in its way also intended to have the force and authority of imperial legislation—a clear statement that summarized the fundamental principle of the inviolability of ecclesiastical asylum, regardless of the crime committed. The slightly later legislation of Constantine VII (Zepos, *Jus*, 1:232–34; see further discussion in Macrides, "Justice under Manuel I," 190ff.) shows that whatever the date of the letter and the motive for Leo's inclusion of this paragraph in Const. 20, there were important issues about the right of sanctuary after this time which needed to be addressed, and according to Arethas's letter, these concerned the relationship between local civil and military officials, the church and those who sought asylum. Following the legislation of Constantine VII, which seems to have resolved some of the issues in favor of the asylum seeker, a number of cases of individuals fleeing to churches and claiming asylum from the first half of the eleventh century, preserved in the *Peira* (66.24–27) suggest that on the whole the right of asylum was respected; see Macrides, "Killing, Asylum and the Law" (note: the pagination for this article in the first Dennis edition, 561 n. 51 is incorrect), esp. 517–19; and on asylum more generally: Hermann, "Asylsrecht" and Siems, "Asyl in der Kirche." Further discussion of Leo's ecclesiastical legislation: Troianos, "'ἐκκλησιαστικές' Νεαρές"; and idem, "Canons of the Trullan."

344-54   §71 An important passage throwing much light on the nature of Byzantine provincial society, fiscal arrangements and the relationship between the military and the state. See on Const. 4§1.10–11 above (and also on Const. 8§§26.82–87).

344   Ἀγγαρείας ἁπάσης ἰδιωτικῆς   See Stavridou-Zaphraka, "Ἀγγαρεία." *Munera* of various sorts could be imposed either individually, as in this case, or on communities.

345   τῆς λεγομένης ἐξατορίας   Almost certainly the word *exatoria* is a corruption or a demoticization for the technical term ἐξ ἀδορείας. In the legislation of Romanos I and Constantine VII *adoreia* referred to an exemption from military obligations of an impoverished *stratiotes* (and his transferral to other duties) and to the consequent attribution of this holding by the state (first, according to a list of priorities outlined in the novel in question, or, when the sequence of persons of

stratiotic status liable for the fiscal dues of the property in question failed to provide a suitable tenant) to a person not registered as of military status, in order to ensure the continued (fiscal) productivity of the property in question and to ensure the continued contribution of such properties, as partial *strateiai*, to the maintenance of a soldier. See Constantine VII's novel of 947, "On Soldiers," in Svoronos, *Novelles*, 104–26, at §A9 (121.70–122.75) (= Dölger, *Regesten*, no. 673); Engl. trans. in McGeer, *Land Legislation*, 71–76, esp. p. 74, 1§9; and cf. *De cerimoniis*, 696.1–9; with Haldon, "Military Service," 30–32. It had as an effect the extension of military obligations (as opposed to active military service) to nonmilitary subjects of the state. The *adoreia* functioned similarly to the regulation known as *sympatheia*, by which civilian landholders within a fiscal community, who had fallen on hard times, had their properties temporarily relieved of fiscal burdens until the owner could once more bring them back into good order (a maximum period of thirty years was usually granted). After this time, such a property was normally declared a *klasma*, and was detached from the fiscal community and attributed to a new owner/holder by the state. In the case of military holdings the second stage—that of the *klasma*—did not apply. Instead, the state attempted first to maintain the holder of the *strateia* through the appointment of contributors (*syndotai*). If this did not work, then the military version of the *sympatheia* was invoked, and the land in question was placed in *adoreia*, by which it was granted to another. For clasmatic land, see Lemerle, *Agrarian History*, 81–82, 162–63; and especially Oikonomidès, "Verfalland."

346–47 τούς τε δημοσίους φόρους καὶ τὰ ἐπικείμενα αὐτοῖς ἀερικὰ   The "public taxes" were chiefly the land tax and related regular impositions such as the *kapnikon* on households and the *ennomion* on pasturage/flocks and herds; see Oikonomidès, *Fiscalité*, 46–76. The *aerikon* seems to have been a tax in cash raised on the basis of, or in proportion to, the value of the other regular taxes (the word derives from Lat. *aes/aeris*, copper/bronze), although it later (by the middle of the eleventh century) referred to a judicial fine. Each *thema* had its own treasury, under the *chartoularioi* (sometimes called *exochartoularioi*) *ton arklon*, responsible to the general *logothesion* (see *De* *cerimoniis* 694.18; Philotheos, *Kletorologion*, 113.29) and the cash raised from the *aerikon* was retained by the thematic treasury, to be disbursed on local needs such as those listed in the next lines (347–50): fortress construction, shipbuilding, bridge building, road building, and so forth; see Haldon, "Aerikon/aerika."

350–54 This refers to a general imposition upon all nonmilitary households—Leo is particularly keen to emphasize that the wealthy as well as the ordinary peasantry should be subject to such impositions, harking back both to what was said in 4§1 and in 18§123. See on §205 below.

355–62 §72 See on 14§§31–32.204–10 (and Onasander, *Strategikos*, 36). See also Const. 16§11.

363–65 §73 Cf. 11.231–38; 7.8–20; Onasander, *Strategikos*, 9.324–33 (9§§2–3); 10.337–41 (10§1); and see also Consts. 6§§1, 19; 11§41.

366–68 §74 See 12.409–20 and 13.12–19 and commentary, above; also 14§101; and in a naval context, see 19§35.

369–72 §75 Cf. 14§59.412–22 and 4§66, where Leo (following Maurice, *Strategikon*, 12.B.8) repeats the ancient arrangements for infantry files of 16, but also repeats Maurice in building into his account (4§65 = Maurice, *Strategikon*, 12.B.8.12–16) the possibility that numbers may not permit this arrangement. The infantry file depth of 10 men appears to be the standard alternative and may represent Leo's own attempt to rationalize. Alternatively it may be intended to include also archers or light troops. Cf. Maurice, *Strategikon*, 12.B.12.2–7 (*Taktika* 14§60.429–32), where archers are to be stationed behind the heavy infantry in proportion of 8 *skoutatoi* to 2 archers, making a line 10 men deep. It is not clear whence Leo derives a 10-wide infantry column. For the positioning of the cavalry, see Const. 7§42.

373–77 §76 Probably based on Polyaenus, *Strategika*, 3.9.10. Cf. Const. 18§134, where Leo suggests a similar drill for cavalry horses; Const. 19§29, suggested for naval forces.

378–85 §77 Cf. *Rhetorica militaris*, 40–41; and see Consts. 2§§22–24; 14§1; 19§24; cf. also 19§58. The motif "God helps those who help themselves" is ancient, of course: cf. Xenophon, *Cyropaedia*, 1.6.5–6.

386–90 §78 See on 14§101.703–7 above; and cf. 12§57; 20§§10 and 141; with Onasander, *Strategikos*, 10.10 (and Frontinus, *Stratagems*, 1.11.14–16).

391–94 §79 Cf. 20§14 and commentary, above.

395–409 §80   Polyaenus, *Strategika*, 8.16.1–2 (*Hypotheseis*, 3.8)

410–21   §81   See on 6§5 and cf. 11§41.

422–26   §82   Cf. 7§43.304–12 and commentary (and see 13§15), and cf. Onasander, *Strategikos*, 10.2.381–91 (10§2). At l. 543 translate σπουδαῖον as "zealous" or "diligent," rather than "serious," which makes little sense in the context, as also at Const. 2§18; ep.§29.

427–31   §83   See on 8§26.82–87 and 19§19.119–23 with further literature and discussion; and cf. part 3 of Constantine VII's novel of 947: καὶ οὕτως οἱ ἄρχειν τοῦ στρατοῦ λαχόντες ὥσπερ ὄντος νόμου τὰ τούτων εἶναι τῶν στρατηγῶν, ἅπαντες ἦγον ἄνω τε καὶ κάτω τοὺς στρατιώτας, οἳ δῶρα λαμβάνοντες ἀντεδίδωσαν αὐτοῖς ἀστρατείαν, ἄνθρωποι ὤνιοι, ἀμελεῖς, ἀπόλεμοι, μυρμήκων ἀγενέστεροι καὶ λύκων ἁρπακτικώτεροι δι᾽ ὧν τοὺς ἐχθροὺς δασμολογεῖν οὐκ ἔχοντες ἠργυρολόγουν τοὺς ὑπηκόους. Ὅθεν οὐκ ἐδεήθησαν χρόνου πρὸς τὸ πᾶν ἀνατρέψαι ἀλλ᾽ ἐκ τῆς σφῶν καταφορᾶς εἰς ἔσχατον κίνδυνον τὴν τῶν Ῥωμαίων ἤλασαν ἀρχήν ("and so those who received command of the army, as though there were a law that commanders should be of this sort, all led the *stratiotai* hither and thither, and taking gifts they handed out exemptions from service in return. Venal, negligent, unwarlike, these men were baser than ants and more ravenous than wolves. Unable for these reasons to exact tribute from the enemy, they forced money from our subjects, with the result that they needed no time to achieve complete disruption, and by their indolence they drove the Roman state into the utmost danger"): Svoronos, *Novelles*, 124.121–25.129; Engl. trans. in McGeer, *Land Legislation*, 75–76, with commentary at 68–70.

432–37   §84   Maurice, *Strategikon*, 8.2.26.

438–40   §85   Maurice, *Strategikon*, 8.2.27 = Vegetius, *Epitoma*, 3.26.30; cf. Const. 16§§3–4.

441–43   §86   Maurice, *Strategikon*, 8.2.28 = Vegetius, *Epitoma*, 3.26.32. The idea is also given expression in Consts. 12§§107–108; 17§2, §4.

444–49   §87   Maurice, *Strategikon*, 8.2.29. See on Const. 17§89 with references, and cf. Onasander, *Strategikos*, 10.393–400 (10§9), with 10.427–35 (10§15). See Rance, *Strategikon*, 8.2.29 and note 59. See Const. 17§91.

450–55   §88   An expanded version of Maurice, *Strategikon*, 8.2.31 with 8.1.5. The quotation of Alexander the Great is Leo's addition; see also Consts. 2§26; 3§1, §8.

456–63   §89   See Const. 12§90. Based on *Strategikon*, 8.2.80, but with Leo's own additional remarks on the possible treachery of allied forces: cf. Const. 18§40 and 20§62. The best example is from considerably later, during the opening stages of the battle of Mantzikert in 1071, when some of Romanos IV's Pecheneg allies deserted the Romans and went over to the Seljuk commander Alp Arslan; cf. Attaleiates, *Historia*, 157.20–158.3 (trans. Kaldellis and Krallis, 287). Misidentified in Dennis, *Taktika* (1st ed.), as *Strategikon*, 8.2.38 (= Const. 20§99; see below)

464–67   §90   The concept, which Leo takes from Maurice, *Strategikon*, 8.2.60, was ancient and well-established: cf. also Vegetius, *Epitoma*, 3, pr.8; cf. Syrianos, *Strategy*, 4.9–14; 5.6–10; *Rhetorica militaris*, 3.3–4, 4.1; and see Haase, "Si vis pacem." See also pr.§4; Const. 2§§30–31. *Pace* Dennis, *Taktika*, 569 n. 61 the text here has no connection with Aristotle, *Politics*, 7.1333a35.

468–69   §91   Maurice, *Strategikon*, 8.2.58; cf. Consts. 2§5, §25.

470–74   §92   Maurice, *Strategikon*, 8.2.32; cf. Onasander, *Strategikos*, 13; and cf. ibid., 14.580–90 (14§§1–2).

475–82   §93   Maurice, *Strategikon*, 8.2.35 (and cf. Const. 2§§19, 28); see Onasander, *Strategikos*, 1.82–83 (1§10) and 2.155–57 (2§2). Misidentified by Dennis, *Taktika* (1st ed.), as *Strat.*, 8.2, 33.

483–85   §94   Maurice, *Strategikon*, 8.2.34. See Const. 2§4.

486–88   §95   Maurice, *Strategikon*, 8.2.33.

489–92   §96   Maurice, *Strategikon*, 8.2.37; cf., e.g., Polyaenus, *Strategika*, 1.38.2 (*Hypotheseis*, 47.1)

493–99   §97   Maurice, *Strategikon*, 8.2.36; cf. also Onasander, *Strategikos*, 37 and comm. to Const. 20§27 above; Const. 16§§15–16.

500–504   §98   Maurice, *Strategikon*, 8.2.37; cf. Const. 17§§69–73 and Const. 20§96.

505–14   §99   Maurice, *Strategikon*, 8.2.38. Cf. Consts. 12§34; 17§§71–72.

515–18   §100   Maurice, *Strategikon*, 8.2.73.

519–23   §101   Maurice, *Strategikon*, 8.2.40.

524–26   §102   Maurice, *Strategikon*, 8.2.41. For §§101–2 see Const. 14§§10, 28, 79; and cf. 20§155.

527–31   §103   Maurice, *Strategikon*, 8.2.43; see Const. 14§31.

532–40   §104   Maurice, *Strategikon*, 8.2.44. See Const. 12§§37–38 and commentary to 12.274ff.

541–43 §105 Maurice, *Strategikon*, 8.2.47; cf. Const. 12§104.

544–47 §106 Maurice, *Strategikon*, 8.2.45.

548–51 §107 Maurice, *Strategikon*, 8.2.42 (and cf. Const. 12§71.525–27)

552–55 §108 Cf. ep.§51; this is a well-established tactic: Maurice, *Strategikon*, 8.2.39 = Vegetius, *Epitoma*, 3.14.1–3 and Polyaenus, *Strategika*, 6.38.4 (restored after *Hypotheseis*, 23), cf. Frontinus, *Stratagems*, 2.2.7. See Rance, *Strategikon*, 8.2.39, note 68.

556–59 §109 Maurice, *Strategikon*, 8.2.49; see Const. 12§106 and cf. Vegetius, *Epitoma*, 2.9.11–12.

560–63 §110 Maurice, *Strategikon*, 8.2.74. See *Rhetorica militaris*, 1.3–4 and 4.1 and Onasander, *Strategikos*, 1.97–111 (1§§13–16). See Const. 2§12, §18.126ff.

564–70 §§111–12 Maurice, *Strategikon*, 8.2.51–52; cf. Consts. 15§6 and 16§9.

571–76 §113 Maurice, *Strategikon*, 8.2.53; and cf. Const. 5.

577–78 §114 Maurice, *Strategikon*, 8.2.46; cf. Const. 12§55.

579–81 §115 Maurice, *Strategikon*, 8.2.48. See on Const. 18§118.576–78; and note Rance, *Strategikon*, 8.2.48, note 72.

582–84 §116 Maurice, *Strategikon*, 8.2.54; cf. Const. 2§§15–16 and Onasander, *Strategikos*, 2.129–53 (1§§21–25)

585–87 §117 Maurice, *Strategikon*, 8.2.55 (and cf. Const. 11§12)

588–89 §118 Maurice, *Strategikon*, 8.2.66; cf. Const. 2§2; Onasander, *Strategikos*, 1.59–63 (1§§2–3)

590–97 §119 Maurice, *Strategikon*, 8.2.56; cf. Consts. 17§4, 18§121, 19§36.

598–600 §120 Maurice, *Strategikon*, 8.2.57; cf. Const. 2§14.

601–4 §121 Maurice, *Strategikon*, 8.2.61. Cf. also Vegetius, *Epitoma*, 1.13.6–8 The general importance of "extensive investigation" is examined in Const. 3§§11–13. On l. 604 (οἱ γὰρ τεθνηκότες διεφθάρησαν), see Wheeler, "Πολλὰ κενὰ τοῦ πολέμου," 165 n. 38.

605–16 §§122–23 Maurice, *Strategikon*, 8.2.75–76. Camps are explored in more detail in Const. 11; the topics specifically mentioned here appear in Const. 11§§3–4 (health of the camp), §6 (provisions), §27 (health).

617–20 §124 Maurice, *Strategikon*, 8.2.77; cf. also Onasander, *Strategikos*, 41.1151–54 (41§6), and Const. 17§§3–4.

621–25 §125 Maurice, *Strategikon*, 8.2.78. See Consts. 11§12, 13§7, 16§9, 18§66.

626–28 §126 Maurice, *Strategikon*, 8.2.63. Cf. Const. 11§12.

629–30 §127 Maurice, *Strategikon*, 8.2.62.

633-34   See on 2§32.220–21, from where it is repeated (and cf. 2§13). See Plutarch, *Moralia*, 187D (Chabrias); Dennis, *Taktika*, 37, n. 15; and Rance, *Strategikon*, 8.2.79 and note 85; with Maurice, *Strategikon*, 8.2.93.

635-41   §§129–30   Maurice, *Strategikon*, 8.2.64–65. Cf. Const. 7§2. The advice in §130 appears also in 3§3 and §8.

642-49   §§131–32   Maurice, *Strategikon*, 8.2.67–68; cf. Const. 3§1.

650-52   §133   Maurice, *Strategikon*, 8.2.72; cf. Const. 3§7.

653-56   §134   Maurice, *Strategikon*, 8.2.71; cf. Const. 19§21.

657-61   §135   Maurice, *Strategikon*, 8.2.69; Onasander, *Strategikos*, 1.112–15 (1§§17–18); Polyaenus, *Strategika*, 1.pr.3; and cf. Const. 2§6, §32.

662-67   §136   Cf. Aelian, *Tactica theoria*, 1.pr.3. From this point until the end of Const. 20 Leo relies heavily on Polyaenus, or the *Hypotheseis* based on his *Strategika*, and Onasander. Cf. Const. 12§4.

668-72   §137   Onasander, *Strategikos*, 2.131–48 (1§§21–23).

672   ἄχρι καὶ αὐτοῦ τοῦ ὀνόματος εἶναι τῆς νίκης σημαντικήν   Compare this paragraph with Leo's (i.e., Onasander's, ultimately Aristotle's) reflections in Const. 2§15; see on 2§32.220–21 above (and cf. also 2§9). Is this a not-so-veiled reference to Leo's successful general *Nikephoros* Phokas?

673-78   §138   The source of this story is not clear; but see Polyaenus, *Strategika*, 7.6.5; Onasander, *Strategikos*, 21.695–99 (21§3).

679-84   §139   See 18§§132–33 and cf. Polyaenus, *Strategika*, 3.11.9–10; 12 (*Hypotheseis*, 57.10).

685-90   §140   The general sentiments in this paragraph parallels those expressed throughout Basil I, Κεφάλαια παραινετικά. A specific military example Leo may have had in mind from his own experience may have been the quarrel between the commanders Leo Apostypes and Prokopios in S. Italy in 880 (*PmbZ* 24341, 26758), as a result of which Apostypes failed to reinforce Prokopios's wing in battle with the Saracens and Prokopios was killed, although the Byzantines were victorious: *Vita Basilii imperatoris*, 66 (pp. 228–32). An earlier case of which he may have read in the chronicles at his disposal occurred at the battle of Versinikia in 813, when the general Leo (later Leo V: *PmbZ* 4244) reportedly failed to support another commander, and as a result the battle was lost: *Theophanis Chronographia*, 500 (trans.

Mango and Scott, 684); *Scriptor incertus* (ed. Iadevaia) 42.78–81; and Theophanes continuatus, 13–15. Cf. Const. 3§3.

691–96   §141   See on Const. 14§101.703–7 above.

697–701   §142   Cf. Onasander, *Strategikos*, 27.

702–6   §143   Cf. *Rhetorica militaris*, 10 and 14.4–5; and cf. *Taktika*, pr.§4.

707–13   §144   See Polyaenus, *Strategika*, 1.34.1 (= *Hypotheseis*, 57.1), 5.16.3 (= *Hypotheseis*, 58.2), 5.19 (= *Hypotheseis*, 54.17), 5.40 (= *Hypotheseis*, 58.3), 5.41 (= *Hypotheseis*, 58.4), 5.44.5 (= *Hypotheseis*, 28.3), 6.12; for this stratagem see also, for example, Frontinus, *Stratagems*, 2.5.10; 2.9.9–10; 3.2.3–4, 2.7–11; and 4.7.12 and 23.

714–21   §145   Polyaenus, *Strategika*, 1.40.3 (= *Hypotheseis*, 25.1); and cf. Const. 14§30.

722–28   §146   See on Const. 17.117–19 above; and cf. Polyaenus, *Strategika*, 2.1.4 (*Hypotheseis*, 45.2); 3.9.2 (*Hypotheseis*, 45.3), with 1.39.3 and 1.40.7.

726–28   See Polyaenus, *Strategika*, 1.40.7 (*Hypotheseis*, 22.2) and cf. 1.39.3 (= *Hypotheseis*, 22.1).

733–37   Cf. Polyaenus, *Strategika*, 1.39.2 and *Parangelmata poliorcetica*, 11.18–21 for wooden supports or pattens under the soles of the soldiers' boots. The somewhat obscure last sentence means that, even if some, infantry or cavalry, should wish to protect themselves from falling onto the caltrops by other means, this method will still protect them. On sole-plates for horses as protection against caltrops, see on Const. 5.33; 14.277–92.

738–44   §148   A tale borrowed from Numbers 25:6–14; cf. also Polyaenus, *Strategika*, 8.16.2 and 8.16.6 (*Hypotheseis*, 3.8 and 4.13) on Scipio's restraint and self-discipline. For the self-control to be exercised by the general see Const. 2§§1–3, 18.

745–48   §149   See on 20§14 above; and cf. 20§179 and §213 and 14.703–7; cf. also Onasander, *Strategikos*, 10.486–506 (10§§25ff.) on taking omens before battle or campaign; and see Const. 12§57 and 20§78.

749–54   §150   Refers to a classic Byzantine stratagem, frequently deployed and regularly recounted in contemporary histories and chronicles. See in particular the recommendations in *DAI* 1: chaps. 2–6 and 10–12, for example. For literature on similar earlier Roman policy, with examples: Wheeler, "Methodological Limits," 225 n. 167.

755-57   §151   Leo passes over the difficulties of actually carrying out this maneuver, about which he has more to say in earlier constitutions. See Polyaenus, *Strategika*, 2.1.19 (*Hypotheseis*, 32.3) (and cf. Frontinus, *Stratagems*, 2.6.6); and Consts. 12§61–62; 14§§52, 94.

758-60   §152   Cf. Polyaenus, 3.11.8, although a distant parallel.

761-65   §153   See Onasander, *Strategikos*, 33, and above, on Const. 14.690–99 (§§99–100). See also Consts. 7.276; 12§105; 14§§3, 23; 20§§159, 193; and ep.§34.

766-68   §154   Polyaenus, *Strategika*, 2.1.17 (*Hypotheseis*, 15.2); 4.4.3; 4.19; similar strategem at 2.1.7. This is described as a method employed by the Saracens in Const. 18§107.

769-73   §155   See Consts. 14§§91, 98; 20§§101–2; and cf. Polyaenus, *Strategika*, 2.2.3. The advice is standard in the older treatises, e.g., Frontinus, *Stratagems*, 2.2.5; note also Julius Africanus, *Kestoi*, 1.1.22–25 (Vieillefond 83–86); Ammianus Marcellinus, 25.1.

774-78   §156   See Polyaenus, *Strategika*, 2.3.4, a tale which Leo has edited to fit his own preconceptions here. See also ibid., 1.32.2 (Leonidas warns his Spartan army of a forthcoming thunderstorm, thus preparing them for the shock to follow) for similar tales of generals turning such events to their own advantage, and on §198 below.

779-81   §157   It may be symptomatic of the already multiethnic composition of Byzantine field armies that Leo incorporates this bit of advice. His original source is Polyaenus, *Strategika*, 2.3.6 (*Hypotheseis*, 14.5), a story about Epaminondas of Thebes.

782-86   §158   Polyaenus, *Strategika*, 2.1.23 (cf. *Hypotheseis*, 34.1); 2.3.11; 7.15.4 (*Hypotheseis*, 34.1). Similar strategems were recommended in order to conceal the number of deserters, for example; see Polyaenus, *Strategika*, 2.1.15 (*Hypotheseis*, 34.2).

787-91   §159   Polyaenus, *Strategika*, 2.3.15 (*Hypotheseis*, 14.7; and cf. 3.9.22 [cf. *Hypotheseis*, 18.4]). An example is given in Const. 14§97. See on 20§153 above.

792-94   §160   See Onasander, *Strategikos*, 24, and on Const. 4.157–67 above.

795-97   §161   Polyaenus, *Strategika*, 4.3.15 (*Hypotheseis*, 8.5); see on §22 above.

802-3   See Polyaenus, *Strategika*, 2.10.5, repeated from Plutarch, *Moralia, Apophthegmata Lakonika*, 229B (Lysander) and idem, *Lives, Lysander*, 7.4.

804-7   §163   Polyaenus, *Strategika*, 2.20, repeated from Aeneas Tacticus, 31.14.

808-15   §164   Leo's summary of a story in Polyaenus, *Strategika*, 3.6.

816-20   §165   Cf. Polyaenus, *Strategika*, 3.9.1 (*Hypotheseis*, 5.2; and see Const. 13§6)

821-24   §166   On weeding out the timorous among the soldiers see Polyaenus, *Strategika*, 3.9.10; and cf. 3.11.8.

825-31   §167   Cf. Polyaenus, *Strategika*, 3.9.13 (*Hypotheseis*, 13.4). Leo appears to suggest the opposite in Const. 18§36, reflecting the contradictions among the sources he exploits.

832-38   §168   See Polyaenus, *Strategika*, 3.9.18 (*Hypotheseis*, 46.4). See Consts. 2§§29-31, 16§§15-16, 19§39.

839-44   §169   Cf. Onasander, *Strategikos*, 4, and see on Const. 2.193-221; for discussion see Stouraitis, "Methodologische Überlegungen," 278ff.; idem, "Jihād and Crusade," 19-23, and on Const. 20§57 above. Note also ep.§§14-17.

845-56   §§170-71   See on Const. 2.193-216; *Rhetorica militaris*, 9ff.; Onasander, *Strategikos*, 4.197-206 (4§§4-5)

857-59   §172   Onasander, *Strategikos*, 5. See Const. 14§1. The naval version is offered at Const. 19§24.

860-65   §173   Onasander, *Strategikos*, 6.218-24 (6§1); see Const. 9§§5, 29.

866-68   §174   Onasander, *Strategikos*, 6.256-60 (6§§7-8); see Const. 4§25 and above on 4.103 and 4.105.

869-75   §175   Cf. Polyaenus, *Strategika*, 3.9.35; see Onasander, *Strategikos*, 10 and *Rhetorica militaris*, 40-41; cf. Consts. 7§§2, 12; 11§5.

876-80   §176   Cf. Maurice, *Strategikon*, 1.8.16 (= Const. 8§20).

881-86   §177   Cf. Polyaenus, *Strategika*, 3.9.32 (*Hypotheseis*, 3.2); cf. Const. 18§15 (and Const. 15§55)

887-91   §178   See Onasander, *Strategikos*, 3; 10.480-85 (10§24); Const. 3.9.

892-97   §179   See §149 and commentary, above.

898-903   §180   Cf. Onasander, *Strategikos*, 11.530-36 (11§4). See also Consts. 14§7; 12§§10, 104.

904-8   §181   This paragraph summarizes Const. 2.58-70 but reflects also the advice scattered through Polyaenus and Onasander. Cf. Consts. 14§2; 18§20.

909-15   §182   Cf. Onasander, *Strategikos*, 15-16 (and see Const. 14.3-7); cf. Consts. 18§§4, 14-15, 73; also briefly mentioned in Const. 1§11.

916-22   §183   Cf. Onasander, *Strategikos*, 20; cf. Consts. 7§§48, 53–54; 14§91.

923-37   §184   Cf. Onasander, *Strategikos*, 21.705–27 (21§§5–8); and see on Const. 19.292ff., with further references (from Polyaenus, *Strategika*) to the crescent formation in naval contexts. It is also described in Const. 12§§63 and 67.

938-42   §185   Cf. Onasander, *Strategikos*, 22.754–61 (22§4). Leo seems to give an example where this is not the case in Const. 12§78, in which he recommends attacking the flanks; see also Const. 12§§26–27.

943-48   §186   Cf. Onasander, *Strategikos*, 25, and see the opening and closing advice in Const. 12§59; for the issue of orders see also Const. 13§4, and in a naval context, Const. 19§26. On the manuscripts Dennis used from this point forward see above, p. 58 n. 47.

949-52   §187   Cf. Onasander, *Strategikos*, 32.899–911 (32§§9–10). See on Const. 18.3–9 above; and cf. Consts. 14§100, 19§58 (and compare with advice on planning at 3§1 and 5§11).

953-59   §188   For shining weaponry and equipment, see on Const. 14.211–28. Cf. Maurice, *Strategikon*, 7.B.15 and Onasander, *Strategikos* 29.834–41 (29§§1–2); also see Const. 14§98. On battle cries see Const. 12§§55 and 83.

960-63   §189   See Polyaenus, *Strategika*, 2.5.2; Onasander, *Strategikos* 31; cf. Const. 14§77 and 20. §64.

964-67   §190   Cf. Onasander, *Strategikos*, 32.884–95 (32§§5–7)

968-83   §§191–92   Cf. Onasander, *Strategikos*, 34; Const. 19§79; and for the general's praise of the soldiers after a victory see also *Rhetorica militaris*, 55.5. Although the issue of dividing the spoils of war has been mentioned already on several occasions (e.g., Const. 16§§3–5), this is the first time that Leo refers specifically to the proportional division to be observed and, more particularly, to the amount reserved for the fisc (*to demosion*). The emperor leaves the distribution largely up to the general's discretion, apart from the fifth designated for the state; although in the *Ecloga*, 18.1 (repeated in the *Procheiros Nomos*, 40.1 = *Epanagoge*, 40§93), and in the *Sylloge tacticorum*, 50.4 the proportion reserved for the state was one sixth; and in practice this varied from campaign to campaign and from commander to commander, according to circumstances; see Dain, "Partage du butin"; Dagron and Mihăescu, *Traité sur la guérilla*, 231ff.; *ODB* 1:309. The spoils of war were an important incentive for the soldiers, as clearly noted by Leo at Const. 16§§2.6–5.32 (see also commentary, above); and even more explicitly mentioned at *Rhetorica militaris*, 45.6. The regulation in the *Ecloga*, repeated in the *Procheiros Nomos*, also permits the general to take from the sixth reserved for the fisc additional rewards for officers who have distinguished themselves.

984–92 §193 Body metaphors in which the army represented one element in the state or society were common. Constantine VII noted that "the army is to the state as the head is to the body; neglect it, and the state is in danger": Svoronos, *Novelles*, 118.1–2; Dölger, *Regesten*, no. 673; and note Const. 11. §9, where Leo VI describes peasants and soldiers as the two pillars upon which the polity was founded. Here the different elements of an army themselves are represented by the limbs, head and trunk of the body. See also above, commentary to Const. 4.10–11; and note 20§159 (the enemy general as the head of a viper—once destroyed the enemy force will be as a headless body). Polyaenus, *Strategika*, 3.9.22 records that Iphikrates (an early fourth-century BCE mercenary commander) likened the battle line to the parts of a body—the main phalanx to the trunk, the light troops to the hands, the cavalry to the feet, and the head to the general: "for if any of the lesser parts are wanting the army was defective; but if it lacked the general, then it lacked everything." This body metaphor also had important implications for political theory, of course. Since the state or polity could be seen metaphorically and allegorically in terms of the human body, the latter was a microcosm of the wider world, and explanations for sickness and its causes could thus have a very specific political dimension; see esp. Benakis, "Stellung"; Krüger, "Anthropologie"; and in general Gahbauer, *Anthropologische Modell*. The question of the application of the correct cure could thus become a particularly acute issue—differences of opinion on how the human body became open to illness and disease, and what treatment was best prescribed, when transferred to the political level, would result in radically different suggestions for dealing with the maladies of state and society.

990–92 διὰ τοῦτο . . . σωτηρίας See on 14§§99.690–100.699 and cf. Consts. 7.276; 12§105; 14§§3, 23; and 20§§153, 159.

993–1000   §194   Cf. Polyaenus, *Strategika*, 3.9.19 (*Hypotheseis*, 15.3).

995   καμάρδαν   See on 10§11.50 above.

1001–6   §195   Cf. Onasander, *Strategikos*, 10.362–75 (10§4); Polyaenus, *Strategika*, 3.9.32 (*Hypotheseis*, 3.2). On the issue of noises, see Const. 18§134. Preparing the soldiers to deal with this through practice is described throughout Const. 7, but the most significant section is Const. 7§36, where Leo advises the general to "perform these drills as though you were actually at war."

1007–13   §196   Polyaenus, *Strategika*, 3.9.38; see also Pryor and Jeffreys, *Age of the Δρόμων*, 516 n. 65, who seem puzzled by this advice. In the original text of Polyaenus, however, the passage makes sense: Ἰφικράτης ... πλέων ἐπ᾽ Αἰγύπτου, τῆς χώρας οὔσης ἀλιμένου, παρήγγειλε τοῖς τριηράρχοις "τεσσαράκοντα σάκκους ἕκαστος ἐχέτω." προσορμιζομένων δὲ τοὺς σάκκους ἄμμου πλήσας [...] ἐξῆπτεν ἑκάστης νεὼς καὶ οὕτως ἀνείλκυσεν αὐτὰς τεταρσωμένας ("Iphikrates had carried the war into Egypt.... But since there were no harbors in that part of the world, he commanded that each trierarch should have forty sacks. When they made landfall, he filled the sacks with sand [and suspended them] from the bows of each ship and thus they were safely moored, oars and all" [author's trans.]). Krentz and Wheeler, *Polyaenus*, 261 translate the last section as ". . . fastened them to the bow of each ship, and in this way he dragged the ships up complete with their oars." While ἀνέλκω usually refers to dragging up a vessel, it can also have the sense of "to moor," which fits the context of this anecdote better—there was no beach or harbor in which to draw up the vessels, so the filled sandbags acted as a fixed mooring for each ship. On this passage in particular, see Wheeler, "Notes."

1014–17   §197   Polyaenus, *Strategika*, 3.9.47.

1018–23   §198   Polyaenus, *Strategika*, 3.10.2, and see on §156 above.

1024–29   §199   Cf. Polyaenus, *Strategika*, 3.11.2.

1030–36   §200   Cf. Onasander, *Strategikos*, 2.154–60 (2§§2–5); and cf. Const. 4§3.26–29.

1037–43   §201   See Pryor and Jeffreys, *Age of the Δρόμων*, 401, and on Const. 19.432–35 above.

1044–46   §202   See §151 above; and Const. 14§23. Cf. also. Consts. 12§§10, 104; 14§7.

1047–51 §203 Onasander, *Strategikos*, 7; Polyaenus, *Strategika*, 3.11.4 (*Hypotheseis*, 46.5 and more distantly 3.9.49, 54).

1052–55 §204 Cf. Onasander, *Strategikos*, 29.834–41 (29§§1–2); and cf. Const. 12§§54–55.

1056–59 §205 See on Const. 4§1.3–14 and on ll. 345–54 in this constitution.

1056–57 τοῖς εὐπόροις μέν, μὴ στρατευομένοις δέ, κέλευε, ἐὰν μὴ βούλωνται στρατεύεσθαι Read: "order those who are well-provided for but not serving actively themselves that, if they do not wish to serve. . . ." In the tenth-century legislative documents the term *strateuomenos* meant technically those who were active soldiers, the verb *strateuesthai* to serve actively or to be enrolled on the military register for service, whether to serve actively under arms or to be registered fiscally. By the same token, *stratiotes* meant (in the technical documents, and apart from its generic meaning of "soldier") a person whose land was registered and who contributed to supporting an active soldier. Here Leo appears to be referring to the practice of commuting the active service owed on "military" land for cash. The best-known example is for the year 921, when the thematic soldiers who were registered were obliged to raise a total of 100 lbs. in gold at the rate of 5 *nomismata* per head, or 5 *nomismata* between two registered persons for the less well-off, in addition to 1,000 horses with their tack: *DAI* 1: chap. 51.199–204; chap. 52.12–15. The total of cash to be raised amounted to 7,200 gold *nomismata*, so divided by 5 per head would suggest a maximum of 1,440 soldiers in the Peloponnese, and no more than 2,880 if every single registered *stratiotes* could afford only 2.5 *nomismata*. Since this is unlikely, the total thematic army from the Peloponnese, at least in terms of those registered for military service, can have been only about 2,000; see also Oikonomidès, "Middle Byzantine Provincial Recruits." The demand for 1,000 horses would suggest that the empire expected to raise just 1,000 cavalry soldiers from the Peloponnese.

Note that this demand is not the same as the general imposition, for particular military undertakings, on the population of a particular province or group of provinces, including the Church and monastic foundations, of demands for cash and materials, a potential which the state seems always to have exercised when it needed to do so, and which was also imposed on the church of the Peloponnese in the same year: *DAI* 1: chap. 52, 4–10. Such impositions were not, of course, regular in the sense that yearly taxes were. See, for example, Nicholas I, *Letters* 92.10–26, 94.31–40 (in which the patriarch supports the state's extraordinary impositions in view of the Bulgarian war), 150, 183 (in which he opposes the imposition of military burdens upon individual clerics, in the first case, and the renewed general imposition of extraordinary levies on Church lands) and *CPTT* C.103–12, for horses and mules with pack-saddles to be contributed by metropolitans and bishops throughout the empire to imperial expeditionary forces (a text which probably dates from the time of Basil I).

1060–65   §206   This seems to reflect Leo's remarks on the marching order of the Arabs of Tarsos at Const. 18§110.

1066–70   §207   A passage which reflects Leo's and the fisc's concern with the abuse of patronage enabling individuals to avoid their obligations regarding the fisc or receive exemption from service. See on Const. 20§83 above.

1071–75   §208   Cf. Onasander, *Strategikos*, 41.1129–34 (42§2); and Const. 15.36ff.; cf. also Const. 2§§10, 13. Similar injunctions regarding enemy subordinates: Const. 15§§34, 39.

1076–80   §209   Cf. commentary on §193 above, and see on Const. 11.45–53 above (and 9§§16–18).

1081–88   §210   See *Rhetorica militaris*, 5.1–2 and esp. 36.3–10. On the well-being of subjects, see Const. 14§25; and on the safety of those under the general's command, Const. 1§12.

1089–97   §211   See on Const. 14§§80.547–83.569; and cf. Maurice, *Strategikon*, 8.2.58; also Onasander, *Strategikos*, 1.59–63 (1§§2–3); Const. 2§§1, 3, 5, 18, 25.

1096–97   ὀξεῖς δὲ … συνισταμένου   Seems oddly out of place here—does it refer to specific events in the recent past? On the dangers of alcohol for a ruler or someone wielding authority, see also Basil I, Κεφάλαια παραινετικά (ed. Emminger), p. 58, §25.

1098–1106   §212   See Const. 18§133 and commentary.

1102–03   δέον τοὺς πλωΐμους … καταλαβόντας   Read: "it is necessary for the fleet generals with their naval forces to reach/arrive at Cyprus …" (rather than "to occupy Cyprus": καταλαμβάνω can also have the meaning "to reach/arrive at."). See Vasiliev, *Byzance et les Arabes*, 2.1, 211, followed by Pryor and Jeffreys, *Age of the Δρόμων*, 386–87, suggesting this paragraph—because of the reference to Cyprus as an assembly point—must be a later addition, after Himerios's expedition of 910–11 but before his subsequent defeat in April 912. See also Dolley, "Forgotten Byzantine Conquest," arguing for an earlier and otherwise unrecorded attack on Cyprus. This need not necessarily be the case, however—naval forces could rendezvous at or near a Cypriot port or a coastal point, and this makes more sense both linguistically and in the context.

1107–18   §213   Compare Basil I, *Κεφάλαια παραινετικά* (ed. Emminger), 60, §30, where the ruler is compared to a physician who cures (or prevents) ailments among the people. Further comparisons between the general and a physician at Const. 2§§12 and 26. Cf. Onasander, *Strategikos*, 30 and *Rhetorica militaris*, 5.1–2, 36.2–3, 53.1–2 on the paternal care and respect to be shown by the commander for his troops and the corresponding discipline and obedience on the part of the soldiers; and ibid., 57.8. See Const. 20§148 for warnings about fornication, and on moral rectitude cf. Polyaenus, *Strategika*, 8.16.2 (*Hypotheseis*, 3.8; and Maurice, *Strategikon*, 8.2.57–59).

1119–27   §214   See Dennis, *Taktika*, 615, note 122 (and note Tougher, *Reign of Leo VI*, 44): Basil I took his son Constantine on the campaign of 878, described in the account in the *Vita Basilii imperatoris* (46.16–20 [164]) as γενναῖον σκύλακα, exactly the same term used of the sons of offficers and soldiers in this passage of the *Taktika* (indeed the similarities between the description in these two passages is striking). Less fortunate were the sons of the officers who accompanied the emperor Nikephoros I on his Bulgarian campaign of 811, most of whom are reported to have died in the panic and ensuing massacre when the Bulgarian troops attacked the imperial and thematic encampments; see *Scriptor incertus* (ed. Iadevaia), 27.4–10, 31.126–32.139. On a related note, Leo mentions that the children of the general can become valuable colleagues in Const. 2§11.

1128–34   §215   Appears to be Leo's own words, based on other passages he has employed from both Onasander and Polyaenus. See §§200, 208. See also 2§10 (and cf. §18), 15§16.

1135-41   §216   Polyaenus, *Strategika*, 3.13.1 (*Hypotheseis*, 7.1), 5.28.2 (*Hypotheseis*, 7.2); and see for similar advice 17§89, and comm. to 17.517–33 above.

1142-46   §217   Polyaenus, *Strategika*, 3.11.1; see also *Rhetorica militaris*, 35, although the whole of this text is an exercise in and demonstration of the sort of rhetoric to be employed in encouraging amd motivating the soldiers. Cf. Const. 2§12.

1147-53   §218   Cf. Polyaenus, *Strategika*, 4.3.8; cf. 2.3.2 (*Hypotheseis*, 14.4); Onasander, *Strategikos*, 33.939–42 (33§6).

1154-60   §219   Advice taken from a tale in Polyaenus, *Strategika*, 4.6.2.

1161-66   §220   For sealed orders in naval expeditions, cf. Polyaenus, *Strategika*, 5.2.12 (*Hypotheseis*, 6.2); with a close parallel at 4.7.2 (*Hypotheseis*, 57.13).

1167-78   §221   Refers back to both Const. 8 and Leo's advice in Const. 18§§123–35.

EPILOGUE

The epilogue consists of seventy-three paragraphs summarizing precepts intended to encapsulate the key elements of the advice set out in the twenty constitutions of the treatise and the prologue, beginning with seventeen paragraphs exhorting the commander to bear in mind the unique relationship between God and mankind, to trust in God, to accept that all reward comes from God and that both rulers and their subordinates are appointed from God or owe their authority ultimately to Him, and to show respect for all things holy, including the Church and those committed to the monastic calling. Leo supports his advice with quotations from the Old and New Testaments, and in all refers to divine support some twenty-two times (compared with twenty-eight references in Const. 20 and nineteen in Const. 18); see discussion in the Introduction. In many respects it reflects also the advice Leo received from his father Basil (albeit framed and expressed by Photios) in the Κεφάλαια παραινετικά (ed. Emminger), 50, §2 (on faith in Christ), 55, §18 (on counsel), 57, §22 (on offices and venality), 73, §66 (reading scripture and the wisdom of the ancients); see the discussion in Čičurov, "Gesetz und Gerechtigkeit," 40–44. As noted in the commentary to the prologue, above, the concepts and wisdom embodied in Basil's *paraineseis* are reflected likewise in the advice given to the general in Consts. 1 and 2. The epilogue is addressed directly to the general and takes up the points developed in Const. 2 regarding both the qualities of the general and his response to the situations with which he may be confronted in war. For the most part each paragraph from §18 onward refers back to one or more sections of the preceding twenty constitutions, with §§18–37 consisting of general exhortations to be prepared for war, to follow the advice and guidance of ancient authorities, to rely on God and to do everything possible to bolster the morale and commitment of the soldiers. Paragraphs 38–41 review the qualities and attributes needed for a general, including the ability to speak publicly, to be morally upright, and to display a high level of personal integrity, and the following paragraphs, §§42–52, summarize key strategic and tactical precepts. Particularly interesting is Leo's survey of a series of technical skills upon which the general will need to draw and which he should be sure to have available: armaments, logistics, "architectonics" (camp design, fortifications, and so forth), astronomy, and priestly and medical services and expertise (§§53–69). The final paragraphs, §§70–73, close the epilogue with an exhortation to pay close attention to the precepts set out in the *Taktika*, to remain flexible in their interpretation and application, and to devote oneself to prayer, since victory is God-given; and of course §71 notes the key motif behind the compilation of the treatise, the threat from the Saracens "now causing us trouble." The whole text ends with the standard invocation to the one "true God, eternal emperor of all, to whom be the glory and power for the ages. Amen," thus ending in effect as a prayer and thereby emphasizing both Leo's legislative and his divinely inspired authority.

6–8   §2   Leo both insists on the primacy of the divinity and recalls his own prooemium (line 1); see also 2§§22–24 and 20§47.

11–13   καὶ κριτικός . . . αὐτοῦ   Quoted from Paul, Hebrews 4:12–13.

15–16   Ἕπεται γὰρ . . . πατέρα   Cf. the paternal relationship between the general and his soldiers, on the one hand, and the emperor and his armies, on the other: Const. 4§1.12, 20§213; *Rhetorica militaris*, 53.1–2.

18   ἐν αὐτῷ . . . ἐσμεν   Acts 17:28.

18–19   καὶ τοσοῦτον . . . βασιλεῖ   Leo sets out another representation of society here, complementing his earlier analogy of the army and the general with parts of the human body (along with the peasantry and the emperor: Const. 11.11; and cf. 20§193, §209), underlining the relationships which should ideally pertain between the soldiers and their officer, between a good master (with the emphasis on the adjective) and his dependents, and between those who hold an office—*archontes* (all military and civilian post- and title-holders)—and the emperor (the last recalling also Leo's remarks in 1§13; see commentary to 1§13.44–49 above).

23–24   τοῦ καλοῦ. . . . ἡμέτερον   A paraphrase of John 10:11.

25–26   Καὶ οὐδεὶς . . . ὑπάρχει   The reader was no doubt intended also to recall Leo's comments on Islamic beliefs alluded to in 18§105.

26–27   Psalm 13 (14):1.

32-33　Καὶ αὐτὸς. . . . βασιλεύουσι　Cf. Basil I, Κεφάλαια παραινετικά
(ed. Emminger), 53, §10 and 59, §27. That the general should entrust
his life to God's providence is called for in Const. 2§§22–24. The
allusion to Proverbs 8:15 is reinforced in many Byzantine texts.
Indeed the assumption of divine appointment of rulers and leaders in general provided also the framework for a causal logic which
explained bad rulers or leaders as punishment for the sins of the
people involved, a motif which becomes especially obvious during the late seventh and early eighth century, and during the second Iconoclastic period and thereafter. See esp. Beck, "Senat und
Volk"; Enßlin, "Gottkaiser"; and for the late seventh century see also
Haldon, "Ideology and Social Change." Here Leo is especially keen
to reinforce the point that generals are also appointed by God, and
should therefore show due respect and reverence through prayer
and entrusting his life to divine providence, points made at greater
length in Const. 2§§18, 21–24, and 32, for example.

37-41　§8　See 20§70 and commentary, above.

42-60　§§9–12　Leo is generally recognized to have been concerned with the
rights and authority of the church, something instilled in him since
his youth under the tutelage of the patriarch Photios and very apparent
in the "book of advice" addressed to him from his father Basil I, generally assumed to have been composed largely by the patriarch. See,
in this connection, Basil I, Κεφάλαια παραινετικά (ed. Emminger),
51, §3. Leo's warning that military commanders and especially the
strategoi of the *themata* must prevent their subordinates—whether
military or civil (ll. 47–48)—from mistreating church property and
personnel is echoed in complaints from contemporary or near-contemporary churchmen about additional fiscal burdens caused by military requirements, about the conscription of priests into the army,
or about other similar abuses committed by provincial fiscal or military officials: Nicholas I, *Letters*, nos. 150 (denouncing the imposition of military burdens upon individual clerics) and 183 (opposition
to a general imposition of extraordinary levies on Church lands). A
group of three letters, probably of the early tenth century, represents a
short correspondence between a fiscal official and a metropolitan, in
which the former complains that the Church authorities whose lands

he is sent to assess refuse to hand over their tax dues. Darrouzès, *Épistoliers*, no. 22, with nos. 23 and 24. The reason for the ecclesiastical non-cooperation is said to be the heavy demands in fees in kind (corn, wine) made by the official, and in addition his extending his stay at the Church's expense beyond a reasonable length of time (in this context, stated to be three days). The problem was not new. In the early ninth century the bishop Ignatios complained to the thematic *proto-notarios* likewise about additional burdens on the tenants of church lands: Ignatios the Deacon., *Ep.*, nos. 7 and 8. By the same token, as noted already, soldiers might equally be the victims of persecution by more powerful individuals or institutions. While the general situation had changed somewhat in the intervening period, a case preserved in the early eleventh-century *Peira* (66.27 [250]) is indicative, and records how a soldier was chased off his holding, which was on Church land, at the instigation of a *kourator* of the Hagia Sophia, and eventually murdered. The official in question was brought to justice, and compensation was awarded. But the case illustrates the sort of treatment lowlier soldiers may have received at the hands of more powerful officials. See on Consts. 4.10–11 and 20§71 above. Concern for Church property and personnel is expressed likewise in the *Procheiros Nomos*, 31.62 (concerning the abuse of nuns and others devoted to God) and *Epanagoge*, 9.16 (emphasizing that church lands were to be free of all extraodinary fiscal impositions); and Leo's concern more generally with matters of ecclesiastical discipline and respect for the church from the laity is expressed in the first seventeen of his collection of 113 novels and a further nineteen scattered throughout the collection (see Troianos, Νεαρές, 27f.), and in his concern in the *Taktika* for the rights of asylum; see on Const. 20§70.340–43 above with discussion in Troianos, "Kirchenrechtliche Novellen."

61–64 §13 A further reminder that the *strategos* had authority also over the *krites* or *praitor*, the judge of his *thema*, and was thus the ultimate representative of imperial law and the fair treatment of all the emperor's subjects: cf. Const. 1§§11–12, 4§33; note also Const. 2§22.

65–69 §14 Regarding weaponry, see Const. 5, specifically 5§11 as well as Const. 20§113.

65–78 §§14–17 For the initiation of just wars, peace, etc., see 2§§29–31; 16§§15–16; 19§§32, 39; 20§58 and note Stouraitis, "Methodologische Überlegungen," 278–79; "Jihād and Crusade," 19–23; and on Const. 20§57 above with further references. For §17 in particular see Basil I, *Κεφάλαια παραινετικά* (ed. Emminger), 53, §11.

81–85 §19 Cf. 17§69.

81–93 §§19–21 Cf. for very similar sentiments *CPTT* B.18–22. This text, B, was in its original form compiled at Leo's order by the *magistros* Leo Katakylas (on whom see commentary to Const. 18§134.675–78 above); see ibid., 45–53, and it may be that the similarities in this particular paragraph reflect a common source, or that Leo had seen the original. The advice is traditional; cf. Frontinus, *Stratagems*, 1.3.

87 καὶ διά . . . χρημάτων Cf. §30. This is one of the few occasions Leo refers directly to the cost of warfare and the expenses incurred in launching campaigns of any kind, perhaps surprising in light of the near-constant need to keep armies in the field during the second half of his reign.

89–93 §21 Cf. 12§106, 13§3, 20§109.

94–100 §22 Cf. 2§12, 12§§75–76.

101–8 §§23–24 Cf. 14§§1–3; and for §23 cf. 12§57, 20§110.

109 §25 See 17§5; and commentary to 17§§2.8–9.63 above; cf. also 16§15.

110–14 §26 Cf. Const. 15§39; and see the very similar insistence on humility and the dangers of arrogance and haughtiness in Basil I, *Κεφάλαια παραινετικά* (ed. Emminger), 73, §65; and cf. Onasander, *Strategikos*, 42.1246–55. On not being content with what has been achieved, see Const. 14§§21 and 23. For the advice about not letting the enemy see one's suffering, see 18§20. See also 20§§10, 32, 92.

113–14 Cf. Onasander, *Strategikos*, 13.

115–18 §27 See 14§§21–22.

119–22 §28 Cf. 20§170.

123–24 §29 Translate σπουδαῖον as "zealous" or "diligent" here, rather than "serious," which makes little sense in the context; and cf. 2§18, 19§82.

125–26 §30 See on l. 87, above and cf. 2§§5, 25; 16§8. The expression τὰ ἄλλα τὰ κοινὰ may refer either to "other public funds," as in the translation, or to public affairs and state resources more generally.

127-29   §31   Another reference to Leo's ancient sources, esp. Polyaenus, Onasander, and his "modern" ones, i.e., Maurice and Syrianos. Cf. pr.§6; 18§150; 20§1. Explicit discussions of the ideas of modern and ancient authors that the general should keep in mind abound. See, for example, 7§67 and 14§98, among many.

130-34   §§32-33   See 3§1 and §5; with Onasander, *Strategikos*, 3.171-74 (3§1)

135-208   §§34-50   Summarizes some of the key passages from earlier constitutions, esp. 2, 3, 7, 12, and 14.

135-37   §34   Cf. 12§105; 14§§3, 99; also 20§§2, 153.

138-40   §35   See 18§1; also 15§55, 18§150.

146-49   §38   The need for a strong constitution is summarized in 2§18.

146-61   §§38-40   See also Onasander, *Strategikos*, 1.55-105 (1§§1-14).

150-57   §39   Cf. 2§18; and for ll. 150-51, see also Basil I, Κεφάλαια παραινετικά (ed. Emminger), 57, §22; 58, §24; 61, §34.

158-61   §40   See 2§12 (and, in brief, 2§1); 20§§110, 217; and cf. the story of the general who claimed the enemy commander had fallen: 14§97.

162-64   §41   Cf. 14§31, 16§11, 20§72.

165-67   §42   The general, if wealthy, is instructed to share his resources in 4§3.

173-74   §44   Cf. 18§§1, 15; 17§4. The opposite advice appears to be given in 18§10, where Leo claims that "the formation and drill customary for the Romans is, in our opinion, *suitable for use against any people*." On the identification of specific enemies, see pp. 33-34 n. 73-74 above.

175-78   §45   Cf. 19§30.

179-83   §46   Examples abound—see, for instance, 12§4.

184-95   §47   Continuation of §46. Regarding the conditions for naval battle, see 19§57.

196-99   §48   On commands, see 7§16.

200-206   §49   Cf. 15§55.

207-8   §50   See on 18§119.579-83 above.

209-11   §51   Cf. 20§108 and below, on ep.§61.

212-13   §52   Leo often talks about the general's speaking skills, but except for this occasion, never specifically his voice.

214-303   §§53-68   Here Leo lists a series of specialist areas with which the general ought to be familiar and take account of in his plans, a list which bears some similarities in concept to the much longer and more detailed catalogue of occupations and tasks at at the beginning of Syrianos's *Strategy*. See ibid., chaps. 1–3. The list at this point is not complete. To armament, logistical, architectonic, astronomic, priestly, and medical skills he later adds tactics and mechanics.

216–25   §§54–56   Refers back to 5§2 and 6§§1–3. For §55 see also 7§§2–6.

226–37   §57   See Const. 4, and ep.§64. For Leo, logistics is effectively organization and management, rather than the more restricted sense of the management of specific sets of resources. Some of this information is new—Leo has not previously discussed the selection of soldiers for garrison duty, for example, nor has he discussed the process of selecting soldiers according to fitness, age, and health, except very cursorily in 4§§1 and 65. Much of the work of organizing the individual units—*banda*—fell to the thematic *chartoularios* and his staff, who were responsible for maintaining the provincial military registers in collaboration with the fiscal officials under the *protonotarios*, and upon whose information the *strategos* and his senior officers would depend for an assessment of how many men, of what condition and with what equipment, were available for different tasks. See on Const. 4§33.127–33 above. The term, in its military connotation, was not adopted as a *terminus technicus* until many centuries later, of course, and (as noted in the introduction, p. 5, n. 7) appears first in the work of Joly de Maïzeroy, who translated Leo's *Taktika*, and subsequently in that of J. W. von Bourscheid, both writing in the second half of the eighteenth century. Von Bourscheid wrote *Kurs der Taktik und Logistik* (Vienna, 1780–81) in which the term has the meaning of tactical maneuvering to win an advantage. Baron Antoine de Jomini's summary treatise on warfare, published in 1838, shows no awareness of Leo, but it is unlikely that he was unaware of either Joly de Maïzeroy or Bourscheid. Unlike the two eighteenth-century writers, Jomini seems to have derived the term from the French *logis* (and *loger*), the lodging or billet of the soldiers, responsibility for which had been assigned to special officers since the seventeenth century at least. See Jomini, *Précis de l'Art de la guerre*, ch. 6.

238–43   §58   See Consts. 6 and 7 (and cf. 1§2). Tactics here is seen by Leo as an aspect of logistics, i.e., of organization and management of the army in the field. Although very brief, Leo's paragraph bears some similarity in topic and treatment to Syrianos, *Strategy*, 14 (Περὶ τακτικῆς) and especally to Aelian, *Tactica theoria*, 2.1–4. Cf. Also *Rhetorica militaris*, 41.1–4, with commentary, Eramo, *Siriano*, 169–71.

244-50 §59 Refers to Const. 11.

251-56 §60 Refers to Const. 15.

257-65 §61 Cf. Onasander, *Strategikos*, 10.502–6 (10§28), whom Leo may have paraphrased here. Using the weather to one's advantage was important in Leo's estimation; see Const. 18§§27, 118–20; 19§§2, 31; 20§§108, 115. See Dennis, *Taktika*, comments at 639 n. 17.

262-65 περὶ δὲ σεισμῶν … ἀστρονομίας εἰσίν The word ἀστρονομίας should be replaced by the manuscript reading ἀστρολογίας: the final two sentences denote a change of subject. The editor's emendation of the text from "astrology" to "astronomy," presumably on the basis of the introductory sentence in the paragraph and the absence of astrology from Leo's earlier list of skills in §53, is not justified, the more so since in two other places in this section he expands the original "skill" to include supplementary elements (thus §58 "tactics" is added to §57 "logistics," and §60 "mechanics" is added to §59 "architectonics"). The interpretation of earthquakes and "other signs" and turning them to appropriate use did indeed belong to the general sphere of predictive science and thus to astrology, as the manuscript tradition has it, although it is certainly true that for Byzantines the difference between the two was not necessarily clearly drawn (see the edition, 639 n. 17; Hunger, *Profane Literatur*, 2:232–39). For predictive skills and the books which offered advice upon and interpretations of natural phenomena, see on Consts. 19.9–15 with §31; 20§156; and *CPTT* C.200–202; and see Magdalino, *L'orthodoxie des astrologues*.

266-72 §62 See 11§19; 13§1 and 14§1, with commentary, above, and see 19§24; 20§172. Cf. *Rhetorica militaris*, 10.1, 26.1. The phrase διά τε ἱερολογιῶν καὶ ἱερουργιῶν refers to both the divine liturgy and to a range of sacred rites, such as blessing the standards and soldiers: cf. Suidas, 1:2.179 (p. 616); 193 (p. 617) and Trapp, *Lexikon*, 4:703, 705.

273-79 §63 Apart from his comments on the medical orderlies (*depoutatoi*) at 4§7, §17; 12§§37–39 and 96, whom he likens to doctors (*iatroi*), and his emphasis on establishing camps in healthy situations (see on Const. 11§§2.10–3.22 above), Leo offers very little information about medical care. It is not clear to what extent, if at all, the basic system of medical care which had evolved in the earlier Roman army was maintained beyond the sixth century. For a brief survey of Roman practice and arrangements see Dixon and Southern, *Roman Cavalry*, 98–106 and see Davies, *Service in the Roman Army*, 212–15, 218–25 and Jackson, *Doctors and Diseases*, 134–36. Given what is known of Byzantine medical pracice and hospitals in general it is likely that there was a good degree of continuity; see *ODB* 2:951, "Hospital" for a brief account and further literature. See esp. the useful discussion, with sources and statistics, in Birkenmeier, *Komnenian Army*, 206–30. In the late tenth-century treatise on *Campaign Organization*, 31.20–23, the wounded were to be taken back, when the army was returning to imperial territory, with a section of the rearguard, transported on the pack animals no longer required for the army's supplies. Kinnamos records that during the battle with the Hungarians for the Danube fortress of Zeugminon in 1165 the Roman wounded were taken off by warship and replacements were ferried in by the same means: Kinnamos, *Epitome*, 238. The treatment of wounds, like the study of diseases, had certainly been seen until the seventh century as an important aspect of medical knowledge. The Alexandrian physician Paul of Aegina, writing in the middle of the seventh century, devoted part of a medical treatise to the problems of extracting arrowheads (including also a substantial amount of detail on types of arrowhead and shaft), dealing with fractured or broken bones, and related injuries that were typical of fighting: Paul of Aegina, *Epitome iatrike*, 6.88: Περὶ βελῶν ἐξαιρέσεως ("On the extraction of arrows"). Procopius describes in detail how military surgeons treated various wounds (this included the extraction of an arrowhead, which had pierced the soldier's face between the nose and the right eye; the withdrawal of a javelin head from the skull, resulting in the death of the soldier from infection; a third case concerned a soldier who had received a series of deep cuts in the back and thigh, eventually died from blood loss, in spite of the efforts of the physicians. Several other historians also give descriptions of the sorts of wounds incurred by soldiers in battle (and see Birkenmeier, *Komnenian Army*, 207–11, 220–27) or of the treatment of wounds (ibid., 211–13), up to the twelfth century. Byzantine surgical instruments have been excavated from various contexts, and suggest—in combination with what can be gleaned from the medical treatises themselves—that surgical practice was quite advanced relative to much of the medieval world; see Temkin, "Byzantine Medicine"; and the entries "Medicine" and "Disease" in *ODB* (2:1327–8; 1:638); and see in particular Salazar, *Treatment of War Wounds*. There is no evidence of specifically military hospitals, and if soldiers were not cared for in the general hospitals established at Constantinople and in some provincial centers, they were most probably looked after by their families, assuming they survived their wounds, the journey, and that they had families who could look after them. Choniates mentions that Manuel issued cash from the imperial treasury to cover the expenses of medical treatment for soldiers wounded in the wars with the Turks in Asia Minor in the 1170s. How exceptional this was is impossible to say: Choniates, *Historia*, 191 and cf. Kinnamos, *Epitome*, 62.

The rate of attrition from wounds received in battle is impossible to quantify. For armies of the middle Roman republic an average rate of some 8.8% of soldiers involved in documented battles—ranging from 4.2% in victories and 16% in defeats—has been calculated, across a period when warfare was relatively common, in contrast with a much lower probable rate for the Principate in the first two centuries CE. But these are gross calculations per year rather than per battle; and in the Byzantine situation defeats may often have generated much higher rates, although the sources, predisposed as they generally are to exaggerate numbers in both victories and defeats, can hardly be relied upon to offer any guidance. In general it is thought that mortality in battle was fairly high in the medieval period, although clearly this depended to some extent on context, the degree to which an army held together or broke up, and so forth. Studies of mortality rates in the crusades suggest that across a two- to four-year period some 15%–20% of nobles (the only specific group for whom there is documentary evidence) may have died from wounds received in battle, averaged across victories, defeats, skirmishes, and raiding activities; see Mitchell, *Medicine in the Crusades* and Mitchell et al., "Weapon Injuries." For useful brief discussions see Scheidel, "Marriage, Families and Survival," esp. 427–28; France, "Casualty Rates"; and Mitchell, "Medical Treatment," 585–89.

280–84   §64   See on §§57–58 above. "Logistics" here seems to mean "accounting" as much as the management of resources as such, although of course dividing up the booty and other money which fell to the army on a campaign was an important function, over which the commander had nominal control; see on Const. 20 §192 above. But this paragraph must also refer to the regular fiscal revenues from the *thema* set aside for military purposes, since the armies in the *themata* certainly could not depend upon anything other than regular and predictable revenues for major expenses associated with campaigning. Again, collaboration between the commander and his staff on the one hand and the fiscal authorities in the provinces was a prerequisite here. For campaign financing see *CPTT* 167, 236; Haldon, "Chapters II, 44 and 45," 285–302 with sources and literature; and idem, *Warfare, State and Society*, 143–45.

286   τεχνῖται   In the context, this term is probably much better translated as "craftsmen." This is the only time in the treatise that Leo uses the term ἐπιστημόνως, "scientific/-ally," with the sense of having a full or complete knowledge of something, see *LSJ*, s.v.

287–89   §66   Leo summarizes the three classes of individual involved in "logistics": craftsmen/artisans, clerical and administrative staff, and the military.

287–88   Τὴν μὲν . . . τεχνῖται   On the production of armaments, weaponry, and military equipment, which was assured through a combination of state-imposed corvées on provincial craftsmen and workers, on the one hand, and imperial arms manufactories, on the other, see on 5§1.3–8 above, with references, and commentary to ep.§64 above. At least some armorers must have been involved in full-time production, since a group of them were classed as *exkoussatoi*, exempt from extraordinary and additional fiscal burdens (like soldiers), in a mid-tenth-century letter of the archbishop Basil; see Cantarella, "Basilio Minimo. II" (letter to the emperor Constantine VII). There is no reason to think that the partially exempt status of such persons in the tenth century did not derive directly from their similar situation in the late Roman period; see, for example, *CJ* 12.40.4; 11.10.1–6; 12.40.4 for the *fabricenses* (and cf. *Basilika* 57.5.4, 8).

290–98   §67   See the editor's comments on this paragraph in the edition, 641 notes 19–23 with sources. On Ptolemy see further Hunger, *Profane Literatur*, 1:511–14; *Der kleine Pauly*, 4:1224–32; on Aratos, Hunger, *Profane Literatur*, 2:227 and n. 24; *Der kleine Pauly*, 1:488–89 (Aratos 4) and on Lydos, Hunger, *Profane Literatur*, 1:250–51; 2:234; *Der kleine Pauly*, 3:801–2. The notion that one's fate can be predicted from the circumstances, time and date of birth was challenged by both Christian and non-Christian thinkers already in the late Roman period, as exemplified in the work of the fifth-century neoplatonist Hierocles, excerpted in Photios, *Bibliotheca*, cod. 214 (172b).

302   στρατηγικῆς ... ἀφηγήσεως   This is the only time Leo refers explicitly to his treatise as a work on generalship.

304–11   §69   Cf. 15§1; 18§§1, 15.

312–17   §70   See on pr.§6.68–69 and 75–76 above; and cf. 18§150.

318–38   §§71–73   See 19§58 and cf. 18§150. For §73, see also 2§§22–23. Leo paraphrases Maurice, *Strategikon*, 11.4.227–39, rearranging the ideas at ll. 331–33 from their position in the text of the *Strategikon* (Maurice, *Strategikon*, 11.4.237–39) and adding the comment on the Saracen threat as the motive for the composition of the treatise at ll. 319–20. On this see above, on Const. 18.686–95. For the unpredictability of events more generally and the fickleness of chance, see also Basil I, *Κεφάλαια παραινετικά* (ed. Emminger), 63, §38.

By way of a conclusion, Leo once again underlines the importance of the general being pious and placing his trust in God, both as the means of securing victory (but in Christian terms presented as preserving the general himself and those under his command from harm) and spiritual salvation. The whole treatise ends with the "Amen," stressing once more the emperor's wish to see the text understood as both imperial legislation and as divinely inspired and guided.

CONCORDANCE OF PARAGRAPHS TO THE EDITIONS

The numbers in the columns below are paragraph or section numbers, ordered by constitution. The exceptions are in the prooemium in the Vári edition, where the attribution and title are not given a paragraph number, so the references for this section are by page and line number; likewise, in the Meursius/Lami edition in PG no paragraph numbers are given for the prooemium, so column references only are given.

DENNIS	VÁRI	PG
Prooemium		
1	3.1–8	672C–D
2	1	672D–673B
3	2	673B
4	3	673B–C
5	4	673C–676B
6	5	676B–D
7	6	676D–677A
8	7	677A–B
9	8	677B–C
10	9	677C–D
11	10	680A
Const. 1		
1–10	1–10	1–10
11	11–13	11–13
12	14	14
13	15	15

DENNIS	VÁRI	PG
Const. 2		
1–8	1–8	1–8
9	9	9–10
10	10	11
11	11	12
12	12	13–14
13	13	15–16
14	14–17	17–20
15	18–19	21–23
16	20	24
17	21	25
18	22–30	26–34
19	31–32	35–36
20	33	37
21	34	38
22	35	39
23	36	40

DENNIS	VÁRI	PG	DENNIS	VÁRI	PG
Const. 2 (CONTINUED)			11	11	9
24	37	41	12	12	10
25	38–40	42–44	13	13	11
26	41	45	14	14	12
27	42	46	15	15	13
28	43	47	16	16	14
29	44	48	17	17	15
30	45	49	18	18	16
31	46	50	19	19	17
32	47	51	20	20	18
33	48	52	21	21	19
34	49	53	22	22	20
Const. 3			23	23	21
1	1	1–2	24	24	22
2	2	3	25	25	23
3	3	4	26	26	24
4	4	5	27	27	25
5	5	6	28	28	26
6	6	7	29	29	27
7	7	8	30	30	28
8	8	9	31	31	29
9	9	10	32	32	30
10	10	11	33	33	31
11	11	12	34	34	32
12	12	13	35	35	33
13	13	14	36	36	34
14	14	15	37	37	35
15	15	16	38	38	36
16	16	17	39	39	37
17	17	18	40	40	38
Const. 4			41	41	39
1–5	1–5	1–5	42	42	40
6	6	6	43	43	41
7	7	6	44	44	42
8	8	7	45	45	43
9	9	8	46	46	44
10	10	8	47	47	45

DENNIS	VÁRI	PG	DENNIS	VÁRI	PG
48	48	46	8	8	9
49	49	47	9	9	10
50	50	48	10	10	11
51	51	49	11	11	12
52	52–53	49–50	12	12	13
53	54	51	13	13	14
54	55	52	*Const. 6*		
55	56	53	1–14	1–14	1–14
56	57	54	15	15–17	15–17
57	58	55	16	18–20	18–20
58	59	56	17	21	21
59	60	57	18	22	22
60	61	58	19	23	23
61	62	59	20	24	24
62	63	60	21	25	25
63	64	61	22	26	26
64	65	62	23	27	27–28
65	66	63	24	28	29
66	67	64	25	29	30
67	68	65	26	30	31
68	69	66	27	31	32
69	70	67	28	32	33
70	71	68	29	33	34
71	72	69	30	34	34–35
72	73	70	31	35	36
73	74	71	32	36	37
74	75	72	33	37	38
75	76	73	34	38	39
76	77	74	35	39	39
Const. 5			*Const. 7*		
1	1	1–2	1–2	1–2	1–2
2	2	3	3	3	2–6
3	3	4	4	4–6	7–10
4	4	5	5	7	11–12
5	5	6	6	8	13
6	6	7	7	9–10	14–15
7	7	8	8	11	16

DENNIS	VÁRI	PG	DENNIS	VÁRI	PG
Const. 7 (CONTINUED)			45	56	63
9	12	17	46	57	64
10	13–14	18–19	47	58	65
11	15	20	48	59	66
12	16	21	49	60	67
13	17	22	50	61	67–68
14	18	23	51	62	69
15	19–22	24–27	52	63	70
16	23–24	28–29	53	64–65	71–72
17	25	30	54	66	73
18	26	31	55	67	74
19	27	32	56	68	75
20	28	33	57	69	76
21	29	34	58	70	77
22	30	35	59	71	78
23	31	36	60	72	79
24	32	37	61	73	80
25	33	38	62	74	81
26	34	39	63	75	82
27	35	40	64	76	83
28	36	41	65	77	84
29	37	42	66	78	85
30	38	43	67	79	86
31	39	44	68	80	87
32	40	45	69	81	88–89
33	41	46	70	82	90
34	42	47–48	*Const. 8*		
35	43	49–50	Paragraph numbering is the same		
36	44	51	in all three editions		
37	45–46	52–53	*Const. 9*		
38	47	54	1–10	1–10	1–10
39	48	55	11	11	11
40	49–50	56–57	12	12	11
41	51	58	13	13	12
42	52–53	59–60	14	14	13
43	54	61	15	15	14
44	55	62	16	16	15

DENNIS	VÁRI	PG	DENNIS	VÁRI	PG
17	17	16	54	56	56
18	18	17	55	57	57
19	19	18	56	58	58
20	20	19	57	59	59
21	21	20	58	60	60
22	22	21	59	60	61
23	23	22	60	61	62
24	24	23	61	62	63
25	25	24	62	63	64
26	26	25	63	64	65
27	27	26	64	65	66
28	28	27	65	66–68	67–69
29	29	28	66	69	70–72
30	30	29	67	70	73–74
31	31	30	68	71	75
32	32	31	69	72	76
33	33	32	70	73	77
34	34	33	71	74	78
35	35	34	72	75	79
36	36	35	73	76	80
27	37–38	36–37	74	77	81
38	39	38	75	78	82
39	40	39	76	78	82
40	41	40	*Const. 10*		
41	42	41	1	1–2	1–2
42	43	42	2	3	3
43	44	43	3	4	4
44	45–46	44–45	4	5	5
45	47	46	5	6	6
46	48	47	6	7	7
47	49	48	7	8	8
48	50	49	8	9	9
49	51	50	9	10	10
50	52	51	10	11	11
51	53	52	11	12	12
52	54	53–54	12	13–14	13–14
53	55	55	13	15	15

DENNIS	VÁRI	PG	DENNIS	VÁRI	PG
Const. 10 (CONTINUED)			38	47	46
14	16	16	39	48	47
15	17–18	17–18	40	49	48
16	19	19	41	50–51	49
17	20	19	*Const. 12*		
Const. 11			1–4	1–4	1–4
1–7	1–7	1–7	5	5–6	5
8	8–10	8–10	6	7–8	6–7
9	11	11	7	9	8–9
10	12	12	8	10	10
11	13	13	9	11	11
12	14	14	10	12	12
13	15	15	11	13–14	13–14
14	16	16	12	15–16	15–16
15	17	17–18	13	17–18	17–18
16	18	18	14	19	19
17	19	19–20	15	20	20
18	20	20	16	21–22	21–22
19	21	21	17	23	23
20	22–23	22–23	18	24	24
21	24–25	24–25	19	25–26	25–26
22	26	26	20	27	27
23	27	27	21	28	28
24	28	28	22	29–30	29–30
25	29	29	23	31	31
26	30	30	24	32–33	32–33
27	31	31	25	34	34
28	32	32	26	35	35
29	33–36	33–36	27	36–37	36–37
30	37	37	28	38	38
31	38	38	29	39	39
32	39–40	39–40	30	40	40
33	41–42	41	31	41	41
34	43	42	32	42–43	42–44
35	44	43	33	44	45
36	45	44	34	45–47	46–48
37	46	45	35	48	49

DENNIS	VÁRI	PG	DENNIS	VÁRI	PG
36	49	50	73	93	94
37	50	51	74	94	95
38	51	52	75	95	96
39	52	53	76	96	97
40	53	54	77	97	98
41	54	55	78	98–99	99–100
42	55	56	79	100–101	101–2
43	56	57	80	102–3	103
44	57	58	81	104	104
45	58	59	82	105	105
46	59	60	83	106	106
47	60	61	84	107	107
48	61	62	85	108	108
49	62	63	86	109	109
50	63	64	87	110	110
51	64	65	88	111	111
52	65	66	89	112	112
53	66	67	90	113	113
54	67	68	91	114	114
55	68–69	69–70	92	115	115
56	70	71	93	116	116
57	71	72	94	117	117
58	72	73	95	118	118
59	73–74	74–75	96	119	119
60	75	76	97	120	120
61	76–78	77–79	98	121	121
62	79	80	99	122	122
63	80	81	100	123	123
64	81	82	101	124	124
65	82	83	102	125	125
66	83	84	103	126	126
67	84–85	85–86	104	127	127
68	86	87	105	128	128
69	87	88	106	129–38	129–36
70	88	89	107	139	137
71	89–90	90–91	108	140	138
72	91–92	92–93	109	141	138

DENNIS	VÁRI	PG	DENNIS	VÁRI	PG
Const. 13			12	13	13
1	1	1	13	14–15	14–15
2	2	1	14	16	16
3	3	2	15	17–18	17–18
4	4	3	16	19	19
5	5	4	17	20	20
6	6	5	18	21	21
7	7	6	19	22	22
8	8	7	20	23–24	23–24
9	9	8	21	25	25
10	10	9	22	26	26
11	11	10	23	27	27
12	12	11	24	28	28
13	13	12	25	29	29
14	14	13	26	30	30
15	15	14	27	31	31
16	16	15	28	32	32
17	17	16	29	33	33
18	18	16	30	34	34
Const. 14			31	35	35
1–8	1–8	1–8	32	36	36
9	9–10	9–10	33	37	37
10	11	11	34	38	38
11	12	12			

The Vári edition ends here

DENNIS	PG	DENNIS	PG	DENNIS	PG
Const. 14 (CONTINUED)		71	80	25	26
35	39	72	81	26	27
36	40	73	82	27	28–32
37	41	74	83	28	33
38	42	75	84	29	34–35
39	43	76	85	30	36
40	44	77	86	31	37
41	45	78	87	32	38
42	46	79	88	33	39
43	47	80	89	34	40
44	48	81	90	35	41
45	49	82	91	36	42
46	50	83	92	37	43
47	51	84	93	38	44
48	52	85	94	39	45
49	53	86	95–97	40	46
50	54	87	98–99	41	47
51	55	88	100	42	48
52	56	89	101	43	49
53	57	90	102	44	50
54	58	91	103	45	51–53
55	59	92	104–5	46	54
56	60–61	93	106–7	47	55
57	62	94	108	48	56
58	63	95	109	49	57–59
59	64–68	96	110	50	60
60	69	97	111	51	61
61	70	98	112–13	52	62–63
62	71	99	114	53	64
63	72	100	115	54	65
64	73	101	116	55	66
65	74	*Const. 15*		56	67
66	75	1–20	1–20	57	68
67	76	21	21–22	58	69–70
68	77	22	23	59	71
69	78	23	24	60	72
70	79	24	25	61	73

DENNIS	PG
Const. 15 (CONTINUED)	
62	74
63	75–76
64	77
65	78
Const. 16	
1–2	1 -2
3	3–4
4	5–6
5	7
6	8
7	9
8	10
9	11
10	12
11	13
12	14
13	15
14	16
15	17–19
16	20
17	21
Const. 17	
1–3	1–3
4	4–5
5	6–11
6	12
7	13
8	14
9	15
10	16
11	17
12	18
13	19
14	20
15	21
16	22

DENNIS	PG
17	23–25
18	26
19	27
20	28
21	29
22	30
23	31
24	32
25	33
26	34
27	35
28	36
29	37
30	38
31	39–40
32	41–43
33	44
34	45–46
35	47
36	48–49
37	50
38	51
39	52
40	53
41	54
42	55
43	56
44	57
45	58
46	59
47	60
48	61
49	62
50	63–64
51	65
52	66
53	67

DENNIS	PG
54	68–69
55	70
56	71
57	72–73
58	74–75
59	76
60	77
61	78
62	79
63	80
64	81
65	82–83
66	84
67	85–86
68	87
69	88
70	89
71	90
72	91
73	92
74	93
75	94
76	95
77	96–97
78	98
79	99
80	100
81	101
82	102
83	103
84	104
85	105
86	106
87	107
88	108
89	109–11
90	112–13

DENNIS	PG	DENNIS	PG	DENNIS	PG
91	114–15	59	61	96	102–3
92	116	60	62	97	104
Const. 18		61	63	98	105
1–25	1–25	62	64	99	106
26	26–27	63	65	100	107
27	28–29	64	66	101	107
28	30	65	67	102	108
29	31	66	68	103	109
30	32	67	69	104	110
31	33	68	70	105	111
32	34–35	69	71	106	112
33	35	70	72	107	113
34	36	71	73	108	114
35	37	72	74	109	114
36	38	73	75–77	110	115
37	39	74	78	111	116
38	40	75	79	112	117
39	41	76	80	113	118
40	42	77	81	114	119
41	43	78	82	115	120
42	44	79	83–84	116	121–22
43	45	80	85	117	123
44	46	81	86	118	124
45	47	82	87	119	125
46	48	83	88	120	126
47	49	84	89	121	127
48	50	85	90	122	128
49	51	86	91	123	129
50	52	87	92	124	130
51	53	88	93	125	131
52	54	89	94	126	132
53	55	90	95	127	133
54	56	91	96	128	134
55	57	92	97–98	129	135–36
56	58	93	99	130	137
57	59	94	100	131	138
58	60	95	101	132	139

DENNIS	PG	DENNIS	PG	DENNIS	PG
Const. 18 (CONTINUED)		21	19	53–55	47
133	140	22–23	20	56	48
134	141	24	21	57	49
135	142	25	22	58	50
136–41	143	26	23	59	51
142	144–45	27	24	60	52–53
143	146	28–29	25	61	54
144	147	30	26	62	55
145	148	31	27	63	56
146	149	32	28	64	57
147	150	33	29	65	58
148	151	34	30	66	59
149	152–53	35	31	67	60
150	154	36	32	68	61
Const. 19		37–38	33	69	62
1–6	1–6	39	34	70–71	63
7	7	40	35	72	64
8	7–8	41	36	73	65
9	9	42	37	74	66
10	10	43	38	75	67
11	11	44	39	76	68
12	12	45	40	77	69
13–14	13	46–47	41	78	70
15	14	48	42	79	71
16	15	49	43	80	72
17	16	50	44	81	73–74
18–19	17	51	45	82–83	75
20	18	52	46		

Const. 20 and the epilogue share exactly the same paragraph enumeration in both editions

ABBREVIATIONS

AB	*Analecta Bollandiana*
ActaAntHung	*Acta Archaeologica,* Academiae Scientiarum Hungaricae
AASS	*Acta Sanctorum*
AnnalesESC	*Annales: Economies, sociétés, civilisations*
BAGB	Bulletin de l'Association Guillaume Budé
BBA	Berliner byzantinistische Arbeiten
BBS	Berliner byzantinistische Studien
BCH	*Bulletin de correspondance Hellénique*
Byz	*Byzantion*
ByzF	*Byzantinische Forschungen*
BGA	*Bibliotheca Geographorum Arabicorum,* ed. M.-J. de Goeje (Leyden, 1873–1939)
BHO	*Bibliotheca Hagiographica Orientalis*
BMGS	*Byzantine & Modern Greek Studies*
BSCAbstr	*Byzantine Studies Conference, Abstracts of Papers*
BSl	*Byzantinoslavica*
BSOAS	*Bulletin of the School of Oriental and African Studies*
BZ	*Byzantinische Zeitschrift*
CEFR	Collection de l'École française de Rome
CFHB	Corpus Fontium Historiae Byzantinae
CJ	*Codex Justinianus* (= *CJC* 2)
CJC	*Corpus Juris Civilis,* vol. 1, *Institutiones; Digesta,* ed. P. Krüger and T. Mommsen; vol. 2, *Codex Justinianus,* ed. P. Krüger; vol. 3, *Novellae,* ed. R. Schöll and W. Kroll (Berlin, 1892–95, repr. 1945–63)
CPTT	J. Haldon, ed. and trans., *Constantine Porphyrogenitus, Three Treatises on Imperial Military Expeditions,* CFHB 28 (Vienna, 1990)
CSHB	Corpus Scriptorum Historiae Byzantinae
CTh	*Theodosiani libri 16 cum constitutionibus Sirmondianis,* ed. T. Mommsen, P. Meyer, et al. (Berlin, 1905)

DAI   *Constantine Porphyrogenitus: De Administrando Imperio,* vol. 1, Greek
        text ed. G. Moravcsik, Eng. trans. R. J. H. Jenkins, CFHB 1 = DOT 1
        (Washington D.C., 1967); vol. 2, *Commentary,* ed. R. J. H. Jenkins
        (London, 1962; repr. Washington, D.C., 2012).
DOP   *Dumbarton Oaks Papers*
DOS   Dumbarton Oaks Studies
DOSeals E. McGeer, J. Nesbitt, and N. Oikonomidès, *Catalogue of Byzantine
        Seals at Dumbarton Oaks and in the Fogg Museum of Art* (1991–)
ΕΕΒΣ   Ἐπετηρὶς Ἑταιρείας Βυζαντινῶν Σπουδῶν
EHB   Laiou, A. E., et al., eds. *The Economic History of Byzantium from the
        Seventh through the Fifteenth Century* (Washington, D.C., 2002)
EHR   *English Historical Review*
EI   *Encyclopaedia of Islam,* 2nd ed. (Leyden and London, 1960–)
EO   *Échos d'Orient*
FHG   *Fragmenta Historicorum Graecorum,* ed. C. and T. Müller, 5 vols. (Paris
        1874–85)
GRBS   *Greek, Roman, and Byzantine Studies*
HMSO   Her Majesty's Stationery Office
JAOS   *Journal of the American Oriental Society*
JHS   *Journal of Hellenic Studies*
JÖB   *Jahrbuch der österreichischen Byzantinistik*
JÖBG   *Jahrbuch der österreichischen byzantinischen Gesellschaft*
JRMES   *Journal of Roman Military Equipment Studies*
JRS   *Journal of Roman Studies*
JThS   *Journal of Theological Studies*
LSJ   H. G. Liddell, R. Scott, H. S. Jones, et al., *A Greek-English Lexicon*
        (Oxford, 1968)
MGH Ep *Monumenta Germaniae Historica, Epistolae,* 8 vols. (Berlin 1887–1939)
Νέος Ἑλλ. *Νέος Ἑλληνομνήμων*
OCP   *Orientalia Christiana Periodica*
ODB   *The Oxford Dictionary of Byzantium,* ed. A. P. Kazhdan et al. (Oxford
        and New York, 1991)
PBE   *Prosopography of the Byzantine Empire, 641–886,* ed. J. R. Martindale,
        CD-ROM/online version (London, 2000)
PG   Patrologiae Cursus completus, series Graeco-Latina, ed. J.-P. Migne
        (Paris, 1857–1866, 1880–1903)
PL   Patrologiae Cursus completus, series Latina, ed. J.-P. Migne (Paris,
        1844–1974)
PLBH   A. G. Savvides, B. Hendrickx, et al., *Encyclopaedic Prosopographical
        Lexicon of Byzantine History and Civilization,* 2 vols. (Turnhout,
        2007–8)

PmbZ	R.-J. Lilie, C. Ludwig, T. Pratsch, I. Rochow, et al., *Prosopographie der mittelbyzantinischen Zeit*, vol. 1, *641–867*, 6 parts (Berlin and New York, 1999–2002); vol. 2, *867–1025*, 8 parts (Berlin and New York, 2013)
PSAS	*Proceedings of the Society of Antiquaries of Scotland*
RB	*Reallexikon der Byzantinistik*, ed. P. Wirth (Amsterdam 1968–)
RbK	*Reallexikon der byzantinischen Kunst*, ed. K. Wessel and M. Restle (Stuttgart, 1963–)
REArm	*Revue des Études Arméniennes*, n.s.
REB	*Revue des Études Byzantines* (vols. 1–3: *Études Byzantines*).
REG	*Revue des Études Grecques*
RESEE	*Revue des Études Sud-Est Européennes*
RH	*Revue Historique*
ROC	*Revue de l'Orient chrétien*
RSBN	*Rivista di studi bizantini e neoellenici*
SbB	*Sitzungsberichte der bayerischen Akademie der Wissenschaften, phil.-hist. Klasse*
SBN	*Studi bizantini e neoellenici*
SbWien	*Sitzungsberichte der österreichischen Akademie der Wissenschaften, phil.-hist. Klasse*
SJ	*Saalburg Jahrbuch*
SPBS	Society for the Promotion of Byzantine Studies
TIB	*Tabula Imperii Byzantini*, ed. H. Hunger (Vienna, 1976–)
TM	*Travaux et Mémoires*
VV	*Vizantiiskii Vremmenik*
WByzSt	Wiener Byzantinistische Studien
ZRVI	*Zbornik Radova Vizantološkog Instituta*
ZV	G. Zacos and A. Veglery, *Byzantine Lead Seals*, vol. 1, parts 1–3 (Basel, 1972)

SOURCES

Below are listed ancient and medieval textual sources referred to by shorthand in the commentary, indicating significant editions, translations, or commentaries.

Acts of Iviron	Lefort et al., *Actes d'Iviron*
Acts of Lavra	Lemerle et al., *Actes de Lavra*
Acts of Pantokrator	Kravari, *Actes de Pantokrator*
Acts of Vatopedi	Bompaire et al., *Actes de Vatopédi*
Acts of Xeropotamou	Bompaire, *Actes de Xéropotamou*
Aelian, *Tactica theoria*	Köchly and Rüstow, *Griechische Kriegsschriftsteller*, 2.1:241–471 (crit. ed.); Devine, "Aelian's Manual" (Eng. trans.)
Aeneas Tacticus	Dain and Bon, *Poliorcétique* (crit. ed.); Oldfather, *Aeneas Tacticus* (older ed. with Eng. trans.); Bettali, *Enea tattico* (ed. and Ital. trans.); Whitehead, *Aineias the Tactician* (comm.)
Agathias	Keydell, *Historiarum libri* (crit. ed.); Frendo, *Agathias* (Eng. trans.)
Akropolites, George	Heisenberg, *Georgii Acropolitae Opera*
Ammianus Marcellinus	Rolfe, *Ammianus Marcellinus: History*
Anastasius of Sinai	Richard and Munitiz, *Anastasii Sinaitae*
Anna Komnene	*See* Komnena, Anna
Apollodorus, *Poliorketika*	Schneider, *Griechische Poliorketiker* (crit. ed.); Wescher, *Poliorcétique* (older ed.); Whitehead, *Apollodorus Mechanicus* (Eng. trans.)
Apparatus bellicus	Zuckerman, "Chapitres," 366–69
Appendix Eclogae	Burgmann and Troianos, "Appendix Eclogae"
Appian, *Spanish Wars*	White, *Appian's Roman History*
Arethas	Westerink, *Scripta minora*
Aristotle, *Meteorologica*	Lee, *Aristotle: Meteorologica* (ed. and Eng. trans.)
——, *Physics*	Wicksteed and Comford, *Aristotle: The Physics* (ed. and Eng. trans.)

———, *Politics* Rackham, *Aristotle: The Politics* (ed. and Eng. trans.)

Arrian, *Ektaxis* (Ἔκταξις κατὰ Ἀλανῶν) Roos and Wirth, *Flavii Arriani*, 2:177–85 (crit. ed.); Gilliver, *Roman Art*, 178–80 (Eng. trans.)

———, *Techne taktike* Roos and Wirth, *Flavii Arriani*, 2:129–76 (crit. ed.)

Asclepiodotus, *Tactica* Poznanski, *Asclépiodote* (ed. and Fr. trans.); Oldfather, *Aeneas Tacticus*, 229–340 (ed. and Eng. trans.)

Attaleiates, Michael, *Historia* Pérez Martín, *Historia* (ed. and Span. trans.); Kaldellis and Krallis, *History* (Eng. trans.); Bekker, *Historia* (crit. ed.)

Basil I, *Basilica* Scheltema and Van Der Wal, *Basilicorum libri LX;* Brokaar et al., *Oracles* (ed. and Eng. trans.)

———, Κεφάλαια παραινετικά Emminger, *Studien* (ed. of "Basilii imperatoris Romanorum exhortationum capita sexaginta sex ad Leonem filium"; supersedes PG 107:21–56); PG 107:57–60 (ed. "Basilii imperatoris Romanorum altera exhortatio ad filium suum Leonem imperatorem")

———, *Naumachika* Dain, *Naumachica*, 61–68; Pryor and Jeffreys, *Age of the Δρόμων*, 522–44; Dimitroukas, *Naumachika*, 152-73.

Book of Ceremonies See *De cerimoniis*

Bryennios, Nikephoros Gautier, *Historiarum libri*

Campaign Organization Dennis, *Three Byzantine Military Treatises*, 241–335 (crit. ed. and Eng. trans.)

Choniates, *Historia* van Dieten, *Nicetae Choniatae Historia* (crit. ed.); Magoulias, *O City* (Eng. trans.)

Chronicon Paschale Dindorf, *Chronicon Paschale* (crit. ed.); Whitby and Whitby, *Chronicon Paschale* (Eng. trans.)

Constantine Porphyrogennetos Ahrweiler, "Discours inédit"; Haldon, *Constantine Porphyrogenitus* (= *CPTT*); McGeer, "Two Military Orations". *See also De administrando imperio, De cerimoniis, De thematibus*

De administrando imperio (= DAI) Jenkins et al., *Constantine Porphyrogenitus* (ed., Eng. trans., and commentary)

De cerimoniis Reiske, *Constantini Porphyrogeniti imperatoris De cerimoniis*

De expugnatione Thessalonicae Böhlig, *Ioannis Caminiatae De expugnatione Thessalonicae*

De militari scientia Müller, "Ein griechisches Fragment"; Bertazzoli, "Il *De militari scientia*"

Demosthenes	Demosthenes, *Works*
De obsidione toleranda	van den Berg, *Anonymous* (crit. ed.); Sullivan, "Byzantine Instructional Manual" (Eng. trans.)
De rebus bellicis	Ireland, *De rebus bellicis*
De thematibus	Pertusi, *Costantino Porfirogenito, De Thematibus*
De velitatione bellica	*See* Skirmishing
De XLII martyribus Amoriensibus	Wassiliewski and Nikitine, "Narrationes et carmina sacra"
Digenes Akritas	Trapp, *Digenes Akrites*
Dio, *Roman History*	Cary et al., *Dio's Roman History* (ed. and Eng. trans.)
Ecloga	Burgmann, *Ecloga*
Eisagoge (Epanagoge)	Zepos, *Jus*, 2:229–368
Euripides, *Scholia*	Smith, *Scholia metrica anonyma*
Eustathios of Thessalonike	van der Valk, *Eustathii archiepiscopi Thessalonicensis commentarii*
Evagrius	Bidez and Parmentier, *Ecclesiastical History* (crit. ed.); Whitby, *Ecclesiastical History* (Eng. trans.)
Excerpta Polyaeni	*See* Hypotheseis
Frontinus, *Stratagems*	McElwain, *Frontinus* (ed. and Eng. trans.)
Genesios	Lesmüller-Werner and Thurn, *Iosephi Genesii Regum libri* (crit. ed.); Kaldellis, *Genesios* (Eng. trans.)
George Akropolites	*See* Akropolites, George
George of Pisidia	Pertusi, *Giorgio di Pisidia*
George the Monk	de Boor, *Georgii Monachi Chronicon*
Georgius Monachus continuatus	Bekker, *Theophanes continuatus*, 763–924; Istrin, *Chronika*, 1–65
al-Harthamī, Abū Saʿīd al-Shaʿrānī	al-Harthamī, *Mukhtasar*
Herodian	Whittaker, *Herodian*
Hypotheseis	Melber, *Polyaeni Strategematon*, 427–500; Krentz and Wheeler, *Polyaenus*, 2:851–1003 (ed. and Eng. trans.)
Ibn al ʿAwwām, *Kitāb al-filāha*	Clément-Mullet, *Livre*
Ibn al-Fakīh	Brooks, "Description"
Ibn Hawqal Abu'l-Qasīm	Kramers and Wiet, *Kitāb*
Ibn al-Kalbi, *Livres*	Levi della Vida, "*Livres des Chevaux*"
Ibn Khurradādhbīh	BGA 6:76–85 ("Abū'l-Kāsim ʿUbayd Allāh b. ʿAbd Allāh b. Khurradādhbīh, Kitāb al-Masālik wa'l-Mamālik")

Ibn Rusta	de Goeje, *Kitāb al-A'lāq*
Ignatios the Deacon, *Ep.*	Mango and Efthymiadis, *Correspondence*
Isocrates, *Ad Demonicum*	Mandilaras, *Isocrates*
John of Damascus, *De haeresibus*	Kotter, *Die Schriften* (crit. ed.); PG 94:677–780 (older ed.); Chase, *St. John of Damascus* (Eng. trans.)
John of Nikiu	Zotenberg, *Chronique* (ed. and Fr. trans.); Charles, *Chronicle* (Eng. trans.)
John of Salisbury	Nederman, *Policraticus*
Joshua the Stylite	Wright, *Chronicle* (ed. and Eng. trans.); Trombley and Watt, *Chronicle* (Eng. trans.)
Julius Africanus, *Kestoi*	Vieillefond, *Cestes* (ed. and Fr. trans.); Wallraff, ed., *Julius Africanus, Cesti* (Eng. trans. Adler)
Justinian, *Novellae*	*CJC* 3
Kedrenos, George	Bekker, *Cedrenus*
Kekaumenos, *Strategikon*	Litavrin, *Soveti* (ed. and Russ. trans.); Spadaro, *Raccomandazioni* (ed. and It. trans.)
——, *Logos nouthetetikos pros Basilea*	Litavrin, *Soveti i rasskazi Kekavmena*
Kinnamos, *Epitome*	Meineke, *Ioannis Cinnami Epitome* (crit. ed.); Brand, *Deeds* (Eng. trans.)
Komnene, Anna, *Alexiad*	Reinsch and Kambylis, *Annae Comnenae Alexias* (crit. ed.); Sewter, revised Frankopan, *Alexiad* (Eng. trans.)
Kudāma ibn Ja'far	*BGA* 6:196–99 ("Abū'l-Faraj al-Kātib al-Bagdādī Kudāma ibn Ja'far, Kitāb al-Harāj")
Leo the Deacon, *Historia*	Hase, *Leonis diaconi* (crit. ed.); Talbot and Sullivan, *History* (Eng. trans.)
Leo VI, Ἐκ τοῦ κύρου Λέοντος τοῦ βασιλέως	Dain, *Naumachica*, 37–38; Pryor and Jeffreys, *Age of the Δρόμων*, 516–18; Dimitroukas, *Naumachika*, 84–87
——, Ἐκ τῶν τακτικῶν τοῦ βασιλέως Λέοντος τοῦ σόφου	Dain, L'"extrait tactique," 84–100 (crit. ed. and Fr. trans.)
——, *Guidance*	Papadopoulos-Kerameus, *Varia*, 213–53 ("Οἰακιστικὴ ψυχῶν ὑποτύπωσις")
——, *Homilies*	Antonopoulou, *Homiliae*
——, *Naumachika* (Ναυμαχικὰ Λέοντος βασιλέως)	Dain, *Naumachica*, 19–33; Pryor and Jeffreys, *Age of the Δρόμων*, 484–516; Dimitroukas, *Naumachika*, 38–77
——, *Novellae*	Troianos, Νεαρές (ed. and mod. Grc. trans.); Noailles and Dain, *Novelles* (ed. and Fr. trans.)

———, *Problemata*   Dain, *Problemata*

———, *Taktika*   Dennis, *Taktika* (crit. ed. and trans.); Potamianos, *Αὐτοκράτορος Λέοντος, Τακτικά* (PG ed. and mod. Grk. trans.); Vári, *Leonis imperatoris tactica* (older, partial ed. [vol. 1: prologue, const. 1–11; vol. 2: const. 12–14.38]); PG 107:672–1120. See also: Cheke, *De bellico apparatu liber* (1554, 1595); Meursius, *Leonis imperatoris Tactica* (1612); Meursius, *Claudii Aeliani et Leonis imp. Tactica* (1613)

Leo Grammaticus   Bekker, *Leonis Grammatici Chronographia*, 1–331

Liudprand of Cremona   Wright, *Liudprand of Cremona* (Eng. trans.); Becker, *Liutprandi episcopi Cremonensis opera* (crit. ed.)

Malalas, John, *Chronographia*   Thurn, *Chronographia* (crit. ed.); Jeffreys et al., *Chronicle* (Eng. trans.)

Mas'ūdī   Carra de Vaux, *Livre de l'avertissement*

Maurice, *Strategikon*   Dennis and Gamillscheg, *Strategikon* (crit. ed. and Ger. trans.); Rance, *Strategikon* (Eng. trans. with extensive commentary and notes); Dennis, *Maurice's Strategikon* (older Eng. trans.)

———, *Ἐκ τοῦ Μαυρικίου*   Dain, *Naumachica*, 41–42; Dimitroukas, *Naumachika*, 94–99

Menae patricii, de scientia politica dialogus   Mazzucchi, *Menae patricii*; Bell, *Three Political Voices*, 123–89 (Eng. trans. and comm.)

Menander, *Sententiae*   Jaekel, *Menandri Sententiae*

Menander Prot., *Frg.*   Blockley, *History of Menander*

Michael Attaleiates   *See* Attaleiates, Michael

Michael Syr.   Chabot, *Chronique*

Miracula S. Demetrii   Lemerle, *Les plus anciens recueils*

Miracula S. Georgii   Aufhauser, *Drachenwunder*

Miracula S. Theodori tironis   Delehaye, *Légendes grecques*, 183–201

Miracula S. Therapontis   Deubner, *De Incubatione*, 120–34

Mukhtasar siyāsat al-Hurūb li'l Harthamī Sāhib al-Ma'mun   *See* al-Harthamī, Abū Sa'īd al-Sha'rānī

[Cornelius] Nepos, *Hannibal*   Rolfe, *On Great Generals*

Nicholas I   Jenkins and Westerink, *Nicholas I Patriarch*

Nikephoros Patr., *Antirrhetici i–iii adversus Constantinum Copronymum*   PG 100:205–533; Mondzain-Baudinet, *De notre bienheureux père*, 57–296

———, *Apologeticus maior*   PG 100:533–832

———, *History*	Mango, *Nikephoros* (ed. and Eng. trans.); de Boor, *Nicephori*, 1–77 (older ed.)
Nikephoros, *Praecepta*	McGeer, *Sowing*, 3–78
Nikephoros Ouranos, *Taktika*	McGeer, *Sowing*, 88–163 (chaps. 56–65 only); de Foucault, "Douze chapitres inédits" (chaps. 63–74); Dain, *"Tactique"*
Nomos Mosaikos	Burgmann and Troianos, "Nomos Mosaikos"
Nomos stratiotikos	Korzensky, "Leges poenales militares"; Ashburner, "Byzantine Mutiny Act"
Notitiae Episcopatuum Ecclesiae Constantinopolitae	Darrouzès, *Notitiae*
Onasander, *Strategikos*	Korzensky and Vári, *Sylloge*, 155–63 (crit. ed.; line numbers cited here); Oldfather, *Aeneas Tacitus,* 343–527 (ed. and Eng. trans.; book, paragraph references cited here, in parentheses)
On Strategy	See under Syrianos
Parangelmata poliorcetica	Sullivan, *Siegecraft,* 26–113 (ed. and Eng. trans.), 153–248 (commentary); Wescher, *Poliorcétique* (older ed.); Schneider, *Griechische Poliorketiker* (older ed.)
Parekbolai	de Foucault, *Strategemata*, 69–120
Paul of Aegina, *Epitome iatrike*	Heiberg, *Paulus Aegineta*
Piera sive Practica ex Actis Eustathii Romani	Zepos, *Jus*, 4:1–260
Philo of Byzantium, *Poliorcetica*	Garlan, *Recherches*, 279–327 (ed.), 329–404 (comm.)
Philotheos, *Kletorologion*	Oikonomidès, *Listes de préséance*, 81–235
Photios, *Bibliotheca*	Wilson, *Photios*
———, *Epistulae*	Laourdas and Westerink, *Epistulae*
———, *Homilies*	Mango, *Homilies* (Eng. trans.)
Plutarch, *Sayings of Spartans*	Babbitt, *Plutarch's Moralia*, 3:243–421
———, *Lysander*	Perrin, *Plutarch's Lives*, vol. 4
———, *Moralia*	Babbitt, *Plutarch's Moralia*
———, *Pyrrhus*	Perrin, *Plutarch's Lives*, vol. 9
Polyaenus, *Strategika*	Melber, *Polyaeni Strategematon*, 1–425; Krentz and Wheeler, *Polyaenus*
Polybius, *Hist.*	Paton, *Polybius*
Procheiros Nomos	Zepos, *Jus*, 2:107–228
Procopius, *Wars*	Dewing, *Procopius, History*
———, *Secret History*	Dewing, *Procopius, Anecdota*

Psellos, Michael, *Chronographia*   Renauld, *Michel Psellos*; Sewter, *Michael Psellos* (Eng. trans.)

———, *Letters*   Sathas, Ἱστορικοὶ λόγοι

Regino of Prüm   MacLean, *History and Politics*

Rhetorica militaris   See under *Syrianos*

Scriptor incertus de Leone Armenio   Iadevaia, *Scriptor incertus* (crit. ed. and It. trans.); Bekker, *Leonis* (older ed.)

Skirmishing   Dagron and Mihăescu, *Traité*, 28–135; Dennis, *Three Byzantine Military Treatises*, 137–239; Hase, *Leonis Diaconi*, 179–258

Skylitzes, *Synopsis historiarum*   Thurn, *Ioannis Scylitzae synopsis*; Wortley, *John Skylitzes*

Strategemeta Ambrosiana   de Foucault, *Strategemata*

Suidas   Adler, *Suidae Lexicon*

Sylloge tacticorum   Dain, *Sylloge tacticorum*

Symeon *magistros*   Wahlgren, *Symeonis magistri*

pseudo-Symeon   Bekker, *Theophanes continuatus*, 603–760

Syrianos (*magistros*), *Naumachiai*   Dain, *Naumachica*, 43–55; Pryor and Jeffreys, *Age of the Δρόμων*, 456–81; Dimitroukas, *Naumachika*, 152–73

———, *Rhetorica militaris*   Eramo, *Siriano*; Köchly, *Rhetorica militaris*

———, *On Strategy* (*De re strategica*)   Dennis, *Three Byzantine Military Treatises*, 1–136

Tabarī, *Ta'rikh al-rusul wa'l-muluk*   Yar-Shater, *History*

Theodore the Studite, *Epistulae*   Fatouros, *Theodori Studitae Epistulae*

Theognis, *Elegies*   Young, *Theognis*

Theophanis Chronographia   de Boor, *Theophanis Chronographia* (crit. ed.); Mango and Scott, *Chronicle*

Theophanes continuatus   Bekker, *Theophanes continuatus*, 1–481

Th. Sim.   de Boor, *Theophylacti Simocattae Historiae* (crit. ed.); Whitby and Whitby, *Theophylact Simocatta* (Eng. trans.)

Thucydides, *History*   Foster Smith, *Thucydides*

Vegetius, *Epitoma rei militaris*   Reeve, *Vegetius, Epitoma rei militaris*; Milner, *Vegetius*; Lang, *Epitoma rei militaris*

Vita Basilii imperatoris   Ševčenko, *Chronographiae*

Vita Euthymii Iunioris   Petit, "Vie"

Vita Euthymii patr. CP (BHG 651)   Karlin-Hayter, "Vita Euthymii"

Vita S. Ioannicii (BHG 935)   *AASS Nov.* 2.1: 332–83

Vita S. Lazari monachi in monte Galesio	*AASS Nov.* 3:508–88; Greenfield, *Life*
Vita S. Nili	PG 120:16–165
Vita S. Philareti	Fourmy and Leroy, "Vie"
Vita S. Philareti iunioris	Caruso, "Bios"; *AASS Apr.* 1:603–18; Caietanus, *Vitae*, 2:112–27
Vita S. Stephani iunioris	Auzépy, *Vie*
Vita Theodori Studitae	Mai, *Nova Patrum Bibliotheca*, 6.2:291–363
Xenophon, *Cyropaedia*	Miller, *Xenophon*
al-Yaqʿūbī, *Kitāb al-Buldān*	*BGA* 7:231–360; Wiet, *Kitāb al-Buldān* (Fr. trans.)
Yahya ibn al-Antaki	Kratchkovsky et al., *Histoire*
Zonaras	Büttner Wobst, *Ioannis Zonarae*

BIBLIOGRAPHY

* denotes primary sources (see Sources, above)

Adams, J. N. *Pelagonius and Latin Veterinary Terminology in the Roman Empire*. Leiden, 1995.

* Adler, A., ed. *Suidae Lexicon*. 5 volumes. Leipzig, 1928–38.

Aerts, W. J. "Zu einer neuen Ausgabe der 'Revelationes' des Pseudo-Methodius (syrisch-griechisch-lateinisch)." In *XXIV. Deutscher Orientalistentag: Ausgewählte Vorträge*. Edited by W. Diem and A. Falaturi. 123–30. Stuttgart, 1990.

Agius, D. A. *Classic Ships of Islam*. Leiden and Boston, 2008.

Ahrweiler, H. "L'Asie mineure et les invasions arabes." *Revue Historique* 227 (1962): 1–32.

———. *Byzance et la mer: La marine de guerre, la politique at les institutions maritimes de Byzance aux VII–XV siècles*. Paris 1966.

* ———. "Un discours inédit de Constantin VII Porphyrogénète." *TM* 2 (1967): 393–404.

———. "Recherches sur l'administration de l'empire byzantin aux IX–XI siècles." *BCH* 84 (1960): 1–109.

Aik, L. E., and Z. Zainuddin, "Curve Analysis for Real-Time Crowd Estimation System." *European Journal of Scientific Research* 38, no. 3 (2009): 441–53.

Alexakis, A. "Leo VI, Theophano, a *Magistros* called Slokakas, and the *Vita Theophano* (BHG 1794)." *ByzF* 21 (1995): 45–56.

Allmand, C. *The* De re militari *of Vegetius: The Reception, Transmission and Legacy of a Roman Text in the Middle Ages*. Cambridge, 2011.

Ambaglio, D. "Il Trattato 'Sul Comandante' di Onasandro." *Athenaeum* 59 (1981): 353–77.

Amatuccio, G. *Peri Toxeias: L'arco da guerra nel mondo bizantino e tardoantico*. Bologna, 1996.

Amedroz, H. F., and D. S. Margoliouth. *The Eclipse of the 'Abbasid Caliphate: Original Chronicles of the Fourth Islamic Century*, 6 vols. Oxford, 1920–21.

Anastasiadis, M. P. "On Handling the Menavlion." *BMGS* 18 (1994): 1–10.

Anderson, J. G. C. "The Campaign of Basil I against the Paulicians in 872." *Classical Review* 10 (1896): 138–39.

———. "The Road System of Eastern Asia Minor with the Evidence of Byzantine Campaigns." *JHS* 17 (1897): 22–30.

Andrés, G. "Nota al 'Inventaire raisonné des cents manuscrits des *Constitutions tactiques* de Leon VI le Sage'." *Scriptorium* 11 (1957): 261–63.

Andriotes, N. P. Ἐτυμολογικὸ λεξικὸ τῆς κοινῆς νεοελληνικῆς. Athens, 1951.

Anonymous. *Cavalry Training (Horsed)*. HMSO. London, 1937.

Anonymous. *Manual of Horse and Stable Management*. HMSO. London, 1904.

Anonymous. *Manual of Horsemanship, Equitation and Animal Transport*. HMSO. London, 1937.

Antoniadis-Bibicou, H. *Études d'histoire maritime de Byzance: À propos du "thème des Caravisiens."* Paris, 1966.

———. *Recherches sur les douanes à Byzance, l'"octava," le "kommerkion" et les commerciaires*. Paris, 1963.

Antonopoulou, T. "Homiletic Activity in Constantinople around 900." In *Preacher and Audience: Studies in Early Christian and Byzantine Homiletics*. Edited by M. B. Cunningham and P. Allen. Leiden, 1998.

———. *The Homilies of the Emperor Leo VI*. The Medieval Mediterranean: Peoples, Economies and Cultures, 400–1453, no. 14. Leiden, 1997.

———. "Les manuels militaires byzantins: La version byzantine d'un 'chef romain'." Βυζαντιακά 14 (1994): 97–104.

* ———, ed. *Leonis VI Sapientis Imperatoris Byzantini Homiliae*. CCSG 63. Turnhout, 2008.

Argyriou, A. "Perception de l'Islam et traductions du Coran dans le monde byzantin grec," *Byzantion* 75 (2005): 25–69.

* Ashburner, W. "The Byzantine Mutiny Act." *JHS* 46 (1926): 80–109.

* Aufhauser, J. B. *Das Drachenwunder des Heiligen Georg*. Leipzig, 1913.

Aussaresses, F. *L'Armée byzantine à la fin du VIᵉ siècle d'après le "Strategicon" de l'empereur Maurice*. Bibliothèque des Universités du Midi, fasc. 14. Paris, 1909.

Auzépy, M.-F. "Manifestations de la propagande en faveur de l'orthodoxie." In Brubaker, *Byzantium in the Ninth Century*, 85–100.

———. "State of Emergency (700–850)." In Shepard, *Cambridge History*, 251–91.

* ———. *La Vie d'Étienne le Jeune par Étienne le diacre*. Aldershot, 1997.

Avramea, A. "Land and Sea Communications." In *EHB* 1:57–90.

Baatz, D. "Quellen zur Bauplanung römischer Militärlager." In *Bauplanung und Bautheorie der Antike*. 315–25. Berlin, 1994.

* Babbitt, F. C., ed. and trans. *Plutarch's Moralia*. Loeb Classical Library. 15 volumes. London and Cambridge, Mass., 1927–.

Babuin, B. "Standards and Insignia of Byzantium." *Byz* 71 (2001): 5–59.

Bacharach, J. L. "African Military Slaves in the Medieval Middle East: The Cases of Iraq (869–955) and Egypt (868–1171)." *IJMES* 13 (1981): 476–95.

Bachrach, B. S. "Animals and Warfare in Early Medieval Europe" in *L'Uomo di fronte al mondo animale nell'Alto Medioevo*. 707–51. Settimane di Studio del Centro Italiano di Studi sull'Alto Medioevo 31. Spoleto, 1985. Reprinted in *Armies and Politics in the Medieval West*. No. XVII. London, 1993.

———. "Charlemagne's Cavalry: Myth and Reality." *Military Affairs* 47, no. 4 (December 1983): 181–87.

———. *Early Carolingian Warfare*. Philadelphia, 2001.

———. *Merovingian Military Organization 481–751*. Minneapolis, 1972.

———. "Verbruggen's 'Cavalry' and the Lyon-Thesis." *Journal of Medieval Military History* 5 (2006): 137–63.

Bachrach, D. S. *Warfare in Tenth-Century Germany*. Woodbridge, 2012.

Baldwin, B. "On the Date of the Anonymous Περὶ Στρατηγικῆς." *BZ* 81 (1988): 290–93.

Bálint, C. *Die Archäologie der Steppe*. Vienna and Cologne, 1989.

———. "Der Schatz von Nagyszentmiklós in der bulgarischen archäologischen Forschungen." *Acta Archaeologica Academiae Hungaricae* 51 (1999–2000): 429–38.

Banfi, E. "Problemi di lessico balcanico: Di alcune continuazioni del lessico militare nel neogreco e nelle lingue balcaniche." In *Studi Albanologici, Balcanici, Bizantini e Orientali in onore di Giuseppe Valentini, S.J.* 1–29. Studi Albanesi, Studi e Testi 6. Florence, 1986.

Barker, E. *Social and Political Thought in Byzantium: From Justinian to the Last Paleaologus*. Oxford, 1957.

Barford, P. M. *The Early Slavs: Culture and Society in Early Medieval Eastern Europe*. Ithaca, 2001.

Barker, E. *Social and Political Thought in Byzantium: From Justinian to the Last Palaeologus*. Oxford, 1957.

Barthélemy, D., and J.-Cl. Cheynet, eds. *Guerre et société au moyen âge: Byzance—Occident (VIIIᵉ—XIIIᵉ siècle)*. Paris, 2010.

Bartikian, H. "Ἡ λύση τοῦ αἰνίγματος τῶν Μαρδαΐτων." In *Byzantium: Tribute to Andreas N. Stratos*. 1:17–39. Athens, 1986.

Bartosiewicz, L. "Lovak a Kárpát-medencében a honfoglalás előtt" [Horses in the Carpathian Basin prior to the Hungarian Conquest]. *História* (2005): 1–2, 10–12. [Hungarian with Engl. summary.].

———. "Phenotype and Age in Protohistoric Horses: A Comparison between Avar and Early Hungarian Crania." In *Recent Advances in Ageing and Sexing Animal Bones*. Edited by D. Ruscillo. 204–15. Oxford, 2006.

Bartosiewicz, L., and J. Dirjec, "Camels in Antiquity: Roman Period Finds from Slovenia." *Antiquity* 75 (2001): 279–85.

Bartusis, M. *The Late Byzantine Army: Arms and Society, 1204–1453*. Philadelphia, 1992.

Baun, J. *Tales from Another Byzantium: Celestial Journey and Local Community in the Medieval Greek Apocrypha*. Cambridge, 2007.

Beck, H.-G. "Byzantinisches Gefolgschaftswesen." *Sitzungsber. d. Bayer. Akad. d. Wiss., phil.-hist. Kl.* 5 (1965). Reprinted in H.-G. Beck, *Ideen und Realitäten in Byzanz* (London, 1972), XI.

———. *Geschichte der byzantinischen Volksliteratur*. Handbuch der Altertumswissenschaft xii, 2.3 = Byzantinisches Handbuch 2.3. Munich, 1959.

————. *Kirche und theologische Literatur im byzantinischen Reich.* Handbuch der Altertumswissenschaft xii, 2.1 = Byzantinisches Handbuch 2.1. Munich, 1959.

————. "Senat und Volk von Konstantinopel: Probleme der byzantinischen Verfassungsgeschichte." *Sitzungsber. d. Bayer. Akad. d. Wiss., phil.-hist. Kl.* 6 (1966): 1–75. Reprinted in H.-G. Beck, *Ideen und Realitäten in Byzanz* (London, 1972), XII.

————. *Vorsehung und Vorherbestimmung in der theologischen Literatur der Byzantiner.* Rome, 1937.

* Becker, J., ed. *Liutprandi episcopi Cremonensis opera.* Leipzig, 1915.

* Bekker, I., ed. *Cedrenus: Compendium historiarum.* 2 volumes. CSHB. Bonn, 1838–39.

* ————, ed. *Leonis Grammatici Chronographia.* CSHB, Bonn, 1842.

* ————, ed. *Michaelis Attaliotae Historia.* CSHB. Bonn, 1853.

* ————, ed. *Theophanes continuatus, Ioannes Caminiata, Symeon Magister, Georgius Monachus continuatus.* CSHB. Bonn, 1825.

Belke, K. "Communications: Roads and Bridges." In Jeffreys, Haldon, and Cormack, *Oxford Handbook,* 295–308.

————. "Einige Überlegungen zum Sigillion Kaiser Nikephoros I. für Patrai." *JÖB* 46 (1996): 81–96.

————. "Verkehrsmittel und Reise- bzw. Transportgeschwindigkeit zu Lande im byzantinischen Reich." In *Handelsgüter und Verkehrswege: Aspekte der Warenversorgung im östlichen Mittelmeerraum (4. bis 15. Jahrhundert).* Edited by E. Kislinger, J. Koder and A. Külzer. 45–58. Vienna, 2010.

————. "Von der Pflasterstrasse zum Maultierpfad? Zum kleinasiatischen Wegenetz in mittelbyzantinischer Zeit," In *Byzantine Asia Minor (6th–12th cent.).* Edited by N. Oikonomidès. 267–84. Athens, 1998.

* Bell, P. N. *Three Political Voices from the Age of Justinian: Agapetus, "Advice to the Emperor"; Dialogue on Political Science; Paul the Silentiary, "Description of Hagia Sophia."* Translated texts for historians 52. Liverpool, 2009.

Benakis, L. "Die Stellung des Menschen im Kosmos in der byzantinischen Philosophie," in L'homme et son univers au moyen âge (Louvain, 1986), 56–76.

Bendall, S., and C. Morrisson. "Protecting Horses in Byzantium: A Bronze Plaque from the Armamenton, a Branding Iron and a Horse Brass." In *Byzantium, State and Society, in Memory of Nikos Oikonomides.* Edited by A. Avramea, A. Laiou, and E. Chrysos. 31–49. Athens, 2003.

Benecke, N. "Zur Kenntnis der völkerwanderungszeitlichen und frühmittelalterlichen Pferde aus den Pferdegräbern Nordost-Polens." *Zeitschrift für Archäologie* 19 (1985): 197–205.

Bennett, M. "The Medieval Warhorse Reconsidered." In *Medieval Knighthood.* Edited by S. Church and R. Harvey. 19–40. Woodbridge 1995.

Berger, A. "Das apokalyptische Konstantinopel: Topographisches in apokalyptischen Schriften der mittelbyzantinischen Zeit." In Brandes and Schmider, *Enzeiten,* 135–55.

* Bertazzoli, P. *Il* De militari scientia: *Introduzione, testo, traduzione e commento storico.* University of Bologna, 2006.

Beston, P. "Hellenistic Military Leadership." In *War and Violence in Ancient Greece*. Edited by H. Van Wees. 321–29. London, 2000.

* Bettali, M., ed. *Enea tattico: La difesa di una città assediata*. Pisa, 1990.

* Bidez, J., and L. Parmentier. *The Ecclesiastical History of Evagrius*. London, 1898; repr. Amsterdam, 1964.

Bidwell, P. "The Systems of Obstacles on Hadrian's Wall: Their Extent, Date and Purpose." *Arbeia Journal* 8 (2005): 53–76.

Biraben, J.-N. "Diseases in Europe: Equilibrium and Breakdown of the Pathocenosis." In *Western Medical Thought from Antiquity to the Middle Ages*. Edited by M. Grmek and B. Fantini. 319–53. Cambridge, Ma., 1998.

Birkenmeier, J. W. *The Development of the Komnenian Army, 1081–1180*. Leiden, 2002.

Bishop, M. C., and J. C. Coulston. *Roman Military Equipment from the Punic Wars to the Fall of Rome*. 2nd edition. London, 2006.

Bivar, A. D. H. "Cavalry Equipment and Tactics on the Euphrates Frontier." *DOP* 26 (1972): 271–91.

* Blockley, R. C., ed. and trans. *The History of Menander the Guardsman : Introductory Essay, Text, Translation and Historiographical Notes*. Classical and Medieval Texts, Papers and Monographs 17. Liverpool, 1985.

Blysidou, B. *Ἐξωτερική πολιτική καί ἐσωτερικές ἀντιδράσεις τήν ἐποχή τοῦ Βασιλείου Α´*. Athens, 1991.

———. "Nochmals zum Brief des papstes Stephan V. an den Kaiser Basileios I. als Zeugnis für die Datierung des Feldzuges Nikephoros Phokas des älteren in Kalabrien; Mit addenda in der zweiten Auflage Dölgers Regesten von 867–1025." *Byzantion* 78 (2008): 9–33.

Blysidou, B., E. Kountoura-Galaki, et al. *Η Μικρά Ασία των Θεμάτων: Έρευνες πάνω στην γεωγραφική φυσιογνωμία και προσωπογραφία των Βυζαντινών θεμάτων της Μικράς Ασίας (7ος–11 ος αι.)*. Athens 1998.

Blysidou, B., S. Lampakes, M. Leontsini, and T. Loungis. *Βυζαντινά στρατεύματα στην Δύση (5ος—11ος αι.)*. Athens, 2008.

* Böhlig, G., ed. *Ioannis Caminiatae De expugnatione Thessalonicae*. CFHB 4. Vienna, 1973.

Böhlig, H. *Untersuchungen zum rhetorischen Sprachgebrauch der Byzantiner*. BBA 2. Berlin 1956.

Böhnen, K. "Zur Herkunft der frühmittelalterlichen Spangenhelme." In *Actes du XIIe Congrès International des Sciences préhistoriques et protohistoriques*. 4:199–207. Bratislava, 1993.

Bökönyi, S. *History of Domestic Mammals in Central and Eastern Europe*. Translated by L. Halápy, revised by R. Tringham. Budapest, 1974.

* Bompaire, J., ed. *Actes de Xéropotamou*. Archives de l'Athos. Paris, 1964.

* Bompaire, J., J. Lefort, V. Kravari and C. Giros, eds. *Actes de Vatopédi*. Volume 1, *Des origines à 1329*. Archives de l'Athos. Paris, 2001.

Bonner, M. *Aristocratic Violence and Holy War: Studies in the Jihad and the Arab-Byzantine Frontier.* New Haven, Conn., 1996.

———. *Jihad in Islamic History: Doctrines and Practice.* Princeton, 2006.

———. "The Naming of the Frontier: 'Awāsim, Thughūr and the Arab Geographers." *BSOAS* 57 (1994): 17–24.

———. "Some Observations Concerning the Early Development of Jihad on the Arab-Byzantine Frontier." *Studia islamica* 75 (1992): 5–31. Reprinted in M. Bonner, ed., *Arab-Byzantine Relations in Early Islamic Times,* Formation of the Classical Islamic World 8 (Aldershot, 2004), 402–27.

Borca, F. "Towns and Marshes in the Ancient World." In *Death and Disease in the Ancient City.* Edited by V. M. Hope and E. Marshall. 74–84. London and New York, 2000.

Bosworth, A. B. "Arrian and the Alani." *Harvard Studies in Classical Philology* 81 (1977): 217–55.

Bosworth, C. E. "The City of Tarsus and the Arab-Byzantine Frontiers in Early and Middle Abbasid Times." *Oriens* 33 (1992): 268–86.

Bourdara, C. A. "Quelques cas de *damnatio memoriae* à l'époque de la dynastie macédonienne." *JÖB* 32, no. 2 (1982): 337–46.

Bowersock, G. "Arabs and Saracens in the Historia Augusta." In *Bonner Historia-Augusta-Colloquium 1984–1985.* 71–80. Bonn, 1987.

Bowlus, C. R. "Die militärische Organisation des karolingischen Südostens 791–907." *Fruhmittelalterliche Studien* 31 (1997): 46–69.

Bowlus, C. R. "Franks, Moravians and Magyars: The Struggle for the Middle Danube, 788–907." Philadelphia, 1995.

* Brand, C. M. *The Deeds of John and Manuel Comnenus.* New York, 1976.

Brandes, W. "Der frühe Islam in der byzantinischen Historiographie: Anmerkungen zur Quellenprobelmatik der Chronographia des Theophanes." In *Jenseits der Grenzen: Beiträge zur spätantiken und frühmittelalterlichen Geschichtsschreibung.* Millennium-Studien 25. Edited by A. Goltz, H. Leppin, and H. Schlange-Schöningen. 313–43. Berlin and New York, 2009.

———. "Kaiserprophetien und Hochverrat: Apokalyptische Schriften und Kaiservaticinien als Medium antikaiserlicher Propaganda." In Brandes and Schmieder, *Endzeiten,* 157–200.

Brandes, W., and F. Schmieder, eds. *Endzeiten: Eschatologie in den monotheistischen Weltreligionen.* Millennium Studies in the Culture and History of the First Millennium C.E. 16. Berlin and New York, 2008.

Breccia, G. "Grandi imperi e piccole guerre: Roma, Bisanzio e la guerriglia," *Medioevo Greco* 7 (2007): 13–68; 8 (2008): 49–131.

Briggs, C. F. *Giles of Rome's De regimine principorum: Reading and Writing Politics at Court and University, c. 1275–c. 1525.* Cambridge, 1999.

Brizzi, G. *I sistemi informativi dei Romani.* Stuttgart, 1982.

Brock, S. "From Antagonism to Assimilation: Syriac Attitudes to Greek learning." In *East of Byzantium: Syria and Armenia in the Formative Period*, ed. N. G. Garsoïan, T. F. Matthews and R. W. Thomson. 17–34. Washington, DC, 1982.

Brokaar, W. G. "Basil Lecapenus. Byzantium in the Tenth Century." In *Studia Byzantina et Neohellenica Neerlandica*, ed. W. F. Bakker, A. F. Van Gemert, and W. J. Aerts. 199–234. Byzantina Neerlandica 3. Leiden, 1972.

* Brokaar, W. G., et al., eds. *The Oracles of the Most Wise Emperor Leo and the Tale of the True Emperor: Amstelodamensis Graceus VI E8*. Amsterdam, 2002.

Brooks, E. W. "Arabic Lists of Byzantine Themes." *JHS* 21 (1901): 67–77.

———. "Byzantines and Arabs in the Time of the Early Abbasids, 1," *EHR* 15 (1900): 737–39.

* ———. "Ibn al-Fakīh al-Hamadānī, Description of the Land of the Byzantines." In idem, "Arabic lists," 72–77.

Brown, T. S. "Byzantine Italy (680–876)." In Shepard, *Cambridge History*, 433–64.

———. *Gentlemen and Officers: Imperial Administration and Aristocratic Power in Byzantine Italy, A.D. 554–800*. Rome, 1984.

Browning, R. "The Language of Byzantine Literature." In *Byzantina kai Metabyzantina*, vol. 1, *The "Past" in Medieval and Modern Greek Culture*. Edited by S. Vryonis. 103–33. Malibu, 1978. Reprinted in idem, *History, Language and Literacy in the Byzantine World* (Northampton, 1989), XV.

Brubaker, L., ed. *Byzantium in the Ninth Century: Dead or Alive?* Aldershot, 1998.

———. *Vision and Meaning in Ninth-Century Byzantium: Image as Exegesis in the Homilies of Gregory of Nazianzus*. Cambridge, 1999.

Brubaker, L., and J. F. Haldon. *Byzantium in the Iconoclast Era, c. 680–850: A History*. Cambridge, 2011.

Bruhn-Hoffmeyer, A. *Middelalderens tvaeggede svaerd*. Copenhagen, 1954.

———. "Military Equipment in the Byzantine Manuscript of Scylitzes in the Biblioteca Nacional in Madrid." *Gladius* 5 (1966): 1–160.

Brunel, C. "David d'Ashby, auteur méconnu des Faits des tartares." *Romania* 79 (1958): 39–46.

Bryer, A. "Rural Society in Matzouka." In *Continuity and Change in Late Byzantine and Early Ottoman Society*. Edited by A. Bryer and H. Lowry. 53–95. Birmingham and Washington, DC, 1986.

Bryer, A., and D. Winfield. *The Byzantine Monuments and Topography of the Pontos*. DOS 20. Washington, DC, 1985.

Büttner-Wobst, T. "Die Anlage der historischen Encyklopädie des Konstantinos Porphyrogennetos." *BZ* 15 (1906): 88–120.

Buora, M. "Nuovi studi sulle plumbatae (= mattiobarbuli?): A proposito degli stanziamenti militari nell'Illirico occidentale e nell'Italia orientale nel IV e all'inizio del V secolo." *Aquileia Nostra* 68 (1997): 237–46.

* Burgmann, L., ed. *Ecloga: Das Gesetzbuch Leons III. und Konstaninos' V.* Forschungen zur byzantinischen Rechtsgeschichte 10. Frankfurt a. M., 1983.

———. "Die Nomoi stratiotikos, georgikos und nautikos." *ZRVI* 46 (2009): 53–64.

* Burgmann, L., and S. Troianos, eds. "Appendix Eclogae." In *Fontes Minores.* 3:24–125. Frankfurt am Main, 1979.

* ———. "Nomos Mosaikos." In *Fontes Minores.* 3:126–67. Frankfurt am Main, 1979.

Burliga, B. "Aeneas Tacticus between History and Sophistry: The Emergence of the Military Handbook." In *The Children of Herodotus: Greek and Roman Historiography and Related Genres.* Edited by J. Pigoń. 92–101. Newcastle, 2008.

Bury, J. B. "The Ceremonial Book of Constantine Porphyrogennetos." *EHR* 22 (1907): 207–27.

———. "The Naval Policy of the Roman Empire in Relation to the Western Provinces from the 7th to the 9th Century." In *Centenario della nascita di Michele Amari.* 2:21–34. Palermo, 1910.

———. "The Treatise De Administrando Imperio." *BZ* 15 (1906): 517–77.

* Büttner Wobst, Th., ed. *Ioannis Zonarae epitomae historiarum libri XIII usque ad XVIII.* CSHB. Bonn, 1897.

Cahen, C. "Réflexions sur le waqf ancien." *Studia Islamica* 14 (1961): 37–56.

* Caietanus, O., ed. *Vitae Sanctorum Siculorum.* 2 volumes. Palermo, 1657.

Čalkin, V. I. *Drevneyi životnovodstvo plenyon vostočnoy Evropu i sredniye Azya.* Moscow, 1966.

Cameron, Av., ed. *States, Resources and Armies: Papers of the Third Workshop on Late Antiquity and Early Islam.* Princeton, 1995.

———. "The Theotokos in Sixth-Century Constantinople: A City Finds Its Symbol." *JThS* 29, no. 1 (1978): 79–108.

Campagnolo-Pothitou, M. "Les échanges de prisonniers entre Byzance et l'Islam aux IXᵉ et Xᵉ siècles." *Journal of Oriental and African Studies* 7 (1995): 1–55.

Campbell, J. B. *Greek and Roman Military Writers: Selected Reading.* London and New York, 2004.

Canard, M. "Quelques observations sur l'introduction géographique de le Bughyat at't'alab de Kamal ad-Din d'Alep." *Annales de l'Institut des Études orientales* 15 (1957): 41–53.

———. "Les sources arabes de l'histoire byzantin aux confins des Xᵉ et XIᵉ siècles." *REB* 19 (1961): 284–314.

Canfora, L. "Le 'cercle des lecteurs' autour de Photius: une source contemporaine." *REB* 56 (1998): 269–73.

Cankova-Petkova, G. "Der erste Krieg zwischen Bulgarien und Byzanz unter Simeon und die Wiederaufnahme der Handelsbeziehungen zwischen Bulgarien und Konstantinopel." *ByzF* 3 (1968): 80–113.

Cantarella, R. "Basilio Minimo. II." *BZ* 26 (1926): 3–34.

Caracausi, G. *Lessico greco della Sicilia e dell'Italia meridionale (secoli X–XVI).* Palermo, 1990.

* Carra de Vaux, B., trans. *al-Mas'ūdī, Kitāb al-Tanbīh w'al-Ishraf: Le livre de l'avertissement et de la revision*. Paris, 1897.

* Caruso, S. "Il bios di S. Filareto il Giovane. XI sec. e la Calabria tardo-bizantina." In *Sant'Eufemia d'Aspromonte*. Edited by S. Leanza. Atti del Convegno di Studi per il bicentenario dell'autonomia, Sant'Eufemia d'Aspromonte 14–16 dicembre 1990. Bibliotheca Vivariensis 5. Squillace, 1997.

* Cary, E., G. P. Goold, and J. Henderson, ed. and trans. *Dio's Roman History*. Loeb Classical Library. London and Cambridge, Mass., 1914.

Caseau, B., and J.-C. Cheynet. "La communion du soldat et les rites religieux sur le champ de bataille." In *Pèlerinages et lieux saints dans l'Antiquité et le Moyen Âge: Mélanges offerts à Pierre Maraval*. Edited by B. Caseau, J.-Cl. Cheynet and V. Déroche. 101–16. Paris, 2006.

Cassis, M. "Çadır Höyük: A Rural Settlement in Byzantine Anatolia." In *Archaeology of the Countryside in Medieval Anatolia*. Edited by T. Vorderstrasse and J. Roodenberg. 1–24. Leiden, 2009.

Cassis, M., et al., "The Byzantine Period at Çadır Höyük: A Rural Community in the Byzantine Hinterland: Report on the 2008 Season." http://www.doaks.org/research/byzantine/doaks_eid_2422.html.

Castellvi, G. "Clausurae (les cluses des Pyrénées orientales): Forteresses-frontières du Bas-Empire Romain." In *Frontières terrestres, frontières célestes dans l'Antiquité*. Edited by . A. Rousselle. 81–117. Perpignan, 1995.

Cavallo, G. *Lire à Byzance*. Paris, 2006.

* Chabot, J. B., ed. and trans. *La Chronique de Michel le Syrien, Patriarche Jacobite d'Antioche*. 4 volumes. Paris, 1899–1924.

Charles, M. B. *Vegetius in Context: Establishing the Date of the Epitoma Rei Militaris*. Historia: Einzelschriften 194. Stuttgart, 2007.

* Charles, R. H., trans. *The Chronicle of John, c. 690 A.D., Coptic Bishop of Nikiu*. London, 1916.

* Chase, F. H., Jr., trans. *St. John of Damascus, Writings*. Fathers of the Church 37. New York, 1958.

el Cheikh, N. M. *Byzantium Viewed by the Arabs*. Cambridge, Mass. and London, 2004.

* Cheke, J. *De bellico apparatu liber, e Graeco in Latinum conversus, Joann. Checo Cantabrigiensi interp: Accessit libellus Modesti De vocabulis rei militaris, ad Tacitum Augustum*. Basel, 1554. Reprinted 1595.

Chevedden, P. "The Hybrid Trebuchet: The Halfway Step to the Counterweight Trebuchet." In *On the Social Origins of Medieval Institutions: Essays in Honor of Joseph O'Callaghan*. Edited by D. J. Kagay and T. M. Vann. 179–222. Leiden 1998.

Chevedden, P. E. "Artillery in Late Antiquity: Prelude to the Middle Ages." In *The Medieval City under Siege*. Edited by I. A. Corfis and M. Wolfe. 131–73. Woodbridge, 1995.

Cheynet, J.-C. *The Byzantine Aristocracy and Its Military Function*. Aldershot, 2006.

———. "La guerre sainte à Byzance au Moyen Âge: Un malentendu." In *Regards croisés sur la guerre sainte: Guerre, religion et idéologiedans l'espace méditerranéen latin (XIᵉ-XIIIᵉ siècle)*. Edited by D. Baloup and P. Josserand. 13–32. Toulouse, 2006.

———. "Les Phocas." In Dagron and Mihăescu, *Traité sur la guérilla*, 289–315. Reprinted in J.-C. Cheynet, *La société byzantine à l'apport des sceaux*, 2 vols. (Paris, 2008), 2:473–97.

* Cheynet, J.-C., and D. Theodoridis. *Sceaux byzantins de la collection D. Theodoridis: Les sceaux patronymiques*. Paris, 2010.

Christensen, A. *L'Iran sous les Sassanides*. Copenhagen, 1944.

Christides, V. "Arab-Byzantine Struggle in the Sea: Naval Tactics (7th–11th c. AD): Theory and Practice." In *Aspects of Arab Seafaring: An Attempt to Fill in the Gaps in Maritime History*. Edited by Y. Yousef al-Hijjri and V. Christides. 87–106. Athens, 2002.

———. *The Conquest of Crete by the Arabs (ca. 824): A Turning Point in the Struggle between Byzantium and Islam*. Athens, 1984.

———. "The Image of the Sudanese in Byzantine Sources." *Byzantinoslavica* 43 (1982): 8–17.

———. "Military Intelligence in Arabo-Byzantine Naval Warfare." In Oikonomidès, *Τό εμπόλεμο Βυζάντιο*, 269–81.

———. "The Names ΑΡΑΒΕΣ, ΣΑΡΑΚΗΝΟΙ etc. and Their False Byzantine Etymologies." *BZ* 65 (1972): 329–33.

———. "The Naval Engagement at Dhat as-Sawari AH 34/AD 655–56: A Classical Example of Naval Warfare Incompetence." *Βυζαντινά* 13 (1985): 1331–45.

———. "Naval Warfare in the Eastern Mediterreanean (6th–14th centuries): An Arabic Translation of Leo VI's Naumachica." *Graeco-Arabica* 3 (1984): 137–48.

———. "New Light on Navigation and Naval Warfare in the Eastern Mediterranean, the Red Sea and the Indian Ocean (6th–14th Centuries AD)." *Nubica* 3, no. 1 (1994): 3–42.

———. "Raid and Trade in the Eastern Mediterranean: A Treatise by Muhammad bn. ʿUmar, the *Faqīh* from occupied Moslem Crete, and the Rhodian Sea Law, Two Parallel Texts." *Graeco-Arabica* 5 (1993): 63–102.

———. "The Raids of the Moslems of Crete in the Aegean Sea: Piracy and Conquest." *Byzantion* 51 (1981): 76–111.

———. "The Second Arab Siege of Constantinople (717–18?): Logistics and Naval Power." In *Bibel, Byzanz und christlicher Orient: Festschrift für Stephen Gerö zum 65. Geburtstag*. Edited by D. Bumazhnov, E. Grypheou, T. B. Sailors, and A. Toepel. 511–33. Leuven, 2011.

———. "Two Parallel Naval Guides of the Tenth Century: Qudāma's Document and Leo VI's Naumachica; A Study on Byzantine and Moslem Naval Preparedness." *Graeco-Arabica* 1 (1982): 51–103.

Christophilopoulou, A. *Ἐκλογή, ἀναγόρευσις καὶ στέψις τοῦ βυζαντινοῦ αὐτοκράτορος*. Athens, 1956

Chrysos, E. "Νόμος πολέμου." In Tsiknakes, *Εμπόλεμο Βυζάντιο*, 201–11.

————. "The Title Basileus in Early Byzantine International Relations." *DOP* 32 (1978): 29–75.

Čičurov, I. "Gesetz und Gerechtigkeit in den byzantiniscehn Fürstenspiegeln des 6.–9. Jahrhunderts." In *Cupido legum.* Edited by L. Burgmann, M.-T. Fögen, and A. Schminck. 33–45. Frankfurt a. M., 1985.

Cirolot, V. "Techniques guerrières en Catalogne féodale: le maniement de la lance." *Cahiers de civilisation médiévale* 28 (1985): 36–43.

* Clément-Mullet, J. J., ed. and trans. *Le livre de l'agriculture d'Ibn al-Awam (Kitāb al-felahah).* Paris, 1864. Revised by M. el Faiz, Paris, 2000.

Cohn, L. "Bemerkungen zu den konstantinischen Sammelwerken." *BZ* 9 (1900): 154–60.

Collinet, P. "Sur l'expression ΟΙ ΕΝ ΤΟΙΣ ΤΟΥΛΔΟΙΣ ΑΠΕΡΧΟΜΕΝΟΙ 'ceux qui partent dans les bagages'." In *Mélanges Charles Diehl.* 1:49–54. Paris, 1930.

Colson, B., and H. Coutau-Bégari, eds. *Pensée stratégique et humanisme: De la tactique des Anciens à l'éthique de la stratégie.* Paris, 2000.

Conrad, L. "Theophanes and the Arabic Historical Tradition: Some Indications of Intercultural Transmission." *ByzF* 15 (1990): 1–44.

Cormack, R. "Interpreting the Mosaics of S. Sophia at Istanbul." *Art History* 4 (1981): 138–41.

Corrigan, K. "The Ivory Scepter of Leo VI: A Statement of Post-Iconoclastic Imperial Ideology." *Art Bulletin* 60 (1978): 407–16.

Cosentino, S. "Per una nuova edizione dei Naumachica ambrosiani: Il De fluminibus traiciendis (Strat. XII. B.21)." *Bizantinistica: Rivista di studi bizantini e slavi* 3 (2001): 63–107.

————. "The Syrianos's 'Strategikon': A Ninth-Century Source?" *Bizantinistica* 2 (2000): 243–80.

————. "Writing about War in Byzantium." *Revista de História das Ideias* 30 (2009): 83–99.

Coulston, J. "Arms and Armour of the Late Roman Army." In Nicolle, *Companion,* 3–24.

Coulston, J. C. N. "The Draco Standard." *JRMES* 2 (1991): 101–14.

Coupland, S. "Carolingian Arms and Armour in the Ninth Century." *Viator* 21 (1990): 30–50.

Cowan, R. "The Clashing of Weapons and Silent Advances in Roman Battles." *Historia* 56 (2007): 114–17.

Cross, S. H. "Primitive Slavic Culture." In *A Handbook of Slavic Studies.* Edited by L. I. Strakhovsky. 24–43. Cambridge, Mass., 1949.

Crumlin-Pedersen, O. "Schiffe und Schiffahrtswege im Ostseeraum während des 9.–12. Jahrhunderts." In *Oldenburg-Wolin-Staraja Ladoga-Novgorod-Kiev: Handel und Handelsverbindungen im südlichen und östlichen Ostseeraum während des frühen Mittelalters; Bericht der Römisch-Germanisch Kommission* 69 (1988): 536–42.

Csernus, S., and K. Korompay, eds. *Les Hongrois et l'Europe: Conquête et intégration.* Paris and Szeged, 1999.

Curta, F. "The Earliest Avar-Age Stirrups, or the 'Stirrup Controversy' Revisited." In Curta and Kovalev, *Other Europe*, 297–326.

——. *Southeastern Europe in the Middle Ages 500–1250.* Cambridge, 2006.

Curta, F., and R. Kovalev, eds. *The Other Europe in the Middle Ages: Avars, Bulgars, Khazars and Cumans.* Leiden, 2008.

Dąbrowa, E. "Dromedarii in the Roman Army: A Note." In *Roman Frontier Studies 1989.* Edited by V. A. Maxfield and M. J. Dobson. 364–66. Exeter, 1991.

Dagron, G. "Apprivoiser la guerre: Byzantins et arabes ennemis intimes." In Tsiknakes, *Ἐμπόλεμο Βυζάντιο*, 37–49.

——. "Byzance entre le djihad et la croisade: Quelques remarques." In *Le concile de Clermont de 1095 et l'appel à la croisade.* 325–37. Clermont-Ferrand, 1997.

——. "Byzance et le modèle islamique au Xe siècle: A propos des consitutions tactiques de l'empereur Léon VI." *Comptes Rendus de l'Académie des Inscriptions et Belles-Lettres* (1983): 219–43.

——. "'Ceux d'en face': Les peuples étrangers dans les traités militaires byzantins." *TM* 10 (1987): 207–32.

——. "Le combattant byzantin à la frontière du Taurus: Guérilla et société frontalière." In *Le combattant au Moyen Âge.* Papers from the 18th Congrès de la société des historiens médiévistes de l'enseignement supérieur public, Montepellier. 37–43. Montpellier, 1991.

——. *Emperor and Priest: The Imperial Office in Byzantium.* Translated by J. Birrell Cambridge, 2003.

——. "Das Firmament soll christlich werden: Zu zwei Seefahrtskalendern des 10. Jahrhunderts." In *Fest und Alltag in Byzanz.* Edited by G. Prinzing and D. Simon. 145–56, 210–15. Munich, 1990.

——. "Lawful Society and Legitimate Power." In *Law and Society in Byzantium: Ninth–Twelfth Centuries.* Edited by A. E. Laiou and D. Simon. 38–46. Washington, DC, 1994.

——. "Légitimer la guerre à Byzance." *Mélanges de l'université Saint-Joseph* 62 (2009): 1–19.

——. "Le saint, le savant et l'astrologue: Étude de thèmes hagiographiques à travers quelques recueils de 'Questions et réponses' des Ve–VIIe siècles." In *Hagiographie, cultures et sociétés (IVe–VIIe s.).* 143–55. Paris, 1981. Reprinted in idem, *La romanité chrétienne en Orient: Héritages et mutations* (London, 1984), no. IV.

Dagron, G., and D. Feissel. *Inscriptions de Cilicie.* TM Monographies 4. Paris, 1987.

* Dagron, G., and M. Mihăescu, eds. *Le traité sur la guérilla: De velitatione de l'empereur Nicéphore Phocas (963–69).* Paris, 1986.

Dain, A. "Les cinq adaptations byzantines des 'Stratagèmes' de Polyen." *REArm* 33 (1931): 321–45.

——. "L'encyclopédisme de Constantin Porphyrogénète." BAGB, suppl. [ser. 3] 4. *Lettres d'Humanité* 12 (1953): 64–81.

——. "L'Extrait nautique' tiré de Léon VI." *Eranos* 54 (1956): 151–59.

* ———. *L'"extrait tactique" tiré de Léon VI le Sage*. Paris, 1942.

———. *L'histoire du texte d'Élien le tacticien des origines à la fin du Moyen Âge*. Paris, 1946.

———. "Inventaire raisonné des cents manuscrits des 'Constitutions tactiques' de Léon VI le Sage." *Scriptorium* 1 (1946–47): 33–49.

* ———, ed. *Leonis VI sapientis Problemata*. Paris, 1935.

———. "Les manuscrits des Traités tactiques d'Arrien.' In *Mélanges Bidez*. 1:157–84. Brussels, 1934.

———. *Les manuscrits d'Onésandros*. Paris, 1930.

* ———, ed. *Naumachica*. Paris, 1943.

———. "Le partage du butin du guerre d'après les traités juridiques et militaires." *Actes du 6ᵉ congrès int. des études byzantines*. 1:347–52. Paris, 1950.

———. "Σοῦδα dans les traits militaires." *AIPHOS* 5 (1937): 233–41.

* ———, ed. *Sylloge Tacticorum, quae olim "inedita Leonis Tactica" dicebatur*. Paris, 1938.

* ———, ed. *La "Tactique" de Nicéphore Ouranos*. Paris, 1937.

———. "'Touldos' et 'touldon' dans les traités militaires." In *Mélanges Henri Grégoire*. 2:161–69. Brussels, 1950.

———. "La transmission des textes littéraires classiques de Photius à Constantin Porphyrogénète." *DOP* 8 (1954): 33–47. Reprinted in *Griechische Kodikologie und Textüberlieferung*, ed. D. Harlfinger (Darmstadt 1980), 206–24.

* Dain, A., and A. M. Bon, eds. *Poliorcétique*. Paris, 1967.

Dain, A., and J.-A. de Foucault. "Les stratégistes byzantins." *TM* 2 (1967): 317–92.

Darkó, E. "Die Glaubwürdigkeit der Taktik des Leo Philosophus." *Ungarische Rundschau für historische und soziale Wissenschaften* 5 (1916–17): 129–46.

———. "Influences touraniennes sur l'évolution de l'art militaire des Greces, des Romains et des Byzantins." *Byzantion* 10 (1935): 443–69.

* Darrouzès, J. *Épistoliers byzantins du Xᵉ siècle*. Archives de l'Orient chrétien 6. Paris, 1960.

———. "Inventaire des épistoliers byzantins du Xᵉ siècle." *REB* 18 (1960): 109–35. Reprinted in *Littérature et histoire des textes byzantins* (London, 1972), art. V.

* ———, ed. *Notitiae Episcopatuum Ecclesiae Constantinopolitanae*. Paris, 1981.

Davidson, D. P. *The Barracks of the Roman Army from the 1st to the 3rd Centuries AD*. Oxford, 1989.

Davies, R. W. "The Roman Military Diet." Britannia 2 (1971): 122–42.

———. *Service in the Roman Army*. Edinburgh, 1989.

Dawson, T. "'Fit for the Task': Equipment Sizes and the Transmission of Military Lore, Sixth to Tenth Centuries." *BMGS* 31 (2007): 1–12.

———. "Kremasmata, Kabadion, Klibanion: Some Aspects of Middle Byzantine Military Equipment Reconsidered." *BMGS* 22 (1998) 38–50.

———. "Suntagma Hoplôn: The Equipment of Regular Byzantine Troops, c. 950–c. 1204." In Nicolle, *Companion*, 81–90.

* de Boor, C., ed. *Georgii Monachi Chronicon*. 2 vols. Leipzig, 1904.

* ———, ed. *Nicephori Archiepiscopi Constantinopolitani Opuscula Historica*. Leipzig, 1880.

———. "Suidas und die konstantinische Excerptsammlung." *BZ* 21 (1912): 381–424; 23 (1914–19): 1–127.

* ———, ed. *Theophanis Chronographia*. 2 volumes. Leipzig, 1883–85.

* ———, ed. *Theophylacti Simocattae Historiae*. Leipzig, 1887. Rev. P. Wirth, Stuttgart, 1972.

* de Foucault, J.-A. "Douze chapitres inédits de la *Tactique* de Nicéphore Ouranos." *TM* 5 (1973): 281–312.

* ———. *Strategemata*. Paris, 1949.

* de Goeje, M. J., ed. *Abū ʿAlī b. ʿUmar Ibn Rusta: Kitāb al-Aʾlāq an-Nafīsa*. Leiden, 1892.

* ———, ed. *Bibliotheca Geographorum Arabicorum*. 8 vols. Leyden, 1873–1939.

Delbrück, H. *History of the Art of War*. Translated by W. J. Renfroe, Jr. 4 vols. Wesport, Conn., and London, 1975–85. Reprinted Lincoln and London, 1990.

* Delehaye, H., ed. *Les légendes grecques des saints militaires*. Paris 1909.

Delogu, P. "Lombard and Carolongian Italy." In McCitterick, *New Cambridge Medieval History*, 290–319.

Demetrakos, D. B. *Μέγα λεξικὸν ὅλης τῆς ἑλληνικῆς γλωσσῆς*. Edited by I. Zerbos. 2nd edition. 9 vols. Athens, 1953.

de Meun, J., and C. de Pisan. *L'art de [la] chevalerie selon Végèce*. Paris, 1488.

* Demosthenes. *Works*. 7 vols. Loeb Classical Library. London and Cambridge, Mass., 1926–49.

Dennis, G. T. "Byzantine Battle Flags." *ByzF* 8 (1982): 51–59.

———. "The Byzantines in Battle." In Tsiknakes, *Εμπόλεμο Βυζάντιο*, 165–78.

———. "Defenders of the Christian People: Holy War in Byzantium." In *The Crusades from the Perspective of Byzantium and the Muslim World*. Edited by A. E. Laiou and R. P. Mottahedeh. 31–39. Washington D.C., 2001.

———. "Flies, Mice and the Byzantine Crossbow." *BMGS* 7 (1981): 1–5.

* ———. *Maurice's Strategikon. Handbook of Byzantine Military Strategy*. Philadelphia, 1984.

———. "Perils of the Deep." In *Novum Millenium: Studies on Byzantine History and Culture Dedicated to Paul Speck*. Edited by C. Sode and S. Takacs. 81–88. Aldershot, 2001.

———. "Religious Services in the Byzantine Army." In *Ευλόγημα: Studies in Honor of Robert Taft, S. J.* Edited by E. Carr et al. 107–17. Studia Anselmiana 110. Rome, 1993.

* ———, ed. and trans. *The Taktika of Leo VI: Text, Translation and Commentary*. CFHB 49. DOT 12. Washington DC, 2010.

* ———, ed. and trans. *Three Byzantine Military Treatises*. CFHB 25 = DOT 9. Washington D.C., 1985.

* Dennis, G. T., ed., and E. Gamillscheg, trans. *Das Strategikon des Maurikios*. CFHB 17. Vienna, 1981.

Detorakis, T., and J. Mossay, "Un office byzantin inédit pour ceux qui sont morts à la guerre, dans Cod. Sin. Gr. 734–35." *Muséon* 101 (1988): 185–211.

* Deubner, L. *De Incubatione capita quattuor.* Leipzig, 1900.

De Vajay, S. "Über die Wirtschaftsverhältnisse der landnehmenden Ungarn." *Ungarn Jahrbuch* (1979): 9–19.

* Devine, A. M. "Aelian's Manual of Hellenistic Military Tactics." *Ancient World* 19 (1989): 31–64.

Devos, P. "La translation de s. Jean Chrysostome BHG 877h: Une oeuvre de l'empereur Léon VI." *AB* 107 (1989): 5–29.

* Dewing, H. B., ed. and trans. *Procopius, The Anecdota or Secret History.* Loeb Classical Library. London and Cambridge, Mass., 1935.

* ——, ed. and trans. *Procopius, History of the Wars.* Loeb Classical Library. 5 volumes. London and Cambridge, Mass. 1914–28.

Diethart, J.-M. "Lexikalische Rara in drei byzantinischen Mitgift- und Heiratsgutlisten des 6.–8. Jahrhunderts aus der Wiener Papyrussammlung." *JÖB* 33 (1983): 7–14.

Dietz, K. "Cohortes, ripae, pedaturae: Zur Entwicklung der Grenzlegionen in der Spätantike." In *Klassisches Altertum, Spätantike und frühes Christentum: Adolf Lippold zum 65. Geburtstag gewidmet.* Edited by D. Hennig and H. Kaletsch. 279–329. Würzburg, 1993.

* Dimitroukas, I. C. Λέοντος στ᾽, Μαυρικίου, Συριανοῦ μαγίστρου, Βασιλείου πατρικίου, Νικηφόρου Οὐρανοῦ Ναυμαχικά. Athens, 2005.

——. *Reisen und Verkehr im byzantinischen Reich vom Anfang des 6. Jhr. bis zur Mitte des 11. Jhr.* 2 vols. Athens, 1997.

Dimitrov, H. "Bulgaria and the Magyars at the Beginning of the Tenth Century." *Études balkaniques* 22, no. 2 (1986): 61–77.

* Dindorf, L., *Chronicon Paschale.* CSHB. Bonn, 1832.

Ditten, H. *Ethnische Verschiebungen zwischen der Balkanhalbinsel und Kleinasien vom Ende des 6. bis zur zweiten Hälfte des 9. Jahrhunderts.* Edited by R.-J. Lilie, I. Rochow, F. Winkelmann, and I. Ševčenko. BBA 59. Berlin 1993.

Dixon, K. R., and P. Southern. *The Roman Cavalry: From the First to the Third Century AD.* London, 1992.

Dölger, F. *Beiträge zur Geschichte der byzantinischen Finanzverwaltung besonders des 10. und 11. Jahrhunderts.* Byzantinisches Archiv 9. Munich, 1927. Reprinted Hildesheim, 1960.

——. *Regesten der Kaiserurkunden des oströmischen Reiches 565–1453.* Corpus der griechischen Urkunden des Mittelalters und der neueren Zeit, Reihe A, Abt. 1. 5 volumes. Munich and Berlin, 1924–65. [Volume 2: 2nd ed. P. Wirth (Munich, 1977); volume 1: 2nd ed. A. Müller (Munich, 2003).]

Dolley, R. H. "The Date of the St Mokios Attempt on the Life of the Emperor Leon VI." *Mélanges Henri Grégoire. Annuaire de l'Institut de Philologie et d'Histoire Orientales et Slaves* 10 (1950): 231–38.

——. "A Forgotten Byzantine Conquest of Kypros." In *Acad. royale de Belgique. Bulletin de la classe des letters et des sciences morales et politiques* 35 (1949): 209–24.

———. "The Lord High Admiral Eustathios Argyros and the Betrayal of Taormina to the African Arabs in 902." *SBN* 7 (1953): 340–53.

———. "Naval Tactics in the Heyday of the Byzantine Thalassocraty." *SBN* 7 (1953): 323–39.

Doyen-Higuet, A.-M. "Contribution à l'étude du lexique hippiatrique grec." In *Le cheval dans les sociétés antiques et médiévales*. Edited by S. Lazaris. 213–22. Strasbourg, 2012.

Du Cange, C. Du Fresne. *Glossarium ad scriptores mediae et infimae Graecitatis*. Lyon, 1688. Reprinted Paris, 1943.

———. *Glossarium ad scriptores mediae et infimae Latinitatis*. Edited by G. A. L. Henschel. 7 vols. Paris 1840–50.

Ducellier, A. "Byzance, juge cruel dans un environnement cruel? Notes sur le 'musulman cruel' dans l'empire byzantine entre VIIeme et XIIIeme siècles." In *Crudelitas: The Politics of Cruelty in the Ancient and Medieval World*. Edited by T. Viljamaa, A. Timonen, and C. Krötzl. 148–80. Medium Aevum Quotidianum, Sonderband II. Krems, 1992.

Dufrenne, S. "Aux sources des gonfanons." *Byzantion* 43 (1973): 51–60.

Dujcev, I. "La chronique byzantine de l'an 811," *TM* 1 (1965): 205–54.

Dunn, A. W. "Heraclius' "Reconstruction of Cities" and Their Sixth-Century Balkan Antecedents," in *Acta XIII. Congressus Internationalis Archaeologiae Christianae*. 795–806. Studi di Antichità Cristiana 54. Vatican City and Split, 1998.

———. "The Kommerkiarios, the Apotheke, the Dromos, the Vardarios, and the West." *BMGS* 17 (1993): 3–24.

Dvornik, F. *Origins of Intelligence Services*. New Brunswick, 1974.

Eickhoff, E. *Seekrieg und Seepolitik zwischen Islam und Abendland*. Berlin, 1966.

Elbern, V. H. "Leuchterträger für byzantinische Soldaten." *Aachener Kunstblätter* 50 (1982): 148–49.

Elton, H. *Warfare in Roman Europe, AD 350–425*. Oxford 1996.

* Emminger, K. *Studien zu den griechischen Fürstenspiegeln*. Volume 3, *Basileiou kephalaia parainetika*. Munich, 1913.

Engels, D. *Alexander the Great and the Logistics of the Macedonian Army*. Berkeley, 1978.

England, A. W., J. Eastwood, C. N. Roberts, R. Turner, and J. F. Haldon. "Historical Landscape Change in Cappadocia (Central Turkey): A Paleoecological Investigation of Annually-Laminated Sediments from Nar Lake." *The Holocene* 18, no. 8 (2008): 1229–45.

Enßlin, W. "Gottkaiser und Kaiser von Gottes Gnaden." In *Sitzungsber d. Bayer. Akad. D. Wiss., phil.-hist. Kl.* (Munich, 1943), vol. 6. Reprinted in H. Hunger, ed., *Das byzantinische Herrscherbild*, Wege der Forschung 341 (Darmstadt, 1975), 54–85.

Eramo, I. "Composition and Structure of Syrianus Magister's Military Compendium." *Classica et Christiana* 7, no. 1 (2012): 97–116.

———. "Disegni di guerra: La tradizione dei diagrammi tattici greci nell'Arte della guerra di Niccolo Machiavelli." In *Scienza antica in età moderna: Teoria e immagini*. Edited by V. Marglino. 35–62. Bari, 2012.

———. "Ὦ ἄνδρες στρατιῶται: Demegorie protrettiche nell'Ambrosianus B 119 sup." *Annali Bari* 50 (2007): 127–65.

———. "Omero e i Maccabei: Nella biblioteca di Siriano Μάγιστρος." *Annali Bari* 51 (2008): 123–47.

———. "Retorica militare fra tradizione protrettica e pensiero strategico." *Talia Dixit* 5 (2010): 25–44.

———. "'Ρωμαῖοι e Ἄραβες a battaglia? Nota al De re strategica di Siriano Μάγιστρος." *Invigilata Lucernis* 31 (2009): 95–104.

* ———, ed. *Siriano: Discorsi di guerra; Testo, traduzione e commento.* Bari 2010.

———. "Sul compendio militare di Siriano Magistro." *Rivista storica dell'Antichità* 41 (2011): 201–22.

———. "'Un certo tractatello de l'officio del buon capitanio': Ludovico Carbone traduttore di 'opere pellegrine'." *Paideai* 61 (2006): 153–95.

Erdkamp, P. *Hunger and the Sword: Warfare and Food Supply in Roman Republican Wars (264–30 B. C.).* Amsterdam, 1998.

* Fatouros, G., ed. *Theodori Studitae Epistulae.* CFHB 31. Berlin, 1992.

Featherstone, J. M. "Further Remarks on the De Cerimoniis." *BZ* 97 (2004) 113–21.

———. "Preliminary Remarks on the Leipzig Manuscript of De Cerimoniis." *BZ* 95 (2002): 457–78.

Featherstone, J. M., J. Gruškova, and O. Kresten. "Studien zu den Palimpsestenfragmenten des sogenannten Zeremonienbuches. I. Prolegomena." *BZ* 98 (2005): 423–30.

Felix, W. *Byzanz und die islamische Welt im früheren 11. Jahrhundert.* Byzantina Vindobonensia 14. Vienna, 1981.

Ferluga, J. "Niže vojno-administrativne jedinice tematskog uredenja." *ZRVI* 2 (1953): 61–94.

Fettich, N. *Das Kunstgewerke der Avarenzeit in Ungarn.* Archaeologia Hungarica 1. Budapest, 1926.

Fiedler, U. "Bulgars in the Lower Danube Region: A Survey of the Archaeological Evidence and the State of Current Research." In Curta and Kovalev, *Other Europe,* 151–236.

Fischer, K.-D. "Ancient Veterinary Medicine: A Survey of Greek and Latin Sources and Some Recent Scholarship." *Medizinhistorisches Journal* 23 (1988): 191–209.

Fischer, W. "Zu Leo und Alexander als Mitkaiser von Byzance." *BZ* 5 (1896): 137–39.

Fleckenstein, J. "Adel und Kriegertum und ihre Wandlung im Karolingerreich." In *Nascita dell'Europa ed Europa carolingia, un'equzione da verificare.* 67–94. Settimane di Studio del Centro Italiano di Studi sull'alto Medioevo 27. Spoleto, 1981.

Fledelius, K. "Woman's Position and Possibilities in Byzantine Society with Particular Reference to the Novels of Leo VI." *JÖB* 32, no. 2 (1982): 425–32.

Flori, J. "Encore l'usage de la lance: La technique du combat chevaleresque vers l'an 1000." *Cahiers de civilisation médiévale* 31 (1988): 213–40.

Flusin, B. "Les Excerpta constantiniens: Logique d'une anti-histoire." In *Fragments d'historiens grecs: Autour de Denys d'Halicarnasse; histoire d'un texte.* Edited by S. Pittia. Rome, 2002.

Fögen, M.-Th. "Legislation und Kodification des Kaisers Leons VI," *Subseciva Groningana* 3 (1989): 23–35.

———. "Leon liest Theophilos: Eine Exegese der Novellen 24–27 des Kaisers Leon VI." *Subseciva Groningana* 4 (1990): 83–97.

———. "Reanimation of Roman Law in the Ninth Century: Remarks on Reasons and Results." In Brubaker, *Byzantium in the Ninth Century*, 11–22.

Foss, C. "Byzantine Malagina and the Lower Sangarios." *Anatolian Studies* 40 (1990): 161–83. Reprinted in idem, *Cities, Fortresses and Villages of Byzantine Asia Minor* (Aldershot, 1996) no. VII.

Foss, C., and D. Winfield. *Byzantine Fortifications.* Pretoria 1986.

* Foster Smith, C. *Thucydides, History of the Peloponnesian War.* Loeb Classical Library. London-Cambridge, Mass., 1953.

* Fourmy, M.-H., and M. Leroy. "La vie de S. Philarète." *Byzantion* 9 (1934): 85–170.

Foxhall, L., and H. A. Forbes. "Sitometreia: The Role of Grain as a Staple Food in Classical Antiquity." *Chiron* 12 (1982): 41–90.

France, J. "Casualty Rates." In Rogers, *Oxford Encyclopaedia*, 1:348–49.

———. "Size." In Rogers, *Oxford Encyclopaedia*, 1:65–67.

Franklin, S., and J. Shepard. *The Emergence of Rus, 750–1200.* London and New York, 1996.

———, eds. *Byzantine Diplomacy.* Aldershot, 1992.

* Frendo, J. D. C., trans. *Agathias, History.* Berlin and New York, 1975.

———. "The Miracles of St. Demetrius and the Capture of Thessaloniki: An Examination of the Purpose, Significance and Authenticity of John Kaminiates' *De expugnatione Thessalonicae.*" *BSl* 58 (1997): 205–24.

Friesinger, E. H. "Waffenkunde des 9. und 10. Jahrhunderts aus Niederösterreich." *Archaeologia austriaca* 52 (1972): 43–64.

Frolow, A. *La relique de la vraie croix: Recherches sur le développement d'un culte.* Paris, 1961.

Fülöp, G. "Awarenzeitliche Fürstenfunde von Igar." *Acta Archaeologica Academiae Scientiarum Hungaricae* 40 (1988): 151–90.

Furse, G. A. *The Art of Marching.* London, 1901.

Gahbauer, F. R. *Das anthroplogische Modell: Ein beitrag zur Christologie der frühen Kirche bis Chalkedon.* Würzburg, 1984.

Gamillscheg, E., and D. Harlfinger. *Repertorium der griechischen Kopisten, 800–1600.* 3 vols. Vienna, 1981–97.

Ganshof, F. L. "L'armée sous les Carolingiens." In *Ordinamenti Militari in Occidente nell'alto Medioevo.* 109–30. Settimane di Studi del centro Italiano di Studi sull'alto Medioevo 15. Spoleto, 1968.

Garam, É. "Bemerkungen zum ältesten Fundmaterial der Awarenzeit." In *Typen der Ethnogenese unter besonderer Berücksichtigung der Bayern.* Edited by H. Friesinger and F. Daim. 2:253–72. Veröffentlichungen der Kommission für Frühmittelalterforschung 13. Vienna, 1990.

———. "Sepolture de Cavalli." In *Gli Avari, un populo d'Europa*. Edited by G. C. Menis. 143–49. Udine, 1995.

* Garlan, Y., ed. *Recherches de poliorcétique grecque*. Bibliothèque des Écoles françaises d'Athènes et de Rome, fasc. 223. Athens, 1974.

Garnsey, P. *Food and Society in Classical Antiquity*. Cambridge, 1999.

Garrood, W. "The Byzantine Conquest of Cilicia and the Hamdanids of Aleppo, 959–65." *Anat Studies* 58 (2008): 127–40.

Garzya, A., G. Giangrande, and M. Manfredini. *Sulla tradizione manoscritta dei Moralia di Plutarco*. Edited by I. Gallo. Salerno, 1988.

Gaudeul, J. M. "The Correspondence between Leo and 'Umar: 'Umar's Letter Rediscovered?" *Islamochristiana* 10 (1984): 109–57.

* Gautier, P., ed. *Nicephori Bryennii Historiarum libri quattuor*. CFHB 9. Brussels, 1975.

Gavrilović, Z. A. "The Humiliation of Leo VI the Wise (the Mosaic of the Narthex at Saint Sophia)." *Cahiers Archéologiques* 28 (1979): 87–94.

Genito, B. "Archaeology of the Early Medieval Nomads in Italy: The Horse-Burials in Molise (7th Century) South-Central Italy." In *Kontakte zwischen Iran, Byzanz und der Steppe im 6.–7. Jahrhundert*. Edited by C. Bálint. 229–47. Varia Archaeologica Hungarica 10. Budapest, 2000.

Gertwagen, R. "Harbours and Facilities along the Eastern Mediterranean Sea Lanes to Outremer." In *The Logistics of the Crusades*. Edited by J. Pryor. 95–118. Sydney, 2006.

Gertwagen, R., and E. Jeffreys, eds. *Shipping, Trade and Crusade in the Medieval Mediterranean: Studies in Honour of John Pryor*. Farnham, 2012.

Giagkoulis, K. G. Θησαυρός Κυπριακής Διαλέκτου: Ερμηνευτικός και ετυμολογικός. Από το ό αι. μέχρι σήμερα. Leukosia, 2002.

Gilbey, W. *Small Horses in Warfare*. London, 1900.

Gilliver, C. M. "Hedgehogs, Caltrops and Palisade Stakes." *JRMES* 4 (1993): 49–54.

* ———. *The Roman Art of War*. Stroud, 1999.

Gillmor, C. "The Brevium Exempla as a Source for Carolingian Warhorses." *Journal of Medieval Military History* 6 (2008): 32–57.

———. "Stirrups and Stirrup Thesis." In Rogers, *Oxford Encyclopaedia*, 3:312–14.

Giuffrida, C. "Disciplina Romanorum: Dall'Epitome di Vegezio allo Strategikon dello Pseudo-Mauricius." In *Le trasformazioni della cultura nella tarda antichità: Atti del convegno tenuto a Catania, Università degli studi, 27 sett.–2 ott. 1982*. Edited by M. Mazza and C. Giuffrida. 2:837–60. Rome, 1985.

Gladitz, C. *Horse Breeding in the Medieval World*. Dublin 1997.

Goetz, H.-W. "Social and Military Institutions." In McKitterick, *New Cambridge Medieval History*, 451–80.

Golden, P. B. "War and Warfare in the Pre-Činggisid Western Steppes of Eurasia." In *Warfare in Inner Asian History (500–1800)*. Edited by N. Di Cosmo. 105–72. Leiden, 2002.

Goldsworthy, A. K. *The Roman Army at War 100 BC–AD 200*. Oxford, 1996.

* Gombos, A. F. *Catalogus fontium historicae Hungaricae*. 3 vols. Budapest, 1937–43.

Górecki, D. "The Strateia of Constantine VII: The Legal Status, Administration and Historical Background." *BZ* 82 (1989): 157–76.

Gorelik, M. "Arms and Armour in South-Eastern Europe in the Second Half of the First Millenium AD." In Nicolle, *Companion,* 127–47.

———. "Oriental Armour of the Near and Middle East as Shown in Works of Art." In *Islamic Arms and Armour.* Edited by R. Elgood. 30–63. London, 1979.

Goria, F. Review of Schminck, *Studien.* In *Studia et documenta historiae et iuris* 55 (1989): 529–54.

Graf, D. F. "Camels, Roads and Wheels in Late Antiquity." In *Donum Amicitiae: Studies in Ancient History.* Edited by E. Dąbrowa. 43–49. Electrum 1. Kraków, 1997.

Graf, D., and M. O'Connor, "The Origin of the Term Saracen and the Rawwāfa Inscriptions." *Byzantinische Studien* 4 (1977): 52–66.

Grafton, A. "The Availability of Ancient Works." In *The Cambridge History of Renaissance Philosophy.* Edited by C. B. Schmitt. 767–91. Cambridge, 1988.

Grafton, A., G. W. Most, and S. Settis, eds. *The Classical Tradition.* Cambridge, Mass., 2010.

Greatrex, G. *Rome and Persia at War: 502–32.* Leeds, 1998.

* Greenfield, R. P. H. *The Life of Lazaros of Mt. Galesion: An Eleventh-Century Pillar Saint.* Washington, D.C., 2000.

Grégoire, H. "Un captif arabe à la cour de l'empereur Alexandre." *Byz* 7 (193?): 666–73.

———. "La carrière du premier Nicéphore Phocas." In Προσφορὰ εἰς Στίλπώνα Π. Κυριακίδην. 232–54. Thessalonike, 1953.

Grierson, P. *Catalogue of the Byzantine Coins in the Dumbarton Oaks Collection and in the Whittemore Collection.* Volume 3, *Leo III to Nicephorus III, 717–1081.* Washington D.C., 1973.

Griffith, S. "What Has Constantinople To Do with Jerusalem? Palestine in the Ninth Century: Byzantine Orthodoxy in the World of Islam." In Brubaker, *Byzantium in the Ninth Century,* 181–94.

Grigoriou-Ioannidou, M. "Το ναυτικό θέμα των Κιβυρραιώτων. Συμβολή στο πρόβλημα της υδρύσεώς του." *Byzantina* 11 (1982): 201–21.

Grmek, M. D. "Les ruses de guerre biologiques dans l'antiquité." *REG* 92 (1979): 141–63.

Grosdidier de Matons, J. "Trois études sur Léon VI. II, Hippocrate et Léon VI: remarques sur l'Οἰακιστικὴ ψυχῶν ὑποτύπωσις." *TM* 5 (1973): 206–28.

———. "Trois études sur Léon VI. III: Les *Constitutions tactiques* et la *damnatio memoriae* de l'empereur Alexandre." *TM* 5 (1973): 229–42.

Grosse, R. "Das römisch-byzantinische Marschlager vom 4.–10. Jahrhundert." *BZ* 22 (1913): 90–121.

Grotowski, P. Ł. *Arms and Armour of the Warrior Saints: Tradition and Innovation in Byzantine iconography (843–1261).* Leiden, 2010.

Grumel, V. *La Chronologie.* Traité d'Études Byzantines 1. Paris, 1958.

———. "Chronologie des événements du règne de Léon VI (886–912)." *EO* 35 (1936): 5–42.

———. "Notes chronologiques: La révolte d'Andronic Doux sous Léon VI: La victorie navale d'Himérius." *EO* 36 (1937): 202–7.

*———. *Les Regestes des actes du patriarcat de Constantinople.* Volume 1, *Les actes des patriarches.* Part 1, *Les regestes de 381 à 715.* Paris, 1972. Part 2, *Les regestes de 715 à 1043.* Chalcedon, 1936; 2nd rev. ed. J. Darrouzès, Paris 1989.

Guemara, R. "La libération et le rachat des captifs. Une lecture musulmane." In *La liberazione dei "captivi" tra cristianità e islam: Oltre la Crociata e il Ğihād; Tolleranza e servizio umanitario; Atti del Congresso interdisciplinare di studi storici.* Edited by G. Cipollone. 333–44. Vatican City, 2000.

Gyftopoulou, S. "Riding and Reserving Equi in the Late Antique/Middle Byzantine Army." *Βυζαντινός Δόμος* 16 (2007–8): 389–410.

Györffy, G. "Landnahme, Ansiedlung und Streifzüge der Ungarn." *Acta Academiae Scientiarum Hungaricae* 31 (1985): 231–70.

———. "Système des résidences d'hiver et d'été chez les nomades et les chefs hongrois au Xe siècle." *Archivum Eurasiae Medii Aevi* 1 (1975): 45–153.

Haase, W. "Si vis pacem, para bellum." In *Akten des XI. Int. Limeskongresses, 1976.* 721–55. Budapest, 1977.

Hahlweg, W. *Die Heeresreform der Oranier und die Antike.* Berlin, 1941.

Haldon, J. F. "Administrative Continuities and Structural Transformations in East Roman Military Organisation ca. 580–640." In *L'Armée romaine et les barbares du IIIe au VIIe siècle.* Edited by F. Vallet and M. Kazanski. 45–53. Paris, 1993.

———. "Aerikon/aerika: A Re-interpretation." *JÖB* 44 (1994): 135–42.

———. "Approaches to an Alternative Military History of the Period ca. 1025–1071." In *Byzantium in the Eleventh Century.* Edited by E. Chrysos. 45–74. Athens, 2003.

———. *Byzantine Praetorians: A Social, Institutional and Administrative History of the Opsikion and Tagmata.* Bonn and Berlin, 1984.

———. *Byzantium in the Seventh Century: The Transformation of a Culture.* Cambridge, 1997.

———. "'Cappadocia Will Be Given Over to Ruin and Become a Desert': Environmental Evidence for Historically-Attested Events in the 7th–10th Centuries." In *Byzantina Mediterranea: Festschrift für Johannes Koder zum 65. Geburtstag.* Edited by K. Belke, E. Kislinger, A. Külzer, and M. Stassinopoulou. 215–30. Vienna, 2007.

———. "Chapters II, 44 and 45 of the Book of Ceremonies: Theory and Practice in Tenth-Century Military Administration." *TM* 13 (2000): 201–352.

*———, ed. and trans. *Constantine Porphyrogenitus, Three Treatises on Imperial Military Expeditions.* CFHB 28. Vienna, 1990.

———. "Fighting for Peace: Attitudes to Warfare in Byzantium." In idem, *Warfare, State and Society,* 13–33.

———. "Ideology and Social Change in the Seventh Century: Military Discontent as a Barometer." *Klio* 68 (1986): 139–90.

———. "Kudāma Ibn Dja'far and the Garrison of Constantinople," *Byzantion* 48 (1978): 78–90.

———. "Military Service, Military Lands and the Status of Soldiers: Current Problems and Interpretations." *DOP* 47 (1993): 1–67. Repr. in idem, *State, Army and Society*, no. VII.

———. "The Organisation and Support of an Expeditionary Force: Manpower and Logistics in the Middle Byzantine Period." In Tsiknakes, *Εμπόλεμο Βυζάντιο*, 111–51.

———. *Recruitment and Conscription in the Byzantine Army c. 550–950: A Study on the Origins of the Stratiotika Ktemata*. Sitzungsber. d. österr. Akad. d. Wiss., phil.-hist. Kl. 357. Vienna, 1979.

———. "Roads and Communications in Byzantine Asia Minor: Wagons, Horses, Supplies." In *The Logistics of the Crusades*. Edited by J. Pryor. 131–58. Sydney, 2006.

———. "Some Aspects of Byzantine Military Technology from the Sixth to the Tenth Centuries." *BMGS* 1 (1975): 11–47.

———. "Some Aspects of Early Byzantine Arms and Armour." In Nicolle, *Companion*, 65–87.

———. *State, Army and Society in Byzantium. Approaches to Military, Social and Administrative History*. Aldershot, 1995.

———. "Strategies of Defence, Problems of Security: The Garrisons of Constantinople in the Middle Byzantine Period." In *Constantinople and Its Hinterland*. Edited by C. Mango. 143–55. Oxford, 1995.

———. *Warfare, State and Society in the Byzantine World 550–1204: An Introductory Survey*. London, 1999.

———. "Why Model Logistical Systems?" In *General Issues in the Study of Medieval Logistics: Sources, Problems and Methodologies*. Edited by J. Haldon. 1–35. Leiden, 2005.

———. "The Works of Anastasius of Sinai: A Key Source for the History of Seventh-Century East Mediterranean Society and Belief." In *The Byzantine and Early Islamic Near East*. Volume 1, *Problems in the Literary Source Material*. Edited by Av. Cameron and L. Conrad. 107–47. Princeton, 1992.

Haldon, J. F., V. Gaffney, G. Theodoropoulos, and P. Murgatroyd. "Marching across Anatolia: Medieval Logistics and Modeling the Mantzikert Campaign." *DOP* 65–66 (2011–12): 209–36.

Haldon, J. F., and H. Kennedy, "The Arab-Byzantine Frontier in the Eighth and Ninth Centuries: Military Organisation and Society in the Borderlands." *ZRVI* 19 (1980): 79–116. Repr. in *Arab-Byzantine Relations in Early Islamic Times*. Edited by M. Bonner. 142–78. Formation of the Classical Islamic World 8. Aldershot, 2004.

Halkin, F. "Trois dates historiques précisés grâce au Synaxaire." *Byzantion* 24 (1954): 14–17.

Halperin, C. "Bulgars and Slavs in the First Bulgarian Empire: A Reconsideration of the Historiography." *Archivum Eurasiae Medii Aevi* 3 (1983): 183–200.

Halsall, G. *Warfare and Society in the Barbarian West 450–900*. London, 2003.

Hamblin, W. "Sassanian Military Science and Its Transmission to the Arabs." *Proceedings of the 1986 International Conference on Middle Eastern Studies*. 99–106. Oxford, 1986.

Hanel, N. "Military Camps, Canabae, and Vici: The Archaeological Evidence." In *Companion to the Roman Army*. Edited by P. Erdkamp. 395–416. Oxford, 2007.

Harris, J. "Wars and Rumours of Wars: England and the Byzantine World in the Eighth and Ninth Centuries." *Mediterranean Historical Review* 14 (1999): 29–46.

al-Harthamī, al-Sh. *Mukhtasar siyāsat al-Hurūb li'l Harthamī Sāhib al-Ma'mun*. Edited by A. al-R'ūf 'Aun. Cairo, 1964.

* Hase, C. B., ed. *Leonis diaconi Caloensis Historiae libri decem*. CSHB. Bonn, 1828.

* Heiberg, J. L., ed. *Paulus Aegineta*. 2 volumes. Corpus medicorum Graecorum 9. Leipzig, 1921–24.

* Heisenberg, A., ed. *Georgii Acropolitae Opera*. Volume 1. Leipzig 1903.

Hen, Y. "Charlemagne's Jihad." *Viator* 37 (2006): 34–51.

———. *Roman Barbarians: The Royal Court and Culture in the Early Medieval West*. Basingstoke, 2007.

Hermann, E. "Zum Asylsrecht im byzantinischen Reich." *OCP* 1 (1935): 204–38.

Hild, F. *Das byzantinische Strassensystem in Kappadokien*. Veröffentlichungen der Kommission der tabula Imperii Byzantini 2. Denkschr. d. Österr. Akad. D. Wiss., phil.-hist. Kl. 131. Vienna, 1977.

Hild, F., and H. Hellenkamper. *Tabula Imperii Byzantini*. Volume 5, *Kilikien und Isaurien*. 2 pts. Denkschr. d. Österr. Akad. d Wiss., phil.-hist. Kl. 215. Vienna, 1990.

Hoffmann, F. G. *Lexicon bibliographicum sive index editionum et interpretationum scriptorum graecorum tam sacrorum tam profanorum*. Leipzig, 1832.

Holmes, C. "Byzantine Political Culture and Compilation Literature in the Tenth and Eleventh Centuries," *DOP* 64 (2010): 55–80.

Holmes, C., and J. Waring, eds. *Literacy, Education and Manuscript Transmission in Byzantium and Beyond*. Leiden, 2002.

Holum, K. G. "Pulcheria's Crusade A.D. 421–22 and the Ideology of Imperial Victory." *GRBS* 18 (1977): 153–72.

Honigmann, E. *Die Ostgrenze des byzantinischen Reiches von 363 bis 1071*. Brussels, 1935. = Vasiliev, *Byzance et les arabes*, vol. 3.

Horden, P., and N. Purcell. *The Corrupting Sea: A Study of Mediterranean History*. Oxford, 2000.

Hose, M. "Das Gnomologion des Stobaios. Eine Landkarte des "paganen" Geistes". *Hermes* 133 (2005): 93–99

Howard-Johnston, J. "Byzantine Anzitene." In Mitchell, *Armies and Frontiers*, 239–90.

———. "The *De administrando imperio*: A Re-examination of the Text and a Re-evaluation of Its Evidence about the Rus." In *Les centres proto-urbains russes entre Scandinavie, Byzance et Orient*. Edited by M. Kazanski, et al. 301–36. Paris, 2000.

Hoyland, R. G. *Seeing Islam as Others Saw It: A Survey and Evaluation of Christian, Jewish and Zoroastrian Writings on Early Islam*. Studies in Late Antiquity and Early Islam 13. Princeton, 1997.

Hunger, H. *Die hochsprachliche profane Literatur der Byzantiner*. 2 vols. Handbuch der Altertumswissenschaft xii, 5.1–2 = Byzantinisches Handbuch 5.1–2. Munich, 1978.

————. *Prooimion: Elemente der byzantinischen Kaiseridee in den Arengen der Urkunden.* Wiener byzantinistische Studien 1. Wien-Graz-Köln, 1964.

Huuri, K. *Zur Geschichte des mittelalterlichen Geschützwesen aus orientalischen Quellen.* Societas Orientalia Fennica, Studia Orientalia 9.3. Helsinki, 1941.

Huxley, G. "Michael III and the Battle of Bishop's Meadow [AD 863]." *GRBS* 16 (1975): 443–50.

————. "A Porphyrogenitan Portulan." *GRBS* 17 (1976): 295–300.

Hyland, A. *Equus: The Horse in the Roman World.* London, 1990.

————. *The Medieval Warhorse from Byzantium to the Crusades.* London, 1994.

* Iadevaia, F., ed. and trans. *Scriptor incertus: Testo critico, traduzione e note.* Messina, 1987.

Ilari, V. *Imitatio, restitutio, utopia: la storia militare antica nel pensiero strategico moderno.* Milan, 2001. http://www.scribd.com/doc/10971682/Storia-Militare-Antica (accessed 5 June 2012).

Iotov, V. "A Note on the 'Hungarian Sabers' of Medieval Bulgaria." In Curta and Kovalev, *Other Europe*, 327–38.

* Ireland, R. I., ed. *De rebus bellicis.* Leipzig, 1982.

Irigoin, J. "Pour une étude des centres de copie byzantins." *Scriptorium* 12 (1958): 208–27; 13 (1959): 177–209.

————. "Survie et renouveau de la littérature antique à Constantinople (IXᵉ siècle)." *Cahiers de civilisation médiévale* 5 (1962): 287–302.

Irmscher, J. "Die Gestalt Leons VI. des Weisen in Volkssage und Historiographie." In *Beiträge zur byzantinischen Geschichte im 9.–11. Jahrhundert.* 205–24. Prague, 1978.

* Istrin, V. M., ed., "Prodolzhenie chroniki Georgija Amartola po Vatikanskomu spisku No. 153." In V. M. Istrin, *Chronika Georgija Amartola v drevnem slavjanorusskom perevode: Tekst, issledovanie i slovar,* vol. 2. Petrograd, 1922.

Ivanišević, V., and I. Bugarski, "Les étriers byzantins: La documentation du Balkan central." In *Le cheval dans les sociétés antiques et médiévales.* Edited by S. Lazaris. 135–42. Strasbourg, 2012.

Jackson, R. *Doctors and Diseases in the Roman Empire.* Norman, 1988.

* Jaekel, S., ed. *Menandri Sententiae: Comparatio Menandri et Philistionis.* Leipzig, 1964.

Jähns, M. *Handbuch einer Geschichte des Kriegswesens von der Urzeit bis zur Renaissance: Bewaffnung, Kampfweise, Befestigung, Belagerung, See.* Leipzig, 1880.

James, S. "Archaeological Evidence for Roman Incendiary Projectiles." *SJ* 39 (1983): 142–43.

————. "Dura-Europos and the Introduction of the 'Mongolian Release'." In *Roman Military Equipment: The Accoutrements of War.* 77–84. BAR International Series 336. Oxford, 1987.

————. "Evidence from Dura Europos for the Origins of Late Roman Helmets." *Syria* 63 (1986): 107–34.

————. "Stratagems, Combat and 'Chemical Warfare' in the Siege Mines of Dura-Europos." *American Journal of Archaeology* 115 (2011): 69–101.

Janniard, S. "Armati, scutati et la catégorisation des troupes dans l'Antiquité tardive." In *L'Armée romaine de Dioclétien à Valentinien Ier*. Edited by Y. Le Bohec and C. Wolff. 371–88. Actes du Congrès de Lyon, 12–14 sept. 2002. Paris, 2004.

———. "Centuriones ordinarii et ducenarii dans l'armée romaine tardive (IIIe–VIe s. apr. J.-C.)." In Lewin and Pelegrini, *Late Roman Army*, 383–93.

———. "Végèce et les transformations de l'art de la guerre aux IVe et Ve siècles après J.-C.," *Antiquité Tardive* 16 (2008): 19–36.

Jeffery, A. "Ghevond's Text of the Correspondence between 'Umar II and Leo III." *Harvard Theological Review* 37 (1944): 269–332.

Jeffreys, E. "The Image of the Arabs in Byzantine Literature." In *The 17th International Byzantine Congress, Major Papers*. 305–23. New Rochelle and York, 1986.

Jeffreys, E., J. F. Haldon, and R. Cormack. *The Oxford Handbook of Byzantine Studies*. Oxford, 2008.

* Jeffreys, E., M. Jeffreys, and R. Scott. *The Chronicle of John Malalas*. Byzantina Australiensia 4. Melbourne 1986.

Jenkins, R. J. H. "The Date of Leo VI's Cretan Expedition." In *Prosphora eis St. P. Kyriakidên*. 277–81. Hellênika 4. Athens, 1953. Repr. in idem, *Studies on Byzantine History of the 9th and 10th Centuries* (London, 1970), no. XIV.

———. "Leo Choerosphactes and the Saracen Vizier." *Mélanges G. Ostrogorsky*, vol. 1. *ZRVI* 9 (1963): 167–75.

———. "Three Documents concerning the 'Tetragamy'," *DOP* (1962): 231–41.

* Jenkins, R. J. H., et al. *Constantine Porphyrogenitus, De Administrando Imperio*. Volume 1. Greek text edited by Gy. Moravcsik, English translation by R. J. H. Jenkins. CFHB 1 = DOT 1. Washington D.C., 1967. Volume 2, *Commentary*. Edited by R. J. H. Jenkins. London, 1962, reprinted 2012.

Jenkins, R. J. H., and Laourdas, B. "Eight Letters of Arethas on the Fourth Marriage of Leo the Wise." Ἑλληνικα 14 (1956): 124–40.

* Jenkins, R. J. H., and L. G. Westerink, ed. and trans. *Nicholas I Patriarch of Constantinople. Letters*. CFHB 6. DOT 2. Washington D.C., 1973.

Johnson, A. *Roman Forts of the 1st and 2nd Centuries AD in Britain and the German Provinces*. London, 1983.

* Joly de Maïzeroy, P.-G. *Institutions militaires de l'empereur Léon le Philosophe*. 2 vols. Paris, 1758–78. Repr. in F. C. Liskenne and J. B.-B. Sauvan, *Bibliothèque historique et militaire* (Paris, 1837), 3:437–552.

Jomini, Antoine de. *Précis de l'Art de la guerre: Des principales combinaisons de la stratégie, de la grande tactique et de la politique militaire*. Brussels, 1838. English trans. O. F. Winship and E. E. McLean. *The Art of War*. New York, 1854. Trans. G. H. Mendell and W. P. Craighill. *The Art of War*. Philadelphia, 1862; reprinted, Westport, CT, 1971; reprinted, with a new introduction by C. Messenger, London, 1992.

Jones, A. H. M. *The Later Roman Empire, 284-602: A Social, Economic, and Administrative Survey*. 2 vols. Oxford, 1964.

Jopson, N. B. "Early Slavonic Funeral Ceremonies." *The Slavonic Review* 6, no. 16 (1927): 59–67.

Kaegi, W. E. "Confronting Islam: Emperors versus Caliphs (641–c. 850)." In Shepard, *Cambridge History*, 365–94.

———. "The Contribution of Archery to the Turkish Conquest of Anatolia." *Speculum* 39 (1964): 96–108.

———. "The Heraclians and Holy War." In Koder and Stouraitis, *Byzantine War Ideology*, 17–26.

———. *Muslim Expansion and Byzantine Collapse in North Africa*. Cambridge, 2010.

———. "Some Seventh-Century Sources on Caesarea." *Israel Exploration Journal* 28 (1978): 177–81.

Kahane, H. and R. "Abendland und Byzanz III. Literatur und Sprache. B. Sprache." In *RB* A. Vol. 1, fasc. 4–6: 345–640.

Kaiser, A.-M. "Die Fahndung nach Deserteuren im spätantiken Ägypten." In *Actes du 26e Congrès international de papyrologie*. Edited by P. Schubert. 381–90. Geneva, 2012.

Kaldellis, A. "Classicism, Barbarism and Warfare." *American Journal of Ancient History*, n.s., 3–4 (2004–2005): 189–218.

* ———. *Genesios on the Reigns of the Emperors*. Canberra, 1998.

— —. Review of Luttwak, *Grand Strategy*. *Bryn Mawr Classical Review* 2010.01.49.

Kaldellis, A., and D. Krallis. *Michael Attaleiates, The History*. Dumbarton Oaks Medieval Library. Cambridge, Mass., 2012.

Kalomenopoulos, N. Ἡ στρατιωτικὴ ὀργάνωση τῆς Ἑλληνικῆς αὐτοκρατορίας τοῦ Βυζαντίου. Athens, 1937.

Kaplan, M. *Les hommes et la terre à Byzance du VIᵉ au XIᵉ siècle: Propriété et exploitation du sol*. Paris, 1992.

Karapli, K. "Βυζαντινόν χελάνδιον—αραβικό shalandi." In *Cultural Relations between Byzantium and the Arabs*. Edited by Y. Y. al-Hijji and V. Christides. 79–84. Athens, 2007.

———. *Κατευόδωσις στρατού: Η οργάνωση και η ψυχολογική προετοιμασία του Βυζαντινού στρατού πριν απο τον πόλεμο (610–1081)*. Athens, 2010.

———. "'Κιβώτια', 'Ξυλόκαστρα', 'Καστέλλωμα'." *Byzantina* 29 (2009): 111–19.

———. "Speeches of Arab Leaders to Their Warriors According to Byzantine Texts." *Graeco-Arabica* 5 (1993): 233–42.

Karayannopoulos, J., and G. Weiss. *Quellenkunde zur Geschichte von Byzanz (324–1453)*. Schriften zur Geistesgeschichte des östlichen Europa 14, nos. 1–2. Wiesbaden, 1982.

Karlin-Hayter, P. "Arethas, Choirosphactes and the Saracen Vizir." *Byz* 35 (1965): 455–81.

———. "Aréthas et le droit d'asile." *Byz* 34 (1964): 613–18.

———. "Clément d'Ochrid, la guerre bulgare de Léon et la prise de Thessalonique en 904." *Byz* 35 (1965): 606–11.

———. "The Emperor Alexander's Bad Name." *Speculum* 44 (1969): 585–96.

———. "L'Hétériarque: L'évolution de son rôle du De Ceremoniis au Traité des Offices." *JÖB* 23 (1974): 101–43.

———. "Le mort de Théophano (10 nov. 896 ou 895)." *BZ* 62 (1969): 13–19.

———. "La 'préhistoire' de la dernière volonté de Léon VI." *Byz* 33 (1963): 483–86.

———. "The revolt of Andronicus Ducas." *BSl* 27 (1966): 23–25.

———. "Le synode à Constantinople de 886 à 912 et le rôle de Nicolas le Mystique dans l'affaire de la tétragamie." *JÖB* 40 (1990): 205–8.

* ———. *Vita Euthymii Patriarchae CP: Text, Translation, Introduction and Commentary.* Brussels, 1970.

———. "'When Military Affairs Were in Leo's Hands': A Note on Byzantine Foreign Policy (886–912)." *Traditio* 23 (1967): 15–40. Repr. in eadem, *Studies in Byzantine Political History* (London, 1981), no. XIII.

Kartsonis, A. *Anastasis: The Making of an Image.* Princeton, 1986.

Kasparek, M. U. "Stand der Forschung über den Hufbeschlag des Pferdes." *Zeitschrift für Agrargeschichte und Agrarsoziologie* 6 (1958): 38–43.

Kazanski, M. "Les armes et les techniques de combat des guerriers steppiques du début du moyen âge. Des Huns aux Avares." In *Le cheval dans les sociétés antiques et médiévales.* Edited by S. Lazaris. 193–99. Strasbourg, 2012.

Kazanski, M., A. Mastykova, and P. Périn. "Byzance et les royaumes barbares d'Occident au début de l'époque mérovingienne." In *Probleme der frühen Merowingerzeit im Mitteldonauraum: Materialien des XI. Internationalen Symposiums "Grundprobleme der frühgeschichtlichen Entwicklung im nördlichen Mitteldonaugebiet.* Edited by J. Tejral. 159–94. Brno, 2002.

Kazanski, M., and J.-P. Sodini, "Byzance et l'art 'nomade': Remarques à propos de l'essai de J. Werner sur le dépôt de Malaja Pereščepina (Pereščepino)." *Revue archéologique* 1 (1987): 71–90.

Kazhdan, A. "Hagiographical Notes 2: On Horseback or On Foot? A 'Sociological' Approach in an Eleventh-Century Saint's Life." *Byzantion* 53 (1983): 544–45.

Kazhdan, A., and I. Ševčenko, "Modesty, Topos of." *ODB* 2:1387.

Keegan, J. A. *History of Warfare.* London, 1993.

Kennedy, H. *The Armies of the Caliphs: Military and Society in the Early Islamic State.* London and New York, 2001.

———. "The Financing of the Military in the Early Islamic State." In Cameron, *States, Resources and Armies,* 361–78.

———. *The Prophet and the Age of the Caliphates.* London, 1986; repr. 2004.

* Keydell, R., ed. *Agathiae Myrinaei Historiarum libri V.* Berlin, 1967.

Khoury, A. T. *Polémique byzantine contre l'Islam (VIIIᵉ–XIIIᵉ s.).* Leiden, 1972.

———. *Les théologiens byzantins et l'Islam. Textes et auteurs (VIIIᵉ–XIIIᵉ s.).* Louvain, 1969.

Khudyakov, Yu. S. *Vooruženie srednevokovykh kočevnikov Južnoy Sibiri i Central'noy Azii.* Novosibirsk, 1986.

Kiechle, F. K. "Die Entwicklung der Brandwaffen im Altertum." *Historia* 26 (1977): 253–56.

Kirpichnikov, A. N. *Drevnierusskoe Oruzhie.* Vol. 3. Leningrad, 1971.

Kislinger, E. "Eudokia Ingerina, Basileios I. und Michael III." *JÖB* 33 (1983): 389–400.

———. "Verkehrsrouten zur See im byzantinischen Raum." In *Handelsgüter und Verkehrswege: Aspekte der Warenversorgung im östlichen Mittelmeerraum (4. bis 15. Jahrhundert).* Edited by E. Kislinger, J. Koder, and A. Külzer. 149–74. Vienna, 2010.

Kiss, A. "Frühmittelalterliche byzantinische Schwerter im Karpatenbecken." *Acta Archaeologica Hungarica* 39 (1987): 193–210.

Kızıltan, Z. ed. *Stories from the Hidden Harbor. Shipwrecks of Yeni Kapı.* Istanbul, 2013.

Der kleine Pauly: Lexikon der Antike in fünf Bänden. Munich, 1979.

Klingshirn, W. "Christian Divination in Late Roman Gaul: The *Sortes Sangallenses.*" In *Mantikê: Studies in Ancient Divination.* Edited by S. I. Johnston and P. T. Struck. 99–128. Leiden, 2005.

———. "Divination and the Disciplines of Knowledge According to Augustine." In *Augustine and the Disciplines: From Cassiciacum to Confessions.* Edited by K. Pollmann and M. Vessey. 113–40. Oxford, 2005.

Kocabaş, U. ed. *The "Old Ships" of the "New Gate."* Yenikapı'nın Eski gemileri 1. Yenikapı shipwrecks 1 / Yenikapı batıkları 1. Istanbul University Yenikapı Shipwrecks Project. Istanbul, 2008.

Kocabaş, I. O. and U. "Technological and Constructional Features of Yenikapı shipwrecks: a preliminary evaluation." In Kocabaş, *"Old ships" of the "New Gate,"* 99–183.

* Köchly, H., ed. *Anonymi Byzantini rhetorica militaris.* Index lectionum in literarum Universitate Triecensi. Turici, 1855–56.

* Köchly, H., and W. Rüstow. *Griechische Kriegsschriftsteller.* 3 volumes. Leipzig, 1853–55.

Koder, J. "Anmerkungen zu γραικόω." *Byzantina* 21 (2000): 199–202.

———. "Aspekte der thalassokratia der Byzantiner in der Ägäis." In *Griechenland und das Meer.* Edited by E. Chrysos, D. Letsios, H. A. Richter and R. Stupperich. 101–9. Mannheim and Möhnsee, 1999.

———. *Tabula Imperii Byzantini.* Volume 10, *Aigaion Pelagos (die nördliche Ägäis).* Denkschr. d. Österr. Akad. d Wiss., phil.-hist. Kl. 259. Vienna, 1998.

Koder, J., and I. Stouraitis, eds. *Byzantine War Ideology between Roman Imperial Concept and Christian Religion.* Denkschr. d. Österr. Akad. d Wiss., phil.-hist. Kl. 452. Vienna, 2012

Koder, J., and T. Weber. *Liutprand von Cremona in Konstantinopel: Untersuchungen zum griechischen Sprachschatz und zu realienkundlichen Aussagen in seinen Werken.* Byz. Vind. 13. Vienna, 1980.

Kolia-Dermitzaki, A. Ὁ βυζαντινός "ἱερός πόλεμος": Ἡ ἔννοια καί ἡ προβολή τοῦ θρησκευτικοῦ πολέμου στό Βυζάντιο. Athens, 1991.

———. "Byzantium at War in Sermons and Letters of the 10th and 11th Centuries: An Ideological Approach." In Tsiknakes, *Ἐμπόλεμο Βυζάντιο,* 213–38.

———. "'Holy War' in Byzantium Twenty Years Later: A Question of Term Definition and Interpretation." In Koder and Stouraitis, *Byzantine War Ideology*, 121–32.

———. "Some Remarks on the Fate of Prisoners of War in Byzantium (9th–10th Centuries)." In *La liberazione dei "captivi" tra cristianità e islam: Oltre la Crociata e il Ġihād; Tolleranza e servizio umanitario; Atti del Congresso interdisciplinare di studi storici.* Edited by G. Cipollone. 583–620. Vatican City, 2000.

Kolias, G. "Περὶ ἀπλήκτου." *ΕΕΒΣ* 17 (1941): 144–84.

Kolias, T. G. "Die byzantinische Kriegsmarine: Ihre Bedeutung im Vertiedigungssystem von Byzanz." In *Griechenland und das Meer.* Edited by E. Chrysos, D. Letsios, H. A. Richter and R. Stupperich. 133–39. Mannheim and Möhnsee, 1999.

———. Byzantinische Waffen: Ein Beitrag zur byzantinischen Waffenkunde von den Anfängen bis zur lateinischen Eroberung. Byz. Vind. 17. Vienna, 1988.

———. "Ein zu wenig bekannter Faktor im byzantinischen Heer: Die Hilfskräfte (παῖδες, πάλληκες, ὑπουργοί . . .)." In *Festschrift für Peter Schreiner.* Edited by C. Scholz and G. Makris. *Byzantinisches Archiv* 19 (2000): 113–24.

———. "Eßgewohnheiten und Verpflegung im byzantinischen Heer." In *BYZANTIOΣ: Festschrift für H. Hunger zum 70. Geburtstag.* Edited by W. Hörandner, J. Koder, O. Kresten and E. Trapp. 193–202. Vienna, 1984.

———. "The Horse in the Byzantine World." In *Le cheval dans les sociétés antiques et médiévales.* Edited by S. Lazaris. 87–97. Strasbourg, 2012.

———. "The *Taktika* of Leo VI the Wise and the Arabs," *Graeco-Arabica* 3 (1984): 129–35.

———. "Τὰ στρατιωτικὰ ἐγκλήματα κατὰ τοὺς Βυζαντινοὺς χρόνους." In Troianos, ed., *Ἔγκλημα καὶ τιμωρία στὸ Βυζάντιο,* 295–316.

———. "Ζάβα, ζαβαρεῖον, Ζαβαρειώτης." *JÖB* 29 (1980): 27–35.

* Konstantopoulos, K. *Βυζαντιακὰ μολυβδόβουλλα τοῦ ἐν Ἀθηναῖς Ἐθνικοῦ Νομισματολογικοῦ Μουσείου.* Athens, 1917.

Köpstein, H. "Profane Gesetzgebung und Rechtssetzung." In *Quellen zur Geschichte des frühen Byzanz (4.–9. Jahrhundert).* Edited by F. Winkelmann and W. Brandes. 134–48. BBA 55. Berlin, 1990.

* Korzensky, E. "Leges poenales militares e codice Laurentiano LXXV." *Egyetemes philologiai közlöny* 54 (1930): 155–63, 215–18.

* Korzenszky, E., and R. Vári, eds. *Sylloge Tacticorum Graecorum.* Volume 1. Budapest, 1935.

* Kotter, B., ed. *Die Schriften des Johannes von Damaskos.* Volume 2. Berlin, 1973.

Koukoules, Ph. *Βυζαντινῶν βίος καὶ πολιτισμός.* 6 vols. Athens, 1946–52.

———. "Διορθωτικὰ καὶ ἑρμηνευτικὰ εἰς τὴν Ἔκθεσιν τῆς βασιλείου τάξεως Κωνσταντίνου τοῦ Πορφυρογεννήτου καὶ τὸ Κλητορολόγιον του Φιλοθέου." *ΕΕΒΣ* 19 (1949): 75–115.

Kourakes, N. E. *Διαχρονικές αρχές βυζαντινής στρατηγικής και τακτικής: Με έμφαση στο έργο Τακτικά του Λέοντος Στ' του Σοφού.* Athens, 2012.

Koutaba-Deliboria, B. *Ο γεωγραφικός κόσμος Κωνσταντίνου του Πορφυρογεννήτου.* Volume 2, *Η είκονα.* (Athens 1993).

Koutrakou, N. "Diplomacy and Espionage: Their Role in Byzantine Foreign Relations, 8th–10th Centuries." *Graeco-Arabica* 6 (1995): 125–44.

———. "The Image of the Arabs in Middle-Byzantine Politics: A Study in the Enemy Principle (8th–10th Centuries)." *Graeco-Arabica* 5 (1993): 213–24.

———. *La propagande impériale byzantine: Persuasion et réaction (VIIIe–Xe siècles).* Athens, 1994.

———. "'Spies of Towns': Some Remarks on Espionage in the Context of Byzantine-Arab Relations (VIIth–Xth Centuries)." *Graeco-Arabica* 7–8 (1999–2000): 243–66.

Kovács, L. "A kalandozások hadművészete és zsákmányának régészeti emlékei" (The art of war of the Hungarian raids and the archaeological evidence for their booty). In *Válaszúton: pogányság-kereszténység, kelet-nyugat.* Edited by L. Kredics. 223–37. Veszprém, 2000.

Kraft, U. "Lat.-griech. Scala als spätantike Bezeichnung für den Steigbügel, und die Benennung des Steigbügels bei anderen Völkern." In *Le cheval dans les sociétés antiques et médiévales.* Edited by S. Lazaris. 155–92. Strasbourg, 2012.

* Kraemer, C. J., ed. *Excavations at Nessana.* Volume 3, *Non-Literary Papyri.* Princeton, 1958.

Kramer, J. "Ein gräzismus gotischer Herkunft im Italienischen: Bando." *Balkan-Archiv,* n.F. 12 (1987): 197–207.

———. "Papyrusbelege für fünf germanische Wörter: ἀρμαλαύσιον, βάνδον, βουρδῶν, βράκιον, σαφώνιον." *Archiv für Papyrusforschung* 42 (1996): 113–26.

* Kramers, J. H., and G. Wiet, ed. and trans. *Ibn Hawqal Abu'l-Qasīm: Kitāb Sūrat al-Ard, Configuration de la terre.* 2 volumes. Beirut and Paris, 1964.

* Kratchkovsky, I., F. Micheau, and G. Troupeau, trans. *Histoire de Yahya ibn Sa'id d'Antioche.* Patrologia Orientalis 42.4 (no. 212). Turnhout, 1997.

Krausmüller, D. "Killing at God's Command: Niketas Byzantios' Polemic against Islam and the Christian Tradition of Divinely-Sanctioned Murder." *Al-Masāq* 16 (2004): 163–76.

Kravari, V., ed. *Actes de Pantokrator.* Archives de l'Athos. Paris, 1991.

Krentz, P. "Deception in Archaic and Classical Greek Warfare." In *War and Violence in Ancient Greece.* Edited by H. van Wees. 167–200. London, 2000.

* Krentz, P., and E. L. Wheeler. *Polyaenus: Stratagems of War.* 2 volumes. Chicago, 1994.

Kresten, O. "Zur angeblichen Heirat Annas, der Tochter Kaiser Leons VI., mit Ludwig III. 'dem Blinden'." *Römische Historische Mitteilungen* 42 (2000): 171–211.

Kriaras, E. *Λεξικὸ τῆς μεσαιωνικῆς ἑλληνικῆς δημώδους γραμματείας (1100–1669).* Thessalonike, 1968–.

Kroll, H. "Groß und stark? Zur Widerristhöhe und Statur byzantinischer Arbeitstiere." In RGZM Conference Proceedings, 2012. Forthcoming.

———. *Tiere im byzantinischen Reich: Archäozoologosche Forschungen im Überblick.* Monographien des Römisch-Germanischen Zentalmuseums Mainz 87. Mainz, 2010.

Krüger, P. "Anthropologie." In *Kleines Wörterbuch des Christlichen Orients.* Edited by J. Aßfalg, P. Krüger. 13–15. Wiesbaden, 1975.

Krumbacher, K. *Geschichte der byzantinischen Litteratur*. 2nd edition. Munich, 1897.

Kučma, V. "Metodika boevoi podgotovki pa 'Taktike L'va': Gospodstvo principov tradicionalizma." In *Vlast', obščestva i cerkov v Vizantii*. Edited by S. Malakhov and N. Barabanov, 89–116. Armavir, 2007.

———. "Militärische Traktate." In *Quellen zur Geschichte des frühen Byzanz (4.–9. Jahrhundert): Bestand und Probleme*. Edited by F. Winkelmann and W. Brandes. 327–35. BBA 55. Berlin, 1990.

———. "Νόμος στρατιωτικός (K voprosu o svyazi trekh pamyatnikov Vizantiiskogo voennogo prava)." *VV* 32 (1971): 276–84. Repr. in idem, *Voennaya organizaciya Vizantiiskoi imperii*, 243–58.

———. "Principy organizacii boevikh predviščeniy (maršei) po 'Taktika L'va'." *Antičnaya drevnost' i Srednie veka* 39 (2009): 123–41.

———. "Principy osady i oborony gorodov v vizantiiskoi polemologičeskoi tradicii." *VV* 70 (2011): 7–24

———. "'Strategikos' Onasandra i 'Strategikon Mavrikiia': Opyt sravnitel'noĭ charakteristiski." *VV* 43 (1982): 35–53; 45 (1984): 20–34; 46 (1986): 109–23. Repr. in idem, *Voennaya organizaciya Vizantiiskoi imperii*, 139–207).

———. "Taktika L'va (glaby III–VI)." *VV* 68 (2009): 280–99.

———. "'Taktika L'va' kak istoričeski istočnik." *VV* 33 (1972) 75–87.

———. "Vizantiiskie voennye traktaty VI–X vv. kak istoričeskii istočnik: Nekotorye voprosy vnutrennei istorii vizantiiskoi imperii po dannym voennykh traktatov." *VV* 40 (1979): 49–75. Repr. in idem, *Voennaya organizaciya Vizantiiskoi imperii*, 37–56.

———. *Voennaya organizaciya Vizantiiskoi imperii*. St. Petersburg, 2001.

Kühn, H.-J. *Die byzantinische Armee im 10. und 11. Jahrhundert*. Vienna, 1991.

Kulakovskiy, J. "Lev Mudryi ili Lev Isavr byl avtorom 'Taktiki'?" *VV* 5 (1898): 398–403.

———. "'Vizantiiskii lager' k. X veka." *VV* 10 (1903): 63–91.

* Kurtz, E. *Zwei griechische Texte über die hl. Theophano, die Gemahlin Kaisers Leo VI*. St. Petersburg, 1898.

Kustas, G. L. *Studies in Byzantine Rhetoric*. Thessalonike, 1973.

Laiou, A. E. "On Just War in Byzantium." In *Tὸ Ἑλληνικόν: Studies in Honor of Speros Vryonis, Jr*. Edited by J. S. Langdon, S. W. Reinert, J. S. Allen, C. P. Ioannides. 153–74. New York, 1993.

———. "The Just War of Eastern Christians and the Holy War of the Crusaders." In *The Ethics of War: Shared Problems in Different Traditions*. Edited by R. Sorabji and D. Rodin. 30–43. Aldershot and Burlington, Vt., 2006.

Laiou, A. E., et al., eds. *The Economic History of Byzantium from the Seventh through the Fifteenth Century*. Washington D.C., 2002.

Lambros, S. "Τρία κείμενα συμβάλλοντα εἰς τὴν ἱστορίαν τοῦ ναυτικοῦ παρὰ Βυζαντινοῖς." *Νέος Ἑλλ.* (1912): 162–77.

Lammert, F. "Die älteste erhaltene Schrift über Seetaktik und ihre Beziehung zum Anonymus Byzantinus des 6. Jahrhunderts, zu Vegetius und zu Aineias' *Strategika*." *Klio* 33 (1940): 271–88.

Lampe, G. W. H. *A Patristic Greek Lexicon*. Oxford 1961–68.

* Lang, C. *Flavii Vegetii Renati Epitoma rei militaris*. Leipzig, 1885.

Langó, P. "Archaeological Research on the Conquering Hungarians: A Review." In *Research on the Prehistory of the Hungarians: A Review*. Edited by B. G. Mende. 177–80. Budapest, 2005.

* Laourdas, B. and G. Westerink, eds. *Photii patriarchae Constantinopolitani Epistulae et Amphilochia*. 2 volumes. Leipzig, 1983–84.

La Salvia, V. "La fabbricazione delle spade delle grandi invasioni: Per la storia del 'processo indiretto' nella lavorazione de ferro." *Quaderni medievali* 44 (1997): 28–54.

László, G. "Contribution à l'archéologie de l'époque des migrations, II: Le carquois d'arc des Hongrois conquérants." *Acta archaeologica Academiae Scientiarum Hungaricae* 7 (1957): 165–98 (172–86).

———. "Études archéologiques sur l'histoire de la société des Avares." *Archaeologia Hungarica* 34 (1955).

———. "Der Grabfund von Koroncó und der altungarische Sattel," *Archaeologica Hungarica* 27 (1943): 159–70.

Latham, M. L. W. *Revised Medieval Latin Word-List from British and Irish Sources*. London, 1965.

* Laurent, V. *Le Corpus des sceaux de l'empire byzantin*. Volume 2, *L'administration centrale*. Paris, 1981.

Lawrence, A. W. "A Skeletal History of Byzantine Fortifications." *Annual of the British School at Athens* 78 (1983): 171–227.

Lazaris, S. ed. *Le cheval dans les sociétés antiques et médiévales* (Strasbourg, 2012).

Lechner, K. *Hellenen und Barbaren im Weltbild der Byzantiner*. Munich, 1954.

Lee, A. D. "The Empire at War." In *The Cambridge Companion to the Age of Justinian*. Edited by M. Maas. 113–33. Cambridge' 2005.

Lee, A. D., and J. Shepard, "A Double Life: Placing the Peri Presbeon." *Byzslav* 52 (1991): 15–39.

* Lee, H. D. P., ed. and trans. *Aristotle: Meteorologica*. Loeb Classical Library. London and Cambridge, Mass., 1952.

* Lefort, J., N. Oikonomidès, and D. Papachryssanthou, eds. *Actes d'Iviron*. Volume 1. Archives de l'Athos. Paris, 1985.

Lemerle, P. *The Agrarian History of Byzantium from the Origins to the Twelfth Century: The Sources and the Problems*. Galway, 1979.

———. *Byzantine Humanism, the First Phase: Notes and Remarks on Education and Culture in Byzantium from its Origins to the 10th Century*. Byzantina Australiensia 3. Translated by H. Lindsay and A. Moffat. Canberra, 1986.

———. "L'histoire des Pauliciens d'Asie Mineure d'après les sources grecques." *TM* 5 (1973): 1–144.

* ———. *Les plus anciens recueils des miracles de S. Démétrius et de la pénétration des Slaves dans les Balkans*. Volume 1, *Le texte*. Paris, 1979.

* Lemerle, P., N. Svoronos, A. Guillou, and D. Papachryssanthou, eds. *Actes de Lavra.* Part 1, *Des origines à 1204.* Archives de l'Athos. Paris, 1970.

Lendle, O. *Schildkröten: Antiken Kriegsmaschinen in poliorketischen texten.* Wiesbaden, 1975.

Lenski, N. "Captivity and Slavery among the Saracens in Late Antiquity (ca. 250–630 CE)." *Antiquité Tardive* 19 (2011): 237–66.

Leone, A. *Animali di trasporto nell'antico Egitto: Una rassegna papirologica dalla dinastia dei Lagidi ai Bizantini.* Naples, 1998.

Leoni, B. *La parafrasi Ambrosiana dello Strategicon di Maurizio: L'arte della guerra a Bisanzio.* Milan, 2003.

Leriche, P. "Techniques de guerre sassanides et romaines à Doura-Europos." In Vallet and Kazanski, *L'Armée romaine et les barbares,* 83–100.

* Lesmüller-Werner, I., and I. Thurn. *Iosephi Genesii Regum libri quattuor.* CFHB 14. Berlin and New York 1978.

Lev, Y. "A Mediterranean Encounter: The Fatimids and Europe, Tenth to Twelfth Centuries." In Gertwagen and Jeffreys, *Shipping, Trade and Crusade,* 131–56.

* Levi della Vida, G. *Les "Livres des Chevaux" de Hisham Ibn al-Kalbi et Muhammad Ibn al-Arabi, publiés d'après le manuscrit de l'Escorial ar. 1705.* Repr. Leiden, 1928.

Lewis, A. R. *Naval Power and Trade in the Mediterranean A.D. 500 to 1100.* Princeton, 1951.

Lewin, A. S., and P. Pelegrini, eds., *The Late Roman Army in the Near East from Diocletian to the Arab Conquest.* 411–19. BAR International Series 1717. Oxford, 2007.

Leyser, K. "The Battle of the Lech, 955: A Study in Tenth-Century Warfare." *History* 50 (1965): 1–25.

Lightfoot, C. S. et al., "The Amorium Project: The 1996 Excavation Season." *DOP* 52 (1998): 323–36.

Lilie, R.-J. *Die byzantinische Reaktion auf die Ausbreitung der Araber.* Miscellanea Byzantina Monacensia 22. Munich, 1976.

———. "'Thrakien' und 'Thrakesion': Zur byzantinischen Provinzorganisation am Ende des 7. Jahrhunderts." *JÖB* 26 (1977): 7–47.

———. "Die zweihundertjährige Reform: Zu den Anfängen der Themenorganisation im 7. und 8. Jahrhundert." *BSl* 45 (1984): 27–39, 190–201.

Lillington-Martin, C. "Archaeological and Ancient Literary Evidence for a Battle near Dara Gap, Turkey, A.D. 530: Topography, Texts and Trenches." In Lewin and Pelegrini, *Late Roman Army,* 299–311.

* Liskenne, F. C., and J. B.-B. Sauvan. *Bibliothèque historique et militaire: Dédiée à l'armée et à la Garde Nationale de France.* Paris, 1837.

* Litavrin, G. G., ed. and trans. *Soveti i rasskazi Kekavmena: Socinenie vizantijskogo polkovodtsa XI veka.* Moscow, 1972.

Löfstedt, L. ed. *Li abregemenz noble honme Vegesce Flave René des establissemenz a chevalerie, traduction par Jean de Meun de Flavii Vegetii Renati viri illustris Epitoma Institutionum Rei Militaris.* Helsinki, 1977.

Lokin, J. H. A. "The Novels of Leo and the Decisions of Justinian." In *Analecta Athenensia ad ius byzantinum spectantia.* Edited by S. Troianos. 1:13–140. Forschungen zur byzantinischen Rechtsgeschichte, Athener Reihe. Athens, 1997.

Long, P. *Openness, Secrecy, Authorship: Technical Arts and the Culture of Knowledge from Antiquity to the Renaissance.* Baltimore, 2001.

Loreto, L. "Il generale e la biblioteca: La trattatistica militare greca da Democrito d'Abdera ad Alessio I Comneno." In *Lo spazio letterario della Grecia antica.* Edited by G. Cambiano, L. Canfora, and D. Lanza. 2:563–89. Rome, 1995.

Loud, G. A. "Byzantine Italy (876–1000)." In Shepard, *Cambridge History,* 560–82.

Lounghis, T. C. "The Adaptability of Byzantine Political Ideology to Western Realities as Diplomatic Message (476–1096)." In *Comunicare e significare nell'alto medioevo.* Settimane di studio della Fondazione Centro italiano di studi sull'alto Medioevo 52. 335–61. Spoleto, 2005.

Luttwak, E. *The Grand Strategy of the Byzantine Empire.* Cambridge, Mass., 2009.

Lyubarskii, I. "Writers' Intrusion in Early Byzantine Literature." In *XVIIIe Congrès int. des études byzantines, Rapports pléniers.* 433–56. Moscow, 1991.

Macartney, C. A. "The Bulgaro-Greek War and the Magyar 'Landnahme'." In idem, *Magyars,* 177–88.

———. *The Magyars in the Ninth Century.* Oxford 1930, reprinted 1968.

MacDonald, M. "Quelques réflexions sur les Saracènes: L'inscription de Rawwāfa et l'armée romaine." In *Présence arabe dans le Croissant fertile avant l'Hégire.* Edited by H. Lozachmeur. 93–101. Paris, 1995.

* MacLean, S. *History and Politics in Late Carolingian and Ottonian Europe: The Chronicle of Regino of Prüm and Adalbert of Magdeburg.* Manchester, 2009.

Macrides, R. J. "Justice under Manuel I Komnenos: Four Novels on Court Business and Murder." *Fontes Minores* 6 (1984): 156–204.

———. "Killing, Asylum and the Law in Byzantium." *Speculum* 63 (1988): 509–38.

———, ed. *Travel in the Byzantine World.* SPBS Publications 10. Aldershot, 2002.

Magdalino, P. "Basil I, Leo VI, and the Feast of the Prophet Elijah." *JÖB* 38 (1988): 193–96.

———. "The Bath of Leo the Wise," *Maistor: Classical, Byzantine, and Renaissance Studies for Robert Browning.* Edited by A. Moffatt. 225–40. Canberra, 1984.

———. "The Bath of Leo the Wise and the 'Macedonian Renaissance' Revisited: Topography, Iconography, Ceremonial, Ideology." *DOP* 42 (1988): 97–118.

———. "The Byzantine Aristocratic *Oikos.*" In *The Byzantine Aristocracy, IX–XIII Centuries.* Edited by M. Angold. 92–111. BAR International Series 221. Oxford, 1984.

———. "The End of Time in Byzantium." In *Endzeiten: Eschatologie in den monotheistischen Weltreligionen.* Edited by W. Brandes and F. Schmieder. 119–33. Millennium Studies in the Culture and History of the First Millennium C.E. 16. Berlin and New York, 2008.

———. "The History of the Future and Its Uses: Prophecy, Policy and Propaganda." In *The Making of Byzantine History: Studies Dedicated to Donald M. Nicol on his Seventieth Birthday.* Edited by R. Beaton and C. Roueché. 3–34. Aldershot, 1993.

———. "Knowledge in Authority and Authorised History: The Imperial Intellectual Programme of Leo VI and Constantine VII." In *Authority in Byzantium*. Edited by P. Armstrong, 187–209. London, 2013.

———. "The Non-juridical Legislation of the Emperor Leo VI." In *Analecta Athenensia ad ius byzantinum spectantia*. Edited by S. Troianos. 1:169–82. Forschungen zur byzantinischen Rechtsgeschichte, Athener Reihe. Athens, 1997.

———. "Observations on the Nea Ekklesia of Basil." *JÖB* 37 (1987): 51–64.

———. *L'orthodoxie des astrologues: La science entre le dogme et la divination à Byzance*. Paris, 2006.

———. "The Road to Baghdad in the Thought-World of Ninth-Century Byzantium." In Brubaker, *Byzantium in the Ninth Century*, 195–213.

———. "Saint Demetrios and Leo VI," *BSl* 51 (1990): 198–201.

———. "The Year 1000 in Byzantium." In *Byzantium in the Year 1000*. Edited by P. Magdalino. 233–70. Leiden, 2002.

Magdalino, P., and M. Mavroudi, eds. *The Occult Sciences in Byzantium*. Geneva, 2006.

Magdalino, P., and R. Nelson. "Introduction." In eidem, *The Old Testament in Byzantium*. 1–38.

———, eds. *The Old Testament in Byzantium*. Washington, D.C., 2010.

Magomedov, M. G. *Obrazhovanie Khazarskogo Kaganata: Po materialam arkheologicheskikh issledovaniya i pis'mennim dannim*. Moscow, 1983.

* Magoulias, H. J., trans. *O City of Byzantium: Annals of Niketas Choniates*. Wayne State, 1984.

* Mai, A., ed. *Nova Patrum Bibliotheca*. 10 volumes. Rome, 1844–1905.

Makris, G. "Ships." In Laiou et al., *Economic history of Byzantium*, 91–100.

Makrypoulias, C. G. "Byzantine Expeditions against the Emirate of Crete c. 825–949." *Graeco-Arabica* 7–8 (1999–2000): 347–62.

———. "The Navy in the Works of Constantine Porphyrogenitus." *Graeco-Arabica* 6 (1995): 152–71.

Malamut, E. "Constantin VII et son image de l'Italie." In *Byzanz und das Abendland im 10. Und 11. Jahrhundert*. Edited by E. Konstantinou. 269–92. Köln, 1997.

———. *Les îles de l'empire byzantin, VIIIᵉ–XIIᵉ siècles*. 2 vols. Byzantina Sorbonensia 8. Paris, 1988.

———. "L'image byzantine des Petchénègues." *BZ* 88 (1995): 105–47.

Maliaras, N. "Die Musikinstrumente im byzantinischen Heer vom 6. bis zum 12. Jahrhundert: Eine Vorstellung der Quellen." *JÖB* 51 (2001): 73–104.

* Mandilaras, B. G., ed. *Isocrates, Opera omnia*. Volume 2. Munich and Leipzig, 2003.

Mango, C. "The Availability of Books in the Byzantine Empire, A. D. 750–850." In *Books and Bookmen in Byzantium*. 29–45. Washington D.C., 1975. Reprinted in idem, *Byzantium and Its Image*, no. VII.

———. *Byzantium and Its Image: History and Culture of the Byzantine Empire and Its Heritage*. London, 1984.

————. "Eudocia Ingerina, the Normans and the Macedonian Dynasty." *ZRVI* 14–15 (1973): 17–27. Reprinted in idem, *Byzantium and Its image*, no. XV.

* ————. *The Homilies of Photios, Patriarch of Constantinople*. DOS 3. Cambridge, Mass., 1958.

————. "The Legend of Leo the Wise." *ZRVI* 6 (1960): 59–93. Reprinted in idem, *Byzantium and Its image*, no. XVI.

* ————, ed. and trans. *Nikephoros Patriarch of Constantinople. Short History*. CFHB 13. DOT 10. Washington D.C., 1990.

* Mango, C., and S. Efthymiadis. *The Correspondence of Ignatios the Deacon*. CFHB 39 = DOT 11. Washington D.C., 1997.

* Mango, C., and R. Scott, ed. and trans. *The Chronicle of Theophanes Confessor*. Oxford, 1997.

Markopoulos, A. "An Anonymous Laudatory Poem in Honour of Basil I." *DOP* 46 (1992): 225–32.

————. "Autour des *Chapitres parénétiques* de Basile Ier." In Ευψυχία: *Mélanges offerts à Hélène Ahrweiler*. 2:469–79. Byzantina Sorbonensia 16. Paris, 1998.

————. "La Chronique de l'an 811 et le Scriptor incertus de Leone Armenia: Problèmes des relations entre l'hagiographie et l'histoire." *REB* 57 (1999): 255–62.

————. "Constantine the Great in Macedonian Historiography: Models and Approaches." In *New Constantines: The Rhythm of Imperial Renewal in Byzantium, 4th–13th Centuries*. Edited by P. Magdalino. 159–70. Aldershot, 1994.

————. "The Ideology of War in the Military Harangues of Constantine VII Porphyrogennetos." In Koder and Stouraitis, *Byzantine War Ideology*, 47–56.

Marsden, E. W. *Greek and Roman Artillery: Historical Development*. Oxford, 1969.

Martin, J.-M. *Guerres, accords et frontières en Italie méridionale pendant le haut Moyen Âge*. Rome, 2005.

————. "Les thèmes italiens : Territoire, administration, population." In *Histoire et culture dans l'Italie byzantine*. Edited by A. Jacob, J.-M. Martin and G. Noyé. 518–58. Rome, 2006.

* Martinez, P. "Gardīzī's Two Chapters on the Turks." *Archivum Eurasiae Medii Aevi* 2 (1982): 109–217.

Mastoropoulos, G. S. "Σίφων—σ(ι)φούνι: ἐπιβίωση ἑνὸς ἀρχαίου (;) ἀγγείου." Ἀρχαιο-λογικὰ ἀνάλεκτα ἐξ Ἀθηνῶν 21 (1988): 158–62.

Mavrodinov, N. *Le trésor protobulgare de Nagyszentmiklós*. Archaeologia Hungarica 29. Budapest, 1943.

Mayor, A. *Greek Fire, Poison Arrows and Scorpion Bombs: Biological and Chemical Warfare in the Ancient World*. Woodstock, 2003.

Mazzucchi, C. M. "Dagli anni di Basilio parakimomenos (Cod. Ambros. B 119 sup.)." *Aevum* 52 (1978): 267–318.

————. "Le καταγραφαὶ dello Strategicon di Maurizio e lo schieramento di battaglia dell'esercito Romano nel VI/VII secolo." *Aevum* 55, no. 1 (1981): 111–38.

* ———, ed. *Menae patricii cum Thoma referendario: De scientia politica dialogus; Iteratis curis quae exstant in codice Vaticano palimpsesto*. Milan, 2002.

McCabe, A. *A Byzantine Encyclopaedia of Horse Medicine: The Sources, Compilation and Transmission of the Hippiatrica*. Oxford, 2007.

McCormick, M. *Eternal Victory: Triumphal Rulership in Late Antiquity, Byzantium, and the Early Medieval West*. Cambridge, 1986.

———. "Western Approaches (700–900)." In Shepard, *Cambridge History*, 395–432.

* McElwain, M. B., ed. *Frontinus, The Stratagems and the Aqueducts of Rome*. Trans. C. E. Bennett. Loeb Classical Library. London and Cambridge, Mass., 1925.

McEvoy, M. *Child-Emperor Rule in the Late Roman West, AD 367–455*. Oxford, 2013.

McGeer, E. "Byzantine Siege Warfare in Theory and Practice." In *The Medieval City under Siege*. Edited by I. A. Corfis and M. Wolfe. 123–29. Woodbridge, 1995.

———. "Infantry Versus Cavalry: The Byzantine Response." *REB* 46 (1988): 135–45.

* ———. *The Land Legislation of the Macedonian Emperors*. Toronto, 2000.

———. "Menaulion—menaulatoi." *Diptycha* 4 (1986–87): 53–57.

———. "Military Texts." In Jeffreys, Haldon, and Cormack, *Oxford Handbook*, 907–14.

* ———. *Sowing the Dragon's Teeth: Byzantine Warfare in the Tenth Century*. DOS 33. Washington D.C., 1995.

———. "The *Syntaxis Armatorum Quadrata*: A Tenth-Century Tactical Blueprint." *REB* 50 (1992): 219–29.

———. "Tradition and Reality in the Taktika of Nikephoros Ouranos." *DOP* 45 (1991): 130–40.

* ———. "Two Military Orations of Constantine VII." In Nesbitt, *Byzantine Authors*, 111–35.

McGeer, E., and A. Cutler. "Battle Standards and Flags." In *ODB*, 1:272.

McKitterick, R., ed. *The New Cambridge Medieval History*. Volume 2, *c. 700–c. 900*. Cambridge, 1995.

McLeod, W. "The Range of the Ancient Bow." *Phoenix* 19 (1965): 1–14.

Mecella, L. "Die Überlieferung der Kestoi des Julius Africanus in den byzantinischen Textsammlungen zur Militärtechnik." In *Die Kestoi des Julius Africanus und ihre Überlieferung*. Edited by M. Wallraff and L. Mecella. 85–144. Berlin and New York, 2009.

Meißner, B. *Die technologische Fachliteratur der Antike: Struktur, Überlieferung und Wirkung technischen Wissens in der Antike (ca. 400 v. Chr.–ca. 500 n. Chr.)* Berlin, 1999.

* Meineke, A., ed. *Ioannis Cinnami Epitome rerum ab Ioanne et Alexio Comnenis gestarum*. CSHB. Bonn, 1836.

* Melber, I., ed. *Polyaeni Strategematon libri octo ex recensione Eduardi Woelfflin*. Leipzig, 1887.

Mende, B. G. ed. *Research on the Prehistory of the Hungarians: A Review*. Budapest, 2005.

Menghin, W. *Das Schwert im frühen Mittelalter*. Stuttgart, 1983.

Mere, J. W., ed. *Medieval Islamic Civilization: An Encyclopaedia*. New York, 2006.

de Meun, J., and C. De Pisan. *L'art de [la] chevalerie selon Végèce*. Paris, 1488. Edited by L. Löfstedt. *Li abregemenz noble honme Vegesce Flave René des establissemenz a chevalerie, traduction par Jean de Meun de Flavii Vegetii Renati viri illustris Epitoma Institutionum Rei Militaris*. Helsinki, 1977.

* Meursius, J. *Claudii Aeliani et Leonis imp. Tactica: Sive, de instruendis aciebus, quorum hic graece primum opera J. Meursius, ille ex Sixti Arcerii nova interpretatione Latina; ambo autem notis et animadversionibus in lucem exeunt*. Leiden, 1613.

* ———. *Leonis imperatoris Tactica: Sive, De re militari liber, Ioannes Meursius graece primus vulgavit, et notas adiecit*. Leiden, 1612.

Meyendorff, J. "Byzantine Views of Islam." *DOP* 18 (1964): 115–32. Reprinted in *Arab-Byzantine Relations in Early Islamic Times*. Edited by M. Bonner. 217–34. Formation of the Classical Islamic World 8. Aldershot, 2004.

Michailidis-Nouaros, G. "Ὁ δίκαιος πόλεμος κατά τα Τακτικά Λέοντος του Σοφού." *Σύμμεικτα Σεφεριάδου*. 411–34. Athens, 1961.

Michel, A. "Sort." In *Dictionnaire de théologie catholique*. Edited by A. Vacant, E. Mangenot, and E. Amann. Paris, 1941.

Mihăescu, H. "Les éléments latins des 'Tactica-Strategica' de Maurice-Urbicius et leur écho en néo-grec." *RESEE* 6 (1968): 481–98; 7 (1969): 155–66, 267–80.

———. "La littérature byzantine, source de connaissance du latin vulgaire." *RESEE* 16 (1978): 195–215; 17 (1979): 39–60, 359–83.

———. "La terminologie d'origine latine dans la littérature byzantine." In *Βυζάντιον: Ἀφιέρωμα στὸν Ἀνδρέα Ν. Στράτο*. 2:587–99. Athens, 1986.

Miller, J. "The Prophetologion: The Old Testament of Byzantine Christianity?" In Magdalino and Nelson, *The Old Testament in Byzantium*, 55–76.

Miller, T. S., and J. S. Nesbitt, eds. *Peace and War in Byzantium*. Washington, DC, 1995.

* Miller, W. *Xenopohon, Cyropaedia*. Loeb Classical Library. 2 volumes. London and Cambridge, Mass., 1947–53.

* Milner, N. P., trans. *Vegetius, Epitome of Military Science*. Liverpool, 1993.

Minkova, M. "Some More Information in Connection with the Meaning of the Word 'Pedatura'." *International Survey of Roman Law/Quaderni camerti di studi romanistici* 20 (1992): 111–23.

Miquel, A. *La géographie humaine du monde musulmane jusqu'au milieu du 11ᵉ siècle*. 2 vols. (Paris, 1967–75).

Mitchell, P. D. "Medical Treatment." In Rogers, *Oxford Encyclopaedia*, 2:585–89.

———. *Medicine in the Crusades: Warfare, Wounds and the Medieval Surgeon*. New York, 2004.

Mitchell, P. D., Y. Nagar, and R. Ellenblum. "Weapon Injuries in the Twelfth-Century Crusader Garrison of Vadum Iacob Castle, Galilee." *International Journal of Osteoarchaeology* 16 (2006): 145–55.

Mitchell, S., ed., *Armies and Frontiers in Roman and Byantine Anatolia*. BAR International series 156. Oxford, 1983.

Mondrain, B. ed. *Lire et écrire à Byzance*. Paris, 2006.

* Mondzain-Baudinet, M.-J. *De notre bienheureux père et archevêque de Constantinople Nicéphore discussion et réfutation des bavardages ignares, athées et tout à fait creux de l'irreligieux Mamon contre l'incarnation de Dieu et le Verbe notre sauveur: Discours contre les iconoclastes.* Paris, 1989.

Mor, H. "The Socio-economic Implications for Ship Construction: Evidence from Underwater Archaeology and the Codex Theodosianus." In *Shipping, Trade and Crusade*, 39–63.

Moravcsik, G., *Byzantinoturcica*. Volume 1, *Die byzantinischen Quellen der Geschichte der Türkvölker*; volume 2, *Sprachreste der Türkvölker in den byzantinischen Quellen*. BBA 10–11. 2nd edition. Berlin, 1958.

———. "A hunok taktikájához," *Kőrösi Csoma-Archivum* 1 (1921–25): 276–80.

———. "Τα συγγράμματα Κωνσταντίνου του Πορφυρογέννητου άπο γλωσσικής απόψεως." *Studi Byzantini e Neoellenici* 5 (1939): 514–20. [= Atti del V Congresso Internazionale di Studi Bizantini, Roma 20–26 sett. 1936, vol. 1.]

Morris, R. "The Powerful and the Poor in Tenth-Century Byzantium: Law and Reality." *Past and Present* 73 (1976): 3–27.

Morrison, J. S., J. F. Coates, and N. B. Rankov. *The Athenian Trireme: The History and Reconstruction of an Ancient Greek Warship.* 2nd edition. Cambridge, 2000.

Mottahedeh, R. P., and R. al-Sayyid, "The Idea of the Jihād in Islam before the Crusaders." In *The Crusades from the Perspective of Byzantium and the Muslim World.* Edited by A. E. Laiou and R. P. Mottahedeh. 23–29. Washington D. C., 2001.

* Müller, K. *Eine griechische Schrift über Seekrieg.* Würzburg, 1882.

* ———. "Ein griechisches Fragment über Kriegswesen." In *Festschrift für Ludwig Urlichs.* 106–38. Würzburg, 1880.

Munitiz, J. A. "War and Peace Reflected in Some Byzantine Mirrors of Princes." In Miller and Nesbitt, eds., *Peace and War in Byzantium*, 50–61.

Muthesius, A. *Byzantine Silk Weaving AD 400 to AD 1200.* Vienna, 1997.

———. "Silken diplomacy." In Franklin and Shepard, *Byzantine Diplomacy*, 237–48.

Napoli, J., and R. Rebuffat, "Clausurae" in *La Frontière: séminaire de recherche.* Edited by Y. Roman. 35–43. Travaux de la Maison de l'Orient 21. Lyons, 1993.

Nederman, C. *John of Salisbury.* Tempe, Ariz., 2005.

* ———, ed. and trans. *John of Salisbury: Policraticus.* Cambridge, 1990.

Nesbitt, J. W., ed. *Byzantine Authors: Literary Activities and Preoccupations.* The Medieval Mediterranean: Peoples, Economies and Cultures, 400–1500, 49. Leiden and Boston 2003.

Nichanian, M. "De la guerre 'antique' à la guerre 'médiévale' dans l'empire romain d'orient." In Barthélemy and Cheynet, *Guerre et société au moyen âge*, 27–41.

Nicolle, D. "Arms of the Umayyad Era: Military Technology in a Time of Change." In *War and Society in the Eastern Mediterranean, 7th–15th Centuries.* Edited by Y. Lev. 9–100. Leiden, 1997.

———. "Byzantine and Islamic Arms and Armour: Evidence for Mutual Influence." *Graeco-Arabica* 4 (1991): 299–325.

———, ed. *Companion to Medieval Arms and Armour*. Woodbridge, 2002.

———. Early Medieval Islamic Arms and Armour. Madrid, 1976.

———. *Medieval Warfare Source Book*. Volume 1, *Warfare in Western Christendom*; volume 2, *Christian Europe and Its Neighbours*. London 1995–96.

Niermeier, J.-F. *Mediae latinitatis lexicon minus*. Leiden, 1976.

Nikonorov, V. P. "The Use of Musical Percussion Instruments in Ancient Near Eastern Warfare: The Parthian and Middle Asian Evidence." In *Studien zur Musikarchäologie*, volume 2, *Musikarchäologie früher Metallzeiten*. Edited by E. Hickmann, I. Laufs, and R. Eichmann. 71–81. Rahden, Westf., 2000.

Nishimura, D. "Crossbows, Arrow-Guides and the Solenarion." *Byz* 58 (1988): 422–35.

* Noailles, P., and A. Dain, eds. *Les novelles de Léon VI le Sage: Texte et traduction*. Paris, 1944.

Nutton, V. *Ancient Medicine*. Abingdon, 2004.

———. "Medical Thoughts on Urban Pollution." In *Death and Disease in the Ancient City*. Edited by V. M. Hope and E. Marshall. 65–73. London and New York, 2000.

Oakshott, E. *The Archaeology of Weapons: Arms and Armour from Prehistory to the Age of Chivalry*. Woodbridge, 1960, reprinted 1994.

Obolensky, D. *The Byzantine Commonwealth: Eastern Europe, 500–1453*. London, 1971.

Odorico, P. *Il prato e l'ape: Il supere sentenzioso del monaco Giovanni*. W ByzSt 17. Vienna, 1986.

Odorico, P. "Lo *Gnomologium Byzantinum* e la recensione del Cod. Bibl. Nat. Athen 1070." *RSBS* 2 (1982): 41–70

Oestreich, G. *Neostoicism and the Early Modern State*. Cambridge, 1982.

Ohnsorge, W. "Zur Frage der Töchter Kaiser Leons VI." *BZ* 51 (1958): 78–81.

Oikonomidès, N. "La dernière volonté de Léon VI au sujet de la tétragamie." *BZ* 56 (1963): 265–70.

———. "L'évolution de l'organisation administrative de l'empire byzantin." *TM* 6 (1976): 125–52.

———. *Fiscalité et exemption fiscale à Byzance (IXᵉ–XIᵉ s.)*. Athens, 1996.

———. "Leo VI and the Narthex Mosaic of Saint Sophia." *DOP* 30 (1976): 151–72.

———. "Leo VI's Legislation of 907 Forbidding Fourth Marriages: An Interpolation in the Procheiros Nomos (IV, 25–27)." *DOP* 30 (1976): 173–93.

* ———. *Les listes de préséance byzantins des IXᵉ-Xᵉ siècles*. Paris, 1972.

———. "Middle Byzantine Provincial Recruits: Salary and Armament." In *Gonimos: Neoplatonic and Byzantine Studies presented to Leendert G. Westerink at 75*. Edited by J. Duffy and J. Peradotto. 121–36. Buffalo, N.Y., 1988.

———. "Poems on the Deaths of Leo VI and Constantine VII in the Madrid Manuscript of Scylitzes." *DOP* 23–24 (1969–70): 185–228.

———. "Silk Trade and Production in Byzantium from the Sixth to the Ninth Century: The Seals of Kommerkiarioi." *DOP* 40 (1986): 33–53.

———. "The Social Structure of the Byzantine Countryside in the First Half of the Xth Century." *Symmeikta* 10 (1996): 105–25.

———. "Das Verfalland im 10.–11. Jahrhundert: Verkauf und Besteuerung." In *Fontes Minores*, 7:161–68. Frankfurt a. M., 1986.

* Oldfather, W. A., ed. and trans. *Aeneas Tacticus, Asclepiodotus, Onasander.* Loeb Classical Library. London and New York, 1923.

Oman, C. W. The *Art of War* in the Middle Ages, A.D. 378–1515. Oxford, 1885. Revised and edited by J. H. Beeler, Ithaca, NY, 1953.

Osiek, C. "The Ransom of Captives: Evolution of a Tradition." *Harvard Theological Review* 74 (1981): 365–86.

Ostrogorsky, G. "L'expédition du prince Oleg contre Constantinople in 907." *Annaly Instituta imeni N. P. Kondakova* 11 (1940): 47–62. Reprinted in idem, *Byzanz und die Welt der Slawen* (Darmstadt, 1974), 17–32.

———. *History of the Byzantine State.* Oxford, 1968.

———. "Das Mitkaisertum im mittelalterlichen Byzanz." In *Doppelprinzipat und Reichsteilung im Imperium Romanum.* Edited by E. Kornemann. 166–78. Leipzig and Berlin, 1930

———. "Observations on the Aristocracy in Byzantium." *DOP* 25 (1971): 2–32.

Pahlitzsch, J. "Zur ideologischen Bedeutung Jerusalems für das orthodoxe Christentum." In *Konflikt und Bewältigung: Die Zerstörung der Grabeskirche zu Jerusalem im Jahre 1009.* Edited by T. Pratsch. 239–55. Berlin and Boston, 2011.

Palme, B. "Spätrömische Militärgerichtsbarkeit in den Papyri." In *Symposion 2003: Vortгräge zur griechischen und hellenistischen Rechtsgeschichte.* Edited by H.-A. Rupprecht. 375–408. Vienna, 2006.

Papadopoulos, A. Ἱστορικὸν λεξικὸν τῆς Ποντικῆς διαλέκτου. 2 vols. Athens, 1961.

* Papadopoulos-Kerameus, A. *Varia Graeca Sacra, Sbornik grečeskich neizdannych bogoslovskich tekstov IV–XV vekov.* St. Petersburg, 1909. Reprinted as Subsidia Byzantina lucis ope iterata 6. Leipzig, 1975.

Papasotiriou, C. "Byzantine Grand Strategy." PhD Diss., Stanford University, 1991.

Parani, M. "Dressed to Kill: Middle Byzantine Military Ceremonial Attire." In *The Byzantine Court: Source of Power and Culture.* Edited by A. Ödekan, N. Neçipoğlu and E. Akyürek. 145–156. Istanbul, 2013.

Parker, G. "The Limits to Revolutions in Military Affairs: Maurice of Nassau, the Battle of Nieuwpoort (1600), and the Legacy." *Journal of military history* 71, no. 2 (2007): 331–72.

Patlagean, E. "Byzantium's Dual Holy Land." In *Sacred Space: Shrine, City, Land.* Edited by B. Z. Kedar and R. J. Zwi Werblowsky. 112–26. London, 1998.

* Paton, W. R., ed. and trans. *Polybius, The Histories.* Loeb Classical Library. London and Cambridge, Mass., 1954.

Patoura, S. Οἱ αἰχμάλωτοι ὡς παράγοντες ἐπικοινωνίας καὶ καὶ πληροφορήσης (4–10 αἰ.). Athens, 1994.

Pedersen, A. "Scandinavian Weaponry in the Tenth Century." In Nicolle, *Companion*, 25–35.

* Pérez Martín, I., ed. and trans. *Miguel Ataliates: Historia.* Nueva Roma 15. Madrid, 2002.

* Perrin, B., ed. and trans. *Plutarch's Lives*. Loeb Classical Library. 11 volumes. London and Cambridge, Mass., 1914–26.

* Perry, B. E. *Aesopica: A Series of Texts Relating to Aesop or Ascribed to Him or Closely Connected with the Literary Tradition that Bears his Name; Collected and Critically Edited, in Part Translated from Oriental Languages, with a Commentary and Historical Essay*. Vol. 1, *Greek and Latin Texts*. Urbana, 1952.

* Pertusi, A., ed. *Costantino Porfirogenito, De Thematibus*. Studi e Testi 160. Vatican City, 1952.

* ——, ed. *Giorgio di Pisidia: Poemi*. Volume 1, *Panegirici epici*. Studia Patristica et Byzantina 7. Ettal, 1959.

—— . "Ordinamenti militari, guerre in Occidente e theoria di guerra dei bizantini [secc. vi–x]." In *Ordinamenti militari in Occidente nell'Alto Medioevo*. 631–700. Settimane di studio del Centro Italiano di studio sull'Altro Medioevo 15. Spoleto, 1968.

—— . "Una acolouthia militare inedita del X secolo." *Aevum* 22–23 (1948–49): 145–68.

* Petit, L. "Vie et office de S. Euthyme le jeune." In *ROC* 8 (1903): 155–205. Reprinted in L. Clugnet, *BHO* 5 (1904): 14–51.

Pétrin, N. "Philological Notes on the Crossbow and Related Missile Weapons." *GRBS* 33 (1992): 265–91.

Petrocelli, C. *Onasandro, Il generale: Manuale per l'esercizio del comando*. Bari, 2008.

Pharmakides, X. P. *Γλωσσάριον Ξεν. Π. Φαρμακίδου*. Edited by Th. D. Kypre. Ὑλικὰ διὰ τὴν σύνταξιν ἱστορικοῦ λεξικοῦ τῆς Κυπριακῆς διαλέκτου 2. Leukosia, 1983.

Photiades, P. J. "A Semi-Greek, Semi-Coptic Parchment." *Klio* 41 (1963): 234–35.

Pitsakis, C. "Guerre et paix en droit byzantin." *Méditerranées* 30–31 (2002): 203–32.

Pizzone, A. "Feeling the Rhythm of the Waves: 'Castaway Rhetoric' in John Eugenikos' Logos eucharisterios." *BMGS* 37, no. 2 (2013): 190–207.

Pletneva, S. A. "Pečenegi, torki i polovcy v yužnorusskikh stepyakh." *Materialy i issledovaniya po arkheologii SSR* 62 (1958): 151–226.

Pohl, W. "Introduction: Ego trouble?" In *Ego trouble: Authors and Their Identities in the Early Middle Ages*. Edited by R. Corradini, M. Gillis, R. McKitterick, and I. Van Renswoude. 9–21. Denkschr. D. österr. Akad. d. Wissenschaften, phil.-hist. Klasse 385. Forschungen zur geschichte des Mittelalters, 15. Vienna, 2010.

Pohl, W., I. Wood, and H. Reimitz, eds. *The Transformation of Frontiers: From Late Antiquity to the Carolingians*. Leiden, 2001.

Pomey, P., Kahanov, Y. and E. Rieth. "Transition from Shell to Skeleton in Ancient Mediterranean Ship-Construction: Analysis, Problems and Future research," *International Journal of Nautical Archaeology* 41 (2012): 235-314.

* Potamianos, K., trans. *Αὐτοκράτορος Λέοντος, Τακτικά: Στρατηγική τακτική*. Athens, 2001.

* Poznanski, L., ed. and trans. *Asclépiodote: Traité de tactique*. Paris 1992.

Preiser-Kapeller, J. "Der Geheimdienst des Kaisers: Spionage und Informationsbeschaffung im byzantinischen Reich." *Karfunkel—Zeitschrift für erlebbare Geschichte, Combat Sonderheft* 7 (2011): 24–29.

Preisigke, F. *Wörterbuch der griechischen Papyrusurkunden.* 5 vols. Berlin, 1925–71.

Pringle, D. *The Defence of Byzantine Africa from Justinian to the Arab Conquest: An Account of the Military History and Archaeology of the African Provinces in the Sixth and Seventh Centuries.* 2nd edition. BAR International Series 99. Oxford, 2001.

Prinzing, F. *Epidemics Resulting from Wars.* Oxford, 1916.

Priskin, K. "A Kárpát-medence avar és honfoglalás kori lóállományának archaeogentikai elemzése" [Archeogenetic analysis of the Avar and early Hungarian horses from the Carpathian Basin]. Diss. Univ. Szeged, 2010. [Includes English summary.]

Pryor, J. *Geography, Technology and War: Studies in the Maritime History of the Mediterranean, 649–1571.* Cambridge, 1988.

———. "Shipping and Seafaring." In Jeffreys, Haldon, and Cormack, *Oxford Handbook,* 482–91.

———. "The Σταδιοδρομικόν of the De Cerimoniis of Constantine VII, Byzantine Warships, and the Cretan Expedition of 949." In *The Greek Islands and the Sea: Proceedings of the First International Colloquium held at The Hellenic Institute, Royal Holloway, University of London, 21–22 September 2001.* Edited by J. Chrysostomides, C. Dendrinos, and J. Harris. 77–108. Camberley, 2004.

———. "Transportation of Horses by Sea during the Era of the Crusades: Eighth Century to 1285 A.D." *Mariners' Mirror* 68 (1982): 9–27, 103–25.

* Pryor, J., and E. Jeffreys. *The Age of the Δρόμων: The Byzantine Navy ca 500–1204.* Leiden and Boston, 2006.

Popov, N. *Imperator Lev VI Mudryj.* Moscow, 1892.

Pulak, C., Ingram, R., Jones, M. and S. Matthews, "The Shipwrecks of Yeni Kapı and Their Contribution to the Study of Ship Construction." In Kızıltan, *Stories,* 22–34.

Purton, P. *A History of the Early Medieval Siege, c. 450–1220.* Woodbridge, 2009.

Putnam, H. *L'église et l'Islam sous Timothée I (720–823): Étude sur l'église nestorienne au temps des premiers ʿAbbāsides avec nouvelle édition et traduction du dialogue entre Timothée et al-Mahdi.* Beirut, 1975.

* Rackham, H., ed. and trans. *Aristotle: The Politics.* Loeb Classical Library. London and Cambridge, Mass., 1932.

Ramsay, W. M. *The Historical Geography of Asia Minor.* Royal Geographical Society, Supplementary Papers 4. London, 1890, reprinted Amsterdam, 1962.

Ramsay, W. M., and G. Bell. *The Thousand and One Churches.* London, 1919.

Rance, P. "Battle." In *The Cambridge History of Greek and Roman Warfare.* Volume 2, *Rome from the Late Republic to the Late Empire.* Edited by P. Sabin, H. van Wees, and L. M. Whitby. 342–78. Cambridge, 2005.

———. "*Campidoctores, Vicarii vel Tribuni*: The Senior Regimental Officers in the Late Roman Army and the Rise of the *Campidoctor.*" In Lewin and Pellegrini, *Late Roman Army,* 395–409.

———. "The Date of the Military Compendium of Syrianus Magister (Formerly the Sixth-Century Anonymus Byzantinus)." *BZ* 100 (2007): 701–37.

————. "The De Militari Scientia or Müller Fragment as a Philological Resource: Latin in the East Roman Army and Two New Loanwords in Greek: palmarium and *recala." *Glotta* 86 (2011): 63–92.

————. "Drungus, Δροῦγγος and Δρουγγιστί—a Gallicism and Continuity in Roman Cavalry Tactics." *Phoenix* 58 (2004): 96–130.

————. "The *Fulcum*, the Late Roman and Byzantine *Testudo*: The Germanization of Late Roman Tactics?" *GRBS* 44 (2004): 265–326.

————. "Noumera or Mounera: A Parallel Philological Problem in De Cerimoniis and Maurice's Strategikon." *JÖB* 58 (2008) 121–29.

————. "The Reception of Aineias' Poliorketika in Byzantine Military Literature." In *Greeks under Siege: Aeneas the Tactician and His World*. Edited by M. Pretzler. Mnemosyne Supplement. Forthcoming.

————. "Simulacra Pugnae: The Literary and Historical Tradition of Mock Battles in the Roman and Early Byzantine Army." *GRBS* 41 (2000): 223–75.

* ————. *The Strategikon of Maurice: Translation and Commentary*. Aldershot, 2014.

————. "'Win but Do Not Overwin'—The History of a Proverb from the Sententiae Menandri, and a Classical Allusion in St. Paul's Epistle to the Romans?" *Philologus* 152, no. 2 (2008): 191–204.

Rapoport, Y. "The View from the South: The Maps of the Book of Curiosities and the Commercial Revolution of the Eleventh Century." In *Histories of the Middle East: Studies in Middle Eastern Society, Economy and Law in Honor of A. L. Udovitch*. Edited by R. E. Margariti et al. 183–212. Leiden, 2010.

Rasi, P. *Exercitus italicus e milizie cittadine nell'alto Medioevo*. Padua, 1937.

* Reeve, M. D., ed. *Vegetius, Epitoma rei militaris*. Oxford, 2004.

* Reinsch, D. R., and A. Kambylis, eds. *Annae Comnenae Alexias*. CFHB 40.1–2. Berlin 2001.

* Reiske, J. J. *De cerimoniis aulae Byzantinae libri duo*. CSHB. Bonn, 1829–30.

* Renauld, P., ed. *Michel Psellos, Chronographie*. 2 volumes. Paris, 1926–28.

Reuter, T. "Carolingian and Ottonian Warfare." In *Medieval Warfare: A History*. Edited by M. Keen. 13–35. Oxford, 1999.

————, ed. *The New Cambridge Medieval History*. Volume 3, *c. 900–c. 1024*. Cambridge, 1999.

* Richard, M., and J. A. Munitiz, eds. *Anastasii Sinaitae: Quaestiones et responsiones*. Corpus Christianorum, Series Graeca 59. Brepols, 2006.

Richardot, P. *Végèce et la culture militaire au Moyen Age: V^e–XVe siècle*. Paris, 1998.

Riedel, M. "A Christian Philosophy of Warfare? Internal Evidence for the Shape and Purpose of the Taktika of Leo VI." In *BSCAbstr* 37 (2011): 91–92.

————. "Fighting the Good Fight: The Taktika of Leo VI and Its Influence on Byzantine Cultural Identity." D.Phil. thesis. Oxford, 2010.

Rigolio, A. "From Sacrifice to the Gods to the Fear of God: Omissions, Additions and Changes in the Syriac Translations of Plutarch, Lucian and Themistius." *Studia Patristica*. Forthcoming.

Roberts, I. P. *The Horse*. New York, 1905.

Rogers, C. J., ed. *The Oxford Encyclopaedia of Medieval Warfare and Military Technology*. Oxford, 2010.

Rogers, H. C. B. *The Mounted Troops of the British Army 1066–1945*. London, 1959.

Roland, A. "Secrecy, Technology, and War: Greek Fire and the Defense of Byzantium, 678–1204." *Technology and Culture* 33 (1992): 655–79.

* Rolfe, J. C., ed. and trans. *Ammianus Marcellinus: History*. 3 vols. Loeb Classical Library. London and Cambridge, Mass., 1935–37.

* ———., ed. and trans. *On Great Generals; On Historians*. Loeb Classical Library. Cambridge, Mass., 1925–84.

* Roos, A. G., and G. Wirth, eds. *Flavii Arriani quae exstant omnia*. Leipzig, 1967–68.

Rösch, G. Ὄνομα βασιλείας: *Studien zum offiziellen Gebrauch der Kaisertitel in spätantiker und frühbyzantinischer Zeit*. Byzantina Vindobonensia 10. Vienna, 1978.

Rosenstein, N. *Rome at War: Farms, Families and Death in the Middle Republic*. Chapel Hill, 2004.

Roth, J. *The Logistics of the Roman Army at War (264 B.C.–A.D. 235)*. Leiden, 1999.

Rothenberg, G. "Maurice of Nassau, Gustavus Adolphus, Raimondo Montecuccoli and the 'Military revolution' of the Seventeenth Century." In *Makers of Modern Strategy: From Machiavelli to the Nuclear Age*. Edited by P. Paret, G. Craig and F. Gilbert. Princeton, 1986.

Rotman, Y. "Byzance face à l'Islam arabe, VIIe–Xe siècle: D'un droit territorial à l'identité par la foi." *AnnalesESC* 60 (2005): 767–88.

———. "Captif ou esclave? Entre marché d'esclaves et marché de captifs en Méditerranée médiévale." In *Les esclavages en Méditerranée. Espaces et dynamiques économiques*. Edited by F. Guillén and S. Trabelsi. 25-46. Madrid, 2012.

Rubin, Z. "The Reforms of Khusro Anushirwan." In Cameron, *States, Resources and Armies*, 227–97.

Runciman, S. *The Emperor Romanus Lecapenus: A Study on Tenth-Century Byzantium*. Cambridge, 1929.

Sahas, D. *John of Damascus on Islam: The "Heresy of the Ishmaelites."* Leiden, 1972.

Şahin, S. *Katalog der antiken Inschriften des Museums von Iznik (Nikaia)*. Volume 2, part 3. Bonn, 1987.

Salazar, C. F. *The Treatment of War Wounds in Greco-Roman Antiquity*. Leiden, 2000.

al-Sarraf, S. "Close Combat Weapons in the Early 'Abbasid Period." In Nicolle, *Companion*, 149–78.

* Sathas, K. N. Μιχαὴλ Ψελλοῦ ἱστορικοὶ λόγοι, ἐπιστολαὶ καὶ ἄλλα ἀνέκδοτα. Vol. 5 of Μεσαιωνικὴ Βιβλιοθήκη. Paris, 1876.

Saunders, W. "Qal'at Seman: A Frontier Fort of the Tenth and Eleventh Centuries." In Mitchell, *Armies and Frontiers*, 291–303.

Scharf, R. *Foederati: Von der völkerrechtlichen Kategorie zur byzantinischen Truppengattung*. Tyche Supplementband 4. Vienna, 2001.

Scheidel, W. "Marriage, Families and Survival: Demographic Aspects." In *A Companion to the Roman Army*. Edited by P. Erdkamp. 417–34. Oxford, 2007.

Schellenburg, H. M. "Einige Bemerkungen zum Strategikos des Onasandros." In *The Impact of the Roman Army (200 BC–AD 476)*. Edited by L. de Blois and E. Lo Cascio. 181–91. Leiden, 2007.

* Scheltema, H. J., and N. van der Wal, eds. *Basilicorum libri LX*, ser. A. Groningen, 1953–.

Schettino, M. T. *Introduzione a Polieno*. Pisa, 1998.

Schilbach, E. *Byzantinische Metrologie*. Handbuch d. Altertumswiss. 12.4. Byzantinisches Handbuch 4. Munich, 1970.

Schindler, F. *Die Überlieferung der Strategemata des Polyainos*. Österreichische Akademie der Wissenschaften phil.-hist. Klasse Sitzungsberichte 284.1. Vienna, 1973.

Schminck, A. "Bermerkungen zum sog. 'Nomos Mosaïkos'." *Fontes Minores* 11 (2005): 249–68.

———. "'Frömmigkeit ziere das Werk': Zur Datierung der 60 Bücher Leons VI." *Subseciva Groningana* 3 (1989): 79–114.

———. "'Novellae extravagantes' Leons VI." *Subseciva Groningana* 4 (1990): 195–209.

———. "Probleme des sog. Ἐνόμος Ῥοδίων Ναυτικός'." In *Griechenland und das Meer*. Edited by E. Chrysos, D. Letsios, H. A. Richter, and R. Stupperich. 171–78. Mannheim/Möhnesee, 1999.

———. *Studien zu mittelbyzantinischen Rechtsbüchern*. Forschungen zur byzantinischen Rechtsgeschichte 13. Frankfurt a. M., 1986.

Schmitt, O. "From the Late Roman to the Early Byzantine Army: Two Aspects of Change." In Lewin and Pelegrini, *Late Roman Army*, 411–19.

———. "Untersuchungen zur Organisation und zur militärischen Stärke oströmischer Herrschaft im Vorderen Orient zwischen 628 und 633." *BZ* 94 (2001): 211–16.

* Schneider, R. *Griechische Poliorketiker*. Volume 2. Abh. d. königl. Gesellschaft d. Wiss. zu Göttingen, phil.-hist. Kl., n.F., 10. Berlin, 1908. No. 1.

Schreiner, P. "Zur Ausrüstung des Kriegers in Byzanz, im Kiewer Rußland und in Nordeuropa nach bildlichen und literarischen Quellen." In *Les pays du nord et Byzance (Scandinavie et Byzance): Actes du colloque nordique et international de Byzantinologie tenu à Upsal 20–22 avril 1979*. Edited by R. Zeitler. 215–36. Acta Universitatis Upsaliensis, Figura. Nova series 19. Upsala 1981.

Schulten, C. "Une nouvelle approche de Maurice de Nassau (1567–1625)." In *Le soldat, la stratégie, la mort: Mélanges André Corvisier*. Edited by P. Chaunu. 42–53. Paris, 1989.

Schwarzer, J. K. "Arms from an Eleventh-Century Shipwreck," *Graeco-Arabica* 4 (1991): 327–50.

Searby, D. M. *The Corpus Parisinum: A Critical Edition of the Greek Text with Commentary and English Translation*. Lewiston NY, 2007.

Sergi, G. "The Kingdom of Italy." In Reuter, *New Cambridge Medieval History*, 346–71.

Serikoff, N. "Leo VI Arabus? An Unknown Fragment from the Arabic Translation of Leo VI's Taktika." *Acta Orientalia Vilnensia* 4 (2003): 112–18.

Seston, W. "Fahnenflucht." *RAC* 7 (1995): 284–86.

* Ševčenko, I., ed. *Chronographiae quae Theophanis continuati nomine fertur Liber quo Vita Basilii imperatoris amplectitur.* CFHB. Berlin 2011.

———. "Hagiography of the Iconoclast Period." In *Iconoclasm: Papers Given at the Ninth Spring Symposium of Byzantine Studies.* Edited by A. A. M. Bryer and J. Herrin. 113–31. Birmingham, 1977.

———. "Levels of Style in Byzantine Prose." *XVI Internationaler Byzantinistenkongreß* = *JÖB* 31, no. 1 (1981): 289–312.

———. "Re-reading Constantine Porphyrogenitus." In Franklin and Shepard, *Byzantine Diplomacy,* 167–95.

de Sérouville, N. Volkyr. *De l'ordre et instruction des batailles.* Paris, 1536.

* Sewter, E. R. A., trans. *The Alexiad of the Princess Anna Comnena.* Revised with introduction and notes by P. Frankopan. London and New York, 2009.

* ———, trans. *Michael Psellos, Chronographia.* New Haven 1953.

Shaban, M. A. *Islamic History: A New Interpretation.* Volume 2, *A.D. 750–1055 (A.H. 132–448).* Cambridge, 1976.

Shatzmiller, M. "The Crusades and Islamic Warfare—A Re-evaluation." *Der Islam* 69 (1992): 247–88.

Shboul, A. "Byzantium and the Arabs: The Image of the Byzantines as Mirrored in Arabic Literature." In *Byzantine Papers: Proceedings of the First Australian Byzantine Studies Conference Canberra, 17–19 May 1978.* 43–68. Canberra, 1981.

Shean, J. F. *Soldiers for God: The Roman Army and Christianity.* Leiden, 2010.

Shepard, J. "Aspects of Byzantine Attitudes and Policy towards the West in the Tenth and Eleventh Centuries." In *Byzantium and the West.* Edited by J. D. Howard-Johnston. 67–118. Amsterdam, 1988.

———. "Bulgaria: The Other Balkan 'Empire'." In Reuter, *New Cambridge Medieval History,* 567–85.

———. "Byzantine Writers on the Hungarians in the Ninth and Tenth Centuries." *The Annual of Medieval Studies at Central European University* 10 (2004): 97–123. Reprinted in idem, *Emergent Élites,* no. VII.

———. "Byzantium and the Steppe-Nomads: The Hungarian Dimension." In *Byzanz und Ostmitteleuropa 950–1453: Beiträge zu einer table-ronde des XIX International Congress of Byzantine Studies, Copenhagen 1996.* Edited by G. Prinzing and M. Salamon. 55–83. Mainzer Veröffentlichungen zur Byzantinistik 3. Wiesbaden, 1999. Reprinted in idem, *Emergent Élites,* no. VIII.

———, ed. *Cambridge History of the Byzantine Empire ca. 500–1492.* Cambridge, 2008.

———. *Emergent Élites and Byzantium in the Balkans and East-Central Europe.* Farnham, 2011.

———. "Equilibrium to Expansion (886–1025)." In idem, *Cambridge History,* 493–536.

———. "Information, Disinformation and Delay in Byzantine Diplomacy." *Byzantinische Forschungen* 10 (1985): 233–93.

———. "The Ruler as Instructor, Pastor and Wise: Leo VI of Byzantium and Symeon of Bulgaria." In *Alfred the Great: Papers from the Eleventh-Centenary Conferences.* Edited by T. Reuter. 339–58. Aldershot, 2003. Reprinted in idem, *Emergent Élites,* no. IV.

———. "Slavs and Bulgars." In McKitterick, *New Cambridge Medieval History,* 228–48. Reprinted in idem, *Emergent Élites,* no. II.

———. "Symeon of Bulgaria—Peacemaker." In *Annuaire de l'Université de Sofia "St. Kliment Ohridski." Centre de Recherches Slavo-Byzantines "Ivan Dujčev"* 83 (1989): 9–48. Reprinted in idem, *Emergent Élites,* no. III.

———. "The Uses of the Franks in Eleventh-Century Byzantium." *Anglo-Norman Studies* 15 (1993): 275–305.

———. "Western Approaches (900–1025)." In idem, *Cambridge History,* 537–59.

Siems, H. "Asyl in der Kirche ? Wechsellagen des Kirchenasyls im Mittelalter." In *Das antike Asyl: Kultische Grundlagen, rechtliche Ausgestaltung und politische Funktion.* Edited by M. Dreher. 263–99. Köln, 2003.

Signes Codoñer, J., and F. J. A. Santos. *La introducción al derecho (eisagoge) del patriarca Focio.* Nueva Roma 28. Madrid, 2007.

Simelides, C. "The Byzantine Understanding of the Qurʾanic Term al-Samad and the Greek Translation of the Qurʾan." *Speculum* 86 (2011): 887–913.

Simeonova, L. "In the Depths of Tenth-Century Byzantine Ceremonial: The Treatment of Arab Prisoners-of-War at Imperial Banquets." *BMGS* 22 (1998): 75–104.

Simon, D. "Legislation as Both a World Order and Legal Order." In *Law and Society in Byzantium: Ninth–Twelfth Centuries.* Edited by A. E. Laiou and D. Simon. 19–25. Washington, DC, 1994.

———. "Princeps legibis solutus." In *Gedächtnisschrift für Wolfgang Kunkel.* Edited by D. Nörr and D. Simon. 449–92. Frankfurt a. M., 1984.

Sinclair, T. "Byzantine and Islamic Fortification in the Middle East: The Photographic Exhibition." In Mitchell, *Armies and Frontiers,* 305–36.

Sinor, D. "The Greed of the Northern Barbarians." In *Aspects of Altaic Civilization.* Edited by L. V. Clark and P. A. Draghi. 171–82. Bloomington, 1978.

Sinor, D. "Horse and Pasture in Inner Asian History." *Oriens extremus* 19 (1972): 171–83.

———. "The Inner Asian Warriors." *JAOS* 101 (1981): 133–44.

Skopelite, B. "Οι ναυτικές δυνάμεις του Βυζαντίου την εποχή της Άλωσης της Θεσσαλονίκης απο τους Άραβες (904)." *Byzantina* 23 (2002–3): 95–115.

Smail, R. C. *Crusading Warfare, 1097–1193.* Edited by C. Marshall. Cambridge 1995.

* Smith, O. L. *Scholia metrica anonyma in Euripidis Hecubam, Orestem, Phoenissas.* Opuscula Graecolatina 10. Copenhagen, 1977.

Sophocles, E. A. *Greek Lexikon of the Roman and Byzantine Periods.* 3rd edition. Cambridge, 1914.

Soustal, P., with J. Koder. *Tabula Imperii Byzantini.* Volume 3, *Nikopolis und Kephallenia.* Denkschr. d. Österr. Akad. d Wiss., phil.-hist. Kl. 150. Vienna, 1981.

Souther, P., and K. R. Dixon. *The Late Roman Army.* London, 1996.

* Spadaro, M. D. *Raccomandazioni e consigli di un galantumomo: Στρατηγικόν.* Alessandria, 1998.

Spanoudes, C. "Παρατηρήσεις για τον Δρόμωνα-Shīni." In *Cultural Relations between Byzantium and the Arabs.* Edited by Y. Y. al-Hijji and V. Christides, 147–52. Athens, 2007.

Speck, P. *Das geteilte Dossier: Beobachtungen zu den Nachrichten über die Regierung des Kaisers Herakleios und seine Söhne bei Theophanes und Nikephoros.* Poikila byzantina 9. Berlin and Bonn, 1988.

———. "Byzantium: Cultural Suicide?" In Brubaker, *Byzantium in the Ninth Century*, 73–84.

———. *Kaiser Konstantin VI: Die Legitimation einer fremden und der Versuch einer eigenen Herrschaft.* Munich, 1978.

———. *Understanding Byzantium: Studies in Byzantine Historical Sources.* Edited by S. Takács. Aldershot, 2003.

———. "Weitere Überlegungen und Untersuchungen über die Ursprünge der byzantinischen Renaissance." In *Varia.* 2:253–83. Poikila Byzantina 6. Bonn, 1987. [English version in idem, *Understanding Byzantium*, no. XIV.]

Spieser, J.-M. "Inventaires en vue d'un recueil des inscriptions historiques de Byzance, I: Les inscriptions de Thessalonique." *TM* 5 (1973): 145–80.

Stadler, P. "La chronologie de l'armement des Avars du VIᵉ au VIIIᵉ siècle." In *L'Armée romaine et les barbares du 4e au 7e siècle, Colloque du CNRS.* Edited by F. Vallet and M. Kazanski. 445–57. Paris, 1993.

Stadter, P. A. "The Ars tactica of Arrian: Tradition and Originality." *Classical Philology* 73 (1978): 117–28.

Stanton, C. D. *Norman Naval Operations in the Mediterranean.* Woodbridge, 2011.

Stavridou-Zaphraka, A. "Η ἀγγαρεία στο Βυζάντιο." *Βυζαντινά* 11 (1982): 23–54.

———. "Vodena, a Byzantine City-Fortress in Macedonia." In *Edessa and Its Region: History and Culture.* Edited by G. Kioutoutskas. 165–78. Edessa, 1995. [In Greek.]

Stephanus, H. *Thesaurus linguae graecae.* Edited by C. B. Hase, G. Dindorf, and L. Dindorf. 8 volumes. Paris, 1831–1865.

Stephenson, P. *Byzantium's Balkan Frontier: A Political Study of the Northern Balkans, 900–1204.* (Cambridge, 2000).

Stouraitis, I. "Byzantine War against Christians—an Emphylios Polemos?" *Βυζαντινά Σύμμεικτα* 20 (2010): 85–110.

———. "Jihād and Crusade: Byzantine Positions towards the Notions of 'Holy War'." *Βυζαντινά Σύμμεικτα* 21 (2011): 11–63.

———. "'Just War' and 'Holy War' in the Middle Ages." *JÖB* 62 (2012): 227–64.

———. *Krieg und Frieden in der politischen und ideologischen Wahrnehmung in Byzanz (7.–11. Jahrhundert).* Byzantinische Geschichtsschreiber, Ergänzungsband 5. Vienna, 2009.

———. "Methodologische Überlegungen zur Frage des byzantinischen 'heiligen' Krieges." *Byzantinoslavica* 67 (2009): 269–90.

Stoyanov, Y. *Defenders and Enemies of the True Cross: The Sasanian Conquest of Jerusalem in 614 and Byzantine Ideology of Anti-Persian Warfare.* SbWien 819. Vienna, 2011.

Strano, G. "Potere imperiale e γένη aristocratici a Bisanzio durante il regno di Leone VI." *Bizantinistica* 4 (2002) 79–99.

———. "La Vita di Teofano (*BHG* 1794) fra agiografia e propaganda." *Bizantinistica* 2nd ser., 3 (2001): 47–61.

Strässle, P. M. *Krieg und Kriegführung in Byzanz: Die Kriege Kaiser Basileios' II. gegen die Bulgaren (976–1019).* Köln, 2006.

———. "*To Monoxylon* in Konstantinos VII. Porphyrogennetos' Werk *De administrando imperio.*" *Études balkaniques* 2 (1990): 99–106.

* Sullivan, D. F. "A Byzantine Instructional Manual on Siege Defense: The *De obsidione toleranda.*" In Nesbitt, *Byzantine Authors*, 139–266.

———. "Byzantine Military Manuals: Prescriptions, Practice and Pedagogy." In *The Byzantine World.* Edited by P. Stephenson. 149–61. London and NY, 2010.

———. "Byzantium Besieged: Prescription and Practice." In *Byzantium, State and Society, in Memory of Nikos Oikonomides.* Edited by A. Avramea, A. Laiou, and E. Chrysos. 509–21. Athens, 2003.

* ———. *Siegecraft: Two Tenth-Century Instructional Manuals by "Heron of Byzantium."* DOS 36. Washington, DC, 2000.

———. "Tenth-Century Byzantine Offensive Siege Warfare: Instructional Prescriptions and Historical Practice." In Tsiknakes, Εμπόλεμο Βυζάντιο, 179–200.

———. "Was Constantine VI 'Lassoed' at Markellai?" *GRBS* 35, no. 3 (1994): 287–91.

* Svoronos, N. *Les novelles des empereurs macédoniens concernant la terre et les stratiotes.* Edited by P. Gounaridis. Athens, 1994.

Syvänne, I. *The Age of Hippotoxotai: Art of War in Roman Military Survival and Disaster (491–636).* Acta Universitatis Tamperensis 994. Tampere, 2004.

Szentpéteri, J. "Archäologische Studien zur Schicht der Waffenträger des Awarentums im Karpatenbecken." *ActaAntHung* 45 (1993): 165–246; 46 (1994): 231–306.

al-Tahir, G. M. "The Nubian Archers in pre-Islamic and Islamic Periods." *Graeco-Arabica* 5 (1993): 139–52.

* Tafel, G. L. Fr. and G. M. Thomas, eds. *Urkunden zur älteren Handels- und Staatsgeschichte der Republik Venedig, mit besonderer Beziehung auf Byzanz und die Levante. Vom neunten bis zum Ausgang des fünfzehnten Jahrhunderts. Fontes rerum Austriacarum.* 2. Abt. *Diplomataria et acta*, 12-14. 3 vols. Vienna, 1856-1857.

* Talbot, A.-M., ed. *Byzantine Defenders of Images: Eight Saints' Lives in English Translation.* Washington D.C., 1998.

———. "Byzantine Pilgrimages to the Holy Land from the Eighth to the Fifteenth Century." In *The Sabaite Heritage in the Orthodox Church from the Fifth Century to the Present.* Edited by J. Patrich. 97–110. Leuven, 2001.

* Talbot, A.-M., and D. Sullivan. *The History of Leo the Deacon.* Washington, D.C., 2005.

Tallett, F. *War and Society in Early Modern Europe, 1495–1715.* London, 1992.

Taragna, A. M. "Λόγος e Πόλεμος: Eloquenza e persuasione nei trattati bizantini di arte militare." *Siculorum Gymnasium* 57 (2004): 797–810. *Atti del VI Congresso nazionale dell'Associazione Italiana di Studi Bizantini.* Edited by T. Creazzo and G. Strano.

Tartaglia, L. "Il Saggio su Plutarco di Teodoro Metochita." In *Talariskos: Studia graeca Antonio Garzya sexagenario a discipulis oblata.* 339–62. Naples, 1987.

Tejeda, J. "Warfare, History and Literature in the Archaic and Classical Periods: The Development of Greek Military Treatises." *Historia* 53 (2004): 129–46.

Temkin, O. "Byzantine Medicine, Tradition and Empiricism." *DOP* 16 (1962): 95–115.

Thatcher, O. J., and E. H. McNeal, eds. *A Source Book for Medieval History.* New York, 1905.

Theis, L. "Ist Frieden darstellbar? Byzantinische Bildlösungen." In *Friede: Eine Spurensuche.* Edited by M. Meyer. 95–110. Vienna, 2008.

Thompson, E. A. *A History of Attila and the Huns.* Oxford, 1948.

* Thurn, H., ed. *Ioannis Malalae Chronographia.* CFHB 35. Berlin and New York, 2000.

* ———, ed. *Ioannis Scylitzae synopsis historiarum.* CFHB 5. Berlin, 1973.

Tougher, S. "The Imperial Thought-World of Leo VI, the Non-Campaigning Emperor of the Ninth Century." In Brubaker, *Byzantium in the Ninth Century*, 51–60.

———. *The Reign of Leo VI (886–912): Politics and People.* Leiden, 1997.

———. "The Wisdom of Leo VI." In *New Constantines: The Rhythm of Imperial Renewal in Byzantium, 4th–13th Centuries.* Edited by P. Magdalino. 171–79. Aldershot, 1994.

Toul, Ch. "Περὶ τῆς νοθογενείας του Λέοντος τοῦ Σόφου." *Παρνασσός* 21 (1979): 15–35.

Toynbee, A. *Constantine Porphyrogenitus and His World.* London, 1973.

Traina, G. *Paludi e bonifiche nel mondo antico.* Rome, 1988.

* Trapp, E., ed. *Digenes Akrites: Synoptische Ausgabe der ältesten Versionen.* WByzSt 8. Vienna, 1971.

———, ed. *Lexikon zur byzantinischen Gräzität, besonders des 9.–12. Jahrhunderts.* Vienna, 1994–.

Trapp, E., et al. *Studien zur byzantinischen Lexikographie.* Byzantina Vindobonensia 18. Vienna, 1988.

Treadgold, W. T. *The Byzantine Revival 780–842.* Stanford, 1988.

———. *Byzantium and Its Army, 284–1081.* Stanford, 1995.

———. "The Chronological Accuracy of the Chronicle of Symeon the Logothete for the Years 813–45." *DOP* 33 (1979): 157–97.

———. "The Macedonian Renaissance." In *Renaissances before the Renaissance: Cultural Revivals of Late Antiquity and the Middle Ages.* Edited by W. Treadgold. 75–98. Stanford, 1984.

———. "Remarks on Al-Jarmi." *Byzantinoslavica* 44 (1983): 205–12.

———. "Standardized Numbers in the Byzantine Army." *War in History* 12 (2005): 1–14.

Trempelas, P. N. *The Three Liturgies according to the Athens Codices.* Texte und Forschungen zur byzantinisch-neugriechischen Philologie. Athens, 1935. [In Greek.]

Tritle, L. "Tatzates' Flight and the Byzantine-Arab Peace Treaty of 782." *Byzantion* 47 (1977): 279–300.

Troianos, S. N. "The Canons of the Trullan Council in the Novels of Leo VI." In *The Council in Trullo Revisited*. Edited by G. Nedungatt and M. Featherstone. 189–98. Rome, 1995.

———. "Οι 'εκκλησιαστικές' Νεαρές του Λέοντος ϛ' και οι πηγές τους." In *Οι νεαρές*. 445–67.

———. "Die kirchenrechtlichen Novellen Leons VI. Und ihre Quellen." *Subseciva Groningana* 4 (1990): 233–47.

———. "Λέων ϛ' ο Σόφος: Νομική σκέψη και κοινωνική συνείδηση." In idem, *Νεαρές*, 415–25.

*   ———. *Οι νεαρές Λέοντος ϛ' του σόφου: Προλεγόμενα, κείμενο, απόδοση στη νεοελληνική και επίμετρο*. Athens, 2007.

———. "Die Novellen Leons VI." In *Analecta Athenensia ad ius byzantinum spectantia*. 141–54. Forschungen zur byzantinischen Rechtsgeschichte, Athener Reihe. Athens, 1997.

———. "Οἱ ποινὲς στὸ Βυζαντινὸ δίκαιο." In *Ἔγκλημα καὶ τιμωρία στὸ Βυζάντιο*. 13–65. Athens, 2001.

Trombley, F. "The *Taktika* of Nikephoros Ouranos and Military Encyclopaedism." In *Pre-Modern Encyclopedic Texts: Proceedings of the Second COMERS Congress, Groningen, 1-4 July 1996*. Edited by P. Binkley, 261 74. Leiden, 1997.

———. "War and Society in Rural Syria c. 502–613 A.D.: Observations on the Epigraphy." *BMGS* 21 (1997): 154–209.

*   Trombley, F. R., and J. W. Watt. *The Chronicle of Ps.-Joshua the Stylite*. Liverpool, 2000.

Tsiknakes, K. *Το εμπόλεμο Βυζάντιο* [Byzantium at war]. Athens, 1997.

Tsougarakis, D., "The Byzantine Seals of Crete," *Studies in Byzantine Sigillography* 2 (1990): 137–52.

Tsurtsumia, M. "The Evolution of Splint Armour in Georgia and Byzantium: Lamellar and Scale Armour in the 10th–12th centuries." *Βυζαντινά Σύμμεικτα* 21 (2011): 65–99.

———. "Τρίβολος: A Byzantine Landmine." *Byz* 82 (2012): 415–22.

Turkey, I. *Naval Intelligence Division, Geographical Handbook Series, B. R. 507*. London, 1942.

Udovitch, A. L. "Time, the Sea and Society: Duration of Commercial Voyages on the Southern Shores of the Mediterranean during the High Middle Ages." In *La navigazione mediterranea nell'alto medioevo: Settimane di studio del Centro italiano di studi sull'alto medioevo* 25 (1978): 2:503–46.

Vallet, F., and M. Kazanski, eds. *L'Armée romaine et les barbares du IIIᵉ au VIIᵉ siècle*. Mémoires de l'Association Française d'Archéologie Mérovingienne 5. Paris, 1993.

van Bochove, T. E. "Some Byzantine Law Books: Introducing the Continuous Debate concerning Their Status and Their Date." In *Introduzione al diritto bizantino: Da Giustiniano ai Basilici*. Edited by J. H. A. Lokin and B. H. Stolte. 239–66. Pavia, 2011.

———. *To Date and Not To Date: On the Date and Status of Byzantine Law Books*. Groningen, 1996.

Van Crefeld, M. *Supplying War: Logistics from Wallerstein to Patton*. Cambridge, 1977.

* van den Berg, H., ed. *Anonymous De obsidione toleranda: Editio critica*. Leyden, 1947.

Vanderheyde, C. "La monture des saints cavaliers dans l'art byzantin." In *Le cheval dans les sociétés antiques et médiévales*. Edited by S. Lazaris. 202–11. Strasbourg, 2012.

* van der Valk, M., ed. *Eustathii archiepiscopi Thessalonicensis commentarii ad Homeri Iliadem pertinentes*. 4 volumes. Leiden, 1971–91.

* van Dieten, J. A., ed. *Nicetae Choniatae Historia*. 2 volumes. CFHB 11.1–2. Berlin and New York, 1975.

Vári, R. *Bölcs Leó császárnak "A hadi taktikáról" szóló munkája*. Értekezések a történeti tudományok köréből 17.10. Budapest, 1898.

———. "Bölcs Leo hadi taktikájának XVIII. fejezete." In *A Magyar Honfoglalás kútföi* [The sources of the Hungarian conquest]. Edited by G. Pauler and S. Szilágyi. 3–89. Budapest, 1900.

———. "Desiderata der byzantinischen Philologie auf dem Gebiete der mittelgriechischen kriegswissenschaftlichen Literatur." *BNJ* 8 (1931): 225–32.

* ———, ed. *Leonis imperatoris tactica*. 2 volumes. Sylloge Tacticorum Graecorum 3. Budapest, 1917–22.

* ———. "Zum historischen Exzerptenwerke des Konstantinos Porphyrogennetos." *BZ* 17 (1908): 75–85.

———. "Zur Überlieferung mittelgriechischer Takitiker." *BZ* 15 (1906): 47–87.

Vasiliev, A. *Byzance et les Arabes*. Volume 1, *La dynastie d'Amorium (820–67)*; volume 2, *Les relations politiques de Byzance et des Arabes à l'époque de la dynastie macédonienne (Les empereurs Basile I, Léon le Sage et Constantin VII Porphyrogénète) (867–959)*. Edited by H. Grégoire and M. Canard. Corpus Bruxellense Hist. Byz. 1–2. Brussels, 1950–68.

———. "Harun ibn Yahya's Description of Constantinople." *Seminarium Kondakovianum* 5 (1932): 149–63.

Vasmcr, M. *Russisches etymologisches Wörterbuch*. Heidelberg, 1980.

———. *Die Slaven in Griechenland*. Abh. d. Preussischen Akad. d. Wissenschaftern, phil.-hist. Klasse 12. Leipzig, 1941.

* Vereeken, J., and L. Hadermann-Misguich. *Les oracles de Léon le Sage illustrés par Georges Klonzas*. Oriens graecolatinus 7. Venice, 2000.

Vernadsky, G. "'The Tactics' of Leo the Wise and the Epanagoge." *Byz* 6 (1931): 333–35.

Verri, P. *Le leggi penali militari dell'impero bizantino nell'alto medioevo*. Rassegna della Giustizia militare, Suppl. 1–2. Rome, 1978.

* Vieillefond, J.-R., ed. and trans. *Les Cestes de Julius Africanus*. Florence and Paris, 1970.

* Vigneron, P. *Le cheval dans l'antiquité gréco-romaine*. 2 vols. Nancy, 1968.

Viguera, M. J., and T. Sobredo. "Hippology." In *Medieval Islamic Civilization: An Encyclopaedia*. Edited by J. W. Mere. 325–26. New York, 2006.

Vogt, A. "Le jeunesse de Léon le sage." *Revue Historique* 174 (1934): 389–428.

Vogt, A., and I. Hausherr. *Oraison funèbre de Basile Ier par son fils Léon le Sage*. Rome, 1932.

Volkyr de Sérouville, N. *De l'ordre et instruction des batailles*. Paris, 1536.

von Bourscheid, J. W. *Kaisers Leo des Philosophen Strategie und Taktik.* 5 vols. Vienna, 1777–81.

———. *Kurs der Taktik und Logistik, in allen dem Dienste der Strategie schuldigen Pflichten dieser zwey Künste, gemäss dem Sistem der gegenwärtigen Zeit.* Vienna, 1780–81, rev. 1782.

von Dobschütz, E. "Sortes apostolorum or sanctorum." In *The New Schaff-Herzog Encyclopaedia of Religious Knowledge.* Edited by S. M. Jackson et al. London and New York, 1911.

von Falkenhausen, V. *La dominazione bizantina nell'Italia meridionale dal IX al XI secolo.* Bari, 1978.

———. *Untersuchungen über die byzantinische Herrschaft in Süditalien vom 9. bis ins 11. Jahrhundert.* Wiesbaden, 1967.

von Petrikovits, H. *Die Innenbauten römischer Legionslager während der Prinzipatszeit.* Opladen, 1975.

Vryonis, S. "Θάλασσα καὶ ὕδωρ: The Sea and the Water in Byzantine Literature." In *The Greeks and the Sea.* 113–33. Athens, 1993.

* Wahlgren, S., ed. *Symeonis magistri et logothetae Chronicon.* CFHB 44.1. Berlin and New York, 2006.

Wallace Hadrill, D. S. *The Greek Patristic View of Nature.* Manchester, 1968.

* Wallraff, M., ed., *Julius Africanus, Cesti: The Extant Fragments.* Trans. W. Adler. *Griechische christliche Schriftsteller,* n.s. 18. Berlin and Boston, 2012.

* Wassiliewsky, B., and P. Nikitine, eds. "De XLII martyribus Amoriensibus Narrationes et carmina sacra." *Mémoires de l'Acad. impériale de St. Petersburg,* classe phil.-hist., 8th ser., 7 (1905).

Watson, R., and P. Yip, "How Many Were There When It Mattered?" *Significance* 8 (2011): 104–7.

Webster, G. *The Roman Imperial Army.* London, 1969.

Wellhausen, J. "Die Kämpfe der Araber mit den Römäern in der Zeit der Umaijiden." *Nachrichten von der königl. Gesellschaft d. Wissenschaften zu Göttingen. Phil.-hist. Kl.* 4 (1901): 414–47. Translation by M. Bonner, "Arab Wars with the Byzantines in the Umayyad Period." In *Arab-Byzantine Relations in Early Islamic Times.* 31–79. Formation of the Classical Islamic World 8. Aldershot, 2004.

Welwei, K.-W. *Unfreie im antiken Kriegsdienst.* Volume 3, *Rom.* Stuttgart, 1988.

Werner, J. "Ein byzantinischer 'Steigbügel' aus Caričin Grad." In *Caričin Grad.* Edited by N. Duval and V. Popović. 1:147–55. CEFR 75. Belgrade and Rome, 1984.

———. *Der Grabfund von Malaja Pereščepina und Kuvrat, Kagan der Bulgaren.* Abh. D. Bayerischen Akad. der Wissenschaften, phil.-hist. Kl., n.F. 91. Munich, 1984.

Werner, K. F. "Heeresorganisation und Kriegsführung im deutschen Königreich des 10. und 11. Jahrhunderts." In *Ordinamenti Militari in Occidente nell'alto Medioevo.* 791–843. Settimane di Studi del centro Italiano di Studi sull'alto Medioevo 15. Spoleto, 1968.

Wesch-Klein, G. "Hochkonjunktur für Deserteure? Fahnenflucht in der Spätantike." In *L'Armée romaine de Dioclétien à Valentinien Ier (Actes du Congrès de Lyon, 12–14 sept. 2002)*. Edited by Y. Le Bohec and C. Wolff. 475–87. Paris, 2004.

* Wescher, C. *Poliorcétique des grecs: Traités théoriques, récits historiques.* Paris, 1867.

* Westerink, L. G., ed. *Arethae archiepiscopi Caesariensis Scripta minora.* 2 vols. Leipzig 1968–1972.

Wheeler, E. L. "The Army and the *Limes* in the East." In *Companion to the Roman Army.* Edited by P. Erdkamp. 235–66. Oxford, 2007.

———. "Firepower: Missile Weapons and the 'Face of Battle'." In *Roman Military Studies.* Edited by E. Dąbrowa. 169–84. Electrum 5. Cracow, 2001.

———. "The General as Hoplite." In *Hoplites.* Edited by V. Hanson. 121–70. New York, 1991.

———. "Hugo Grotius and Aristotle's lost Δικαιώματα πόλεμων: History's First Monograph on Just War." *Politica Antica* 1 (2012): 141–69.

———. "Land Battles." In *The Cambridge History of Greek and Roman warfare.* Volume 1, *Greece, the Hellenistic World and the Rise of Rome.* Edited by P. Sabin, H. van Wees and L. M. Whitby. 186–223. Cambridge, 2007.

———. "The Legion as Phalanx in the Late Empire, Part I." In *L'Armée romaine de Dioclétien à Valentinien Ier (Actes du Congrès de Lyon, 12–14 sept. 2002).* Edited by Y. Le Bohec and C. Wolff. 309–58. Paris, 2004.

———. "The Legion as Phalanx in the Late Empire, Part II." *Revue des études militaires anciennes* 1 (2004): 147–75.

———. "Methodological Limits and the Mirage of Roman Strategy, Part II." *Journal of Military History* 57 (1993): 215–40.

———. "Military Treatises." In *Oxford Encyclopaedia of Ancient Greece and Rome.* Edited by M. Gagarin. 4:434–38. Oxford, 2010.

———. "Notes on a Stratagem of Iphicrates in Polyaenus and Leo Tactica." In *The Greek World in the Fourth and Third Centuries* bc = *Electrum* 19. Edited by E. Dabrowa, 157–63. Cracow, 2012.

———. "The Occasion of Arrian's Ars Tactica." *GRBS* 19 (1978): 351–65.

———. "Πολλὰ κενὰ τοῦ πολέμου: The History of a Greek Proverb." *GRBS* 29 (1988): 153–84.

———. "Polyaenus: Scriptor Militaris." In *Polyainos: Neue Studien/Polyaenus: New Studies.* Edited by K. Brodersen. 7–54. Berlin, 2010.

———. Review of Charles, *Vegetius.* In *Bryn Mawr Classical Review* 2008.06.42.

———. Review of Whitehead, *Apollodoros.* In *Scholia Reviews* n.s. 20 (2011), no. 11.

———. Review of Dennis, *Taktika.* In *The Medieval Review* 11.11.13.

———. "Rome's Dacian Wars: Domitian, Trajan and Strategy on the Danube, Part 1." *Journal of Military History* 74 (2010): 1185–227.

———. *Stratagem and the Vocabulary of Military Trickery.* Leiden, 1988.

Whitby, L. M. *The Emperor Maurice and His Historian: Theophylact Simocatta on Persian and Balkan Warfare.* Oxford, 1988.

* Whitby, M. *The Ecclesiastical History of Evagrius Scholasticus.* Liverpool, 2000.

———. "Recruitment in Roman Armies from Justinian to Heraclius (ca. 565–615)." In Cameron, *States, Resources and Armies*, 61–124.

* White, H., ed. and trans. *Appian's Roman History*. 4 vols. Loeb Classical Library. London and Cambridge, Mass., 1912-1913.

* Whitby, M., and M. Whitby. *Chronicon Paschale, 284–628 A.D.* Liverpool, 1989.

* ———. *Theophylact Simocatta, History*. Oxford, 1986.

White, M. *Military Saints in Byzantium and Rus, 900-1200*. Cambridge, 2013

* Whitehead, D. *Aineias the Tactician*. 2nd edition. Bristol, 2002.

* ———. *Apollodorus Mechanicus, Siege-Matters: Translated with Introduction and Commentary*. Historia, Einzelschriften 216. Stuttgart, 2010.

———. "Fact and Fantasy in Greek Military Writers." *Acta Antiqua Academiae Scientiarum Hungaricae* 48 (2008): 139–55.

* Whittaker, C. R., ed. and trans. *Herodian, History*. Loeb Classical Library. London and Cambridge, Mass., 1969.

Whittow, M. *The Making of Orthodox Byzantium, 600–1025*. London, 1996.

Wickham, C. "Ninth-Century Byzantium through Western Eyes." In Brubaker, *Byzantium in the Ninth Century*, 245–56.

* Wicksteed, P. H., and F. M. Comford, ed. and trans. *Aristotle: The Physics*. Loeb Classical Library. London and Cambridge, Mass., 1929.

Wierschowski, L. "Kriegsdienstverweigerung im römischen Reich." *Ancient Society* 26 (1995): 205–39.

* Wiet, G., and Al-Yaqʿūbī, *Kitāb al-Buldān, Le livre des pays*. Cairo 1937.

Wiita, J. *The Ethnika in Byzantine Military Treatises*. Ann Arbor and London, 1977.

Wijn, J. W. *Het krijswezen in den tijd van prins Maurits*. Utrecht, 1934.

Wilson, N. G. "The Libraries of the Byzantine World." In *Griechische Kodikologie und Textüberlieferung*. Edited by D. Harlfinger. 276–309. Darmstadt, 1980.

* ———. *Photios: The Bibliotheca; A Selection Translated with Notes*. London, 1994.

———. *Scholars of Byzantium*. London, 1983.

Winkelmann, F. *Byzantinische Rang- und Ämterstruktur im 8. und 9. Jahrhundert*. BBA 53. Berlin, 1985.

———. "Probleme der Informationen des al-Ǧarmi über die byzantinischen Provinzen." *Byzantinoslavica* 43 (1982): 18–29.

Winkelmann, F., and W. Brandes, eds. *Quellen zur Geschichte des frühen Byzanz (4.-9. Jahrhundert: Bestand und Probleme*. BBA 55. Berlin, 1990.

Wood, D. M. *Leo VI's Concept of Divine Monarchy Illustrated in a Cave Chapel*. London, 1964.

Woolliscroft, D. J. "Excavations at Garnhall on the Line of the Antonine." *PSAS* 138 (2008): 129–76.

* Wortley, J. *John Skylitzes: A Synopsis of Byzantine History, 811–1057*. Cambridge, 2010.

———. "Military Elements in Psychophelitic Tales and Sayings." In *Peace and War in Byzantium*. Edited by T. S. Miller and J. S. Nesbitt. 89–105. Washington, D.C. 1995.

* Wright, E. A. *The Chronicle of Joshua the Stylite*. Cambridge, 1882.

* Wright, F. A., trans. *Liudprand of Cremona, Works.* London, 1930.

Yannopoulos, P. A. "Cibyrra et Cibyrrhéotes." *Byzantion* 61 (1991): 520–29.

* Yar-Shater, E., ed. *The History of al-Tabarī.* 38 Volumes. SUNY series in Near Eastern studies. Albany, NY, 1985–2007.

* Young, D., ed. *Theognis: Sententiae elegiacae.* Stuttgart, 1998.

* Zacos, G., and A. Veglery. *Byzantine Lead Seals.* Volume 1, parts 1–3. Basel, 1972.

Zakharov, A., and W. Arendt. "Studia Levedica, 2: Türkische Säbel aus den vii–ix Jarhunderten." *Archaeologia Hungarica* 16 (1935).

Zaki, A. R. "Medieval Arab Arms." In *Islamic Arms and Armour.* Edited by R. Elgood. 202–12. London, 1979.

———. "Military Literature of the Arabs." *Cahiers d'histoire égyptienne* 7 (1955): 149–60.

Zástěrová, B. "Ethnika in den Werken frühbyzantinischer Historiker." In Winkelmann and Brandes, *Quellen zur Geschichte,* 180–89.

———. "Zur Problematik der ethnographischen Topoi." In *Griechenland—Byzanz—Europa: Ein Studienband.* Edited by J. Herrmann, H. Köpstein and R. Müller. 16–19. Berlin, 1985.

* Zepos, J. and P. *Jus graecoromanum.* Athens, 1931. Reprinted, 1962.

Ziegler, K. *Die Überlieferungsgeschichte der vergleichenden Lebensbeschreibungen Plutarchs.* Leipzig, 1907.

Zilliacus, H. *Zum Kampf der Weltsprachen im oströmischen Reich.* Helsinki, 1935.

Živković, T. "Uspenskij's Taktikon and the Theme of Dalmatia." *Symmeikta* 17 (2005–7): 49–85.

* Zotenberg, H., ed. and trans. *Chronique de Jean, évêque de Nikiou: Texte éthiopien.* Paris, 1883.

* Zuckerman, C. "Chapitres peu connus de l'*Apparatus Bellicus*." *TM* 12 (1994). 359–89.

———, ed. *La Crimée entre Byzance et le Khaganat Khazar.* Paris, 2006.

———. "Les Hongrois au pays Lébédia: une nouvelle puissance aux confins de Byzance et de la Khazarie ca. 836–89." In Tsiknakes, *Εμπόλεμο Βυζάντιο,* 51–74.

———. "The Military Compendium of Syrianos Magister." *JÖB* 40 (1990): 209–24.

.

INDEX OF SOURCES

GENERAL INDEX

Written works are found under the author's name, unless anonymous or otherwise specified.